General Relativity and John Archibald Wheeler

Astrophysics and Space Science Library

For other titles published in this series, go to
http://www.springer.com/series/5664

Ignazio Ciufolini • Richard A. Matzner
Editors

General Relativity and John Archibald Wheeler

Springer

Editors
Ignazio Ciufolini
Università del Salento
Dipto. Ingegneria
dell'Innovazione
Via per Monteroni
73100 Lecce
Italy
ignazio.ciufolini@unile.it

Richard A. Matzner
University of Texas, Austin
Dept. Physics
Center for Relativity
Austin TX 78712-1081
USA
matzner2@physics.utexas.edu

ISSN 0067-0057
ISBN 978-90-481-3734-3 e-ISBN 978-90-481-3735-0
DOI 10.1007/978-90-481-3735-0
Springer Dordrecht Heidelberg London New York

Library of Congress Control Number: 2010930053

Cover design: eStudio Calamar S.L.

Printed on acid-free paper

Springer is part of Springer Science+Business Media (www.springer.com)

John Archibald Wheeler (seated, first on the left) at the 1970 Clifford Centennial Meeting at the Institute for Advanced Studies, Princeton. In the picture can be recognized, among others, Nicola Cabibbo, Robert Dicke, Eugene Wigner, Stephen Hawking, Brandon Carter, Freeman Dyson, Cecile Dewitt, Charles Misner and Tullio Regge

John Archibald Wheeler in his office at the University of Texas at Austin in 1985. Left to right, Arkady Kheyfets, William Wootters, Ignazio Ciufolini, seated, with a model showing the use of the nodes of two LAGEOS satellites to measure frame-dragging, John Archibald Wheeler, Wolfgang Schleich, Philip Candelas, Roberto Bruno and Warner Miller

John Archibald Wheeler at Princeton Univesity in 1990. Left to right, Eugene P. Wigner, recipient of the 1990 Eugene P. Wigner Reactor Physicist Award, Ignazio Ciufolini and John Archibald Wheeler

We dedicate this book to John Archibald Wheeler, our mentor and teacher

Contents

Part III Gravitational Waves

Part IV Frame Dragging and Gravitomagnetism

Acknowledgment

We gratefully acknowledge the support and encouragement of Antonino Zichichi and the support of the Ettore Majorana Foundation and Centre for Scientific Culture (EMFCSC) of Erice, Sicily, during the First International School of Astrophysical Relativity "John Archibald Wheeler". We especially thank each author of "General Relativity and John Archibald Wheeler" and Giampiero Sindoni for his help in editing this book. Finally, we express our appreciation to Loreta, Svetlana, Sirio and Leonardo Ciufolini, for their patience in the completion of this work.

Introduction to General Relativity and John Archibald Wheeler

Ignazio Ciufolini and Richard Matzner

John Archibald Wheeler was born on July 9, 1911, in Jacksonville, Florida, and passed away on April 13, 2008, in Hightstown, New Jersey; his influence on gravitational physics and science in general will remain forever. Among his many and important contributions to physics, he was one of the fathers of the renaissance of General Relativity. After a golden starting age of General Relativity, a few years after the Einstein's papers of 1915–1916, Einstein's gravitational theory was for many years, to quote the preface of a 1960 book of General Relativity [1], confined to "an ivory tower… and no doubt many a relativist looks forward to the day when the governments will seek his opinion on important questions".

Today we live in Wheeler's renaissance of General Relativity. Indeed, among many other applications of General Relativity, the satellites of the Global Positioning System constellation have relativistic corrections that need to be taken into account and carefully calculated to allow the system to function. If we now live this exciting day in which Einstein's gravitational theory is an important ingredient not only of our understanding of the Universe but also of aerospace engineering, we owe this to John Archibald Wheeler who back in the 1950s transformed General Relativity in an active field of investigation of astrophysics and physics.

Several of the ideas that Wheeler developed or that he inspired are covered in this book: from black holes and wormholes to the initial value problem in General Relativity, from cosmology to space research and from the theory and search for gravitational waves to frame-dragging and its measurement [2]. Most of the papers of this book are based on the First International John Archibald Wheeler School on Astrophysical Relativity that was held in Erice, Italy in May–June 2006.

I. Ciufolini (✉)
Dip. Ingegneria dell'Innovazione, Università del Salento and INFN Sezione di Lecce, via Monteroni, 73100 Lecce, Italy
e-mail: ignazio.ciufolini@unisalento.it

R. Matzner
Center for Relativity, University of Texas at Austin, Austin, Texas 78712, USA
e-mail: matzner2@physics.utexas.edu

I. Ciufolini and R.A. Matzner (eds.), *General Relativity and John Archibald Wheeler*, Astrophysics and Space Science Library 367, DOI 10.1007/978-90-481-3735-0_1,

1 John Archibald Wheeler and General Relativity

This chapter is devoted to the state of the art of some of the research fields that John Wheeler investigated and some of the ideas that he inspired. First, Charles Misner describes the key role that John Wheeler had in "the Recertification of General Relativity as True Physics", then Kip Thorne and Wojeiech Zurek describe some of the important contributions of "Johnny" to General Relativity and to physics in general. Igor Novikov and collaborators then present an update of the theory of wormholes in the modern astrophysics; a wormhole is a "bridge" that connects different regions of spacetime, the terms "wormhole" and "black hole" were coined by Wheeler [3].

2 General Relativity and Its Tests

This chapter describes some fundamental issues in General Relativity, such as the initial value problem and cosmology, and the experimental tests of Einstein's Gravitational theory.

During the past century General Relativity gave rise to remarkable experimental triumphs. On the one hand, a number of key predictions of Einstein's gravitational theory have been experimentally confirmed with impressive accuracy [4]. On the other hand, General Relativity today has practical applications in space research [15], space geodesy, astronomy and navigation in the Solar System, from the orbital modeling of Global Positioning System and laser-ranged satellites, and of spacecraft in the Solar System, to the technique of Very Long Baseline Interferometry (VLBI), and is a fundamental component for the understanding of many astrophysical and cosmological observations. General Relativity has so far passed all the tests it has been put to.

Nevertheless, despite all the observational tests passed by General Relativity, there are still some deep problems to be solved. On the theoretical ground, Einstein's gravitational theory predicts spacetime singularities, events in which the known physical theories cease to be valid and the spacetime curvature diverges, and we are still seeking the merging of General Relativity with quantum mechanics which might solve the problem of the singularities. On the observational ground, we have recently discovered supernovae that appear to accelerate away from us [5, 6]. This can be explained by "dark energy" that is accelerating the expansion of the Universe, and may imply a non-zero value to Einstein's "cosmological constant", or an exotic new form of energy known as "quintessence", or perhaps a modification of General Relativity [6, 7]. Including dark matter, approximately 96% of the mass–energy of the Universe is of a form unknown to us. A relativistic theory of gravity, combining General Relativity at some level with quantum mechanics, may help to reveal what dark energy is and whether it might be related to dark matter. Therefore, new effects are being sought on very small length scales, on very large scales and even in the Solar System for what they might reveal about theories such as quantum gravity, string and brane-world models of the Universe. Furthermore, even though a breakdown of General Relativity should occur at the quantum level, some viable modifications of Einstein's gravitational theory already give different predictions at the classical level [8]. Therefore, every aspect of General Relativity should be directly tested and the accuracy of the present measurements of Einstein's theory and of the foundations of gravitational theories should be further improved. Today, among the biggest remaining experimental challenges of General Relativity are the direct detection of gravitational waves (chapter "Gravitational Waves"), although we already know from the decrease of the orbital period of the binary pulsar PSR B1913 + 16 that this system emits gravitational waves at the predicted level [4, 8], and improved tests of a number of general relativistic phenomena (this chapter "General Relativity and Its Tests") including gravitomagnetism and frame-dragging (chapter "Frame-Dragging and Gravitomagnetism").

First, in this chapter, James York describes a unified form of the initial value conditions in General Relativity; John Wheeler inspired this research and considered the solution of the initial value problem in General Relativity as one of the most important research achievements of Einstein's gravitational theory. Then Clifford Will presents a comprehensive description of the experimental tests of General Relativity and John D. Anderson and Eunice L. Lau report an update of the very accurate measurements of space curvature due to the solar mass. Finally Vladimir Lukash presents a paper on cosmology investigating the riddle of the nature of dark matter.

3 Gravitational Waves

One of the earliest predictions of Einstein's study of gravitation was the prediction of gravitational radiation. Although his analysis correctly describes the physics, it generated a long running controversy over the reality of gravitational radiation (confusion with waves in gauge quantities, due to confusion in the meaning of coordinate conditions which play the role of gauge in general relativity), whether the waves carry energy, and even whether the energy they carry is positive. Bondi gave physical arguments that gravitational waves carry (positive) energy [9, 10] but only in 1984 were the last technical details settled [11]. The article in this volume by Matzner provides a summary introduction into the subject.

Our universe contains a mixture of constituents: baryon matter (plus electrons), photons, neutrinos, dark energy, dark matter, gravitational radiation originating from the big bang (*primordial* gravitational radiation), and gravitational radiation from astrophysical sources (e.g., binary neutron stars). Leonid Grischuk discuses the evolution of primordial gravitational radiation in the early universe and their potentially detectable effects on the cosmic background radiation.

After almost two decades of concentrated effort, the community now has successful computational simulations of compact binary mergers (black hole binaries, neutron star binaries, and black hole – neutron star binaries), and the gravitational radiation they produce. This field is not complete, because extremely large computations are needed, so surprises are still being found in new simulations. For instance it has been found that very large residual velocities can be induced in the final merged black hole in a binary encounter and merger of spinning black holes [12, 13]. The "kick" arises because of asymmetrical radiation of gravitational waves, and can be big enough to eject a black hole from the potential well of a galaxy.

Beginning with the ideas of Weber in the early 1960s [14], an effort has been underway to build direct detectors of gravitational radiation. Weber's were bar detectors, in which passing gravitational waves induce vibrations in a large low loss (high Q) mass. These are prototypes of *resonant* detectors. The paper below by Pizzella gives an update on the current state of that aspect of experimental relativity. Bar detectors are sensitive to waves arriving broadside, but new designs call for *spherical* detectors, which are omnidirectional, and which are sensitive to modes other than the traceless transverse modes that constitute gravitational radiation in Einstein's general relativity. For instance there may be volume breathing modes, or transverse breathing modes, as are present in some alternate descriptions from Einstein's. The paper by Giazotto in this volume discusses these detectors and also introduces the laser-interferometer detectors.

The currently operating laser-interferometer detectors are LIGO, Virgo, GEO, and TAMA. They use interferometry in Michelson interferometers to detect the subtle change of length between end-masses induced by passing gravitational waves. For this to be effective, the arms need to be quite long. For the detectors listed above, arm lengths (in order) are 4, 3, 0.6, and 0.3 km. These detectors have been in operation for approximately two-to-three cumulative years, and the LIGO detectors have accumulated 1 year of simultaneous observing time. To date there have been no convincing events. Giazotto provides a very complete description of the functioning of these detectors and of their data analysis. He also describes a proposed spacebourne detector LISA, which will have three five-million kilometer arms in an equilateral triangle arrangement. Finally, Gaetano Vilasi describes some mathematical solutions of Einstein field equations representing gravitational waves with spin-1. John Wheeler inspired his research some decades ago.

John A. Wheeler was an enthusiastic supporter of Weber's research, but his real contribution in this field was to drive the theoretical understanding of relativity and of gravitational radiation and its generation and propagation.

4 Frame-Dragging and Gravitomagnetism

Some of the results on frame-dragging and gravitomagnetism presented in this chapter are also a heritage of John Wheeler; among other things, he was fascinated by the problem of the origin of inertia and inertial frames in General Relativity [15] and by the Mach principle [16].

Inertial frames are systems in which any isolated body, not subject to any interaction, is at rest or moving with constant velocity. However, what determines an inertial frame? And, in general, what is the origin of inertia? In the mechanics of Galileo Galilei and Isaac Newton an inertial frame has an absolute existence, uninfluenced by the matter in the Universe. In Einstein's gravitational theory, the inertial and centrifugal forces are due to our accelerations and rotations with respect to the local inertial frames which, in turn, are determined, influenced and dragged by the distribution and flow of mass-energy in the Universe [17]; John Wheeler used to describe this as: "mass and mass flow there rule geometry – and inertia – here" [15]. In particular, the local inertial frames are dragged by the motion and rotation of nearby matter; this general relativistic phenomenon is called frame-dragging (ref. [18], and see page 544 of ref. [2] for example) and may be thought as representing in Einstein's theory a weak remnant of the ideas of Ernst Mach on the origin of inertia [15]. Mach thought that centrifugal and inertial forces are due to rotations and accelerations with respect to all the masses in the Universe; this is known as "Mach's principle".

In this chapter, first Robert O'Connell and Sergei Kopeikin present enlightening descriptions of rotation, spin, frame-dragging and gravitomagnetism [19] in General Relativity, and Neil Ashby discusses geodetic precession and the so-called Lense–Thirring effect. Then the main experiments to detect frame-dragging and gravitomagnetism are described. Ciufolini et al. describe the details of the 2004 measurement of Earth's gravitomagnetism and Lense–Thirring effect, achieved with an accuracy of approximately 10% using the data of the laser-ranged satellites LAGEOS and LAGEOS 2 [20] together with the accurate description of the Earth's gravity field by the dedicated spacecraft GRACE. In 2008 John Ries and the US GRACE team independently confirmed this analysis, with an accuracy of about 12%, validated against a number of GRACE Earth's gravity models [21]. Then, Sasha Buchman, for the GP-B collaboration, discusses the Gravity Probe B (GP-B) experiment aimed to measure frame-dragging effects [22, 23]. This relativity satellite, launched in 2004, has been the most complex gravitational space experiment to date and has so-far achieved a measurement of the geodetic precession with an accuracy of about 1%. Finally, the LARES team presents the LARES (LAser RElativity Satellite) space experiment consisting in a new laser-ranged satellite to be launched in 2010 by the Italian Space Agency (ASI) and dedicated to the measurement of Earth's gravitomagnetism and Lense–Thirring effect with accuracy of the order of 1%.

John Archibald Wheeler was a strong supporter of LAGEOS 3 since 1984, when this new laser ranged satellite to test frame-dragging was originally proposed [24]. The LAGEOS 3 proposal has since become the LARES space experiment funded by ASI in 2008.

In conclusion, for the present exciting research in gravitational physics and for many of the discoveries described in this book, we all have to say: "thanks Johnny"!

References

1. Synge, J. L., "*Relativity: the General Theory*", (North-Holland, Amsterdam, 1960).
2. Misner, C. W., Thorne, K. S., and Wheeler, J. A., "*Gravitation*", (Freeman, San Francisco, 1973).

3. Wheeler, J. A., "On the Nature of Quantum Geometrodynamics", *Ann. Phys.* **2**, 604–614 (1957).
4. Will, C. M., "The confrontation between general relativity and experiment", *Living Rev. Rel.* **9**, 3 (2006); http://www.livingreviews.org/lrr-20063 or gr-qc/0510072).
5. Riess, A., et al., "Observational evidence from supernovae for an accelerating universe and a cosmological constant", *Astron. J.* **116**, 1009–1038 (1998).
6. Perlmutter, S., "Supernovae, dark energy, and the accelerating universe", *Phys. Today* **56**, 53–59 (2003).
7. Caldwell, R. R., "Dark energy", *Phys. World* **17**, 37–42 (2004).
8. Will, C. M., "*Theory and Experiment in Gravitational Physics*", 2nd edn (Cambridge Univ. Press, Cambridge, UK, 1993).
9. Bondi, H., "Gravitational Waves", in "*Evidence for Gravitational Theories*", Møller, C., Editor (Academic Press, New York, 1961).
10. Morgan, T. and Bondi, H., "Transfer of energy in general relativity", *Royal Society of London Proceedings Series A* **320** 277–287 (1970).
11. Ehlers, J. and Walker, M., "Gravitational radiation and the 'quadrupole' formula - Report of Workshop A1", in "*General Relativity and Gravitation Conference*", Bertotti, B., de Felice, F. and Pascolini, A., Editors, 125–137, (Springer, Berlin, 1984).
12. Herrmann, F., Hinder, I., Shoemaker, D. M., Laguna, P., and Matzner, Richard A., "Binary black holes: Spin dynamics and gravitational recoil", *Phys. Rev.* **D76**, 084032 (2007).
13. Campanelli, M., Lousto, C. O., Zlochower, Y., and Merritt, D., "Maximum Gravitational Recoil", *Phys. Rev. Lett.*, **98** 231102 (2007).
14. Weber, J., "*General relativity and gravitational waves*", Interscience Tracts on Physics and Astronomy (New York: Interscience, (1961).
15. Ciufolini, I. and Wheeler, J.A., "*Gravitation and Inertia*", (Princeton University Press, Princeton, New Jersey, 1995).
16. Wheeler, J. A., "Mach's Principle as Boundary Condition for Einstein's equations", in "*Gravitation and Relativity*", Chiu, H. Y. and Hoffman, W. F., eds. (Benjanmin, New York, 1964) 303–349.
17. Ciufolini, I., "Dragging of Inertial Frames", *Nature*, **449**, 41–47 (2007).
18. Einstein, A., "Letter to Ernst Mach", Zurich, 25 June 1913, in ref. [2] p. 544.
19. Thorne, K. S., Price, R. H. and MacDonald, D. A., "*Black Holes, the Membrane Paradigm*" (Yale University Press, New Haven and London, 1986).
20. Ciufolini, I. and Pavlis, E. C., "A confirmation of the general relativistic prediction of the Lense-Thirring effect", *Nature*, **431**, 958–960, (2004).
21. Ries, J.C., Eanes, R.J., and Watkins., M. M., "Confirming the Frame-Dragging Effect with Satellite Laser Ranging", *16th International Workshop on Laser Ranging*, Poznan, Poland, 13–17 October 2008.
22. Pugh, G. E., "Proposal for a Satellite Test of the Coriolis Prediction of General Relativity", Weapons Systems Evaluation Group Research Memorandum N. 11 (The Pentagon, Washington, 1959).
23. Schiff, L. I., "Motion of a Gyroscope According to Einstein's Theory of Gravitation", *Proc. Nat. Acad. Sci.*, **46**, 871–82 (1960) and "Possible New Test of General Relativity Theory", *Phys. Rev. Lett.* **4**, 215-7 (1960).
24. Ciufolini, I., "Measurement of the Lense-Thirring drag on high-altitude laser-ranged artificial satellites", *Phys. Rev. Lett.* **56**, 278–281 (1986).

Part I
John Archibald Wheeler and General Relativity

John Wheeler and the Recertification of General Relativity as True Physics

Charles W. Misner

Abstract This lightly edited transcript of my 2006 lecture at the Erice "J.A. Wheeler" school introduces Wheeler by describing him from several different viewpoints. There is brief mention of his early work on nuclear physics and its military applications as well as Wheeler's attitude toward defense work. Also, just briefly, his Texas efforts to probe the foundations of quantum mechanics are mentioned. Considerable attention is given to his methods of working, and to his relationship with students and their reactions to him. A further section describes his push to bring general relativity into the main stream of physics and the influence this had on the development of black hole ideas and on the study of gravitational waves.

1 Introduction

The title, "John Wheeler and the Recertification of General Relativity as True Physics" reflects what I think is the most important contribution John Wheeler made: to pull General Relativity out of the discard pile and get large numbers of good people actively working on it. That will be my concentration in this paper.

But note that John Wheeler went through several different phases of interest in Physics, each lasting about 20 years. In the first phase, where I didn't know him except at the very end, he concentrated on a view that the universe can be understood in terms of particles. He defined the S-matrix early in the 1930s and did much other work in atomic and nuclear physics, and contributed very significantly to the Manhattan project, which produced the first nuclear bombs.

The period I'm going to concentrate on is the second one; I worked with him closely in parts of it. That was when he decided it wasn't going to be all particles – it was going to be all fields – and he started out with gravity and electromagnetism.

C.W. Misner (✉)
Department of Physics, University of Maryland, College Park, Maryland, 20742–4111, USA
e-mail: misner@umd.edu

I. Ciufolini and R.A. Matzner (eds.), *General Relativity and John Archibald Wheeler*, Astrophysics and Space Science Library 367, DOI 10.1007/978-90-481-3735-0_2,

Fig. 1 John Archibald Wheeler

This work dominated 20 some years up until he retired. Upon retiring as a professor at Princeton he moved to take a professorship in Texas at Austin and there his interests changed again.

This last change had begun at Princeton; in Texas his concentration was on information. His motto became "It from Bit" and many of his best students at that time were in this area of the interpretation of quantum mechanics and in trying to understand the nature of information relevant to physics.

This paper was written in 2006, two years before John Wheeler's death (Fig. 1). John Archibald Wheeler died on 13 April 2008 at age 96. Janette Wheeler (John's wife) died on 17 October, 2007. She was 99.

2 John Archibald Wheeler

- Born 9 July 1911
- Ph.D. Johns Hopkins (Herzfeld) 1933
- NYU (Breit) 1933–1934, Bohr Institute 1934–1935

- Married Janette Hegner 10 June 1935
- Asst. Prof. Princeton 1938
- Bohr–Wheeler nuclear fission paper 1939
- A-bomb, Hanford Washington 1942–1945
- H-bomb 1950–1952
- Moved to Austin, Texas, 1976
- Retired from Texas 1986
- Died 13 April 2008

John was born in 1911 which means he celebrated his 95th birthday the summer this lecture was given. He got a Ph.D. from Johns Hopkins in 1933. It was amusing for me to read recently an article by Spencer Weart in "Physics Today" summarizing 50 years of the American Institute of Physics: he said imagine someone who was a graduate student 50 years ago, what he would have gone through. Imagine what that now 100-year-old physicist would have seen in the past 50 years. I remarked to Weart that it doesn't have to be a hypothetical graduate student. John Wheeler was a graduate student at age 19 in 1931 when the AIP was founded. In his autobiography John describes his interactions with postdocs and others, what kind of physics he was seeing at those times, and continues on through. So he was actually participating in physics for those entire 50 years.

John was still 21 (just a few weeks before he became 22) when he got his Ph.D. at Johns Hopkins. He had a postdoc with Breit. Then his second postdoc was at the Bohr Institute (it wasn't called the Bohr Institute then) with Niels Bohr, and that was a very important development. He continued to interact with and admire Bohr all his life. Immediately on returning from Copenhagen he got married and went to North Carolina as an Assistant Professor. Then three years later he turned down tenure at North Carolina to move to Princeton as an Assistant Professor because he thought it was a more lively atmosphere for physics.

With the discovery of fission he and Bohr put together a nuclear liquid drop model that allowed one to estimate some of the important parameters of that process. Then things became much more classified and he worked throughout the Second World War on the development of the A-bomb. He was principally responsible for the operation at Hanford which was to produce plutonium in a nuclear reactor. After the war he came back and encouraged the beginnings of some cosmic ray research at Princeton and tried to get back into nuclear physics with things like a collective model which he was late in publishing.

But then the cold war distracted him again as there was a political call to produce the H-bomb. Late during the war, and after the war, he was very much motivated by the fact that he had a brother he admired very much who died fighting in the war. He always felt that if only the A-bomb had come four months earlier his brother might still be alive. He was very much concerned about that. I think he also felt that he didn't pay enough attention to politics when he was in Copenhagen at the Bohr Institute, and that if one had thought more about what was going on and not just let Hitler have his own way for many years, then also the world would've been a better place. Some of the troubles of World War II could have been avoided. So he was always hawkish on strong defense.

I think in particular some of John's colleagues at Princeton and others didn't think he played an admirable role in the Oppenheimer hearings. Oppenheimer was unenthusiastic about the H-bomb, initially on good scientific grounds because all the proposals were impossible and never have worked. Later when the way to do it was found, Strauss, the Chairman of the Atomic Energy Commission (AEC), really didn't like him (and Teller also didn't like Oppenheimer – you can read about that in the book, [3], about Oppenheimer that has recently come out). Their way of getting Oppenheimer out of the councils was not just to say, "well we want an advisory committee that will pursue defense and since you're not enthusiastic about it we'll replace you". That would have been difficult since Oppenheimer was the hero of the World War II A-bomb. Instead they attacked his loyalty and claimed he was not a reliable person to have secrets – he might be favorable to the communists or something. He certainly had a Communist background which was well-known to all authorities when they appointed him director of Los Alamos. Wheeler didn't really make any strong statements in his support and said so little that he was considered to be supporting this approach to just get rid of Oppenheimer on any grounds. That caused number of hard feelings. Wheeler did spend a lot of time on the H-bomb, much of it eventually done in the neighborhood of Princeton just outside the University on Route 1. That came to an end in '52. I arrived in Princeton in '53 just as John's H-bomb work was finishing up and he was free to start doing something else, which turned out to be gravity.

The main part of this memoir will be about those years when John Wheeler was focusing on gravity and related things. I count them as the period from '53, until in 1976 when at age 65 John retired from Princeton. He then moved to Texas and spent 10 years there until he had some health problems (he had a bypass). He then retired again and moved to Meadow Lakes in Hightstown just outside Princeton where he has been living ever since. Just in the past year, instead of he and his wife running their own apartment and taking pretty much care themselves, they've required some of the assisted living and such things that are available there. This summer was the first time he did not get to his house in Maine since he bought in 1955. As you can calculate, he's 95 years old. His wife Janette is actually 98. They are a continuing to live there, but John's hearing is so poor that it's difficult to have a conversation with him and has been for several years. He does hire a secretary to come around a few hours every week to work with him. His former student, coauthor, and close friend Ken Ford says that John did in fact write a chapter in a book which was published this year, so he's not totally out of everything. But he's under very difficult circumstances right now, I would say. Three years ago John received the first Einstein prize. (Peter Bergmann received the prize at the same time as John Wheeler, and Bryce DeWitt received it just before he died last year. That enterprise needs contributors. If there are any highly financially comfortable people reading this, they could consider donating to that prize through the American Physical Society.)

Wheeler was always a teacher. He talks about teaching in order to learn. In fact he said after the H-bomb work was finished and he settled down to full-time physics that the first thing he did was ask to be assigned to teach the relativity course. He says in his autobiography that he wanted to learn relativity and the best way to learn a thing is to teach it. So that's what he did. He taught that course for a few years but the first year, I guess, was the learning experience. When he taught other things – such as the freshman honors physics, probably (I haven't heard of him teaching the regular introductory premed courses but he did teach freshman physics on many occasions) he would always start out with inspiration. He would start out making sure the students felt that there was something they were going to be really happy to learn going on here. He would typically start by describing the current research he was doing or that his students were doing. Then he would gradually move, after a lecture or a lecture and a half on the current forefront physics that he was enthusiastic about, into some of the background needed for that. He would eventually get around to vectors, acceleration, and things of that sort.

With his research students John thinks questions are really at least as important as, or maybe more important than, the answer. If you pick the really right questions then there is some point to a lot of hard work, because you can probably find an answer. It wasn't the right question if it can't be answered in some way. So he was always interested in what are the important questions to be focused on at any given time. When he thought he had the right question he would put a lot of continuing effort into it. But he always worked with students. He has a rather small fraction of his papers that are single author papers. There are also perhaps a number of single author papers where the author was his student, but where John was actually one of the major contributors. So he often would encourage his students to publish whereas many other people would have put their name on the same paper because there was a lot of interaction going on. In many other cases he was involved in work with students and all the names appeared, which I think is a standard practice. But all his papers had authors listed alphabetically, which almost always put his name at the end. He finally met a student in Texas who outranked him for last place – Zurek. He really believes that students not only do the physics but they ask so many questions that you have to think hard. They come at things from different angles, etc. so that he really appreciated the opportunity to work with students and felt that he could get a lot more physics done by letting them interfere and collaborate.

When I set about to prepare this lecture for Erice, the focus was on those 20 years at Princeton. It turned out that just as he went from Princeton to Texas, there was a book prepared as a gift to him by Peter Putnam, one of his students. That book attempted to gather, from anyone who had benefited from their association with Wheeler, some written comments. The request was for letters to John Wheeler to put in the book, with a CV and list of students. That means I just had to pull out my copy of this and let that do a lot of the review of what was going on in the 20 years. That's what I'm going to do next (Fig. 2).

We start with a list of students from the back of the book. They were organized by dates and, of course, some of his early students are known to everybody, such as Richard Feynman. Arthur Wightman was a very prominent field theorist in Princeton for many years and a leader in the field. Of course Aage Bohr was not actually a student of John's. He came with his father to Los Alamos during World War II, and John of course visited Copenhagen many times immediately after the war and saw

Fig. 2 The 1930s and 1940s

The '30's and '40's

Aage Bohr

Katharine Way

Henry H. Barschall

Richard P. Feynman

Mael A. Melvin

Gilbert N. Plass

Arthur S. Wightman

The '50's

David Lawrence Hill	Arthur Komar	Raymond C. Mjolsness
Lawrence Wilets	Marjory Pratt	Dieter Brill
John Toll	Joseph Weber	Robert N. Euwema
Kenneth W. Ford	Tullio Regge	B. Kent Harrison
Richard Bellman	Thomas O. Philips	John R. Klauder
William Grasberger	Charles W. Misner	John G. Fletcher
David M. Chase	Robert Marzke	Peter Putnam
James J. Griffin	Daniel Sperber	James B. Hartle

Fig. 3 The 1950s

a lot of Aage Bohr and Aage felt that he had had a close enough association with John that he wanted to be included in this list of people who were making a tribute to John (Fig. 3).

Then we go on to the mid-50s. We start out here with the early 50s people who are mostly in nuclear physics. John Toll was not, but he did dispersion relations and field theory with John. Ken Ford was one of his later, nuclear physics, students. Later in life Ken Ford with some support from the Sloan foundation was able to spend a lot of time helping John Wheeler put together an autobiography, which was published a few years ago. Ken still lives in Philadelphia. He goes and sees John several times a month and tries to help him with this correspondence and things like that. Also from the 50s is Jim Griffin, who I think was John's last nuclear physics student. He spent some time at Los Alamos before becoming an important member the Physics Department at the University of Maryland for many years. Larry Wilets is also a very well-known nuclear physicist in Washington State. He leads one of the major groups, as important as the one at Maryland. I'm not sure about Hill, but he did very important work at that time. But I don't keep track of nuclear physics so I can't tell you all these things now.

By then John was going into relativity. His first relativity Ph.D. student was Art Komar, who did some things on quantum gravity and had a postdoc with Peter Bergmann; his work was thereafter more associated with Bergmann than with Wheeler. Later Komar served various times at the National Science Foundation as the Gravitation Program officer, usually substituting for a year for somebody who was away on sabbatical. Joe Weber was with John Wheeler and went to the Institute for Advanced Study so he could work with Wheeler as he was getting into general relativity. Wheeler gave him a lot of encouragement on his development of gravitational wave ideas. In 1955 Joe Weber, myself, and Peter Putnam accompanied Wheeler to Leiden in the Netherlands where John was a visiting professor for the spring semester. Tullio Regge came to visit at that time. I'll have some notes from Tullio Regge about that experience later. There are a number of other people, some of whom are well-known at the University of Maryland, like Dieter Brill. John Klauder is now down in Florida, I believe.

Jim Hartle was actually an undergraduate working with Wheeler and with Dieter Brill. There is a very important paper, Brill-Hartle, which came out of Hartle's senior thesis. It was published in the Physical Review. Peter Putnam, this very eccentric and interesting guy responsible for gathering all the stuff together, was a student of John's; first as an undergraduate, and then during the time I was there he came back as a graduate student and finished a Ph.D. His main interest was Eddington's fundamental theory though John got him away from that long enough to do a Ph.D. Putnam worked around Route 128 in Boston and every time he was offered a raise he would say no, give me some stock instead. So he accumulated stock in some of the startup companies of the 1950s on Route 128. Later on he arranged large gifts, I'm sure a substantial amount from himself. (I think also tried to avoid using his mother's money for anything – she was in the family of the publishing company.) He gave all his stock to the Princeton, saying "don't sell until I tell you", and he told them at exactly the right time about 20 years later. They collected their millions and, as he directed, bought fabulous pieces of sculpture by absolutely first-rate people that are scattered around campus at Princeton (Fig. 4).

Moving on to the 60s: David Sharp was very active. I knew him very well and worked with him. He had a long and distinguished career at Los Alamos. Karel Kuchař has done much fundamental work on quantum gravity. I think we all know Jim York, who has had a long career at Chapel Hill and is now at Cornell. His work on the initial value problem was his strongest contribution, but he has also done important work in recent years on formulations of Einstein's equations, and applying these ideas to computational relativity. Joe Redish was an undergraduate thesis student of John Wheeler's. Maybe his interest in teaching was initiated there. Larry Shepley, who retired many years ago from Texas, was a student of John's, and also partly of mine as he finished up just a few years after I came to Maryland from

The '60's		
	Richard Lindquist	Alfonso Campolattaro
David H. Sharp	Fred K. Manasse	David William Meltzer
Georgia S. Witt	Edward Redish	Edwin F. Taylor
Kuk Pyo Chung	Heinrich Leutwyler	Anthony Zee
Robert Fuller	Andris Suna	Roger Penrose
Masami Wakano	Cheuk-Yin Wong	Michael D. Stern
Allen Mills	George H. Brigman	Paul Boynton
R. Bruce Partridge	Lawrence C. Shepley	Ulrich H. Gerlach
F. Bary Malik	Kip S. Thorne	Robert Geroch
Karel Kuchar	Hans Ohanian	Arthur E. Fischer
James W. York, Jr.	J. Peter Vajk	Jeffrey M. Cohen

Fig. 4 The 1960s

The '70's		
	Jacob D. Bekenstein	John A. Wyler
Brendan B. Godfrey	Bahram Mashhoon	Lawrence Ford
Clifford E. Rhoades, Jr.	Yavuz Nutku	G. David Kerlick
Jeffrey M. Greif	James Isenberg	Friedrich Hehl
Demetrios Christodoulou	Claudio Teitelboim	Adam Burrows
Remo Ruffini	Eckehard Mielke	Larry Smarr
Leonard Parker	Orlando Alvarez	Daniel Rohrlich
Terrence J. Sejnowski	Charles Patton	Gary Glenn Miller
Robert Wald	Dragoljub Cetkovic	Gary Horowitz

Fig. 5 The 1970s

Princeton. Of course Kip is probably John's most distinguished student apart from Feynman and I'm sure everybody knows about all the activities and contributions he's made and is still making. Ed Taylor is on the education side. He joined with John to write a couple of books about relativity for the high school student or the general public. Anthony Zee, I'm not sure exactly – maybe he was undergraduate also – and went on to be a particle theorist. Roger Penrose came for a postdoc for a substantial period of time. Dennis Sciama, whose name is not on this list yet, wasn't a postdoc, he was more senior. Then Robert Geroch, who is a professor in Chicago now, and Arthur Fisher who is distinguished mathematician in Santa Cruz (Fig. 5).

And there are all these in the 1970s: Demetrios Christodoulou; he is a very gifted mathematical physicist who has proved things about the existence of singularities; I think contradicting Penrose's earliest formulation of cosmic censorship. Remo Ruffini, who spent 10 years at Princeton as a postdoc, assistant professor, etc., and worked with a huge batch of people. I think I found that he had 61 different collaborators from Princeton during his ten years there. Leonard Parker – I don't know his connection to John [Visiting Fellow 1971–1972], but I do know of his work on particle creation and quantum field theory in the early universe. Bob Wald, who of course you know from an excellent textbook, is professor at Chicago. Jacob Bekenstein, who was famous for his introduction of entropy of a black hole. Yavuz Nutku must have just been an occasional visitor. I mean, he was a student of Chadrasekhar and did a postdoc at Maryland. He greatly admired Wheeler. Jim Isenberg wrote an undergraduate thesis with Wheeler and feels that his approach to physics was strongly influenced by that. Claudio Teitelboim spent many years at the Institute for Advanced Study as a regular visitor and worked with Wheeler. I'll tell you more about John Wyler later. Larry Ford. Larry Smarr spent, I think, a month in Princeton and felt much inspired by it. Well, Smarr is primarily associated with computers, but he came into it through numerical relativity. Isaacson, who was director at NSF, learned how to manipulate the bureaucracy from Marcel Bardon. His first move to get gravitational waves going was to get supercomputers going. So Isaacson was very important in getting the first round of supercomputers financed around the country for use in science (and especially physics) including setting up the Illinois Supercomputers Institute which Larry Smarr led. Smarr started out in Texas with Bryce DeWitt who had done computations on the A-bomb way back when and knew about computing and was interested in numerical relativity. He got Larry Smarr into that and Larry Smarr did some of the earliest successful work on numerical relativity. Smarr then went to Illinois where he encouraged that and lots of other things including the web browser, which was invented there while he was the director. Then he continued mostly in that field, but he was an important player in the early stages of numerical relativity. Gary Horowitz of course is a professor at Santa Barbara and continues to play an important role in physics (Fig. 6).

So all these people, including Bei-Lok Hu at Maryland, interacted strongly. Bill Unruh just celebrated his 65th birthday in the summer. Zerilli played an important role in early studies of perturbations of the Schwarzschild spacetime and therefore

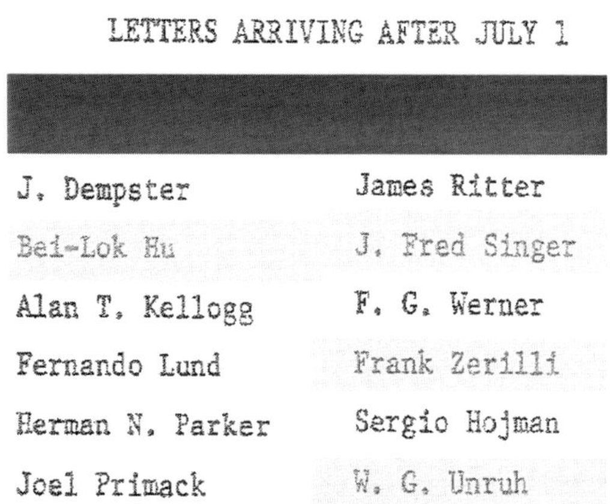
LETTERS ARRIVING AFTER JULY 1

[redacted]

J. Dempster	James Ritter
Bei-Lok Hu	J. Fred Singer
Alan T. Kellogg	F. G. Werner
Fernando Lund	Frank Zerilli
Herman N. Parker	Sergio Hojman
Joel Primack	W. G. Unruh

Fig. 6 Late contributers to the Putnam book

the ring down idea of a signature of black holes which is now very important as a signature in gravitational wave observations. Fred Singer was from John's early stages – cosmic ray physics. He's been a devil's advocate for many years on the unpopular side of numerous questions, currently global warming. He's done all kinds of things beginning with the Van Allen belts of trapped particles around the earth.

So that's a list of the people, a very huge number actually, – large numbers of whom have done very significant things – that wanted to extend their appreciation of John at that time. So let's go through the details of some of their comments.

Fred Singer has this interesting story. His thesis committee included Oppenheimer and Niels Bohr. "I started to explain my work. Oppie interjected, took over, and held the floor. I sat down while he continued, Bohr fell asleep, and [Wheeler] looked vaguely unhappy. I passed the exam."

Here's a note from Tulio Regge (Fig. 7). John had met Regge at a Rochester conference when he was a grad student in the United States and due to go back to Italy at the end. John was impressed by him and invited him to come to Leiden in Holland and found some money to get him there. There they worked on the perturbations of Schwarzschild. John's style, which fits quotes from other people later, is that you should preferably know the answer before you start calculating. So John wrote a paper on the perturbations of Schwarzschild; gave it to Regge, and said "here's the outline of what we expect to find, please fill in the equations". Regge says that wasn't a bad way to approach the problem. He said he learned a lot from it and was up to the task. At that time the nature of the horizon was not understood well and so I don't think they had the boundary conditions exactly right. They laid out all the technology that was needed and pointed out that this was an important problem.

There is a story from someone than you probably don't know (Raymond Mjolsness). He was working as a graduate student with John. He was given John's notebook to borrow for a few days and he thought that was a wonderful experience because he could look in there and see how John goes about organizing his work. John did have this habit for, I guess, all of his life of having bound notebooks. I looked at one at a bookstore the other day; it costs $90 to get one of these books – bound books of blank pages – but John had very nice bound books. They were

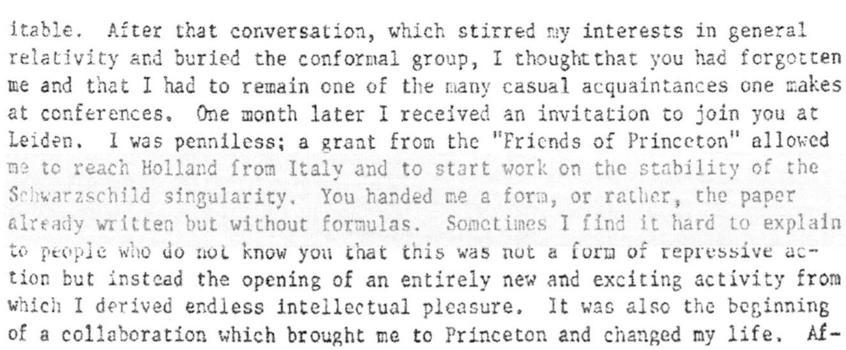
itable. After that conversation, which stirred my interests in general relativity and buried the conformal group, I thought that you had forgotten me and that I had to remain one of the many casual acquaintances one makes at conferences. One month later I received an invitation to join you at Leiden. I was penniless; a grant from the "Friends of Princeton" allowed me to reach Holland from Italy and to start work on the stability of the Schwarzschild singularity. You handed me a form, or rather, the paper already written but without formulas. Sometimes I find it hard to explain to people who do not know you that this was not a form of repressive action but instead the opening of an entirely new and exciting activity from which I derived endless intellectual pleasure. It was also the beginning of a collaboration which brought me to Princeton and changed my life. Af-

Fig. 7 Note from Tulio Regge

always there. When he had a group of students in the office he would sit down and take notes as the discussion went on. He would also make notes to himself about the calculations he was doing, or the work he planned to do. What were the important questions in physics? and so forth. So the student thought he learned a lot by being able to see into John's way of approaching physics. Those notebooks incidentally have been given to the American Philosophical Society in Philadelphia and there they are available to researchers. It's very likely that Joe Weber picked up his similar way of handling things from his association with John in the 1950s.

Another saying of John Wheeler's: "an expert is one who's made every mistake in some field" (Fig. 8) is one version; the version I remember is "the way to be a very good physicist is to be able to make mistakes faster than anybody else". So he sort of assumed there's a quota and that you've got to have a certain number of mistakes before you get to some real progress.

Barry Malik, a student who worked with John: that quote again, "never do a complicated calculation unless you know the answer" (Fig. 9). Another thing I'd

him. The one saying of Professor Wheeler that I quote more than any other to my coworkers is 'An expert is one who has made every mistake in some field.' It is amazing how often this is appropriate."

Fig. 8 Wheeler on expertise. From a note by Bob Euwema

A perfect legacy of Bohr and Einstein. A firm believer in the simplicity and the universality of physics ("Never do a complicated calculation unless you know the answer" or "If one really understands a phenomenon, one can build a simple model and explain to anybody"). A radical conservative, ready to recognize new ideas but only after examining the old one carefully. A worthy carrier of the Bohrian tradition of being

A great devotee to science but never failing to recognize the greater truth, so vividly described by the medieval Bengali poet "Chandidas"

সবার উপরে মানুষ সত্য

(Above everything lies the humanity). A great believer and true practitioner of democracy, human rights and decency,

Fig. 9 Know the answer before you begin. Wheeler was a *Radical Conservative*. From a note by Barry Malik hear so often: "if you really understand a phenomenon you can build a simple model and explain it to anybody". That is certainly one of John's teaching attitudes: if you really understand something you should be able to explain it to less than your collaborators.

As Malik noted, John has this phrase "radical conservative". I remember the phrase "dynamic conservativism" [Dieter Brill recalls "daring conservativism"]. This was a view that John used to justify some of his outrageous work like wormholes and geons and whatnot. Quantum foam. But especially on things like wormholes. His attitude there was: you should not just look to a theory to explain the things you know, you should look to the equations and see what they predict even if you can't imagine why such a thing should be part of physics. He wanted to push the equations to their limit and see what's in them – that's the conservativism. We've got these equations, principally Einstein's, why do we just use them in cases where we already understand what to expect? Why don't we push them and see if they've got anything else to tell us that might be a surprise?

As I was having these pages from this privately published book – there's only a half-dozen copies in existence – as I was having them copied here in our Maryland Physics Department Xerox room, the woman who was running the Xerox looked at this page, Fig. 9 and she said "that's my language!" so one of our staff at Maryland is Bengali.

Now Kip Thorne on John's writing (Fig. 10). John had a style of writing that many people were able to recognize. Kip could do more than recognize it. Kip says one of his students came up with the MTW book and says "This is just too strange a way to write. Why did you let John Wheeler get away with that?" Kip said "Wheeler didn't write that section; I wrote it!" Kip studied John's style of writing. That makes it hard to decode MTW. You can probably find, if things are written in sort of a dull straightforward way, that I had a hand in it, but it's very hard to distinguish Kip from John Wheeler in there. John did the introductions to most of the chapters; made sure there was something like a sentence or two that would be meaningful to anybody to start things out. John's most obvious peculiarity was a sort of Germanic sentence

Style of writing: Nobody else in science writes quite like you, Johnny! (Or almost nobody else. One of my most memorable moments was several years ago when a delegation of my students came to me, MTW in hand, to complain that a certain section was too "Wheeleristic". Why had I permitted it to go through without toning down? With glee I told them "Wheeler did not write that section; I wrote it!") Your style -- elegant and crystal clear, with twists of amazement and amusement; poetry in the midst of prose; the pithy phrase pregnant with insight; parenthetical allusions to whole bodies of knowledge which the reader might have encountered elsewhere; argument by analogy and paradox -- it is a style that creates enthusiasm in some; revulsion in others. But even in those

Fig. 10 John Wheeler had a distinctive writing style. From a note by Kip Thorne structure in which you try, as much as you can in English, to put the verb at the end, having started out with a bunch of other qualifications, until the whole point gets made, with emphasis.

Jim York wrote a long letter talking about how he got into serious physics by getting a chance to meet John and talk to him. Then John simply used his influence to get Jim invited to the Battelle conference. That got York into the real meat of the business and he never slowed down. He continued to work very hard making contributions to primarily the initial value problem, eventually in collaborations with Mme. Choquet-Bruhat with many very rigorous results in that area.

Demetrios Christodoulou was perhaps even younger than John Wheeler as a graduate student. Papapetrou recommended him to John. John was able to meet him in Paris one year when he was there. So John agreed that this was a very outstanding young man, and talked Princeton into accepting him in spite of the fact that he probably hadn't finished high school. He got picked up and learned to do physics at Princeton working with John and very much appreciated the fact that things could be done that way.

Remo Ruffini wrote. As I mentioned, his letter or his CV says that while at Princeton he collaborated with this long list of people. I went through trying to count from his publications and found about 61 different collaborators. He was there during 10 years from 1967 to 1976, I think, during a very active time in the development of the physics of black holes.

Bob Wald says [Wheeler taught him that] "one should always think in a completely down-to-earth manner and decide by physical intuition what ought to be true; then one should obtain a mathematical proof or disproof of one's conjecture." So this again is John's idea that you really should think through a problem in its broad outlines and take all the clues you could have from other aspects of physics and other ways of thinking about it so that you know how to work your way through the problem before you get deep into it.

Sergio Hojman (Fig. 11) points out John's idea of inspiration in saying that physics is going on today even as you attend your first physics class. In that class you meet Newton and learn things that were known 400 years ago, but you should realize that there's real physics going on right now. And of course there are great discoveries being made right now. John was able to get students in the undergraduate classes to realize that physics is really going on now in a serious way.

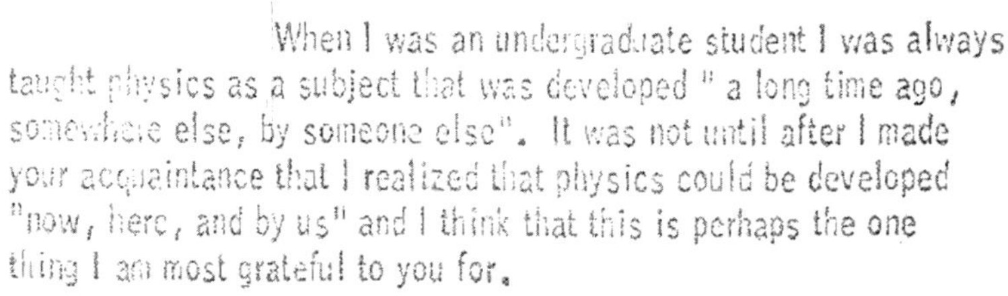
When I was an undergraduate student I was always taught physics as a subject that was developed " a long time ago, somewhere else, by someone else". It was not until after I made your acquaintance that I realized that physics could be developed "now, here, and by us" and I think that this is perhaps the one thing I am most grateful to you for.

Fig. 11 Physics is being developed here and now. From a note by Sergio Hojman the beach where grandson "Wheeler" was playing. (We were at the High Island summer home.) Suddenly John wrote across the top of the page, in large block letters:

"DESLUDGE"

At the time, I had no idea what to make of this strange collection of letters.

Teaching has always been a most important enterprise to John Wheeler [How many full professors want to teach the basic college physics courses, and want to write introductory physics texts?]. And "desludged" teaching is the only kind he does. So it is with the members of the clan. Indeed, one of the clearest

Fig. 12 DESLUDGE! From a note by Jim Isenberg

Dear John:

What clearer evidence of my own scientific debt to you could there be, than that contained in my paper "Rasputin, Science and the Transfiguration of Destiny" (Gen. Rel. Grav. 5, 175 (1974))? I have before me your kind letter of September 1, 1974, expressing appreciation of my small effort. No greater influence, no more important inspiration, no more constant source of ideas, than your own, has any other teacher had on my writings. My existence, indeed, is no less a figment of your conception than of any other. How reasonable to suppose that this influence resulted from a long period of collaboration and association. And yet, nothing could be farther from the truth. There stands out in mind with startling

Fig. 13 John A. Wyler emulated John A. Wheeler's writing style

Jim Isenberg (Fig. 12) wrote about his Ph.D. thesis done at Maryland: he took it up to Maine for John to read and John wrote across the page "DESLUDGE". The question is, what does that mean? Jim found it means that, as you would clean out the garage under your car where it had dripped, you want to get rid of all the junk that's messing things up; clean out unnecessary words, etc. Teaching has been an important enterprise, he says, and de-sludged is the only kind he does. John would work with many of his students with lots of effort on the writing.

Claudio Teitelboim reports the same thing. John would spend a lot of time correcting English. Claudio said "John, why are you spending so much time on that" and John said "That's what I'm paid for. The central idea should always stand out clearly, sharply, just as in 'Cuba si, Yankee no'". That was John's advice to Teitelboim on making sure that there is a clearly understandable point.

Now on John's writing: there is a paper in the GRG Journal by John A. Wyler (Fig. 13) which caused a lot of fun when it first came out. It came out as a Princeton preprint. In those days with no web one had standard little reproduced packages of the typewritten papers which were sent around to your friends. So this came around to everybody. You really had to read for several minutes into that paper before you realized the author was not John A. Wheeler. Wyler was mimicking John Wheeler's style and the paper only slowly got more outrageous than John could be so you begin to see that it was a spoof. But it was such a good spoof that it later got published. You can look it up if you want to.

Bill Press wrote the Wyler paper. He had all the quirks of John's writing down pat. Figure 13 is a letter in the book from John A. Wyler to Wheeler. It is signed: "dictated in Dr. Wyler's absence, proofread and signed by [the secretary] in his presence". This was the style of John Wheeler's known to hundreds of his correspondents. So if you have read many things by Wheeler, you would enjoy reading Wyler.

Again from the book, Larry Smarr talks about how John picked him up and gave him a chance to spend a month at Princeton, during a very active time. Smarr names all these people who are working actively on blackhole related physics. So John, when he found a gifted person, would go really out of his way to make sure they got a leg up to go somewhere.

3 John A. Wheeler and the Renaissance of General Relativity

I will now emphasize less John Wheeler as teacher to talk specifically about his impact on relativity: the recertification of GR as true physics. A brief summary of relativity in the early years: There was a period where cosmology got started. There were some important works in the 30s – the Einstein–Infeld–Hoffman ideas. There was also a paper on inhomogeneous cosmologies that's been very useful for people. Unified Field theories were the bane of GR in those days. Einstein worked on them. Einstein was convinced that physics should be primarily geometry, and he was trying to stretch the ideas of geometry. It's sort of amusing that about 10 years later, maybe 15, Steve Weinberg was convinced that geometry was irrelevant to physics. He wrote a textbook on General Relativity from a field-theory/special-relativity point of view as much as possible saying that. He implies that the geometry is sort of window dressing; it makes things look prettier, but the important stuff is just field theory. A peculiar thing is that it was Weinberg, later, who collaborated in proving that physics really is geometry. Except it's not the geometry of space–time, it's the geometry of the graph paper on which the properties of space–time are conceptually plotted. That's the idea of a curved connection. If you want plot the values of any physical quantity that's of fundamental importance, like a magnetic field, quarks, gluons, etc. you need to plot on curved graph paper. But Einstein wasn't looking there – didn't have that broad an idea of geometry – so he never got on the right track of where geometry really comes predominantly into physics. There were also people who did straight General Relativity, who were just very good physicists, but there were limited numbers of them.

In these early years relativity had a very bad reputation among many physicists. Some Caltech physics professor (reported in Kip's book, [10], but not named by

Kip), as Kip was leaving Caltech as an undergraduate for Princeton as a graduate student, told Kip that "General Relativity has little relevance to physics. You should look elsewhere for interesting physics". Kip was being warned away from relativity. In Russia Novikov was getting interested in relativity and Novikov's wife (also a physicist) was given advice by senior physicists that she should get Ivor out of it or he will never have a career; she was told "relativity is a backwater, Ivor should get out of it". (This is in Kip Thorne's book.)

After World War II there were a number of active people in General Relativity, of the generation ahead of John's students; Post WWII GR leaders:

John Wheeler (Princeton)
Dennis Sciama (Cambridge)
Zel'dovich (Moscow)
Bondi (London)
Lichnerowicz & Choquet–Bruhat (Paris)
Bergmann (Syracuse)
Schild (Austin)
DeWitt (Chapel Hill)

Wheeler went headlong into General Relativity, and his "dynamic conservativism" had these mottos:

- **Mass Without Mass**
  John tried to show the collection of fields, electromagnetic waves, running around in circles and held into circles by gravity, could form an object ('Geon') that had a mass even though there was nothing there but zero mess fields. That was one of the things he was pushing then and later discovered that the whole system is unstable.
- **Brill Waves**
  Proved with Dieter Brill – supervising Brill's thesis – (after I'd been there pushing on initial conditions for some time) that with a rigorous solution of the initial value equations and not just approximations as for geons, but done in a rigorous way, there are bundles of gravitational waves that have mass. It wasn't an attempt to say there was anything stable about them, just that you with nothing but empty space and curvature you can get mass.
- **Charge Without Charge**
  The idea started with John's wormholes. They got turned into good geometry, I guess primarily by me, after I'd read the French book by Lichnerowicz and relying a lot on the differential geometry people were teaching me. Finkelstein, and Beckedorff working with me, got completely geometrical views of the horizon, which was then called the Scharzschild singularity, and so one could make initial conditions corresponding to Wheeler's diagrams of wormholes and show that they really were consistent with the Einstein equations. One later found out that they wouldn't last long enough for anything to pass through them if you had normal kinds of matter.

And so those were the dynamic conservativism kinds of ideas. In the early 1960s John was fighting to understand black holes (as yet unnamed). His concentration was on "what is the ultimate fate of a star?" He expected that there would be something to avoid all the singularities. It took a long time for him to be convinced otherwise. Finkelstein certainly published the first good interpretation of the "Schwarzschild singularity" (the horizon), the real physical interpretation [4]. The [2] results are published in an obscure place, the proceedings of a 1963 conference on Time that Tommy Gold organized at Cornell [8], after the Finkelstein paper. In fact, Finkelstein's paper pushed Wheeler into writing [sic] the Kruskal paper [7]. The Kruskal picture, which goes beyond Finkelstein only in unphysical ways in my view, is anyway mathematically satisfying. So the Beckedorff work came after Kruskal but importantly pointed out how you would match a collapsing star in the Kruskal diagram to the external solution. It showed a physical production of a black hole in a way that you could understand better than Oppenheimer and Snyder [9]. Oppenheimer and Snyder do everything; at least everything exterior to the horizon they had correct. They knew the physics but Oppenheimer never considered the subject again. Did Wheeler really regard extended Schwarzschild (just a vacuum black hole) as mass without mass? I think he eventually did.

John's main theme, "dynamic conservativism" (or as Dieter Brill remembers, "daring conservatism") drove him to look to the physics; get the physics out of the equations, encourage experimental/observational relativity. [*It also, together with desired quantum treatments for cosmology, led him to stressedly support Everett's interpretation of quantum mechanics, which has been the subject, fifty years after Everett's thesis, of two conferences in 2007.*]

John encouraged Weber's work on wave detectors. He worked on the Regge–Wheeler perturbations of Schwarzschild. He influenced Dennis Sciama and Roger Penrose. I think Penrose gives Sciama credit for pulling him into relativity from algebraic geometry. Sciama of course was a strong presence in the astrophysical side, a much better astrophysicist than Wheeler I would think. Sciama was a tremendous influence on Penrose who was also encouraged a lot by Wheeler and made visits in early days to Princeton. New experiments were going on; the Dicke experimental group in fact was right there in Princeton. John saw the beginnings of numerical relativity in which Hahn and Lindquist made an attempt to actually integrate the Einstein equations numerically in a very simple case.

Then as next decade came on, the Princeton group got much bigger with important work being done by many people. And there are new research groups building up various places. Sciama got many people working with him. I came to Maryland and began to have other students and postdocs. Schild went to Texas. Thorne, some years later than me, went to Caltech and began building up a group. Infeld. Chandra came into relativity and attracted people there. Hawking. Ted Newman. Ted Newman is somewhat overlooked, I think, because on the Newman–Penrose formulation, he did the first work which got cleaned up by Penrose. The idea of going to null frames was Newman's, and the idea that doing so would simplify the search for solutions of Einstein's equations was his. The discovery of the Kerr solution would have been much delayed without the work of Ted Newman. Then the real astrophysics began coming in, and a lot of other developments came during the last years of that decade as I have tried to show.

Here are places you can look for some of this history: John Wheeler's autobiography with the assistance of Ken Ford [11], which was published eight years ago. Kip Thorne wrote a book 10 or 11 years ago [10], which has a lot of history of these periods in it. There is the book which Peter Putnam organized – that is, he financed it and pushed people into it, and directed John Wheeler's secretary to send out letters recruiting contributions. This volume, in a really elaborate binding, was given to Wheeler at the GR8 meeting. It was an absolutely beautiful thing; probably cost even in 1977 a thousand or two for the binding. Half a dozen other (less elegantly bound) copies are around. I have one; the American Institute of Physics; Princeton and Texas have copies, so they exist. Then there is a book by Bartusiak [1] containing some interesting history, and there is a book in the press (I've seen a draft [5, 6]) by Kennefick, who is a Caltech physicist who has been doing a lot on the history of physics. And so, you can look forward to that and other sources for the history of this era.

I hope this has given a flavor of the really important influence John Wheeler has had on General Relativity.

After his retirement from Princeton in 1975, John Wheeler moved to The University of Texas at Austin. He established an active and dynamic group with interests in fundamentals of quantum mechanics, quantum measurement, and quantum computing. He also pursued ongoing interests: sum over histories, the structure of field theories, Mach's principle, new formulations of the Einstein system, and astrophysics. Seven people received Ph.D.s from Texas under Wheeler's direction or co-direction: William Wootters, Paul Gleichauf, Ignazio Ciufolini, Warner Miller, David Henry King, Arkady Kheyfets, Benjamin Schumacher. Wheeler retired from The University of Texas at Austin in 1985, but remained a frequent visitor well into the 1990s.

References

1. Bartusiak, M. 2000, Einstein's unfinished symphony : listening to the sounds of space-time (National Academy Press, Washington DC., 2000)
2. Beckedorff, D. L. 1962, Terminal Configurations of Stellar Evolution (Thesis, Princeton University, Department of Mathematics, http://www.physics.umd.edu/grt/cwm/Beckedorff1962.pdf)
3. Bird, K. & Sherwin, M. J. 2005, American Prometheus (Vintage Books, New York, ISBN 0-375-72626-8)
4. Finkelstein, D. 1958, Physical Review, 110, 965
5. Kennefick, D. 2007, Traveling at the Speed of Thought: Proving the Existence of Gravitational Waves ((Princeton University Press) ISBN-13 978-0-691-11727-0)
6. —. 2007, APS Meeting Abstracts, 7001
7. Kruskal, M. D. 1960, Phys. Rev., 1743
8. Misner, C. W. 1967, in: The nature of time, T. Gold and H. Bondi, editors (Ithaca, N.Y., Cornell University Press, http://hdl.handle.net/1903/4280)

9. Oppenheimer, R. & Snyder, H. 1939, Physical Review, 56, 455
10. Thorne, K. S. 1994, Black holes and time warps: Einstein's outrageous legacy (Commonwealth Fund Book Program, New York, NY: W.W. Norton and London: Picador)
11. Wheeler, J. A. & Ford, K. 1998, Geons, black holes and quantum foam : a life in physics (Geons, black holes and quantum foam : a life in physics. Publisher: New York, NY: Norton, 1998. ISBN: 0393046427)

John Archibald Wheeler: A Few Highlights of His Contributions to Physics

Kip S. Thorne and Wojciech H. Zurek

Abstract The following quotations describe in "nutshells" a few highlights of John Archibald Wheeler's contributions to physics. The contributions are arranged in roughly the following order: (i) concrete research results, (ii) innovative ideas that have become foundations for the research of others, (iii) insights that give guidance for the development of physics over the coming decades. Since most of Wheeler's work contains strong elements of two or even all three of these characteristics, the editors have not attempted to delineate the dividing lines between the three categories.

A description of the nucleus which regards the neutrons and protons as spending part of their time in configurations corresponding, for example, to interacting alpha-particles, part of their time in other groupings, already takes into account to a large extent that intimate interaction between nuclear particles which is so entirely different from the situation in atomic structure ... The method of "resonating group structure" builds up a wave function for the whole nucleus out of partial wave functions which describe the close interactions within the individual groups. [10]

[Wheeler's method of "resonating group structure," introduced in this paper, was called "clustering theory" by some later researchers, and was the direct ancestor of the "method of generator coordinates" developed later by Wheeler and D. L. Hill.]

The c's form a unitary [scattering] matrix ... The equality which must exist between the numbers of incoming and outgoing groups in the [given] state ... for arbitrary choices of the a's ... is just the necessary and sufficient condition for unitary character. ... [Time-reversal invariance] shows that $\|c_{mn}\|$ *is a symmetrical matrix, and ... each element of the matrix is determined up to a factor* ± 1. [10]

K.S. Thorne (✉)
Division of Physics and Astronomy, California Institute of Technology, Pasadena, CA 91125
e-mail: kip@caltech.edu

W.H. Zurek
Theoretical Astrophysics, Los Alamos National Laboratory, Los Alamos, NM 87545
e-mail: whz@lanl.gov

I. Ciufolini and R.A. Matzner (eds.), *General Relativity and John Archibald Wheeler*, Astrophysics and Space Science Library 367, DOI 10.1007/978-90-481-3735-0_3,

[The matrix $\|c_{mn}\|$ introduced for the first time in this paper was later given the name "scattering matrix" or "S-matrix" and has come to play a major role in nuclear physics and elementary particle physics.]

On the basis of the liquid drop model of atomic nuclei, an account is given of the mechanism of nuclear fission. In particular, conclusions are drawn regarding the variation from nucleus to nucleus of the critical energy required for fission, and regarding the dependence of the fission cross section for a given nucleus on the energy of the exciting agency." [1]

[This paper, written only a few months after the discovery of nuclear fission, gave the definitive theory of fission. It even enabled Bohr and Wheeler to predict that the yet-to-be-discovered nucleus plutonium-239 would have a very large cross section for fission when bombarded by slow neutrons – a cross section that would make it the chosen fuel for the first American atomic bomb.]

[Describing the first, abortive attempt on September 27, 1944 to operate at full strength the Hanford Washington reactor that was to produce Plutonium-239, Wheeler says]: It had been one of my jobs to consider every possible way that things might go wrong. I was, therefore, very aware that a fission product, when it decayed, could give rise to another one which could absorb the neutrons. When [the reactivity of the reactor mysteriously began to fall] and then, a few hours later, ... began rising again, I was sure that this was what had happened. Now the second nucleus had decayed into a third one which did not absorb the neutrons; ... the culprit had to be Xenon-135. [27]

[Thanks to Wheeler, the Hanford reactor had already been designed to deal with the possibility of such "poisoning" by fission product nuclei with high neutron capture cross sections. Some months earlier he had started to warn of the possibility of cross sections far higher than the typical 10 barns that had been measured up to that time; perhaps even as high as 1–25 million barns. The reactor had been redesigned at the last minute to allow for this possibility; and this permitted it, with the readjustement of the safety rods, to handle Xenon-135 whose cross section turned out to be 3 million barns.]

A quantitative formulation of [the absorber as the mechanism of radiation] is given here on the basis of the following postulates: (1) An accelerated charge in otherwise charge-free space does not radiate energy. (2) The fields which act on a given particle arise only from other particles. (3) These fields are represented by one-half the retarded plus one-half the advanced Lienard–Wiechert solutions of Maxwell's equations. In a system containing particles sufficient in number ultimately to absorb all radiation [these postulates produce the standard radiation-reaction phenomena in the emitter] ... Radiation is [thereby] concluded to be a phenomenon as much of statistical mechanics as of pure electrodynamics. [28]

[Feynman [2] in his Nobel Prize lecture describes the genesis of this work, his interactions with Wheeler during its development, and its considerable influence on Feynman's subsequent Nobel-prize-winning contributions to quantum electrodynamics.]

... by far the dominating type of annihilation is that in which the positron combines with an electron whose spin forms a singlet state with respect to the positron. Associated with this selection of pairs which have zero relative angular momentum, before the annihilation process, is an analogous polarization phenomenon in the two quanta which are left at the end of the process. According to the pair theory, if one of these photons is linearly polarized in one plane, then the photon which goes off in the opposite direction with equal momentum is linearly polarized in the perpendicular plane. [11]

[This paper, in addition to developing the theory of electron–positron atoms, also laid the foundations (in the above quote) for the now-most-familiar variant of the Einstein–Podolsky–Rosen paradox.]

Ignorance of the deeper relation between nucleons and mesons makes it especially appropriate at this time to investigate those features of the behavior of mesons which are largely independent of uncertainties about the nature of elementary particles. Fortunately, a number of conclusions may be drawn about the interaction between the meson and the nucleus when we assume little more than the laws of electrodynamics, elementary notions of nuclear structure, the principle of microscopic reversibility, and the simplest ideas of quantum theory. Thus, it has become clear not only that the meson possesses characteristic Bohr orbits of its own around the nucleus, but also that trapping into these orbits via ordinary atomic interactions is the precursor of any specific reaction with the nucleus. ... Also interesting are the energy levels and interlevel transitions of the meson, the first experimental evidence for which is given by W. Y. Chang in a following paper. Among the possible transitions are not only those in which a photon is emitted or an atomic electron is ejected, but also processes of pair creation and meson-induced fission. ... An attempt is made to analyze these questions in the present paper. [12]

[This paper developed the definitive theory of mu-mesic atoms.]

Analyzing different formulations of conditions for validity of the dispersion formula relating absorption and refraction of light, especially the condition that no wave be scattered before arrival of the primary wave (Kronig) we conclude: it is reasonable to apply the dispersion formula for arbitrarily high frequencies. [9]

[This work showed for the first time that dispersion theory could be extended into the relativistic domain by a proper application of the principle of causality.]

Associated with an electromagnetic disturbance is a mass, the gravitational attraction of which under appropriate circumstances is capable of holding the disturbance together for a time long in comparison with the characteristic periods of the system. Such gravitational-electromagnetic entities, or "geons," are analyzed via classical relativity theory. They furnish for the first time a completely classical, divergence-free self-consistent picture of the Newtonian concept of body... [13]

[This paper introduced the concept of "geon" into physics, and in analyzing geons it introduced the concept of two-lengthscale perturbative expansions into general relativity. Such expansions, in the hands of Wheeler's subsequent students and their students, have become the key foundation for the theory of the interaction of propagating gravitational waves with the "background" curvature of space–time.]

We consider a number of mass concentrations so distributed in space and of such relative magnitudes, that the zone of influence of each can be reasonably approximated by a sphere. Inside each cell [sphere] we replace the actual gravitational potentials by the expressions of Schwarzschild. . . . The mass concentrations on either side of a cell boundary accelerate toward that boundary at such a rate as to nullify the discontinuity in matching of the normal derivative of the gravitational potentials that would otherwise occur. . . . [This] expresses the equation of motion of the mass at the center of a cell as a dynamic condition on the boundary of the cell. . . . The whole of the expansion and subsequent contraction [of the universe] is derived [in this way] from the elementary static Schwarzschild solution. [4]

[This paper introduced into general relativity a technique for matching together two different space–time geometries. This matching technique, in the hands of Wheeler's students and colleagues, has become a foundation for modern analyses of the laws of motion and precession of black holes and other relativistic bodies.]

Space tells matter how to move, . . . and matter tells space how to curve. [5]

. . . we examined the consequences of an equation of state that attempts to span the whole region from normal densities to supranuclear densities. We assume: (1) Cold matter ideally catalyzed to the end point of thermonuclear evolution. . . . (2) General relativity equations of hydrostatic equilibrium. . . . The numerical integrations show for the first time both crushing points [white dwarf and neutron star] on a single curve for mass as a function of central density. [The form of this curve emphasizes that] assembly of an amount of mass that exceeds in order of magnitude the mass of the sun and catalysis of this matter to the endpoint of energy evolution results in a condition which lies at the untamed frontier between elementary particle physics and general relativity. Of all the implications of general relativity for the structure and evolution of the universe, this question of the fate of great masses is one of the most challenging. [3]

[With these words Wheeler began the quest to understand the fates of great masses, a quest that led him a decade later to formulate the concept of a black hole.]

Energy of rotation [of a central neutron star] appears not yet to have been investigated as a source of power [for the Crab nebula]. Presumably this mechanism can only be effective – if then – when the magnetic field of the residual neutron star is well coupled to the surrounding ion clouds. [18]

[This statement, published 1 year before the discovery of pulsars, was the closest anybody ever came before pulsars to the correct explanation of what powers the Crab nebula.]

We write down . . . the linear differential equations for small first-order departures from the Schwarzschild metric. . . . It is shown that a Schwarzschild singularity ["black hole" in the terminology of the 1980s] ... will undergo small vibrations about the spherical form and will therefore remain stable if subjected to a small nonspherical perturbation. [6]

[This was the pioneering analysis of the pulsations and stability of black holes, performed a decade before the concept of a black hole was fully understood.]

... by reason of its faster and faster infall [the surface of a collapsing star] moves away from the observer more and more rapidly. The light is shifted to the red. It becomes dimmer millisecond by millisecond, and in less than a second too dark to see... [The star] like the Cheshire cat fades from view. One leaves behind only its grin, the other, only its gravitational attraction. Gravitational attraction, yes; light, no. No more than light do any particles emerge. Moreover, light and particles incident from outside ... go down the black hole only to add to its mass and increase its gravitational attraction. [19]

[With these sentences, Wheeler coined the name "black hole" thereby catalyzing a major change in viewpoint on something that previously had gone under such names as "Schwarzschild singularity," "collapsed star," and "frozen star." Note that, in typical Wheelerian style, he does not tell us that he is coining a new name for and viewpoint on a fundamental concept in physics; rather, he writes as though this name and viewpoint had always been used – and almost overnight all other researchers in "black-hole physics" embrace them.]

A black hole has no hair. [7]

[This phrase epitomized a result of which hints existed at the time Wheeler coined the phrase; and like so many of Wheeler's pithy phrases, it acted as a stimulus for further research by Wheeler's colleagues, former students and students of students – research that ultimately gave strong validity to the phrase.]

On an atomic scale the metric appears flat, as does the ocean to an aviator far above. The closer the approach, the greater the degree of irregularity. Finally, at distances of the order $L^*[=(\hbar G/c^2) = 1.6 \times 10^{-33} cm]$, *the fluctuations in the typical metric component,* $g_{\mu\nu}$, *become of the same order as* $g_{\mu\mu}$ *themselves. Then the character of the space undergoes an essential change ... Multiple connectedness develops, as it does on the surface of an ocean where waves are breaking.* [14]

[Wheeler's intuitive insight that space–time may be multiply connected (foam-like) on the scale L^* has received considerable substantiation from recent research on the quantum theory of gravity. The lengthscale L^* is called by Wheeler the "Planck length" and by others the "Planck–Wheeler length," in recognition that Wheeler was the first to understand clearly its significance.]

Along with the fluctuations of the metric there occur fluctuations in the electromagnetic field. In consequence the typical multiply connected space ... has a net flux of electric lines of force passing through the "wormhole." These lines are trapped by the topology of space. These lines give the appearance of a positive charge at one end of a wormhole and a negative charge at the other. [14]

The quantum mechanical state of the metric field is a functional of the three-dimensional geometry $\mathcal{G}$ *intrinsic to a spacelike surface* σ

$$\Psi = \Psi(\mathcal{G})$$

The probability amplitude Ψ *typically falls off rapidly for* large scale *departures from a smoothly behaved metric, but falls off very little for departures from smoothness which are of the scale* $L^* = (\hbar G/c^2) = 1.6 \times 10^{-33}$*cm or less.* [15]

[This line of thought led, within the next few years, to the DeWitt–Wheeler formulation of quantum gravity, which in the hands of Hartle, Hawking, and others has recently achieved considerable success in understanding the initial conditions of the universe.]

... one feels that one has, at last in gravitational collapse, a phenomenon where general relativity dramatically comes into its own, and where its fiery marriage with quantum physics will be consummated. [16]

The law of conservation of baryons must be abandoned for a star that exceeds the critical baryon number. [17]

[This conclusion, forced on Wheeler by a long struggle to understand the fate of great masses, became a central issue in the 1970s with the discovery by Hawking that black holes can evaporate. This conclusion also led Wheeler to speculate, as an inspiration for future research, about the mutability of all conservation laws – and indeed of all the laws of physics – in the big bang and big crunch of the universe.]

... These advances in cosmology and black hole physics lend new emphasis to some of the most remarkable consequences of Einstein's standard theory: ... nature conserves nothing; there is no constant of physics that is not transcended; or, in one word, mutability is a law of nature. [20]

Law without Law. *"Physical space–time is not mathematical space–time" is the one lesson of mutability; the other, "Physical law is not ideal mathematical law." Law that comes into being at the beginning of time and fades away at the end of time cannot be forever 100% accurate. Moreover, it must have come into being without anything to guide it into being.* [22]

Not only particles and the fields of force have to come into being at the big bang, but the laws of physics themselves, and this by a process of higgledy-piggledy as genetic mutation or the second law of thermodynamics. [25]

Time is defined so that motion looks simple. [5]

Time ends. That is the lesson of the "big bang." It is also the lesson of the black hole. [23]

Time is not a primary category in the description of nature. It is secondary, approximate, and derived. [25]

[Wheeler is also fond of the phrase "Time is God's way of keeping things from happening all at once" (anonymous graffiti, discovered by him on the wall of Pecan Street Cafe in Austin, Texas).]

No account of existence that presupposes the concept of time can ever account for either time or existence. [25]

No elementary phenomenon is a phenomenon until it is an observed phenomenon. [21]

[This phrase, and variations on its theme, paraphrase the central point of Niels Bohr. They serve Wheeler and his colleagues both as a point of departure and as a motto in the struggle to understand the relation between the quantum and measurements.]

... if one really understood the central point and its necessity in the construction of the world, one ought to be able to state it in one clear, simple sentence. Until we see the quantum principle with this simplicity we can well believe that we do not know the first thing about the universe, about ourselves, and about our place in the universe. [20]

The final story of the relation between the quantum and the universe is unfinished business. We can well believe that we will first understand how simple the universe is when we will recognise how strange it is. [20]

[Wheeler often quotes Gertrude Stein on modern art: "It looks strange and it looks strange and it looks very strange, and then it suddenly does not look strange at all and you can't understand what made it look strange in the first place."]

The elementary quantum phenomenon is the strangest thing in this strange world. It is strange because it has no localisation in space and time. It is strange because it has a pure yes–no character – one bit of meaning. It is strange because it is more deeply dyed with an information-theoretic flavor than anything in all physics. [26]

Quantum mechanics promotes the mere "observer of reality" to the "participater in the defining of reality." It demolishes the view that the universe exists out there. [20]

The universe is a self-excited circuit. As it expands, cools and develops it gives rise to observer participancy. Observer participancy in turn gives what we call tangible reality to the universe. [22]

[A graphic summary of this point can be found, for instance, in Fig. 22.13 of Wheeler [22].]

... The wave function is within the present incomplete marriage of quantum theory and relativity not a very happy device for calculating the probability of real

coincidences. Only when the counters have gone off has the reality of the situation declared itself. [20]

... It is wrong to think of that past [ascribed to a quantum phenomenon] as "already existing" in all detail. The "past" is theory. The past has no existence except as it is recorded in the present. By deciding what questions our quantum registering equipment shall put in the present we have an undeniable choice in what we have the right to say about the past. [22]

[The delayed-choice experiment described by Wheeler in this paper forces one to recognize this unexpected and yet inescapable conclusion about the quantum Universe.]

What we call reality consists ... of a few iron posts of observation between which we fill an elaborate papier-mache of imagination and theory. [22]

Spacetime in the prequantum dispensation was a great record parchment. This sheet, this continuum, this carrier of all that is, was and shall be, had its definite structure with its curves, waves and ripples; and on this great page every event, like a glued down grain of sand, had its determinate place. In this frozen picture a far-reaching modification is forced by the quantum. What we have the right to say of past spacetime, and past events, is decided by choices – of what measurements to carry out – made in the near past and now. [24]

Planck's discovery of the quantum in 1900 drove a crack in the armor that still covers the deep and secret principle of existence. In the exploitation of that opening we are at the beginning, not the end. [25]

Beyond the probability interpretation of quantum mechanics, beyond all the standard analysis of idealised experiments, beyond the principle of indeterminacy and the limits it imposes, lie deep issues on which full agreement has not yet been reached in the physics community. They include questions like these: Does observation demand an irreversible act of amplification such as takes place in a grain of photographic emulsion or in the electron avalanche of a Geiger counter? And if so, what does one mean by "amplification"? And by "irreversible"? Does the quantum theory of observation apply in any meaningful way to the "whole universe"? Or is it restricted, even in principle, to the light cone? And if so, whose light cone? How are the observations made by different observers to be fited into a single consistent picture in space–time? [29]

... the most revolutionary discovery in science is yet to come! And come, not by questioning the quantum, but by uncovering that utterly simple idea that demands the quantum. [26]

WHEELER'S FIRST MORAL PRINCIPLE. Never make a calculation until you know the answer. Make an estimate before every calculation, try a simple physical argument (symmetry! invariance! conservation!) before every derivation, guess the answer to every puzzle. [8]

References

1. Bohr, Niels, and Wheeler, John A. (1939), "The Mechanism of Nuclear Fission," *Phys. Rev.* **56**, 426.
2. Feynman, Richard P. (1966), "The Development of the Space-Time View of Quantum Electrodynamics," *Phys. Today*, August 1966, p. 31.
3. Harrison, Kent, Wakano, Masami, and Wheeler, John A. (1958), "Matter-Energy at High Density; End Point of Thermonuclear Evolution," in *La Structure et L'Evolution de l'Univers*, Proceedings of Eleventh Solvay Congress, Brussels, 9–13 June, 1958 (Stoops, Brussels), p. 124.
4. Lindquist, Richard W., and Wheeler, John A. (1957), "Dynamics of a Lattice Model Universe by the Schwarzschild-Cell Method," *Rev. Mod. Phys.* **29**, 432.
5. Misner, Charles W., Thorne, Kip S., and Wheeler, John A. (1973), *Gravitation* (Freeman, San Francisco), p. 23.
6. Regge, Tullio, and Wheeler, John A. (1957), "Stability of a Schwarzschild Singularity," *Phys. Rev.* **108**, 1063.
7. Ruffini, Remo, and Wheeler, John A. (1971), "Relativistic Cosmology and Space Platforms," in *Proceedings of Conference on Space Physics* (ESRO, Paris).
8. Taylor, Edwin F., and Wheeler, John A. (1966), *Spacetime Physics* (Freeman, San Francisco).
9. Toll, John, and Wheeler, John A. (1951), "Some Pair-Theoretic Applications of the Dispersion Relation," *Phys. Rev.* **81**, 654.
10. Wheeler, John A. (1937), "On the Mathematical Description of Light Nuclei by the Method of Resonating Group Structure," *Phys. Rev.* **52**, 1107.
11. Wheeler, John A. (1946), "Polyelectrons," *Ann. N.Y. Acad. Sci.* **48**, 219–238.
12. Wheeler, John A. (1949), "Some Consequences of the Electromagnetic Interaction between μ^--Mesons and Nuclei," *Rev. Mod. Phys.* **21**, 133.
13. Wheeler, John A. (1955), "Geons," *Phys. Rev.* **97**, 511.
14. Wheeler, John A. (1957), "On the Nature of Quantum Geometrodynamics," *Ann. Phys.* **2**, 604–614.
15. Wheeler, John A. (1960), "Neutrinos, Gravitation, and Geometry," in *Rendiconti della Scuola Internazionale di Fisica "Enrico Fermi," XI Corso* (Zanichelli, Bologna, 1960); reprinted in J. A. Wheeler, *Geometrodynamics* (Academic Press, New York, 1962).
16. Wheeler, John A. (1964a), "Geometrodynamics and the Issue of the Final State," in *Relativity, Groups, and Topology*, C. deWitt and B. DeWitt, eds. (Gordon and Breach, New York), pp. 317–522.
17. Wheeler, John A. (1964b), "The Superdense Star and the Critical Nucleon Number," in *Gravitation and Relativity*, H.-Y. Chiu and W. F. Hoffmann, eds. (Benjamin, New York).
18. Wheeler, John A. (1966), "Superdense Stars," *Ann. Rev. Astron. Astrophys.* **4**, 393.
19. Wheeler, John A. (1968), "Our Universe: The Known and the Unknown," *Am. Sci.* **56**, 1.
20. Wheeler, John A. (1979a), "The Quantum and the Universe," in *Relativity, Quanta and Cosmology*, Vol. II, M. Pantaleo and F. de Finis, eds. (Johnson Reprint Corp., New York).
21. Wheeler, John A. (1979b), *Frontiers of Time* (North-Holland Publishing Co., Amsterdam); also in *Rediconti della Scuola Internazionale di Fisica "Enrico Fermi,"* N. Toraldo di Francia and Bas van Fraassen, eds. (North-Holland Publishing Co., Amsterdam), pp. 395–497.
22. Wheeler, John A. (1980), "Beyond the Black Hole," Chapter 22 in *Some Strangeness in Proportion*, Harry Wolf, ed. (Addison-Wesley, Reading, Massachusetts).
23. Wheeler, John A. (1981a), "The Lesson of the Black Hole," *Proc. Am. Philos. Soc.* **125**, 25.
24. Wheeler, John A. (1981b), "Delayed-Choice Experiments and the Bohr–Einstein Dialogue," The American Philosophical Society and the Royal Society, pp. 25–37, papers read at the Meeting, June 5, 1980 (American Philosophical Society, Philadelphia).
25. Wheeler, John A. (1982), "The Computer and the Universe," *Int. J. Theor. Phys.* **21**, Nos. 6/7.
26. Wheeler, John A. (1984), "Bits, Quanta and Meaning," in *Theoretical Physics Meeting: Commemorative Volume on the Occasion of Eduardo Caianiello's Sixtieth Birthday*, A. Giovanni, M. Marinaro, F. Mancini, and A. Rimini, eds. (Edizioni Scientifici Italiani, Naples), pp. 121–134.

27. Wheeler, John A. (1985), quoted in Jeremy Bernstein, "Retarded Learner," *Princeton Alumni Weekly*, October 9, 1985, p. 28.
28. Wheeler, John A., and Feynman, Richard P. (1945), "Interaction with the Absorver as the Mechanism of Radiation," *Rev. Mod. Phys.* **17**, 157.
29. Wheeler, John A., and Zurek, Wojciech H., eds. (1983), *Quantum Theory and the Measurement* (Princeton University Press, Princeton), Preface, p. i.

Wheeler Wormholes and the Modern Astrophysics

Igor D. Novikov, N.S. Kardashev, and A.A. Shatskiy

Abstract J.A. Wheeler was the main propagandist of thewormhole idea. We consider the hypothesis that some active galactic nuclei and other compact astrophysical objects may be current or former entrances to wormholes. A broad mass spectrum for astrophysical wormholes is possible. We consider various new models of the static wormholes including wormholes maintained mainly by an electromagnetic field. We also discuss observational effects of a single entrance to wormhole and a model for a binary astrophysical system formed by the entrances of wormholes with magnetic fields and consider its possible manifestation.

1 Introduction

First serious paper on wormholes was the 1935 paper by Einstein and Rosen [1]. In this paper the authors used a term "bridge" for the description of the hypothetical objects which we call today "wormholes". They wrote: 'These solutions involve the mathematical representation of physical space by a space of two identical sheets, a particle being represented by a "bridge" connecting these sheets'.

It was a great idea, however, the mathematical model proposed by the authors could not be considered as a correct model.

First real seminal work on the subject was the 1955 paper by J.A. Wheeler [2]. In this paper there is first diagram of a wormhole. The famous 1957, paper [3] by Misner and Wheeler first introduces the term "wormhole" to the physics community.

I.D. Novikov (✉)
Astro Space Center, Lebedev Physical Institute, Russian Academy of Sciences, Profsoyuznaya ul., 84/32, Moscow, 117997, Russia
and
Niels Bohr Institute, Blegdamsvej 17, DK-2100, Copenhagen, Denmark
e-mail: novikov@asc.rssi.ru

N.S. Kardashev and A.A. Shatskiy
Astro Space Center, Lebedev Physical Institute, Russian Academy of Sciences, Profsoyuznaya ul., 84/32, Moscow, 117997, Russia

I. Ciufolini and R.A. Matzner (eds.), *General Relativity and John Archibald Wheeler*, Astrophysics and Space Science Library 367, DOI 10.1007/978-90-481-3735-0_4,

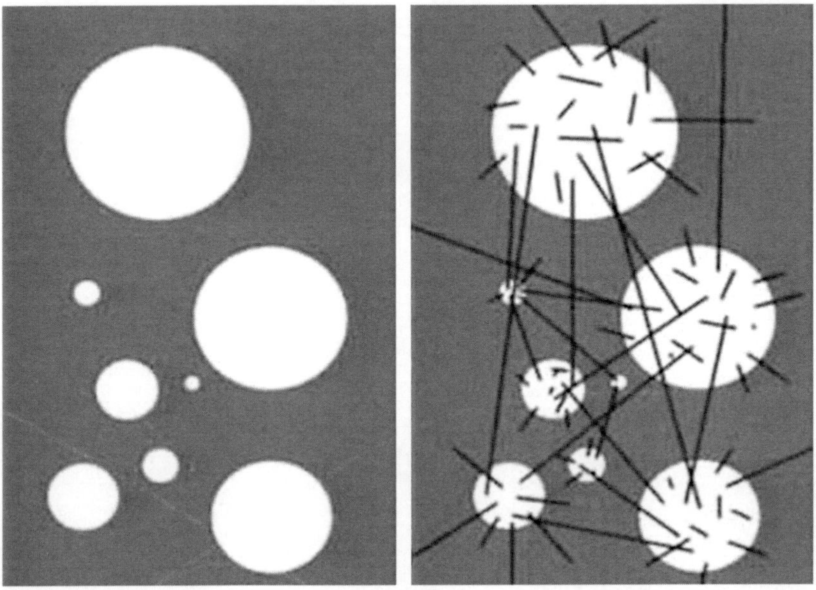

Fig. 1 Model of chaotic inflation in a multi-element Universe without (*left*) and with (*right*) wormholes

The purpose of this paper is to consider a possibility that some astrophysical objects may be current or former entrances to wormholes (WHs). These wormholes may be relic of the inflation epoch of the evolution of the Universe [4–10] and Fig. 1.

It follows from WH models that their existence requires matter with a peculiar equation of state [11–15]. This equation must be anisotropic, and $w_{\parallel} = p_{\parallel}/\varepsilon$ must be smaller than -1, as in the case of phantom matter ($p_{\parallel}$ is the total pressure along the tunnel of the WH, and ε is the total energy density for all components of the matter in the tunnel of the WH). The existence of such matter remains hypothetical [16]. For definiteness, we will use the term "phantom energy" for an isotropic equation of state $p/\varepsilon < -1$, and the term "phantom matter" for an anisotropic equation of state. The units are selected so that $c = 1$ and $G = 1$ (with exceptions for final relations).

In this paper, we consider a model in which the main component of a wormhole having all the necessary properties is a strong magnetic field that penetrates the WH, while phantom matter and phantom energy are required only in small amounts. We also consider a model based on phantom energy, where the equation of state is close to that for a vacuum ($p/\varepsilon = -1$), with some added energy density of the magnetic field.

We do not consider here the dynamics of wormholes. Thus all conditions for the relations between the components of the stress–energy tensor correspond to the static state but not dynamics.

In particular we do not consider the problem of stability of the models of wormholes. It is necessary to emphasize that there are stable models of wormholes (see for example [17]).

We consider the corresponding model of WHs, and their properties in Appendixes. For an external observer, the entrance to the wormhole appears to be a magnetic monopole with a macroscopic size. The accretion of ordinary matter onto the entrance to the wormhole may resulting the formation of a black hole with a radial magnetic field. We consider the possibility that some active galactic nuclei and some of Galactic objects may be current or former entranced to magnetic wormholes. We consider the possible existence of a broad mass spectrum for wormholes, from several billion solar masses to masses of the order of 2 kg. The Hawking effect (evaporation) does not operate in such objects due to the absence of a horizon, making it possible for them to be retained over cosmological time intervals, even if their masses are smaller than 10^{15} g. We also discuss a model for a binary system formed by the entrances of tunnels with magnetic fields, which could be sources of non-thermal radiation and γ-ray bursts.

There is a hypothesis that the primordial wormholes probably exists in the initial state of the expanding Universe [7, 8] and can connect different regions in our Universe and other Universes in the model of Multiverse. It is possible that primordial WHs are preserved after inflation. In this case the search for astrophysical wormholes is a unique possibility to study the Multiverse.

More detail discussion see in [18].

2 Wormholes and Their Remnants in the Universe

As we emphasized in Section 1 our hypothesis suggests that it may be possible existence in the modern Universe entrances to the wormholes with rather strong monopole magnetic fields. Explicit models of them are in Appendix 5.1. Astrophysical accretion of ordinary matter on them may probably converts them into black holes (BHs) with strong monopole magnetic field (wormhole remnants). The monopole magnetic field of the WH remnants differs from ordinary BHs. The last ones nether have strong monopole magnetic field. Probably some of Galactic and extragalactic objects are such WH/BHs. Detecting these WH/BHs is a great challenge.

Figure 2 presents WH/BH models allowing for the possibility that they possess an evolved accretion disk with its own magnetic field (probably dipolar and substantially weaker than the field of the WH/BH themselves). This is a schematic toy model in the vacuum approximation. The presence of a radial magnetic field can be revealed by the law for the increase in the field strength ($H \propto r^{-2}$) and the presence of the same sign of the field on all sides. The rotation of the monopole excites a dipolar electric field, which can provide acceleration for relativistic particles. Note that the dipole electric field (in contrast to the quadruple field for a disk) accelerates electrons towards one of the poles (and protons and positrons towards the other; see Appendix 5.3). Of course such motion of electrons is possible only if there are currents providing the electro neutrality of the system. Something like this may explain the origination of one-sided jets in some sources (for example, the quasar 3C273) [19].

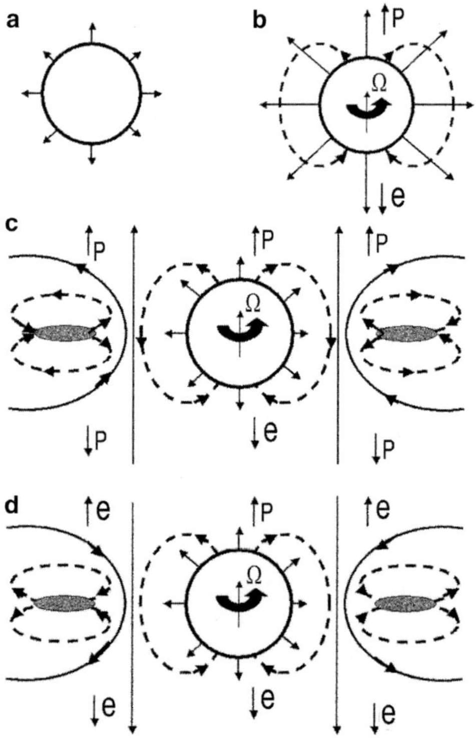

Fig. 2 A schematic toy model of the WH/BH: (**a**) WH or BH with a radial magnetic field. (**b**) Same as (a), with rotation and a dipolar electric field. (**c**) Same as (b), with a rotating accretion disk with its dipole magnetic field and quadrupolar electric field. (**d**) Same as (c), with dipole magnetic and quadrupole electric fields of the opposite direction. The *arrows* (p) and (e) indicate the directions for proton and electron jets, respectively

Actually the general picture should be much more complicated.

First of all the rarefied plasma form a complex magnetosphere around WH/BH (see [20]). In addition the presence of the accretion disk complicates the picture. In a toy model the quadrupolar electric field generates two-sided electron or proton/positron jets (depending on the sign of the quadrupole). As a result, the structure of the jets may be different:

1. Electrons from the magnetic WH/BH are ejected from one pole, while protons (positrons) are ejected from the other.
2. Electrons from the WH/BH and the accretion disk are ejected from one pole, while protons (positrons) from the WH/BH and electrons from the accretion disk are ejected from the other.
3. Electrons from the WH/BH and protons (positrons) from the accretion disk are ejected from one pole, while protons (positrons) from the WH/BH and accretion disk are ejected from the other. The interactions between the electromagnetic fields of the WH/BH and accretion disk are likely to be very strong.

We have to remember that even in the toy model such jets are possible only under the conditions of existence of the electric currents providing the electro neutrality of the system and inducted electric field conserves.

The difference between a WH entrance and a BH may be revealed from observations indicating the absence of a horizon – a source of light falling into a WH should be observed continuously, but with a variable red, and even blue, shift. However, in this case, we must assume the tunnel is transparent. A blue shift can appear if the mass of the further WH entrance (relative to the observer) exceeds that of the closer entrance. If the tunnel is transparent and has accretion disks at both entrances, the redshift of the spectra of these disks will also be different. Thus, two different redshifts from a single source related to aWH could be distinguished.

An observed WH image may display some internal structure, whose angular size could be substantially smaller than that specified by the gravitational diameter.

In this connection, observations of the gravitationally lensed quasar Q0957+561 with the redshift $z = 1.4141$ [21] are of great interest. The observation show that in this active galaxy exist the central compact object with mass $3.6 \cdot 10^9 M_\odot$, strong magnetic moment and without an event horizon. A specific property of WHs, which display strong relativistic effects and also the absence of a horizon, is the possibility of periodic oscillations of a test mass relative to the throat (see Table 1a, b and Appendix 5.1). The redshift will also vary periodically. If the structure of the WH is close to the formation of the horizon, these shifts, the period, and the flux variations can be very large. We point out in this connection the quasi-periodic flux variations observed for so-called IDV sources, such as the BLLac object 0716 + 714 [22].

When sources move along circular orbits around a WH entrance (the laws of motion see in [18] and Appendix 5.4), the flux and redshift of radiation of a compact source will also vary periodically.

Ultimately, an external observer may be able to detect some radiation at the gyrofrequencies and events related to the creation of $e^\pm$ and $\mu^\pm$ pairs (where μ designated a magnetic monopole).

Let us also consider the possibility of binary objects formed by two entrances to the same WH that interact gravitationally and electromagnetically and rotate about their common center of gravity. If the two entrances in the model with a magnetic field represent magnetic monopoles of different signs, then the binary forms a rotating magnetic dipole. Such binaries with closely-spaced entrances are also very likely to originate in the primordial scalar field, and can be preserved until the present (Fig. 3).

Let us denote: d – is the distance between the WH entrances, T – is the period of the circular orbit, and M_∞ – is the asymptotic mass at infinity of each entrance.

As it is shown in Appendix 5.1 in our model the magnetic field H_0 at the throat of the WH is related with M_∞:

$$H_0 \approx \left(c^4/G^{3/2}\right) M_\infty^{-1} \tag{1}$$

Then for the circular orbit (see [18]):

$$V = \sqrt{\frac{GM_\infty}{d}}, \quad T = \pi\sqrt{\frac{d^3}{GM_\infty}}. \tag{2}$$

These relations are derived in the Newtonian approximation and for motion under the action of the gravitational and magnetic fields.

Table 1 (**a**) Parameters of the throats of magnetic WHs with various masses.[a]

$M_\infty = 2M_0$	r_0 (cm)	H_0 (Gs)	$\rho(r_0)$ (g/cm^3)	ν_G (Hz)	ν_H (Hz)	ν_c (Hz)
$6 \cdot 10^{42}$g $= 3 \cdot 10^9 M_\odot$ (Quasar)	$4.5 \cdot 10^{14}$	$7.8 \cdot 10^9$	$2.7 \cdot 10^{-3}$	$7.6 \cdot 10^{-6}$ (1.5 days)	$2.2 \cdot 10^{16}$	$1.16 \cdot 10^{-6}$ (9.8 days)
10^{39}g $= 5 \cdot 10^5 M_\odot$ ($e^\pm$ Pair creation)	$7.4 \cdot 10^{10}$	$4.4 \cdot 10^{13}$	$9.7 \cdot 10^4$	0.045 (22 s)	$1.3 \cdot 10^{20}$	$6.9 \cdot 10^{-3}$ (2.4 min)
$2 \cdot 10^{33}$g $= M_\odot$ (Sun)	$1.5 \cdot 10^5$	$2.3 \cdot 10^{19}$	$2.4 \cdot 10^{16}$	$2.3 \cdot 10^4$	$6.6 \cdot 10^{25}$	$3.5 \cdot 10^3$
$6 \cdot 10^{27}$g $= M_\oplus$ (Earth)	0.45	$7.8 \cdot 10^{24}$	$2.7 \cdot 10^{27}$	$7.6 \cdot 10^9$	$2.2 \cdot 10^{31}$	$1.16 \cdot 10^9$
$5 \cdot 10^{10}$g (Positronium)	$3.5 \cdot 10^{-18}$	10^{42}	$4.4 \cdot 10^{61}$	$9.7 \cdot 10^{26}$	$2.7 \cdot 10^{48}$	$1.5 \cdot 10^{26}$
$1.8 \cdot 10^3$ g ($\mu^\pm$ Pair creation)	$1.3 \cdot 10^{-25}$	$2.6 \cdot 10^{49}$	$3 \cdot 10^{76}$	$2.6 \cdot 10^{34}$	$7.3 \cdot 10^{55}$	$4 \cdot 10^{33}$
$2 \cdot 10^{-5}$ g (Planck mass)	$1.5 \cdot 10^{-33}$	$2.3 \cdot 10^{57}$	$2.4 \cdot 10^{92}$	$2.3 \cdot 10^{42}$	$6.6 \cdot 10^{63}$	$3.5 \cdot 10^{41}$

Table 1 (**b**) Binary WH entrances[a]

M_∞	I_{mag} (erg/s)	d_{em} (cm)	T_{em} (s)	V_{em} (km/s)
$6 \cdot 10^{42}$g $\approx 3 \cdot 10^9 M_\odot$ (Quasar)	$7.0 \cdot 10^{40}$	$3.0 \cdot 10^{19}$	$8.2 \cdot 10^{11}$	1,100
$3.2 \cdot 10^{35}$g $\approx 160 \cdot M_\odot$ ($e^\pm$ Pair creation)	$4.0 \cdot 10^{31}$	$4.2 \cdot 10^{14}$	$1.9 \cdot 10^8$	70
$2 \cdot 10^{33}$g $= M_\odot$ (Sun)	$4.6 \cdot 10^{28}$	$1.4 \cdot 10^{13}$	$1.5 \cdot 10^7$	30
$9 \cdot 10^{18}$g[b] (Micropulsar)	$3.4 \cdot 10^9$	$3.8 \cdot 10^3$	1	120 m/s
$1.8 \cdot 10^3$g ($\mu^\pm$ Pair creation)	$4.0 \cdot 10^{-12}$	$1.3 \cdot 10^{-7}$	$1.4 \cdot 10^{-8}$	30 cm/s
$2.2 \cdot 10^{-5}$g (Planck mass)	$1.1 \cdot 10^{-22}$	$7.0 \cdot 10^{-13}$	$1.6 \cdot 10^{-12}$	1.4 cm/s

[a]See Appendix 5.1.
[b]Binary WH with orbital period 1 s.

The intensity of magnetic-dipole I_{mag} and gravitational I_{grav} radiation is correspondingly [23]:

$$I_{mag} = \frac{2^5 G^3 M_\infty^4}{3c^3 d^4}, \tag{3}$$

$$I_{grav} = \frac{2^9 G^4 M_\infty^5}{5 \cdot c^5 d^5}. \tag{4}$$

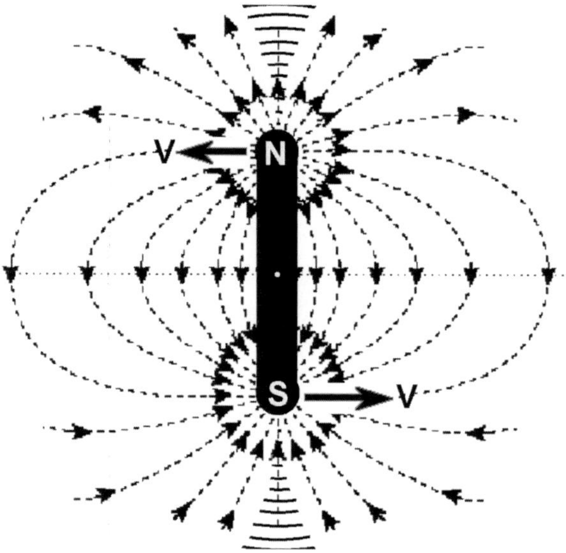

Fig. 3 Schematic model of a binary system of two magnetic WHs/entrances

Gravitational radiation dominates only at very small distances $d < (3/5) \cdot 2^4 \cdot (GM_\infty/c^2) \leq 9.6r_0$, (see Appendix 5.1) so that the losses of the energy and evolution of the system are determined by the magnetic-dipole radiation (3).

The characteristic time of the evolution of the system due to the radiation is $t_{em} = |\mathcal{E}_{total}|/I_{mag}$, where

$$\mathcal{E}_{total} = -M_\infty V^2 \tag{5}$$

Hence the system parameters preserved over time t_{em} are no smaller than

$$d_{em}^3 = \frac{2^5 G^2 M_\infty^2 t_{em}}{3c^3}, \quad T_{em} = \pi \sqrt{\frac{2^5 G M_\infty t_{em}}{3c^3}}. \tag{6}$$

A rotating system of macroscopic magnetic dipole induces an electric field, which accelerate electrons and emit electromagnetic radiation along the axis of the magnetic dipole (similarly to pulsars) or even create of electron–positron pairs. Corresponding critical electric field is $\mathcal{E}_{cr} = 4 \cdot 10^{13}$. Now if we specify t_{em} equal t_U the age of the Universe $t_U \approx 13 \cdot 10^9$ years, we obtain the following characteristic of the system with electric field $\mathcal{E}_{cr}$ and $t_{em} = t_U$:

$$\begin{aligned} M_\infty &= 3.2 \cdot 10^{35}\,\text{g} \approx 160\, M_\odot, \quad d_{em} = 4.2 \cdot 10^{14}\,\text{cm}, \\ T_{em} &= 1.9 \cdot 10^8 s \approx 6\,\text{years}, \quad V_{em} \approx 70\,\text{km/s}. \end{aligned} \tag{7}$$

All binary systems with smaller masses will generate $e^\pm$, forming two-sided jets of relativistic particles in the orbital plane, lose energy, and collapse.

Table 1a, b presents the parameters of magnetic WHs with various masses (notations see in Appendix 5.1) and estimates for the intensity of magnetic-dipole radiation I_{mag}, the component separation d_{em}, the rotational period T_{em} and velocity V_{em} for systems preserved to the present epoch.

In this toy model binaries can be detected as strictly periodical nonthermal sources (similar to pulsars) with two jets of relativistic $e^{\pm}$ in the equatorial plane and a cloud of annihilating low-energy particles $e^{\pm}$.

The approach of the two entrances of the same magnetic WHs ends with its transformation into a BH with a mass of $\sim M_{\infty}$ without a magnetic field, with the radiation of the external magnetic field $\mathcal{E}_{mag,\,total} \approx M_{\infty}c^2$, similar to the collapse of a magnetized body [24, 25]. The electromagnetic impulse accelerates the created $e^{\pm}$ pairs, and may represent the basic mechanism for some type of observable γ-ray bursts. The approach of two closely spaced different WH entrances should be somewhat different: after they merge, probably one entrance or a BH is formed, with its magnetic flux equal to the sum of the original magnetic fluxes, taking into account their signs and the radiation of the difference in the flux. Thus, various merging processes could result in the formation of more massive objects, within which the entrances and magnetic fluxes from many tunnels are combined.

3 Wormholes and the Multi-universe

As we mentioned in Section 1, there is a hypothesis that a wormholes can connect different Universes in the model of multi-universe (see Fig. 1).

In this section briefly consider various aspects of such a possibility. Let us consider the possible observational differences of the two cases:

(a) Both entrances of the wormhole are in our Universe.
(b) One entrance is in our Universe and another entrance is in another Universe.

In case (a) both entrances probably were formed in the very early quantum period of the existence of our Universe. It means that both of them were formed practically at the same moment of time of an external observer which is practically at rest with respect to the entrances. If during the subsequent evolution of the Universe there was not great relative velocity of the motion of the entrances and if there was not great difference in gravitational potential in their surrounding then in our epoch, time near each entrance corresponds to the same cosmological time of our epoch.

Hence an observer looking through the throat of the wormhole sees our Universe at present epoch (but another regions of it). The radiation from the wormhole would be provided mainly by the processes near both entrances rather then the cosmic microwave background which is quite weak.

In case (b) situation may be quite different. In principle the physics in our and other universe can be absolutely different and in principle we can observe it. But even if we suppose that another Universe is similar to our one there is not probably any correlation in times in these universes. Thus in this case we can see through the wormhole any epoch in the evolution of the Hot Universe: (1) very early one, or (2) close to our epoch, or (3) very late epoch.

In case (1) we would see the epoch which corresponds to very hot background radiation. The flux of the radiation through the wormhole into our Universe would rapid decrease the mass and the size of the entrance in our Universe and would lead to rapid conversion of it into (a) black hole or into a naked singularity.

The dynamics of space–time of a wormhole in the case of a strong flux of matter through it needs a special investigation and we will discuss it in a separate paper.

Here we want to mention that such an instability of a wormhole with respect to flux of energy brought it may be very important also in the case (a) at the very early epoch if there is not a good balance of the contrary fluxes of radiation in both entrances.

Case (2) would be similar to the case (a) from the point of view of observations.

In case (3) we would see a very cold Universe without any significant radiation from it.

It is worthy of consideration the following problem. The entrance, in which the flux of radiation is falling down, would increase in mass and in size. As we mentioned above the opposite entrance would shrink down and eventually probably would disappear.

The question arises: what does an observer near first entrance see when the second entrance disappears? This question is a part of the problem about the evolution of the wormhole in the case of a strong flux of energy through it and will be considered in a separate paper.

4 Conclusion

In this paper we propose to analyze the possibility of identifying a new type of primordial cosmological objects – WHs entrances, or BHs formed from WHs – among known Galactic and extragalactic objects that are usually identified with neutron stars or stellar-mass and super massive BHs. The implied presence of a strong radial magnetic field (the "hedgehog" model [25–27]) is very important, and can, together with rotation, bring about the generation of (one- and two-sided) relativistic jets. WH models suggest the specific properties of such objects related to the absence of a horizon, which makes it possible to observe sources of radiation at any point of the tunnel, if the tunnel medium is transparent. A certain behavior is expected for the variations of the spectrum, flux, and polarization of the source of radiating objects making periodic oscillations relative to a WH throat could be detected.

The observer structures of the sources that could be related To WHs entrances (or BHs) with a radial magnetic field could be studied with microarcsecond or better angular resolution, as is proposed in the "Radioastron" and "Millimetron" space-VLBI experiments [28, 29].

It may be possible to detect sources associated with binary WHs entrances, which form systems with strong magnetic-dipole radiation and the ejection of relativistic $e^{\pm}$. The final stage in the evolution of such systems is the formation of a BH, accompanied by strong electromagnetic impulses.

Note also that the possible existence of WHs with strong magnetic fields suggests that theoretically predicted elementary magnetic monopoles [28, 29] could have merged with these objects in the course of their cosmological evolution.

Another direction for future studies is related to spectral and polarization monitoring of these sources.

The detection of tunnels will open the way to studies of the entire Multiverse.

In the Appendixes, we consider the explicit mathematical models of the WHs with magnetic field which are the theoretical basis of our hypothesis.

Acknowledgements This work was supported by the Russian Foundation for Basic Research (project numbers: 05-02-17377, 05-02-16302-*a*, 04-02-16987-*a*, 04-02-17257-*a*) and the Program in Support of Leading Scientific Schools of the Russian Federation (NSH-1653.2003.2) and program "stellar evolution".

The authors are grateful to S.P. Gavrilov [30], B.V. Komberg, M.B. Mensky, D.I. Novikov, V.I. Ritus, and A.E. Shabat for useful discussions and comments.

5 Appendixes

5.1 Spherically Symmetrical Wormhole with Radial Magnetic Field

The metric of a spherical WH can be presented in the form (see [31]):

$$ds^2 = e^{2\phi(r)}\,dt^2 - \frac{dr^2}{1 - b(r)/r} - r^2\,d\Omega^2, \tag{8}$$

where r is the radial coordinate, $\varphi(r)$ the so-called redshift function, and $b(r)$ the form function. The WH neck corresponds to the minimum $r = r_0 = b(r_0)$ and $b_{(r_0)} \le 1$. The presence of a horizon corresponds to the condition $\varphi \to -\infty$ or $e^{\varphi} \to 0$; for WH φ must be finite everywhere.

Let us introduce the mass $M(r)$ of the WH for an external observer:

$$M(r) = M_0 + \int_{r_0}^{r} 4\pi\varepsilon r^2\,dr, \tag{9}$$

where $M_0 = r_0/2$ and $\varepsilon(r)$ – is the energy density. For convenience in graphical representation and calculations, we introduce the variable $x = r_0/r$. The entire interval $r_0 \le r < \infty$ will then be transformed into $0 < x \le 1$, and we obtain the equation for a static WH:

$$\begin{aligned} 8\pi \varepsilon r_0^2 &= -b' x^4/r_0, \\ 8\pi p_{\|} r_0^2 &= -bx^3/r_0 - 2x^3(1 - bx/r_0)\phi', \\ 8\pi p_{\perp} r_0^2 &= (1 - bx/r_0)\left[x^4\phi'' + x^3\phi' + x^4(\phi')^2\right] \\ &\quad + 0.5x^3(xb' + b)(1 - x\phi')/r_0. \end{aligned} \tag{10}$$

where the derivatives are taken with respect to x. It was shown in [32] that, in a spherically symmetrical WH with $w_{\|} = const$ and $w_{\perp} = const$, an inequality specifying possible equations of state and their anisotropy must be satisfied:

$$-2w_{\perp} < w_{\|} < -1 \tag{11}$$

The left-hand side of the inequality specifies the finiteness of the WH mass as $r \to \infty$, while the right-hand side indicates the absence of a horizon.

It is of great interest that condition (11) is "almost" satisfied for a magnetic (or electric) field. If the field is in the r direction, the energy–momentum tensor specifies the equation of state

$$w_{\|} = -1\,, \quad w_{\perp} = 1\,, \quad \varepsilon = (H^2 + E^2)/(8\pi), \tag{12}$$

which satisfies the conditions (11) for a WH to within a small negative increment in $w_{\|}$.

In [33], a model for a phantom-matter WH with an anisotropic equation of state is considered:

$$1 + \delta = -p_{\|}/\varepsilon = p_{\perp}/\varepsilon, \tag{13}$$

and it is shown that even an arbitrarily small δ is sufficient for a WH to exist.

Let us denote $x_h = r_0/r_h > 1$ to be the ratio of the radii of the WH throat and the horizon of the corresponding Reissner–Nordstrem black hole (BH) [34] with magnetic charge $Q = r_h$ (see [18]). ε is specified by the relation

$$\varepsilon = \varepsilon_0\, x^4\ \left[(x_h - 1)/(x_h - x)\right]^{\delta}, \quad \varepsilon_0 = 1/(8\pi r_0^2(1 + \delta)). \tag{14}$$

For an observer far from the throat, the WH mass M_∞ is in the interval

$$M_0 \le M_\infty \le 2M_0 \tag{15}$$

The left-hand side of the inequality follows from (9), and the right-hand side from (14).

In this connection, it may be concluded that the electromagnetic (EM) field may be an appreciable or even predominant part of the WH. Let us consider three types of models:

1. An EM field is the main component of the WH matter. In addition a small amount of the phantom energy is necessary to provide conditions above.
2. An EM field plus phantom energy with an isotropic equation of state.
3. An EM field plus phantom matter with an anisotropic (vector-type) equation of state.

Common to all three models is the assumption that the WH is penetrated by an initial magnetic field, which should display a radial structure for an external observer in the spherically symmetrical case; i.e., it should be correspond to a macroscopic magnetic monopole ($r \to \infty$, $H \propto r^{-2}$, and $\varepsilon_H \propto r^{-4}$). The two entrances to the WH should have opposite signs of the magnetic field.

In model 1 there is the magnetic field plus a small amount of phantom energy or phantom matter. Under real conditions, this model is specified by a strong magnetic field. The fact that δ is small indicates that the configuration is close to a BH. The accretion of normal matter onto the WH entrance results in the growth of $w_{\|}$, and the condition of the absence of a horizon can be violated (the right-hand side of (11)). The WH entrance can then turn into a BH with the radial magnetic field. The opposite is also true: the accretion of phantom energy makes the WH more different from a BH [35]. Overall, model 1 is close to a BH with an extremely strong magnetic field [36], which, however, displays a monopole rather than a dipole structure.

Table 1a, b in Section 2 presents the parameters for the throats of magnetic WHs with various masses M_0. We can use (14) (and the constants c and G) to derive expressions for r_0, H_0, ρ_0 (the mass density), ν_G (the frequency of oscillations with a small amplitude for a sample particle), and ν_H (the gyrofrequency) in the throat and its rest frame; ν_c is the frequency of revolution along the lower stable circular orbit for an external observer. Then,

$$\begin{aligned} r_0 &= (G/c^2) \cdot M_\infty; \\ H_0 &= (c^4)/(G^{3/2}) \cdot M_\infty^{-1}; \\ \rho_0 &= (c^6)/(8\pi G^3) \cdot M_\infty^{-2}; \\ \nu_G &= (c^3)/(2\sqrt{2}\pi G) \cdot M_\infty^{-1}; \\ \nu_H &= (ec^3)/(2\pi m_e G^{3/2}) \cdot M_\infty^{-1}; \\ \nu_c &= (\sqrt{3}c^3)/(32\pi G) \cdot M_\infty^{-1} = \sqrt{(3/128)} \cdot \nu_G. \end{aligned} \tag{16}$$

In Table 1a, b, WH parameters are estimated for quasar cores, as well as for objects with the field (and corresponding WH mass) critical for the creation of electron–positron pairs, with masses of the order of the Sun's and the Earth's, with the magnetic field (and WH mass) critical for stability of the positronium atom, with the magnetic field (and WH mass) critical for the creation of monopole-anti-monopole pairs, and with the Planck mass.

The magnetic field for a model described by (14) with a small δ will be

$$H \approx M_\infty \sqrt{G}/r^2 \tag{17}$$

Assuming that the electric field is small, the constraint associated with the creation of electron–positron pairs is removed (although we present it in the Table 1a, b for the limiting case). This field, $H = m_e^2 c^3/\mu_H \approx 4.4 \cdot 10^{13}\, G$, specifies specific conditions related to the fact that, at large fields, the Landau excitation level exceeds the electron rest energy. The positronium atom becomes stable in fields above $10^{42}\, G$, and the medium becomes filled with such atoms created from the vacuum [37, 38]. In magnetic fields stronger than the critical value H_{max}, a vacuum puncture and the creation of monopole pairs occur [39–42]. If the mass of a stable, colorless monopole [43–45] is $m_\mu \sim 10^{16}\, GeV \sim 10^{-8} g$ and the magnetic charge $\mu = (3/2)\hbar c/e \sim 10^{-7}$, then $H_{max} = m_\mu^2 c^3/\mu\hbar \approx 2.6 \cdot 10^{49}\, G$, and, according to (15), the maximum mass of a magnetic WH with this field in its throat is $M_\infty \approx 1.8$ kg. The created monopoles will be ejected from the WH, decreasing its mass. It is not obvious whether such small WHs are stable against other quantum processes, and the minimum mass of the WH may turn out to be substantially higher than 1.8 kg. The lower limit for the mass of a WH with a composite constitution is obviously even lower. Note also that the absence of a horizon for the WH results in the absence of evaporation (Hawking radiation). Therefore, primordial low-mass WHs could be preserved until the current epoch, unlike primordial BHs (for which the lower limit for the mass is $\sim 10^{15}$ g).

Model 2 corresponds to the case of phantom energy, similar to that used in cosmological models. Since the equation of state is isotropic, phantom energy alone cannot satisfy the left-hand side of (11), and a larger fraction of magnetic field is needed. Let us denote the fraction of the magnetic-energy density $\eta_H = \varepsilon_H/\varepsilon$ and the fraction of the phantom-energy density $\eta_w = \varepsilon_w/\varepsilon$, $\varepsilon = \varepsilon_H + \varepsilon_w$. If the pressure is $p_w = w\varepsilon_w$ and $\eta_H + \eta_w = 1$ for the phantom matter, then $p_\parallel = -\varepsilon_H + w\varepsilon_w$, $p_\perp = \varepsilon_H + w\varepsilon_w$, and it follows from (11) with conditions $w_\parallel = const$ and $w_\perp = const$ that

$$3w/(3w-1) < \eta_H < 1, \quad 0 < \eta_w < 1/(1-3w) \tag{18}$$

As $w \to -1$, the fractions of the magnetic and phantom energies are $0.75 < \eta_H < 1$ and $0 < \eta_w < 0.25$. As $w \to -\infty$, $\eta_H \to 1$, and $\eta_w \to 0$. Thus, for an electromagnetic WH, it is necessary and sufficient to include an arbitrarily small fraction of phantom energy with an isotropic equation of state, with w arbitrarily close, but smaller than -1.

Figure 4 presents the results of calculating equations (10) for all three models. It shows the dependence of the fractions of η_H, η_w, and η_{CDM} on $x = r_0/r$ for the models with a radial magnetic field, phantom energy (with $w \to -1$), and ordinary matter with "zero" pressure (the fraction η_{CDM}). The limiting conditions are $\eta_H = \eta_w = 0.5$ in the throat, $\eta_w \to \eta_{CDM} \to 0$ for $x \to 0$ ($r \to \infty$). Here, in the three-component model, unlike (13), $w_\parallel$ and $w_\perp$ depend on r.

Model 3 can provide any relation between the magnetic and phantom matter density, while, according to [33], δ can be arbitrarily small.

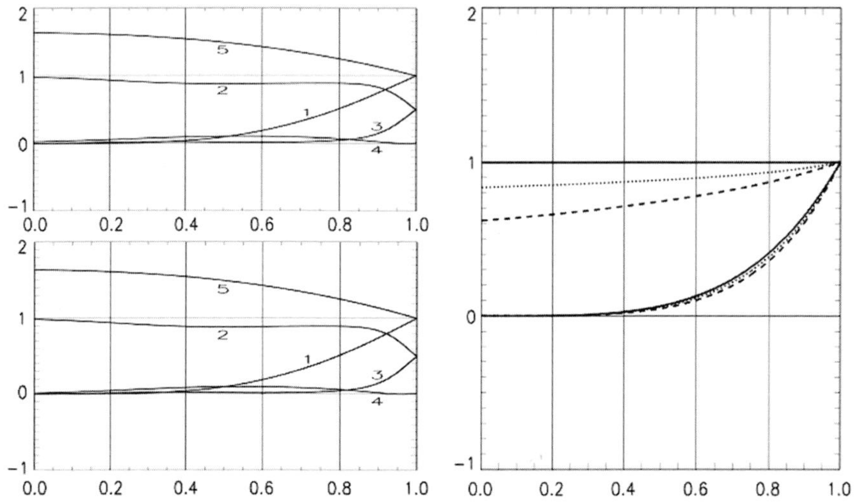

Fig. 4 Basic model parameters as functions WH of $x = r_0/r$. The *top-left plots* show the three-component model with $w_H = 1,\ w = -1$, and $\eta_H(1) = \eta_w(1) = 0.5$. The dependences are indicated for (1) the total density in a throat, (2) the magnetic-field density from the total density, (3) the phantom-energy density, (4) the ordinary-matter density, and (5) the total density in a throat not divided by x^4 (this characterizes the difference of distribution for the density from that for a magnetic monopole). The *bottom-left plots* show the same dependences for the three-component model for $w_H = 1,\ w = 1.02$, and $\eta_H(1) = \eta_w(1) = 0.5$. The *right-hand-side plots* show the model for a WH supported mainly by a magnetic field (or by phantom matter with the same δ). The *three lower curves* represent the dependence of the fractional density on the density in the throat $\eta = \varepsilon/\varepsilon_0$ for $\delta = 0.5$ (*dashed*), $\delta = 0.1$ (*dotted*), and $\delta = 0.001$ (*solid*); the *three upper curves* correspond to the *lower curves* with η/x^4 (characterizing the difference of the structure from a monopole). As $\delta \to 0$ $\eta/x^4 \to 1$. About notations see the end of Section 5.1

We want to emphasize that the relations between p and ε correspond to the static state and we do not consider the dynamics. Particularly we do not consider the problem of stability of wormholes.

5.2 Dipolar Electric Field Induced in a WH

In the case of a rotating WH with a magnetic field, the electric field that is induced should display a dipolar structure, due to the monopolar WH magnetic field.

Characteristic estimates for the induced WH electric field can be obtained from the exact solution corresponding to a BH [20]. Making the substitution "electricity $\longleftrightarrow$ magnetism" in the solution, we obtain for the electric field:

$$\vec{E}(r,\theta) \approx H(r_h) \cdot \frac{a \cdot r_h^2}{r^3} \cdot \left(2\cos\theta \cdot \vec{e}_r + \sin\theta \cdot \vec{e}_\theta\right) \tag{19}$$

Observationally, the existence of such a field in the vacuum approximation in a toy model will result in the ejection of electrons from one of the poles and protons from the other. Standard Blandford–Znajek type [46] hydrodynamical models or models with a quadrupolar electric field predict two-sided jets for BHs. In addition, even in the case of slow rotation, such a jet should display a higher energy, since its acceleration results from a larger electric potential then in the case of a BH: $\varphi_e \sim H(r_0)a$ for a WH, while $\varphi_e \sim H(r_h)a^2/r_h$ for a BH, where a is the rotation parameter in the Kerr metric.

5.3 Observations of Body Oscillating Through a WH Throat

Oscillations of bodies in the vicinity of a WH throat (radial orbits) could give rise to a peculiar observational phenomenon. Signals from such sources detected by an external observer will display a characteristic periodicity in their spectra. All objects (stars, BHs) other than WHs absorb bodies falling onto them irrecoverably. Periodic radial oscillations are a characteristic feature of WHs.

For simplicity, let us consider a test body with zero angular momentum. We will solve the equations of motion using the Hamilton–Jacobi method in a curved space [23].

We then obtain for the velocity of the body:

$$\frac{\partial r}{\partial t} = \pm \exp(\phi) \cdot \sqrt{[1 - b/r] \cdot [1 - \exp(2\phi) \cdot (m_0/E_0)^2]} \tag{20}$$

Here, m_0 and E_0 are the rest mass and total energy of the body.

Expression (20) does not correspond to the physical velocity of the body, since r is not the physical radial coordinate. The physical velocity of the body along the radius, $\dot{l}$ (with respect to time t) is

$$\dot{l} = \pm \exp(\phi) \cdot \sqrt{1 - \exp(2\phi) \cdot (m_0/E_0)^2} \tag{21}$$

The redshift of a signal radiated by the body is specified by the following two factors.

1. The Doppler shift due to the motion of the source yields the factor $\sqrt{1-v^2}/(1 \pm v)$, where $v = \dot{r}\sqrt{|g_{rr}/g_{tt}|}$ is the physical velocity of the body in its proper time (here, the signs $\pm$ correspond to motion of the body from and toward the observer).
2. The gravitational redshift yields the factor $\exp\varphi$.

Thus, the frequency of the signal received by a distant observer will be given by the expression

$$\nu = \nu_0 \cdot \frac{\exp(2\phi) \cdot (m_0/E_0)}{1 \pm \sqrt{1 - \exp(2\phi) \cdot (m_0/E_0)^2}} \tag{22}$$

Here, ν_0 is the frequency of the signal measured on the moving body. It follows that, unlike a BH horizon, the frequency of the signal in the WH throat does not become equal to zero for a distant observer.

In order to find the time dependence of the redshift ν/ν_0 of the body for an external observer, time Δt for the light to propagate from point r to point r_{max} of the body must be added to time t. Figure 5 presents these dependencies.

For extremely small amplitudes $(r_1 - r_0)$, the test-body oscillations become harmonic. This situation is reached when the following inequality is satisfied:

$$r - r_0 \leq r_1 - r_0 \ll r_0 - r_h, \qquad 1 - x \leq 1 - x_1 \ll x_h - 1, \tag{23}$$

where $x_1 = r_0/r_1$. In this case, in the vicinity of the points where the body stops, the components of its velocity $\dot{r}$ (20) can be expanded in a series in $1 - x$ (or $x - x_1$). Restricting the expansion to the main terms yields

$$e^{\Phi} \approx (x_h - 1), \ \left(1 - \frac{b}{r}\right) \approx (x_h - 1)(1 - x), \ 1 - e^{2\Phi}(m_0/E_0)^2 \approx \frac{2(x - x_1)}{x_h - 1}. \tag{24}$$

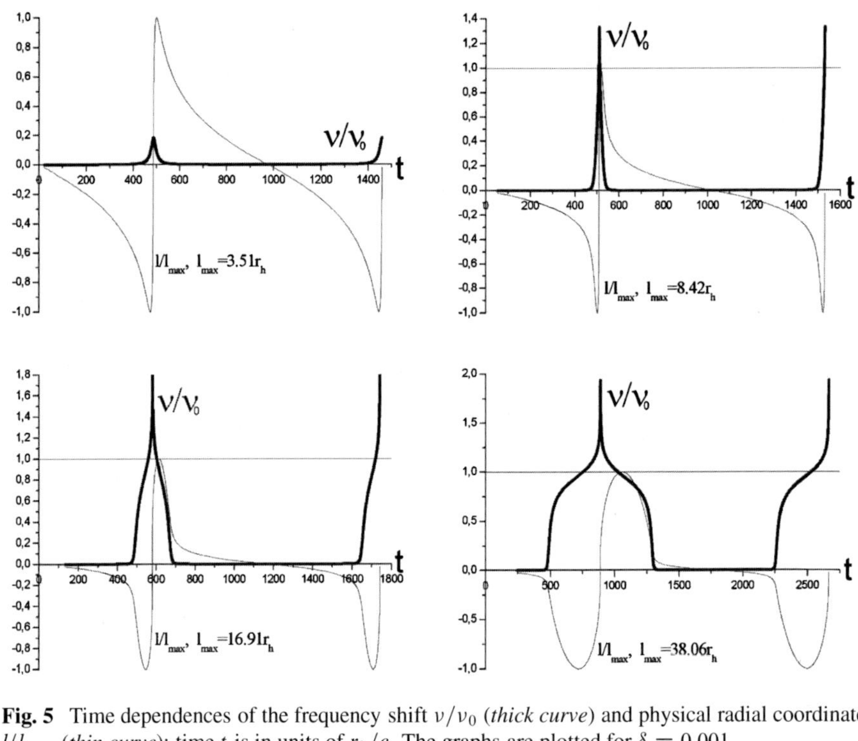

Fig. 5 Time dependences of the frequency shift ν/ν_0 (*thick curve*) and physical radial coordinate l/l_{max} (*thin curve*); time t is in units of r_h/c. The graphs are plotted for $\delta = 0.001$

Hence,

$$(\dot{r})^2 \approx 2(x_h - 1)^2(1 - x)(x - x_1), \quad r_h\ddot{r} = \dot{r}\frac{\partial \dot{r}}{\partial x} = \frac{1}{2}\frac{\partial(\dot{r})^2}{\partial x},$$
$$(1 - x)^{\cdot\cdot} = -\omega_0^2(1 - x), \tag{25}$$

where $w_0 = \sqrt{2}(x_h - 1)/r_h$. This is the equation for harmonic oscillations with the body stopping at $x = 1$ and $x = x_1$; therefore, the period of these oscillations measured by an external observer will be

$$T_1 = \frac{\sqrt{2}\pi r_h}{c(x_h - 1)}. \tag{26}$$

In this case, the oscillations of the physical coordinate l are also harmonic, as follows from (21) and (24). The body oscillates from $-l_1$ to $+l_1$; the condition that l_1 be small, corresponding to (20), is $l_1 \ll r_h$ (thus, in these coordinates, the amplitude does not have to be extremely small).

The oscillations of coordinate l have a period that is twice this value ($T_2 = 2T_1$).

5.4 Circular Orbit Around a WH

The difference of the metric of the limiting Reisner–Nordstrem and our model of the WH is negligible at $r = 2r_h$ (or more). Thus all conclusions about circular orbit around a WH are the same as in the limiting Reisner–Nordstrem geometry at the corresponding distances.

We obtain (see [47,48]) for the period at the circular orbit with respect to time t:

$$\tau = \int_0^{2\pi} \frac{d\varphi}{\dot{\varphi}} = \frac{2\pi r^2 E_0/L}{c^2(1 - r_h/r)^2}, \tag{27}$$

where E_0 – is the energy and L – is the angular momentum.

Hence, the periods of the last stable and unstable circular orbits (according to a distant observer) are

$$\tau(4r_h) = \frac{32\pi r_h}{\sqrt{3}c}, \quad \tau(2r_h) = \frac{8\pi r_h}{c}. \tag{28}$$

References

1. A. Einstein and N. Rosen, Phys. Rev. 48, 73 (1935).
2. J. A. Wheeler, Phys. Rev. **97**, 511 (1955).
3. C. W. Misner and J. A. Wheeler, Ann. Phys. (N.Y.) **2**, 525 (1957).
4. J. A. Wheeler, Ann. Phys. (N.Y.) **2**, 604 (1957).

5. A. Vilenkin, Phys. Rev. D **27**, 2848 (1983).
6. A. Linde, Phys. Lett. B **175**, 395 (1986).
7. S. W. Hawking, *Black Holes and the Structure of the Universe*, Eds. by C. Teitelboim and J. Zanelli (World Sci., Singapore, 2000), p. 23.
8. M. Visser, *Lorential Wormholes: from Einstein to Hawking* (Springer, AIP, 1996).
9. F. S. N. Lobo, Phys. Rev. D **71**, 084011, (2005).
10. H. Shinkai and S. A. Hayward, Phys. Rev. D **66**, 4005, (2002).
11. F. Rahaman, M. Kalam, M. Sarker, and K. Gayen, gr-qc/0512075, (2005).
12. P. K. F. Kuhfittig, gr-qc/0512027, (2005).
13. F. S. N. Lobo, gr-qc/0506001 (2005).
14. M. Visser, S. Kar, and N. Dadhich, gr-qc/0301003, (2003).
15. F. Rahaman, et al, gr-qc/0607061, (2006).
16. H. K. Jassal, J. S. Bagla, and T. Padmanabhan, Phys. Rev. D **72**, 103503, (2005).
17. C. Armendariz-Picon, gr-qc/0201027, (2002).
18. N. S. Kardashev, I. D. Novikov, and A. A. Shatskiy, International Journal of Modern Physics **D**, in press, April, (2007); astro-ph/0610441.
19. L. Stawarz, Astrophys. J. **613**, 119, (2004).
20. Black Holes: *the Membrane Paradigm*, Ed. by K. S. Thorne, R. H. Price, and D. A. Macdonald (Yale Univ. Press, New Haven, 1986, Mir, Moscow, 1988).
21. R. E. Schild, D. J. Leiter, and L. Robertson, astro-ph/0505518, (2005).
22. L. Ostorero, S. J. Wagner, J. Gracia, et al., astro-ph/0602237, (2006).
23. L. D. Landau and E. M. Lifshitz, *Course of Theoretical Physics*, Vol. 2: *The Classical Theory of Fields* (Nauka, Moscow, 1995; Pergamon, Oxford, 1975).
24. V. L. Ginzburg and L. M. Ozernoi, Zh. Eksp. Teor. Fiz. **47**, 1030 (1964) [Sov. Phys. JETP **20**, 689, (1964)].
25. I. D. Novikov, Astron. Tsirk., N290, (1964).
26. N. S. Kardashev, Epilogue to the Russian Eddition of the Monograph by G.P. Burbidge and E.M. Burbidge "Quasars", (Mir, Moscow, 1969) [in Russian].
27. Yu. A. Kovalev and Yu. Yu. Kovalev, Publ. Astron. Soc. Jpn. 52, 1027, (2000).
28. Project "Radioastron", http://www.asc.rssi.ru/radioastron/description/intro_eng.htm.
29. Project "Millimetron", http://www.asc.rssi.ru/millimetron/millim_eng.htm.
30. S. P. Gavrilov, hep-th/0510093, (2005).
31. M. Morris and K. S. Thorne, Am. J. Phys. **56**, 395, (1988).
32. A. A. Shatskiy, Astron. Zh. **81**, 579, (2004) [Astron. Rep. **48**, 525, (2004)].
33. A. A. Shatskiy, Astron. Zh. (2006, in press).
34. C.W. Misner, K. S. Thorne, and J. A. Wheeler, it Gravitation (Freeman, San Francisco, 1973; Mir, Moscow, 1977), Vol. 3.
35. P. F. Gonzalez-Diaz, astro-ph/0510771, (2005).
36. N. S. Kardashev, Mon. Not. R. Astron. Soc. **276**, 515, (1995).
37. A. E. Shabad and V. V. Usov, hep-th/0512236, (2005).
38. A. E. Shabad and V. V. Usov, astro-ph/0601542, (2006).
39. M. Bander and H. R. Rubinstein, Phys. Lett. B **280**, 121, (1992).
40. M. Bander and H. R. Rubinstein, Phys. Lett. B **289**, 385, (1992).
41. R. C. Duncan, astro-ph/0002442, (2000).
42. Qiu-He Peng and Chih-Kang Chou, Astrophys. J. **551**, L23, (2001).
43. G. Hooft, Nucl. Phys. B **79**, 276, (1974).
44. A. M. Polyakov, Zh. Eksp. Teor. Fiz. **20**, 194, (1974).
45. T. W. Kibble, J. Phys. A **9**, 1387, (1976).
46. R. D. Blandford and R. L. Znajek, Mon. Not. R. Astron. Soc. **179**, 433, (1977).
47. V. P. Frolov and I. D. Novikov, *Black Hole Physics. Basic Concepts and New Developments* (Kluwer, 1998).
48. B. Carr, astro-ph/0511743, (2005).

Part II
Foundations and Tests of General Relativity

Unified Form of the Initial Value Conditions*

James W. York

Abstract In this paper both the initial value problem and the conformal thin sandwich problem are written in a unified way.

1 Some Geometry and Notation

The spacetime metric $g_{\mu\nu}$ will be written in a moving frame adapted to the slices of constant time as

$$ds^2 = -N^2(dt)^2 + g_{ij}(dx^i + \beta^i dt)(dx^j + \beta^j dt) \tag{1}$$

where the spatial scalar N is the lapse function and β^i is the (spatial) shift vector. In a *natural* (coordinate) basis, $\beta_i = g_{0i}$ ($\beta_i = g_{ij}\beta^j$ and g_{ij}, g^{kl} are taken as the 3×3 inverses of one another; they are riemannian). From this one can see that β^i is a spatial vector and β_i is a spatial one-form with respect to arbitrary spatial coordinate transformations provided these transformations are *not* time-dependent.

The spacetime cobasis is

$$\theta^0 = dt, \qquad \theta^i = dx^i + \beta^i dt \tag{2}$$

and the dual vector basis is

$$e_0 \equiv \partial_0 = \partial/\partial t - \beta^i \partial/\partial x^i, \qquad e_i \equiv \partial_i = \partial/\partial x^i \tag{3}$$

We see that ∂_0 is a Pfaffian derivative while ∂_t and ∂_i are natural derivatives. The basis vector ∂_0 can be generalized to the operator on spatial tensors

$$\hat{\partial}_0 = \partial_t - \pounds_\beta \tag{4}$$

**I dedicate this paper to John Archibald Wheeler, a wonderful mentor and colleague.*

J.W. York (✉)
Department of Physics, North Carolina State University, Raleigh, NC 27695
e-mail: jaswyork1@mac.com

I. Ciufolini and R.A. Matzner (eds.), *General Relativity and John Archibald Wheeler*, Astrophysics and Space Science Library 367, DOI 10.1007/978-90-481-3735-0_5,

which, it should be noted, *commutes with* ∂_i and propagates orthogonally to $t = const.$ slices. It is obvious that $\partial_i = \partial/\partial x^i$ and $\partial_t = \partial/\partial t$ commute, because they are both natural derivatives. What is sometimes forgotten, but has been known for more than 50 years, is that (for example) the spatial Lie derivative $\pounds_\beta$, for any vector field β^j, *commutes* with $\partial/\partial x^i$ when they act on tensors *and* more general objects such as spatial connections. This result holds in an even more general form. See, for example, Schouten's *Ricci Calculus* [20], p. 105, Eq. (10.17). It is noteworthy that $\hat{\partial}_0$ acts orthogonally to $t = constant$ slices and that it is actually the *only* time derivative that ever occurs in the $3 + 1$ formulation of general relativity based on an exact or locally exact (integrable) "time" basis one-form θ^0, such as dt.

The connection coefficients in our "Cauchy-adapted" frame are given by

$$\omega^\alpha{}_{\beta\gamma} = \Gamma^\alpha{}_{\beta\gamma} + g^{\alpha\delta} C^\epsilon_{\delta(\beta} g_{\gamma)\epsilon} + \frac{1}{2} C^\alpha{}_{\beta\gamma} \tag{5}$$

where Γ denotes an ordinary Christoffel symbol, parentheses around indices denote the symmetric part (so $A_{(\beta\gamma)} = \frac{1}{2}(A_{\beta\gamma} + A_{\gamma\beta})$), and C denotes the coefficients of the commutator

$$[e_\beta, e_\gamma] = C^\alpha{}_{\beta\gamma} e_\alpha \tag{6}$$

Our *spacetime* covariant derivative convention associated with (5) is

$$D_\alpha V^\gamma = \partial_\alpha V^\gamma + \omega^\gamma{}_{\alpha\beta} V^\beta \tag{7}$$

The only non-vanishing C's are

$$C^i{}_{0j} = -C^i{}_{j0} = \partial_j \beta^i \tag{8}$$

For the *spatial* covariant derivative we write

$$\nabla_i V^j = \partial_i V^j + \gamma^j{}_{ik} V^k = \partial_i V^j + \Gamma^j{}_{ik} V^k \tag{9}$$

Because the shift β^i is in the basis, the spacetime metric in the Cauchy basis that we use has no time–space components. A convenient consequence is that there is no ambiguity in writing g_{ij}, g^{kl}, or Γ^i_{jk}. For the spacetime metric determinant we will write $\det g_{\mu\nu} = -N^2 g$, $g = \det g_{ij}$; and g_{ij} is here and hereafter considered as a 3×3 symmetric tensor.

The ω's are given next. Note that [MTW] write $\omega^\alpha{}_{\gamma\beta}$ where we have $\omega^\alpha{}_{\beta\gamma}$ for the same object. (We follow [6] here.) Of course, this convention does not matter in a coordinate basis, where all of the connection coefficients are symmetric. Here, note that only $\omega^i{}_{0j}$ and $\omega^i{}_{j0}$ differ.

$$\omega^i{}_{jk} = \Gamma^i{}_{jk}(g_{\mu\nu}) = \Gamma^i{}_{jk}(g_{mn}) = \gamma^i{}_{jk} \tag{10}$$

$$\omega^i{}_{0j} = -NK^i{}_j + \partial_j \beta^i, \quad \omega^i{}_{j0} = -NK^i{}_j, \quad \omega^0{}_{ij} = -N^{-1} K_{ij} \tag{11}$$

$$\omega^i{}_{00} = N\nabla^i N, \quad \omega^0{}_{0i} = \omega^0{}_{i0} = \nabla_i \log N, \quad \omega^0{}_{00} = \partial_0 \log N \tag{12}$$

The Riemann tensor satisfies the commutator

$$(D_\alpha D_\beta - D_\beta D_\alpha)V^\gamma = (Riem)_{\alpha\beta}{}^\gamma{}_\delta V^\delta \tag{13}$$

where $(Riem)_{\alpha\beta}{}^\gamma{}_\delta$ would be denoted $(Riem)^\gamma{}_{\delta\alpha\beta}$ in [MTW]. Again, we are using here the conventions of [6].

There are a number of possible definitions for the second fundamental tensor or extrinsic curvature tensor K_{ij}. This does not measure curvature in the sense of Gauss or Riemann, where curvature has dimensions of $(length)^{-2}$. The extrinsic curvature is a measure, at a point on a spatial slice, of the curvature of a spacetime geodesic *curve* relative to a spatial geodesic curve to which it is tangent at the point. The dimensions are therefore $(length)^{-1}$. (See, for example, [18] and the Appendix of [19] for a detailed discussion of extrinsic curvature tensors.) This has the same dimension as a connection symbol

$$K_{ij} = -N\omega^0{}_{ij} \tag{14}$$

Also

$$\hat{\partial}_0 g_{ij} = -2NK_{ij} = \partial_t g_{ij} - (\nabla_i \beta_j + \nabla_j \beta_i) \tag{15}$$

where ∇_i is the spatial covariant derivative with connection $\gamma^i{}_{jk} = \omega^i{}_{jk} = \Gamma^i{}_{jk}$; and the final term, apart the sign, is $\mathcal{L}_\beta g_{ij}$.

The Riemann tensor components are in accord with the Ricci identity (13)

$$(Riem)_{\alpha\beta}{}^\gamma{}_\delta = \partial_\alpha \omega^\gamma{}_{\beta\delta} - \partial_\beta \omega^\gamma{}_{\alpha\delta} + \omega^\gamma{}_{\alpha\epsilon}\omega^\epsilon{}_{\beta\delta} - \omega^\gamma{}_{\beta\epsilon}\omega^\epsilon{}_{\alpha\delta} - C^\epsilon{}_{\alpha\beta}\omega^\gamma{}_{\epsilon\delta} \tag{16}$$

The spatial Riemann tensor is denoted $R_{ij}{}^k{}_l$. These curvatures are related by the Gauss–Codazzi–Mainardi equations for codimension one (see, for example, [20])

$$(Riem)_{ijkl} = R_{ijkl} + (K_{ik}K_{jl} - K_{il}K_{jk}) \tag{17}$$

$$(Riem)_{0ijk} = N(\nabla_j K_{ki} - \nabla_k K_{ji}) \tag{18}$$

$$(Riem)_{0i0j} = N(\hat{\partial}_0 K_{ij} + NK_{ik}K^k{}_j + \nabla_i \partial_j N) \tag{19}$$

One can likewise form and decompose the Ricci tensor, which has the definition

$$(Ric)_{\beta\delta} = (Riem)_{\alpha\beta}{}^\alpha{}_\delta \tag{20}$$

Then we can construct

$$(Ric)_{ij} = R_{ij} - N^{-1}\hat{\partial}_0 K_{ij} + KK_{ij} - 2K_{ik}K^k{}_j - N^{-1}\nabla_i \partial_j N \tag{21}$$

$$(Ric)_{0j} = N(\partial_j K - \nabla_l K^l{}_j) \equiv N\nabla_l(\delta^l{}_j K - K^l{}_j) \tag{22}$$

$$(Ric)_{00} = N(\partial_0 K - NK_{ij}K^{ij} + \Delta N) \tag{23}$$

where ΔN denotes the spatial "rough" or scalar Laplacian acting on the lapse function: $\Delta N \equiv g^{ij}\nabla_i\nabla_j N \equiv \nabla^2 N$. The trace of K_{ij} is K, the "mean curvature," of the time slice.
It is important to know the spacetime scalar curvature, which we call $(trRic)$:

$$g^{\alpha\beta}(Ric)_{\alpha\beta} = g^{\alpha\beta} R_{\lambda\alpha}{}^{\lambda}{}_{\beta} \tag{24}$$

in the form

$$N\sqrt{g}(trRic) = N\sqrt{g}(R + K_{ij}K^{ij} - K^2) - 2\partial_t(\sqrt{g}K) + 2\partial_i[\sqrt{g}(K\beta^i - \nabla^i N)] \tag{25}$$

where R is the spatial scalar curvature, because the spacetime scalar curvature density is the lagrangian density of the Hilbert action principle [14], explicitly modified in [24] to conform to the fixation of the metric on the boundary. The scalar curvature itself will be needed later. It is found from (24) and (21), (22), and (23) to be

$$(trRic) = 2N^{-1}\partial_0 K - 2N^{-1}\Delta N + (R + K_{ij}K^{ij} - K^2) \tag{26}$$

2 Einstein's Equations

Einstein used his insights about the principle of equivalence and his principle of general covariance (spacetime coordinate freedom plus a pseudo-riemannian metric not given a priori) in arriving at the final form of his field equations. As is now well known, his learning tensor analysis from Marcel Grossmann was an essential enabling step. The equations, using the Einstein tensor

$$(Ein)_{\mu\nu} \equiv G_{\mu\nu} \equiv (Ric)_{\mu\nu} - \frac{1}{2}g_{\mu\nu}(trRic) \tag{27}$$

are

$$G_{\mu\nu} = \kappa T_{\mu\nu} \tag{28}$$

where $\kappa = 8\pi G$, $c = 1$, $G =$ Newton's constant, and the stress–energy–momentum tensor of fields other than gravity (the "source" tensor) $T_{\mu\nu}$ must satisfy, as Einstein reasoned, in analogy to the conservation laws of special relativity,

$$\nabla_\mu T^{\mu\nu} = 0 \tag{29}$$

corresponding to

$$\nabla_\mu G^{\mu\nu} \equiv 0 \tag{30}$$

which is an identity, the "third Bianchi identity" or the "(twice) contracted Bianchi identity [23]." For purposes of the discussion below, besides $c = 1$ we also take Dirac's form of Planck's constant to be one: $\hbar = 1$. (Mass is now *inverse* length.)

Here we will consider only the vacuum theory. This is non-trivial because the equations are non-linear (gravity acts as a source of itself) and because the global topology and (or) boundary conditions are not prescribed by the equations. We assume that the object of solving the equations is to find the metric. In the 3+1 form of the equations, which is very close to a hamiltonian framework, the object is to obtain g_{ij} and K_{ij}, along with a workable specification of $\alpha = N g^{-1/2}$ and β^i which, as we shall see shortly, are not determined by Einstein's equations. For the vacuum case, one can use in four spacetime dimensions either of these two equations

$$G_{\mu\nu} = 0 \tag{31}$$

or

$$(Ric)_{\mu\nu} = 0 \tag{32}$$

For completeness, we note that the form of equation (32) with "sources" using the Ricci tensor is

$$(Ric)_{\mu\nu} = \kappa \left(T_{\mu\nu} - \frac{1}{2} T g_{\mu\nu} \right) \equiv \kappa \rho_{\mu\nu} \tag{33}$$

where $T = T^{\alpha}_{\alpha}$, the trace of the stress–energy tensor.

We now remark on a couple of points that are sometimes useful to bear in mind. The curvature equations we have given in purely geometric form. They can be converted into physical gravity equations explicitly by assigning the *physical* dimensions length (L), mass (M), and time (T). We have already chosen $c = 1$, so $T = L$. We shall work in terms of L. Next we choose to make action dimensionless by setting $\hbar = 1$, which yields $M = L^{-1}$. Then G (and κ) have dimension L^2. This is a handy viewpoint for the physicist and mathematician even if quantum effects are not considered. It enables us to display rather easily the nonlinear self-coupling that arises even in vacuum from the particular geometric nature of General Relativity.

We take the metric to be dimensionless while t and x^i have dimension L. (Think of natural locally Riemannian normal coordinates to make this view palatable.) The canonical form of the action based on $(2\kappa)^{-1} g^{1/2} N(trRic)$ is given by [1, 2, 24]. It yields in place of K_{ij} the closely related field canonical momentum [2]

$$\pi^{ij} = (2\kappa)^{-1} g^{1/2} (K g_{rs} - K_{rs}) \left(g^{ri} g^{sj} \right) \tag{34}$$

Inverting this expression in *three* dimensions yields

$$K_{ij} = (2\kappa) \left[g^{-1/2} \left(\frac{1}{2} \pi g^{kl} - \pi^{kl} \right) g_{ik} g_{jl} \right] \tag{35}$$

where $\pi = \pi^k_k$ is the trace. We note that from (35) we can obtain for the mean curvature

$$K = \kappa g^{-1/2}\pi = \frac{1}{2}(2\kappa)g^{-1/2}\pi \tag{36}$$

$$(2\kappa)^{-1}g^{1/2}(2\kappa) = \pi \tag{37}$$

the well-known, and vital, integrand of the boundary term of the Hilbert action [14] that can be seen [24] to convert it to the canonical action [1,2].

Denoting the terms in the rectangular brackets in (35) by μ_{ij}, which has dimension L, then

$$K_{ij} = (2\kappa)\mu_{ij} \tag{38}$$

Now we could rewrite the curvature equations with the gravitational interaction explicit. For example, the Gauss–Codazzi–Mainardi equations become

$$(Riem)_{ijkl} = R_{ijkl} + 4\kappa^2(\mu_{ik}\mu_{jl} - \mu_{il}\mu_{jk}) \tag{39}$$

$$(Riem)_{0ijk} = 2\kappa N(\nabla_j\mu_{ki} - \nabla_k\mu_{ji}) \tag{40}$$

$$(Riem)_{0i0j} = 2\kappa N\hat{\partial}_0\mu_{ij} + 4N\kappa^2\mu_{ik}\mu^k{}_j + N\nabla_i\nabla_j N \tag{41}$$

3 The 3 + 1-Form of Einstein's Equations

It is helpful to write out the ten vacuum equations using both (Ric) and (Ein):

$$(Ric)_{ij} = 0, \qquad 2N(Ric)_i^0 = 0, \qquad 2G_0^0 = 0 \tag{42}$$

This form was noted by Lichnerowicz [16] in the case of zero shift as being revealing. First, recall the geometric identity (15)

$$\hat{\partial}_0 g_{ij} = -2NK_{ij}$$

or

$$\partial_t g_{ij} = -2NK_{ij} + £_\beta g_{ij} = -2NK_{ij} + (\nabla_i\beta_j + \nabla_j\beta_i) \tag{43}$$

From the first equation in (42), one can obtain

$$\begin{aligned}\hat{\partial}_0 K_{ij} &= -\nabla_i\partial_j N + N(R_{ij} - 2K_{il}K^l{}_j + KK_{ij})\\ &\equiv \partial_t K_{ij} - £_\beta K_{ij}\\ &\equiv \partial_t K_{ij} - \left(\beta^l\nabla_l K_{ij} + K_{il}\nabla_j\beta^l + K_{lj}\nabla_i\beta^l\right)\end{aligned} \tag{44}$$

The second and third equations in (42) contain no terms $\partial_t K_{ij}$ (i.e., no "accelerations" $\partial_t\partial_t g_{ij}$) and are, therefore, *constraints* on the initial values of g_{ij} and K_{ij}. As previously mentioned, in this "canonical"-like 3 + 1 form, there are no time derivatives of $N = \alpha g^{1/2}$ or of β^i. In a second-order formalism, $\partial_t N$ and $\partial_t\beta^i$ would appear, as we see from (43). To make second order wave operators on *all* components of the spacetime metric, the (original) harmonic coordinate conditions

$(-g)^{-1/2}\partial_\mu[(-g)^{1/2}g^{\mu\nu}] = 0$ in natural coordinates were introduced and shown, along with the constraints, to be conserved by the resulting "reduced" equations if the constraints were assumed to hold at the "initial" time [4]. But no powers of $\dot{N}$ and $\dot{\beta}^i$ appear in (25), the lagrangian of Hilbert's action for the Einstein's equations. We have thus an easy way of seeing that $\dot{N}$ and $\dot{\beta}^i$ are dynamically irrelevant. We find from the second and third equations of (42), respectively,

$$2NR_i^0 \equiv C_i = 2\nabla_j(K_i^j - K\delta_i^j), \tag{45}$$

$$2G_0^0 \equiv C = K_{ij}K^{ij} - K^2 - R \tag{46}$$

These equations were derived in detail and displayed in [25], *without* a 3 + 1 splitting of the basis frames and coframes. An arbitrary spacetime basis was used there in order to remove what some people regarded as the "taint" of using particular coordinates. They are the *standard* 3 + 1 *equations*, wrongly called the standard ADM equations. In regard to evolving g_{ij} and K_{ij}, I do not claim that (43) and (44) are preferred for any other reason other than their maximum simplicity and absolute correctness when written in explicitly canonical form, using the variable π^{ij}, defined below, in place of K_{ij}. I say nothing here about the numerical properties of (43) and (44).

The canonical equations derived by Arnowitt, Deser, and Misner [2] and by Dirac [12, 13], are not equivalent to (44) given above, even when written in the same formalism, that is, with g_{ij} and K_{ij}. This is because their equation of motion is $G_{ij} = 0$ rather than $R_{ij} = 0$. Although $G_{\mu\nu} = 0$ and $R_{\mu\nu} = 0$ are equivalent, this is not true of spatial components. Instead, one has the key identity [1]

$$G_{ij} + g_{ij}G_0^0 \equiv (Ric)_{ij} - g_{ij}g^{kl}(Ric)_{kl} \tag{47}$$

or

$$G_{ij} + \frac{1}{2}g_{ij}C = (Ric)_{ij} - g_{ij}g^{kl}(Ric)_{kl} \tag{48}$$

Therefore, $G_{ij} = 0$ is not the correct equation of motion unless the constraint $C = 0$ also holds.

For the interested reader, I remark that this means the hamiltonian vector field of the ADM and Dirac canonical formalisms is not well-defined throughout the phase space. There is an easy cure in the ADM approach, which is based on a canonical action principle. When the metric g_{ij} is varied, one must hold fixed the "weighted" or "densitized" lapse function $\alpha = g^{-1/2}N$, instead of just the scalar lapse N. Thus, one carries out independent variations of g_{ij}, α, β^i, and $\pi^{ij} = (2\kappa)^{-1}g^{1/2}(Kg^{ij} - K^{ij})$ [1,3].

4 Conformal Transformations

A very useful technique for transforming the constraints C_i and C ((45) and (46)) into a well posed problem involving elliptic partial differential equations is to use conformal transformations of the essential spatial objects g_{ij}, K_{ij}, N, and β^i. Along the way we will again encounter the densitized lapse

$$\alpha \equiv g^{-1/2} N \tag{49}$$

which has coordinate weight (-1): N is a scalar with respect to spatial time-independent coordinate transformations (weight zero by definition).

The conformal factor will be denoted by φ. It will be assumed that $\varphi > 0$ throughout. The conformal transformation is defined by its action on the metric

$$\bar{g}_{ij} = \varphi^4 g_{ij} \tag{50}$$

It is called "conformal" because it preserves angles between vectors intersecting at a given point, whether one constructs the scalar product and vector magnitudes with g_{ij} or $\bar{g}_{ij}$. The power "4" in (50) is convenient for three dimensions. For dimension $n \geq 3$, the "convenient power" is $4(n-2)^{-1}$ for the metric conformal deformation. The neatness of this choice comes out most clearly in the relation of the scalar curvatures $\bar{R} = R(\bar{g})$ and $R = R(g)$ below. From (50) and the fact that the spatial connection is simply the "Christoffel symbol of the second kind $\left\{ {i \atop j\,k} \right\}$" which we denote by $\Gamma^i{}_{jk}$,

$$\Gamma^i{}_{jk} \equiv \left\{ \begin{matrix} i \\ j\;k \end{matrix} \right\} = \frac{1}{2} g^{il} (\partial_j g_{lk} + \partial_k g_{lj} - \partial_l g_{jk}) \tag{51}$$

we find that

$$\bar{\Gamma}^i{}_{jk} = \Gamma^i{}_{jk} + \frac{1}{2} \varphi^{-1} \left(\delta^i_j \, \partial_k \varphi + \delta^i_k \, \partial_j \varphi - g^{il} g_{jk} \, \partial_l \varphi\right) \tag{52}$$

From $\bar{\Gamma}^i{}_{jk}$ we can find the relationship between $\bar{R}_{ij}$ and R_{ij}.

$$\begin{aligned} \bar{R}_{ij} = R_{ij} &- 2\varphi^{-1} \, \nabla_i \partial_j \varphi + 6\varphi^{-2} \, (\partial_i \varphi)(\partial_j \varphi) \\ &- g_{ij} \left[2\varphi^{-1} \Delta\varphi + 2\varphi^{-2} (\nabla^k \varphi)(\partial_k \varphi) \right] \end{aligned} \tag{53}$$

There is no need to derive the transformation of the Riemann tensor, for in three dimensions R_{ijkl} can be expressed in terms of g_{ij} and R_{ij}. One can see that this must be so, for both the Ricci and Riemann tensors have six algebraically independent components. The formula relating them has long been known. It is displayed for example in [27]. This formula can be obtained from the identical vanishing of the Weyl conformal curvature tensor in three dimensions. But riemannian three-spaces are *not* conformally flat, in general. The Weyl tensor in three dimensions is replaced by the Cotton tensor [10], which is conformally invariant and vanishes if the three-space is conformally flat

$$C_{ijk} = \nabla_j L_{ik} - \nabla_k L_{ij} \tag{54}$$

where

$$L_{ik} = R_{ik} - \frac{1}{4} R g_{ik} \tag{55}$$

Its dual [28], found by using the inverse volume form ϵ^{mjk} on the skew pair [jk] and by raising the index i to l is a symmetric tensor $* C^{lm}$ with trace identically zero and covariant divergence identically zero. The Cotton tensor has third derivatives of the metric and is therefore not a curvature tensor, but rather is a *differential curvature tensor* with dimensions $(length)^{-3}$. Therefore, the dual Cotton tensor is

$$* C^{lm} = g^{li} \epsilon^{mjk} C_{ijk} \tag{56}$$

Under conformal transformations, we have that $\bar{g}^{ij} = \varphi^{-4} g^{ij}$, $\bar{\epsilon}_{ijk} = \varphi^6 \epsilon_{ijk}$ (the volume three-form), $\bar{C}_{ijk} = C_{ijk}$, and $\bar{\epsilon}^{ijk} = \varphi^{-6} \epsilon^{ijk}$. Therefore,

$$* \bar{C}^{ij} = \varphi^{-10} (* C^{ij}) \tag{57}$$

The properties of $* C^{ij}$ hold for an entire conformal equivalence class, that is, for all sets of conformally related riemannian metrics as in (50) for all $0 < \varphi < \infty$. The divergence of $* C^{ij}$ is *identically* zero, so it need not be surprising that

$$\bar{\nabla}_j \left(* \bar{C}^{ij} \right) = \varphi^{-10} \nabla_j \left(* C^{ij} \right) \tag{58}$$

using the barred and unbarred connections to form the covariant divergence operators $\bar{\nabla}_i$ and ∇_j. We see that (57) is the natural conformal transformation law for symmetric, traceless type $\binom{2}{0}$ tensors in three dimensions, whose divergence may or may not vanish. Note that in obtaining (58), the scaling (57) was used. But *no* properties of $*C^{ij}$ were employed except the symmetry type of the tensor representation: symmetric with zero trace.

We now pass to the famous formula for the conformal transformation of the scalar curvature R. It was introduced in connection with an early treatment of the constraints in [16]. From (53) it follows that

$$\bar{R} = \varphi^{-4} R - 8 \varphi^{-5} \Delta \varphi \tag{59}$$

So far every transformation has followed from the defining relation $\bar{g}_{ij} = \varphi^4 g_{ij}$. A glance at both the constraints (45) and (46) shows that we must deal with K_{ij}. The method here can be deduced by writing K_{ij} as

$$K^{ij} = A^{ij} + \frac{1}{3} K g^{ij} \tag{60}$$

where A^{ij} is the traceless part of K^{ij}. We treat A^{ij} and K differently because A^{ij} and Kg^{ij} can be regarded as different irreducible types of symmetric two-index tensors. They also have different conformal transformations. Lichnerowicz took $K = 0$ [16]. But this is too restrictive, and even then the simplified "momentum constraint" (45) was not solved. Mme. Choquet-Bruhat first argued that one has to solve the momentum constraint with a second-order operator on a vector potential [5]. A very useful result was given in [26] and it was used for many years to solve the momentum constraint. Its imperfections were noted first by O'Murchadha [17], Isenberg [15], and the author. The inference was that the early method was only an *ansatz*. A better method yet, with no ambiguities, was displayed in [25] and [19]. It is given below.

Suppose an overbar denotes a solution of the constraints and the corresponding object without an overbar denotes a "trial function." The strategy is to "deform" the trial objects conformally into barred quantities, that is, into solutions. Every object we deal with has, in effect, a "conformal dimension," which is not given by its physical dimension or by its tensorial character but by how it transforms under a change of the basis or of the natural coordinates.

5 An Elliptic System

We write (45) and (46) in the barred variables as

$$\bar{\nabla}_j \bar{A}^{ij} - \frac{2}{3}\bar{g}^{ij}\partial_j \bar{K} = 0 \tag{61}$$

$$\bar{A}_{ij}\bar{A}^{ij} - \frac{2}{3}\bar{K} - \bar{R} = 0 \tag{62}$$

Conformal transformations for objects that are purely concomitants of $\bar{g}_{ij}$ (or g_{ij}) are derived as above in a straightforward manner. But the extrinsic curvature variables have to be handled with a modicum of care. The transformations obtained by extending $\bar{g}_{ij} = \varphi^4 g_{ij}$ to all of the spacetime metric variables is not appropriate because the view that spacetime structure is primary is not helpful in a situation, as here, where there is as yet *no* spacetime.

We begin with $\bar{K}$. We hold it fixed because its inverse in the simpler cosmological models is the "Hubble time," without a knowledge of which the epoch is not known. Data astronomers obtain from different directions in the sky, or at different "depths" back in time are basically correlated and they fix $\bar{K}$ implicitly. Therefore, I long ago adopted the rule [24] of fixing the "mean curvature" $\bar{K}$

$$\bar{K} = K \tag{63}$$

under conformal transformations. Thus it is specified *a priori*.

What to do about the symmetric tracefree tensor $\bar{A}^{ij}$? The prior discussion of $*C^{ij}$ indicates the transformation

$$\bar{A}^{ij} = \varphi^{-10} A^{ij} \tag{64}$$

But symmetric tensors "$\bar{T}^{ij}$" have, in a curved space, three irreducible types that are formally L^2-orthogonal [11]. One is the trace ($\bar{g}^{ij}\bar{T}^k_k$), another is like $*C^{ij}$, that is, a part with vanishing covariant divergence. Finally, a symmetric tracefree tensor can be constructed from a vector

$$(\bar{L}X)^{ij} = \left(\bar{\nabla}^i X^j + \bar{\nabla}^j X^i - \frac{2}{3}\bar{g}^{ij}\bar{\nabla}_l X^l\right) \tag{65}$$

the "conformal Killing form" of X^i. (I have not found other constructions that are sufficiently well-behaved under conformal transformations to be useful in this problem.) This expression (65) vanishes if X^i is a conformal killing vector of $\bar{g}_{ij}$. Then, X^i would be a conformal killing vector of every metric conformal to $\bar{g}_{ij}$. Therefore,

$$\bar{X}^i = X^i, \qquad \bar{g}_{ij} = \varphi^4 g_{ij} \tag{66}$$

and

$$(\bar{L}X)^{ij} = \varphi^{-4}(LX)^{ij} \tag{67}$$

which *misses* obeying our "rule" (57) or (64). For a long time, the mismatched powers required a work-around to obtain an *ansatz* for solving the constraints [27], but there is a simple and elegant solution [28]. The vectorial part (65) needs a weight factor and a corresponding change in the measure of orthogonality.

Recall our statement that the densitized lapse α is the preferred undetermined multiplier (rather than the lapse N) in the action principle leading to 3 + 1 (or canonical) equations of motion. See [1] where this is made perfectly clear. This is not to say anything about the "best" form of the $\hat{\partial}_0 K_{ij}$ (or $\hat{\partial}_0 \pi^{ij}$) equation of motion for calculational purposes. In fact, the system for $\hat{\partial}_0 g_{ij}$ and $\hat{\partial}_0 K_{ij}$ is not hyperbolic. But *only* this form gives a hamiltonian vector field well-defined in the entire momentum phase space.

To proceed, we note that one does not scale undetermined multipliers. Therefore

$$\bar{\beta}^i = \beta^i, \qquad \bar{\alpha} = \alpha \tag{68}$$

But because $\bar{\alpha} = \bar{g}^{-1/2}\bar{N}$ and $\bar{g}^{1/2} = \varphi^6 g^{1/2}$, then $\bar{N} = \bar{g}^{1/2}\alpha = g^{1/2}\alpha$ implies

$$\bar{N} = \varphi^6 N \tag{69}$$

I have known that the transformation (69) was useful since 1971. But, thinking that N was an undetermined multiplier – a lowly "C-number," independent of the dynamical variables, in Dirac's well-known parlance – I did not use (69). Then I learned about the densitized lapse and saw its role in the action principle. It then dawned on me that (69) was correct all along. It made its first appearance in the conformal thin sandwich problem [25].

The lapse becomes, thus, a dynamical variable [1, 3, 19]. A look at (69) and at the relation between $\partial_t \bar{g}_{ij}$ and $\bar{K}_{ij}$ gives us the scalar weight factor $(-2\bar{N})^{-1}$ in the decomposition of $\bar{A}^{ij}$

$$\bar{A}^{ij} = \bar{A}^{ij}_{(\delta)} + \left(-2\bar{N}\right)^{-1}\left(\bar{L}X\right)^{ij} \tag{70}$$

The subscript (δ) indicates that the covariant divergence of $\bar{A}^{ij}_{(\delta)}$ vanishes. Note that (70) does *not* mean that the extrinsic curvature is sensitive to N. It is not. What it *does* mean is that the identification of the divergence-free and trace-free part of the extrinsic curvature is, in part, dependent on N. Also note that the two parts of $\bar{A}^{ij}$ are formally L^2-orthogonal both before and after a conformal transformation, with the geometrical *spacetime measure*

$$\sqrt{-g} = Ng^{1/2} \tag{71}$$

instead of the spatial measure $g^{1/2}$. Therefore, we have

$$\begin{aligned} &\int \bar{A}^{ij}_{(\delta)}\left[(-2\bar{N})^{-1}(\bar{L}X)^{kl}\right]\bar{g}_{ik}\bar{g}_{jl}\left(\bar{N}\bar{g}^{1/2}\right)d^3x \\ &= \int A^{ij}_{(\delta)}\left[(-2N)^{-1}(LX)^{kl}\right]g_{ik}g_{jl}\left(Ng^{1/2}\right)d^3x \end{aligned} \tag{72}$$

Upon integration by parts, with suitable boundary conditions, or no boundary, each of the integrals (72) vanishes.

We construct $\bar{A}^{ij}_{(\delta)}$ or $A^{ij}_{(\delta)}$ by extracting from a freely given symmetric tracefree tensor $\bar{F}^{ij} = \varphi^{-10}F^{ij}$ its transverse-tracefree part, which will be our $A^{ij}_{(\delta)}$

$$F^{ij} = A^{ij}_{(\delta)} + (-2N)^{-1}(LY)^{ij} \tag{73}$$

with

$$\nabla_j[(-2N)^{-1}(LY)^{ij}] = \nabla_j F^{ij} \tag{74}$$

The momentum constraints

$$\bar{\nabla}_j \bar{A}^{ij} - \frac{2}{3}\bar{g}^{ij}\partial_j K = 0 \tag{75}$$

become, with $Z^i = X^i - Y^i$,

$$\nabla_j[(-2N)^{-1}(LZ)^{ij}] = \nabla_j F^{ij} - \frac{2}{3}\varphi^6 g^{ij}\partial_j K \tag{76}$$

The solution for Z^i, with given N, will give the parts of $\bar{K}_{ij}$,

$$\bar{A}^{ij} = \varphi^{-10}[F^{ij} + (-2N)^{-1}(LZ)^{ij}] \tag{77}$$

$$\tfrac{1}{3}\bar{K}\bar{g}^{ij} = \tfrac{1}{3}\varphi^{-4}Kg^{ij} \tag{78}$$

However (75) contains φ and is coupled to the "hamiltonian constraint" (62) unless the "constant mean curvature" (CMC) condition $K = constant$ (in space, $\partial_j K = 0$) can be employed, as introduced in [24]. This includes maximal slicing $K = 0$ [16]. (Lichnerowicz did not propose the CMC condition as claimed in [22].)

Gathering the transformations for $\bar{R}$ and $\bar{K}^{ij}$ enables us to write the hamiltonian constraint as the general relativity version of the Laplace–Poisson equation

$$\Delta\varphi - \frac{1}{8}\left[R\varphi + (F_{ij} + (-2N)^{-1}(LZ)_{ij})^2\varphi^{-7} - \frac{2}{3}K^2\varphi^5\right] = 0 \tag{79}$$

Suppose, for example, we choose $N = 1$. Then

$$\bar{N} = \varphi^6 = \bar{g}^{1/2}\left(g^{-1/2}\right) \tag{80}$$

We are certainly entitled to have chosen g_{ij} such that $g^{1/2} = 1$, without loss of generality. Thus we recover Teitelboim's gauge for the lapse equation

$$\bar{N} = \bar{g}^{1/2} \tag{81}$$

in his noted paper on the canonical path integral in general relativity [21].

A bit more generally, if we choose

$$\hat{\partial}_0 g^{1/2} = 0 \tag{82}$$

we see that $\bar{N}$ automatically satisfies the time gauge equation used by Choquet-Bruhat and Ruggeri [8].

The constraints have the *same form* as they do in the thin sandwich formulation: see [25]. Therefore, the space of solutions has the properties obtained in [7]; see also [9].

Write out from the formula for $\partial_t \bar{g}_{ij}$ its tracefree part $\bar{u}_{ij}$, which we can use as the velocity of the conformal metric

$$\bar{u}_{ij} = -2\bar{N}\bar{F}_{ij} + \left[\bar{L}(\bar{Z} + \bar{\beta})\right]_{ij} \tag{83}$$

with $(\bar{Z} + \bar{\beta})_j \equiv \bar{g}_{ij}(Z^i + \beta^i)$. This has the form of the solution in the conformal thin sandwich problem. The choice of shift $\beta^i = -Z^i$ is possible here and renders a simple final form.

Acknowledgement The author thanks Michael Brown for his invaluable assistance in preparing the manuscript.

References

1. A. Anderson and J.W. York, “Hamiltonian time evolution for general relativity,” Phys. Rev. Letters, **81**, 1154-1157 (1998).
2. R. Arnowitt, S. Deser, and C.W. Misner, “The Dynamics of General Relativity”, pp. 227–265 in L. Witten (ed.), *Gravitation* (Wiley, New York, 1962).
3. A. Ashtekar, “New Hamiltonian formulation of general relativity,” Phys. Rev. D **36**, 1587–1602 (1987).
4. Y. Choquet-Bruhat, Théorèm d'existence pour certains systèms d'équations aux derivées partielles non linéaires,” Acta. Math. **88**, 141–255 (1955).
5. Y. Choquet-Bruhat, Seminar, Princeton University, October, 1972.
6. Y. Choquet-Bruhat and C. DeWitt-Morette, *Analysis, Manifolds, and Physics*, (North Holland, Amsterdam, 1977).
7. Y. Choquet-Bruhat, J. Isenberg, and J.W. York, “Asymptotically euclidean solutions of the Einstein constraints,” Phys. Rev. D **61**, 084034 (2000).
8. Y. Choquet-Bruhat and T. Ruggeri, “Hyperbolic form of the Einstein equations,” Commun. Math. Phys. **89**, 269–283 (1983).
9. Y. Choquet-Bruhat and J.W. York, “Well posed reduced systems for the Einstein equation,” Banach Center Publications, I, **41**, 119–131 (1997).
10. E. Cotton, “Sur les variétés à trois dimensions,” Ann. Fac. d. Sc. Toulouse (II) **1**, 385–438, (1899).
11. S. Deser, “Covariant decompositions of symmetric tensors and the gravitational Cauchy problem,” Ann. Inst. Henri Poincaré **A7**, 149–188 (1967).
12. P.A.M. Dirac, “The theory of gravitation in Hamiltonian form,” Proc. Roy. Soc. Lond. **A246**, 333–343 (1958).
13. P.A.M. Dirac, “Fixation of coordinates in the Hamiltonian theory of gravitation,” Phys. Rev. **114**, 924–930 (1959).
14. D. Hilbert, “Die Grundlagen der Physik,” Konigl. Gesell. d. Wiss. Göttingen, Nachr., Math.-Phys. Kl., 395–407 (1915).
15. J.A. Isenberg, private communication (1973).
16. A. Lichnerowicz, “L'intégration des équations de la gravitation problème des n corps,” J. Math. Pures Appl. **23**, 37–63, (1944).
17. N.S. O'Murchadha, private communication (1972).
18. H. Pfeiffer, Ph.D. thesis, Cornell University, 2003.
19. H. Pfeiffer and J.W.York, “Extrinsic curvature and the Einstein constraints,” Phys. Rev. D **67**, 044022 (2003).
20. J.A. Schouten, *Ricci Calculus*, Springer-Verlag (Berlin, 1954).
21. C. Teitelboim, “Quantum mechanics of the gravitational field,” Phys. Rev. D **28**, 297–311 (1983).
22. F.J. Tipler and J.E. Marsden, “Maximal hypersurfaces and foliations of constant mean curvature in general relativity,” Phys. Reports **66**, 109–139 (1980).
23. A. Trautman, “Conservation Laws in General Relativity,” pp. 169–198 in L. Witten (ed.), *Gravitation* (Wiley, New York, 1962).
24. J.W. York, “Role of conformal three-geometry in the dynamics of gravitation,” Phys. Rev. Letters **28**, 1082–1085 (1972).
25. J.W. York, “Conformal thin sandwich problem,” Phys. Rev. Letters **82**, 1350–1353 (1999).
26. J.W. York, “Decomposition of symmetric tensors,” Ann. Inst. Henri Poincaré, **A21**, 319–332 (1974). It should be noted that this method is the correct one for decomposing metric perturbations.
27. J.W. York, “Kinematics and Dynamics of General Relativity,” pp. 83–126 in L. Smarr (ed.), *Sources of Gravitational Radiation*, (Cambridge, 1979).
28. J.W. York, private communication to S. Teukolsky and H. Pfeiffer (2001).

The Confrontation Between General Relativity and Experiment

Clifford M. Will

Abstract We review the experimental evidence for Einstein's special and general relativity. A variety of high precision null experiments verify the weak equivalence principle and local Lorentz invariance, while gravitational redshift and other clock experiments support local position invariance. Together these results confirm the Einstein Equivalence Principle which underlies the concept that gravitation is synonymous with spacetime geometry, and must be described by a metric theory. Solar system experiments that test the weak-field, post-Newtonian limit of metric theories strongly favor general relativity. The Binary Pulsar provides tests of gravitational-wave damping and of strong-field general relativity. Recently discovered binary pulsar systems may provide additional tests. Ongoing and future experiments, such as the Gravity Probe B Gyroscope Experiment, satellite tests of the Equivalence principle, and tests of gravity at short distance to look for extra spatial dimensions could constrain extensions of general relativity. Laser interferometric gravitational-wave observatories on Earth and in space may provide new tests of gravitational theory via detailed measurements of the properties of gravitational waves. Future efforts using X-ray, infrared, gamma-ray and gravitational-wave astronomy may one day achieve one of John Wheeler's visions: to test general relativity in the strong-field regime near black holes.

1 Introduction

During the late 1960s, it was frequently said that "the field of general relativity is a theorist's paradise and an experimentalist's purgatory". To be sure, there were some experiments: Irwin Shapiro, then at MIT, had just measured the relativistic retardation of radar waves passing the Sun (an effect that now bears his name), Robert

C.M. Will (✉)
Department of Physics and McDonnell Center for the Space Sciences Washington University, St. Louis MO, USA
e-mail: cmw@wuphys.wustl.edu; wugrav.wustl.edu/people/CMW

I. Ciufolini and R.A. Matzner (eds.), *General Relativity and John Archibald Wheeler*, Astrophysics and Space Science Library 367, DOI 10.1007/978-90-481-3735-0_6,

Dicke of Princeton was claiming that the Sun was flattened in an amount that would mess up general relativity's success with Mercury's perihelion advance, and Joseph Weber of the University of Maryland was just about to announce (40 years prematurely, as we now know) the detection of gravitational waves. Nevertheless the field was dominated by theory and by theorists. The field *circa* 1970 seemed to reflect Einstein's own attitudes: although he was not ignorant of experiment, and indeed had a keen insight into the workings of the physical world, he felt that the bottom line was the *theory*. As he once famously said, if experiment were to contradict the theory, he would have "felt sorry for the dear Lord".

Since that time the field has been completely transformed, and today experiment is a central, and in some ways dominant component of gravitational physics. To illustrate this, one needs only to cite the first regular article of the 15 June 2004 issue of Physical Review D: the author list of this "general relativity" paper fills an entire page, and the institution list fills most of another. This was one of the papers reporting results from the first science run of the LIGO laser interferometer gravitational-wave observatories, but it brings to mind papers in high-energy physics, not general relativity! The breadth of current experiments, ranging from tests of classic general relativistic effects such as the light bending and the Shapiro delay, to searches for short-range violations of the inverse-square law, to the operation of a space experiment to measure the relativistic precession of gyroscopes, to the construction and operation of gravitational-wave detectors, attest to the ongoing vigor of experimental gravitation.

John Wheeler played a significant role in this transformation. Although he never carried out any experiments, he was one of the first to recognize and to emphasize the central role that general relativity would play in compact astronomical objects. In the early 1960s he helped initiate the emerging field of relativistics astrophysics. As the recognition of general relativity's role in astrophysics began to grow, so too did the recognition that better tests of the theory were needed.

Today, because of its elegance and simplicity, and because of its empirical success, general relativity has become the foundation for our understanding of the gravitational interaction. Yet modern developments in particle theory suggest that it is probably not the entire story, and that modification of the basic theory may be required at some level. String theory generally predicts a proliferation of scalar fields that could result in alterations of general relativity reminiscent of the Brans–Dicke theory of the 1960s. In the presence of extra dimensions, the gravity that we feel on our four-dimensional "brane" of a higher dimensional world could be somewhat different from a pure four-dimensional general relativity. Some of these ideas have motivated the possibility that fundamental constants may actually be dynamical variables, and hence may vary in time or in space. However, any theoretical speculation along these lines must abide by the best current empirical bounds. Decades of high-precision tests of general relativity have produced some very tight constraints. In this article I will review the experimental situation, and assess how well Einstein got it right.

We begin in Section 2 with the "Einstein equivalence principle", which underlies the idea that gravity and curved spacetime are synonymous, and describe its empirical support. Section 3 describes solar system tests of gravity in terms of experimental bounds on a set of "parametrized post-Newtonian" (PPN) parameters. In Section 4 we discuss tests of general relativity using binary pulsar systems. Section 5 describes tests of gravitational theory that could be carried out using future observations of gravitational radiation, and Section 6 describes the possibility of performing strong-field tests of general relativity. Concluding remarks are made in Section 7. For further discussion of topics in this paper, and for references to the literature, the reader is referred to *Theory and Experiment in Gravitational Physics* [1] and to the "living" review articles [2–4].

2 The Einstein Equivalence Principle

The Einstein equivalence principle (EEP) is a powerful and far-reaching principle, which states that

- Test bodies fall with the same acceleration independently of their internal structure or composition (Weak Equivalence Principle, or WEP).
- The outcome of any local non-gravitational experiment is independent of the velocity of the freely-falling reference frame in which it is performed (Local Lorentz Invariance, or LLI).
- The outcome of any local non-gravitational experiment is independent of where and when in the universe it is performed (Local Position Invariance, or LPI).

The Einstein equivalence principle is the heart of gravitational theory, for it is possible to argue convincingly that if EEP is valid, then gravitation must be described by "metric theories of gravity", which state that (i) spacetime is endowed with a symmetric metric, (ii) the trajectories of freely falling bodies are geodesics of that metric, and (iii) in local freely falling reference frames, the non-gravitational laws of physics are those written in the language of special relativity.

General relativity is a metric theory of gravity, but so are many others, including the Brans–Dicke theory. In this narrow sense, superstring theory is not metric, because of residual coupling of external, gravitation-like fields, to matter. Such external fields could be characterized as fields that do not vanish in the vacuum state (in contrast, say, to electromagnetic fields). Theories in which varying non-gravitational constants are associated with dynamical fields that couple to matter directly are also not metric theories.

2.1 *Tests of the Weak Equivalence Principle*

To test the weak equivalence principle, one compares the acceleration of two laboratory-sized bodies of different composition in an external gravitational field. A measurement or limit on the fractional difference in acceleration between two bodies yields a quantity $\eta \equiv 2|a_1 - a_2|/|a_1 + a_2|$, called the "Eötvös ratio", named in honor of Baron von Eötvös, the Hungarian physicist whose experiments carried out with torsion balances at the end of the nineteenth century were the first high-precision tests of WEP [5]. Later classic experiments by Dicke and Braginsky [6, 7] improved the bounds by several orders of magnitude. Additional experiments were carried out during the 1980s as part of a search for a putative "fifth force", that was motivated in part by a reanalysis of Eötvös' original data (the range of bounds achieved during that period is shown schematically in the region labeled "fifth force" in Fig. 1).

In a torsion balance, two bodies of different composition are suspended at the ends of a rod that is supported by a fine wire or fibre. One then looks for a difference in the horizontal accelerations of the two bodies as revealed by a slight rotation of

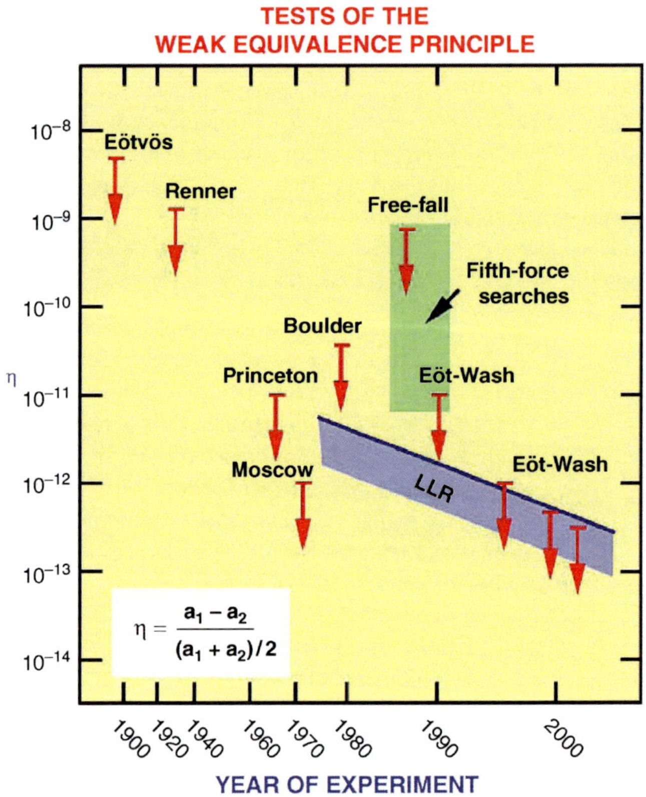

Fig. 1 Selected tests of the Weak Equivalence Principle, showing bounds on the fractional difference in acceleration of different materials or bodies. "Free-fall" and Eöt-Wash experiments, along with numerous others between 1986 and 1990, were originally performed to search for a fifth force. Blue line and shading shows evolving bounds on WEP for the Earth and the Moon from Lunar laser ranging (LLR)

the rod. The source of the horizontal gravitational force could be the Sun, a large mass in or near the laboratory, or, as Eötvös recognized, the Earth itself.

The best limit on η currently comes from the "Eöt-Wash" experiments carried out at the University of Washington, which used a sophisticated torsion balance tray to compare the accelerations of bodies of different composition toward the Earth, the Sun and the galaxy [8, 9]. Another strong bound comes from Lunar laser ranging (LLR), which checks the equality of free fall of the Earth and Moon toward the Sun [10, 11]. The results from laboratory and LLR experiments are:

$$\eta_{\text{Eöt-Wash}} < 3 \times 10^{-13}, \quad \eta_{\text{LLR}} < 3 \times 10^{-13}. \tag{1}$$

In fact, by using laboratory materials whose composition mimics that of the Earth and Moon, the Eöt-Wash experiments [8] permit one to infer an unambiguous bound from Lunar laser ranging on the universality of acceleration of gravitational binding energy at the level of 9×10^{-4} (test of the Nordtvedt effect – see Section 3.2 and Table 1.)

The Apache Point Observatory for Lunar Laser-ranging Operation (APOLLO) project, a joint effort by researchers from the Universities of Washington, Seattle, and California, San Diego, uses enhanced laser and telescope technology, together with a good, high-altitude site in New Mexico, with the goal to improve the Lunar laser-ranging bound by as much as an order of magnitude [11].

High-precision WEP experiments, can test superstring inspired models of scalar–tensor gravity, or theories with varying fundamental constants in which weak violations of WEP can occur via non-metric couplings. The project MICROSCOPE, designed to test WEP to a part in 10^{15} has been approved by the French space agency CNES for a possible 2012 launch. A proposed NASA-ESA Satellite Test

Table 1 Current limits on the PPN parameters

Parameter	Effect	Limit	Remarks
$\gamma - 1$	(i) Time delay	2.3×10^{-5}	Cassini tracking
	(ii) Light deflection	4×10^{-4}	VLBI
$\beta - 1$	(i) Perihelion shift	3×10^{-3}	$J_2 = 10^{-7}$ from helioseismology
	(ii) Nordtvedt effect	2.3×10^{-4}	$\eta = 4\beta - \gamma - 3$ assumed
ξ	Earth tides	10^{-3}	Gravimeter data
α_1	Orbital polarization	10^{-4}	Lunar laser ranging PSR J2317 + 1439
α_2	Solar spin precession	4×10^{-7}	Alignment of Sun and ecliptic
α_3	Pulsar acceleration	2×10^{-20}	Pulsar $\dot{P}$ statistics
η^{a}	Nordtvedt effect	9×10^{-4}	Lunar laser ranging
ζ_1	–	2×10^{-2}	Combined PPN bounds
ζ_2	Binary motion	4×10^{-5}	$\ddot{P}_p$ for PSR 1913 + 16
ζ_3	Newton's 3rd law	10^{-8}	Lunar acceleration
ζ_4	–	–	Not independent

[a]Here $\eta = 4\beta - \gamma - 3 - 10\xi/3 - \alpha_1 - 2\alpha_2/3 - 2\zeta_1/3 - \zeta_2/3$.

of the Equivalence Principle (STEP) seeks to reach the level of $\eta < 10^{-18}$. These experiments will compare the acceleration of different materials moving in free-fall orbits around the Earth inside a drag-compensated spacecraft. Doing these experiments in space means that the bodies are in perpetual fall, whereas Earth-based free-fall experiments (such as the 1987 test done at the University of Colorado [12] indicated in Fig. 1), are over in seconds, which leads to significant measurement errors.

Many of the high-precision, low-noise methods that were developed for tests of WEP have been adapted to carry out laboratory tests of the inverse square law of Newtonian gravitation at millimeter scales and below. The goal of these experiments is to search for additional couplings to massive particles or for the presence of large extra dimensions. The challenge of these experiments is to distinguish gravitation-like interactions from electromagnetic and quantum mechanical (Casimir) effects. No deviations from the inverse square law have been found to date at distances between 10 μm and 10 mm [13–17].

2.2 Tests of Local Lorentz Invariance

Although special relativity itself never benefited from the kind of "crucial" experiments, such as the perihelion advance of Mercury and the deflection of light, that contributed so much to the initial acceptance of general relativity and to the fame of Einstein, the steady accumulation of experimental support, together with the successful merger of special relativity with quantum mechanics, led to its being accepted by mainstream physicists by the late 1920s, ultimately to become part of the standard toolkit of every working physicist. This accumulation included

- The classic Michelson–Morley experiment and its descendents [18–21]
- The Ives–Stillwell, Rossi–Hall and other tests of time-dilation [22–24]
- Tests of the independence of the speed of light of the velocity of the source, using both binary X-ray stellar sources and high-energy pions [25, 26]
- Tests of the isotropy of the speed of light [27–29]

In addition to these direct experiments, there was the Dirac equation of quantum mechanics and its prediction of anti-particles and spin; later would come the stunningly successful relativistic theory of quantum electrodynamics.

Having recently passed the 100th anniversary of the introduction of special relativity, one might ask "what is there to test?" Special relativity has been so thoroughly integrated into the fabric of modern physics that its validity is rarely challenged, except by cranks and crackpots. It is ironic then, that during the past several years, a vigorous theoretical and experimental effort has been launched, on an international scale, to find violations of special relativity. The motivation for this effort is a desire not to repudiate Einstein, but to look for evidence of new physics "beyond" Einstein, such as apparent violations of Lorentz invariance that might result from certain models of quantum gravity. Quantum gravity asserts that there is a fundamental length scale given by the Planck length, $L_p = (\hbar G/c^3)^{1/2} = 1.6 \times 10^{-33}$ cm, but since length is not an invariant quantity (Lorentz–FitzGerald contraction), then there could be a violation of Lorentz invariance at some level in quantum gravity. In brane world scenarios, while physics may be locally Lorentz invariant in the higher dimensional world, the confinement of the interactions of normal physics to our four-dimensional "brane" could induce apparent Lorentz violating effects. And in models such as string theory, the presence of additional scalar, vector and tensor long-range fields that couple to matter of the standard model could induce effective violations of Lorentz symmetry. These and other ideas have motivated a serious reconsideration of how to test Lorentz invariance with better precision and in new ways.

A simple way of interpreting some of these experiments is to suppose that a non-metric coupling to the electromagnetic interactions results in a change in the speed of electromagnetic radiation c relative to the limiting speed of material test particles c_0, in other words, $c \neq c_0$. In units where $c_0 = 1$, this would result in an action for charged particles and electromagnetic fields given, in a preferred reference frame (presumably that of the cosmic background radiation), by

$$I = -\sum_a m_{0a} \int \left(1 - v_a^2\right)^{1/2} dt + \sum_a e_a \int (-\Phi + \mathbf{A} \cdot \mathbf{v_a}) dt + \frac{1}{8\pi} \int (E^2 - c^2 B^2) d^3x dt, \tag{2}$$

where $\Phi = -A_0$, $\mathbf{E} = -\nabla\Phi - \dot{\mathbf{A}}$, and $\mathbf{B} = \nabla \times \mathbf{A}$. This is sometimes called the "c^2" framework [30, 31]; it is a special case of the "$TH\epsilon\mu$" framework of Lightman and Lee [32] for analysing non-metric theories of gravity, and of the "standard model extension" (SME) of Kostelecký and coworkers [33–35]. Such a Lorentz-non-invariant electromagnetic interaction would cause shifts in the energy levels of atoms and nuclei that depend on the orientation of the quantization axis of the state relative to our velocity in the rest-frame of the universe, and on the quantum numbers of the state, resulting in orientation dependence of the fundamental frequencies of such atomic clocks. The magnitude of these "clock anisotropies" would be proportional to $\delta \equiv |c^{-2} - 1|$.

The earliest clock anisotropy experiments were those of Hughes and Drever, although their original motivation was somewhat different [36,37]. Dramatic improvements were made in the 1980s using laser-cooled trapped atoms and ions [38–40]. This technique made it possible to reduce the broading of resonance lines caused by collisions, leading to improved bounds on δ shown in Fig. 2 (experiments labelled NIST, U. Washington and Harvard, respectively).

The SME and other frameworks [41] have been used to analyse many new experimental tests of local Lorentz invariance, including comparisons of resonant cavities with atomic clocks, and tests of dispersion and birefringence in the propagation of high energy photons from astrophysical sources. Other testable effects of Lorentz invariance violation include threshold effects in particle reactions, gravitational Cerenkov radiation, and neutrino oscillations. Mattingly [4] gives a thorough and

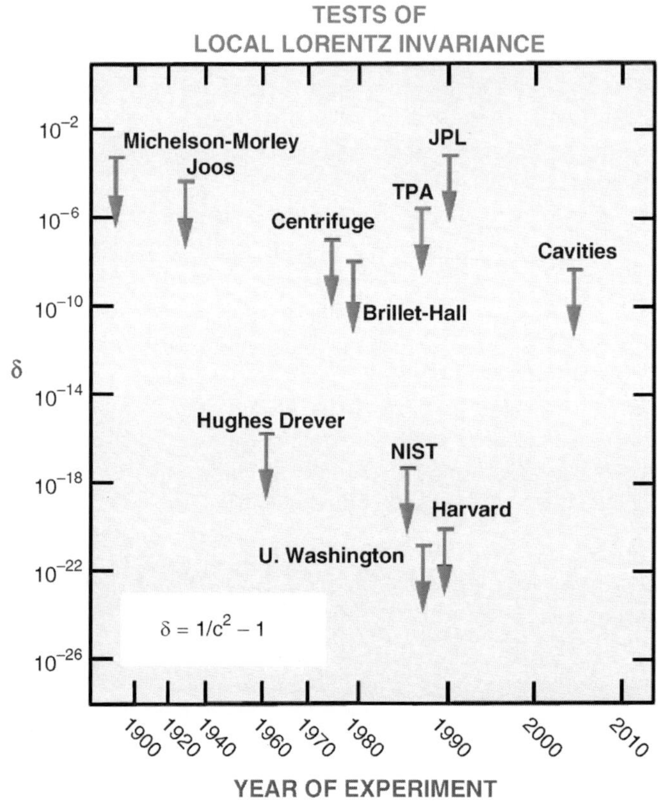

Fig. 2 Selected tests of local Lorentz invariance, showing bounds on the parameter $\delta = c^{-2} - 1$, where c is the speed of propagation of electromagnetic waves in a preferred reference frame, in units in which the limiting speed of test particles is unity

up-to-date review of both the theoretical frameworks and the experimental results, and describes possibilities for future tests of local Lorentz invariance.

2.3 *Tests of Local Position Invariance*

Local position invariance, requires, among other things, that the internal binding energies of atoms be independent of location in space and time, when measured against some standard atom. This means that a comparison of the rates of two different kinds of clocks should be independent of location or epoch, and that the frequency shift of a signal sent between two identical clocks at different locations is simply a consequence of the apparent Doppler shift between a pair of inertial frames momentarily comoving with the clocks at the moments of emission and reception respectively. The relevant parameter in the frequency shift expression $\Delta f/f = (1+\alpha)\Delta U/c^2$, is $\alpha \equiv \partial \ln E_B/\partial(U/c^2)$, where E_B is the atomic or nuclear binding energy, and U is the external gravitational potential. If LPI is valid, the binding energy should be independent of the external potential, and hence $\alpha = 0$. The best bounds come from a 1976 rocket redshift experiment using Hydrogen masers [42], and a series of "null" redshift experiments, in which atomic clocks employing different atoms (Hydrogen, Cesium, Rubidium, Mercury) were intercompared as a function of the varying solar gravitational potential at the Earth [43–46]. The results are

$$\alpha_{\text{Maser}} < 2 \times 10^{-4}, \quad \alpha_{\text{Null}} < 2 \times 10^{-5}. \tag{3}$$

In a similar manner, clock intercomparison experiments were designed to look for possible variations of the fine structure constant on a cosmological timescale. The results showed that the fine structure constant α is constant in time to a few parts in 10^{15} per year [47–50]. Plans are being developed to perform such clock comparisons in space, possibly on the International Space Station.

A better bound on $d\alpha/dt$ comes from analysis of fission yields of the Oklo natural reactor, which occurred in Africa 1.8 billion years ago, namely $(\dot{\alpha}/\alpha)_{\text{Oklo}} < 6 \times 10^{-17}$ per year [51]. These and other bounds on variations of constants, including reports (later disputed) of positive evidence for variations from quasar spectra, are discussed by Martins and others in Ref. [52], and in the extensive review by Uzan [53].

3 Solar-System Tests

3.1 The Parametrized Post-Newtonian Framework

It was once customary to discuss experimental tests of general relativity in terms of the "three classical tests", the gravitational redshift, which is really a test of the EEP, not of general relativity itself (see Section 2.3); the perihelion advance of Mercury, the first success of the theory; and the deflection of light, whose measurement in 1919 made Einstein a celebrity. However, the proliferation of additional tests as well as of well-motivated alternative metric theories of gravity, made it desirable to develop a more general theoretical framework for analysing both experiments and theories.

This "parametrized post-Newtonian (PPN) framework" dates back to Eddington in 1922, but was fully developed by Nordtvedt and Will in the period 1968–1972. When we confine attention to metric theories of gravity, and further focus on the slow-motion, weak-field limit appropriate to the solar system and similar systems, it turns out that, in a broad class of metric theories, only the numerical values of a set of parameters vary from theory to theory. The framework contains ten PPN parameters: γ, related to the amount of spatial curvature generated by mass; β, related to the degree of non-linearity in the gravitational field; ξ, α_1, α_2, and α_3, which determine whether the theory violates local position invariance or local Lorentz invariance in *gravitational* experiments (violations of the Strong Equivalence Principle); and ζ_1, ζ_2, ζ_3 and ζ_4, which describe whether the theory has appropriate momentum conservation laws. For a complete exposition of the PPN framework see Ref. [1].

A number of well-known relativistic effects can be expressed in terms of these PPN parameters:

Deflection of light:

$$\begin{aligned}\Delta\theta &= \left(\frac{1+\gamma}{2}\right)\frac{4GM}{dc^2} \\ &= \left(\frac{1+\gamma}{2}\right)\times 1.7505\frac{R_\odot}{d}\ \text{arcsec},\end{aligned} \tag{4}$$

where d is the distance of closest approach of a ray of light to a body of mass M, and where the second line is the deflection by the Sun, with radius $R_\odot$.

Shapiro time delay:

$$\Delta t = \left(\frac{1+\gamma}{2}\right)\frac{4GM}{c^3}\ln\left[\frac{(r_1+\mathbf{x}_1\cdot\mathbf{n})(r_2-\mathbf{x}_2\cdot\mathbf{n})}{d^2}\right], \tag{5}$$

where Δt is the excess travel time of a round-trip electromagnetic tracking signal, $\mathbf{x}_1$ and $\mathbf{x}_2$ are the locations relative to the body of mass M of the emitter and receiver of the round-trip radar tracking signal (r_1 and r_2 are the respective distances) and $\mathbf{n}$ is the direction of the outgoing tracking signal.

Perihelion advance:

$$\begin{aligned}\frac{d\omega}{dt} &= \left(\frac{2+2\gamma-\beta}{3}\right)\frac{GM}{Pa(1-e^2)c^2} \\ &= \left(\frac{2+2\gamma-\beta}{3}\right)\times 42.98\ \text{arcsec}/100\ \text{year},\end{aligned} \tag{6}$$

where P, a, and e are the period, semi-major axis and eccentricity of the planet's orbit; the second line is the value for Mercury.

Nordtvedt effect:

$$\frac{m_G-m_I}{m_I} = \left(4\beta-\gamma-3-\frac{10}{3}\xi-\alpha_1-\frac{2}{3}\alpha_2-\frac{2}{3}\zeta_1-\frac{1}{3}\zeta_2\right)\frac{|E_g|}{m_I c^2}, \tag{7}$$

where m_G and m_I are the gravitational and inertial masses of a body such as the Earth or Moon, and E_g is its gravitational binding energy. A non-zero Nordtvedt effect would cause the Earth and Moon to fall with a different acceleration toward the Sun.

Precession of a gyroscope:

$$\begin{aligned}
\Omega_{FD} &= -\frac{1}{2}\left(1+\gamma+\frac{\alpha_1}{4}\right)\frac{G}{r^3c^2}(\mathbf{J}-3\mathbf{n}\,\mathbf{n}\cdot\mathbf{J}) \\
&= \frac{1}{2}\left(1+\gamma+\frac{\alpha_1}{4}\right)\times 0.041\,\text{arcsec}\,\text{yr}^{-1}, \\
\Omega_{Geo} &= -\frac{1}{2}(1+2\gamma)\mathbf{v}\times\frac{Gm\mathbf{n}}{r^2c^2} \\
&= \frac{1}{3}(1+2\gamma)\times 6.6\,\text{arcsec year}^{-1},
\end{aligned} \tag{8}$$

where Ω_{FD} and Ω_{Geo} are the precession angular velocities caused by the dragging of inertial frames (Lense–Thirring effect), and by the geodetic effect, a combination of Thomas precession and precession induced by spatial curvature; J is the angular momentum of the Earth, and $\mathbf{v}$, $\mathbf{n}$ and r are the velocity, direction, and distance of the gyroscope. The second line in each case is the corresponding value for a gyroscope in polar Earth orbit at about 650 km altitude (Gravity Probe B, Section 3.3).

In general relativity, $\gamma = 1$, $\beta = 1$, and the remaining parameters all vanish.

3.2 Bounds on the PPN Parameters

Four decades of experiments, ranging from the standard light-deflection and perihelion-shift tests, to Lunar laser ranging, planetary and satellite tracking tests of the Shapiro time delay, and geophysical and astronomical observations, have placed bounds on the PPN parameters that are consistent with general relativity. The current bounds are summarized in Table 1.

To illustrate the dramatic progress of experimental gravity since the dawn of Einstein's theory, Fig. 3 shows a history of results for $(1+\gamma)/2$, from the 1919 solar eclipse measurements of Eddington and his colleagues (which made Einstein a public celebrity), to modern-day measurements using very-long-baseline radio interferometry (VLBI), advanced radar tracking of spacecraft, and orbiting astrometric satellites such as Hipparcos. The most recent results include a measurement of the Shapiro delay using the *Cassini* spacecraft [54], and a measurement of the bending of light via analysis of VLBI data on 541 quasars and compact radio galaxies distributed over the entire sky [55].

The perihelion advance of Mercury, the first of Einstein's successes, is now known to agree with observation to a few parts in 10^3. Although there was controversy during the 1960s about this test because of Dicke's claims of an excess solar oblateness, which would result in an unacceptably large Newtonian contribution to the perihelion advance, it is now known from helioseismology that the oblateness is of the order of a few parts in 10^7, as expected from standard solar models, and too small to affect Mercury's orbit, within the experimental error.

Scalar–tensor theories of gravity are characterized by a coupling function $\omega(\phi)$ whose size is inversely related to the "strength" of the scalar field relative to the

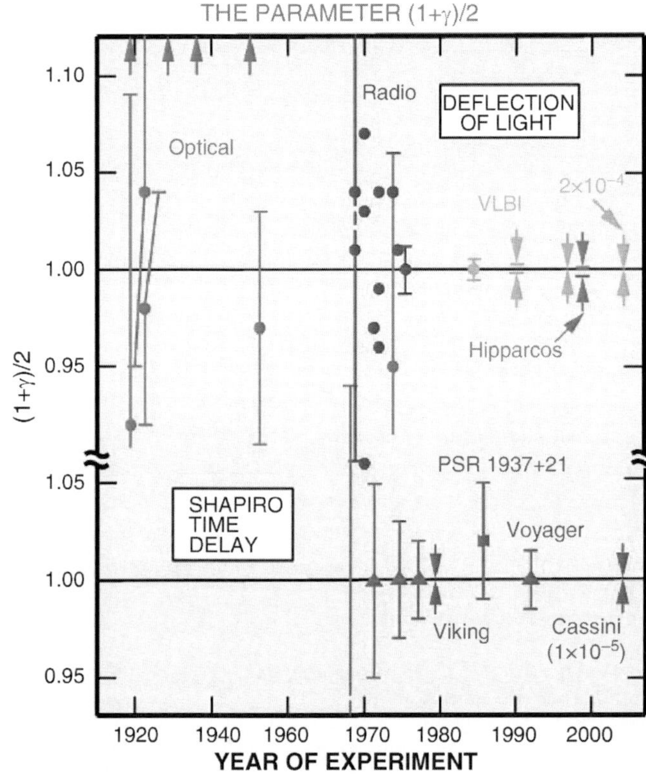

Fig. 3 Measurements of the coefficient $(1 + \gamma)/2$ from observations of the deflection of light and of the Shapiro delay in propagation of radio signals near the Sun. The general relativity prediction is unity. "Optical" denotes measurements of stellar deflection made during solar eclipses, "Radio" denotes interferometric measurements of radio-wave deflection, and "VLBI" denotes Very Long Baseline Radio Interferometry. "Hipparcos" denotes the European optical astrometry satellite. Arrows denote values well off the chart from one of the 1919 eclipse expeditions and from others through 1947. Shapiro delay measurements using the Cassini spacecraft on its way to Saturn yielded tests at the 0.001 percent level, and light deflection measurements using VLBI have reached 0.02%

metric. In the solar system, the parameter $|\gamma - 1|$, for example is equal to $1/(2 + \omega(\phi_0))$, where ϕ_0 is the value of the scalar field today outside the solar system. Solar-system experiments (primarily the Cassini results [54]) constrain $\omega(\phi_0) >$ 40,000.

Proposals are being developed for advanced space missions which will have tests of PPN parameters as key components, including GAIA, a high-precision astrometric telescope (successor to Hipparcos), which could measure light-deflection and γ to the 10^{-6} level [56], the Laser Astrometric Test of Relativity (LATOR), a mission involving laser ranging to a pair of satellites on the far side of the Sun, which could measure γ to a part in 10^8, and could possibly detect second-order effects in light propagation [57], and BepiColombo, a Mercury orbiter mission planned for the

2012 time frame, which could, among other measurements, determine J_2 of the sun to 10^{-8}, and improve bounds on a time variation of the gravitational constant [58]. The ground-based laser ranging program APOLLO could improve the bound on the Nordtvedt parameter η to the 3×10^{-5} level [11].

3.3 Gravity Probe B

The NASA Relativity Mission called Gravity Probe-B completed its mission to measure the Lense–Thirring and geodetic precessions of gyroscopes in Earth orbit[59]. Launched on April 20, 2004 for a 16-month mission, it consisted of four spherical fused quartz rotors coated with a thin layer of superconducting niobium, spinning at 70–100 Hz, in a spacecraft containing a telescope continuously pointed toward a distant guide star (IM Pegasi). Superconducting current loops encircling each rotor measured the change in direction of the rotors by detecting the change in magnetic flux through the loop generated by the London magnetic moment of the spinning superconducting film. The spacecraft orbited the Earth in a polar orbit at 650 km altitude. The proper motion of the guide star relative to distant quasars was measured before, during and after the mission using VLBI. The primary science goal of GPB was a 1% measurement of the 41 milliarcsecond per year frame dragging or Lense–Thirring effect caused by the rotation of the Earth; its secondary goal was to measure to six parts in 10^5 the larger 6.6 arcsecond per year geodetic precession caused by space curvature (Eq. (8)). Final results of the experiment were expected toward the end of 2010.

A complementary test of the Lense–Thirring precession, albeit with about ten times lower accuracy than the GPB goal, was reported by Ciufolini and Pavlis [60]. This experiment measured the precession of the orbital planes of two Earth-orbiting laser-ranged satellites called LAGEOS, using up-to-date models of the gravitational field of the Earth in order to subtract the dominant Newtonian precession with sufficient accuracy to yield a measurement of the relativistic effect. This experiment is described in detail elsewhere in this volume.

4 The Binary Pulsar

The binary pulsar PSR 1913 + 16, discovered in 1974 by Joseph Taylor and Russell Hulse, provided important new tests of general relativity, specifically of gravitational radiation and of strong-field gravity. Through precise timing of the pulsar "clock", the important orbital parameters of the system could be measured with exquisite precision. These include non-relativistic "Keplerian" parameters, such as the eccentricity e, and the orbital period (at a chosen epoch) P_b, as well as a set of relativistic "post-Keplerian" parameters (see Table 2). The first PK parameter, $\langle\dot{\omega}\rangle$, is the mean rate of advance of periastron, the analogue of Mercury's perihelion shift.

Table 2 Parameters of the binary pulsar PSR 1913 + 16

Parameter	Symbol	Value[a] in PSR1913 + 16	Value[a] in J0737-3039
Keplerian parameters			
Eccentricity	e	0.6171338(4)	0.087778(2)
Orbital period	P_b (day)	0.322997448930(4)	0.1022515628(2)
Post-Keplerian parameters			
Periastron advance	$\langle\dot{\omega}\rangle$ ($^{\circ}$yr^{-1})	4.226595(5)	16.900(2)
Redshift/time dilation	γ' (ms)	4.2919(8)	0.39(2)
Orbital period derivative	$\dot{P}_b$ (10^{-12})	−2.4184(9)	−1.20(8)
Shapiro delay ($\sin i$)	s		0.9995(4)

[a]Numbers in parentheses denote errors in last digit.

The second, denoted γ' is the effect of special relativistic time-dilation and the gravitational redshift on the observed phase or arrival time of pulses, resulting from the pulsar's orbital motion and the gravitational potential of its companion. The third, $\dot{P}_b$, is the rate of decrease of the orbital period; this is taken to be the result of gravitational radiation damping (apart from a small correction due to galactic differential rotation). Two other parameters, s and r, are related to the Shapiro time delay of the pulsar signal if the orbital inclination is such that the signal passes in the vicinity of the companion; s is a direct measure of the orbital inclination $\sin i$. According to general relativity, the first three post-Keplerian effects depend only on e and P_b, which are known, and on the two stellar masses which are unknown. By combining the observations of PSR 1913 + 16 with the general relativity predictions, one obtains both a measurement of the two masses, and a test of the theory, since the system is overdetermined. The results are [61]

$$m_1 = 1.4414 \pm 0.0002 M_{\odot}, \quad m_2 = 1.3867 \pm 0.0002 M_{\odot},$$
$$\dot{P}_b^{GR}/\dot{P}_b^{OBS} = 1.0013 \pm 0.0021. \tag{9}$$

The accuracy in measuring the relativistic damping of the orbital period is now limited by uncertainties in our knowledge of the relative acceleration between the solar system and the binary system as a result of galactic differential rotation.

The results also test the strong-field aspects of metric gravitation in the following way: the neutron stars that comprise the system have very strong internal gravity, contributing as much as several tenths of the rest mass of the bodies (compared to the orbital energy, which is only 10^{-6} of the mass of the system). Yet in general relativity, the internal structure is "effaced" as a consequence of the Strong Equivalence Principle (SEP), a stronger version of EEP that includes *gravitationally* bound bodies and local *gravitational* experiments. As a result, the orbital motion and gravitational radiation emission depend *only* on the masses m_1 and m_2, and not on their internal structure. By contrast, in alternative metric theories, SEP is not valid in general, and internal-structure effects can lead to significantly different behavior, such as the emission of dipole gravitational radiation. Unfortunately, in the case of scalar–tensor theories of gravity, because the neutron stars are so similar in PSR

1913 + 16 (and in other double-neutron star binary pulsar systems), dipole radiation is suppressed by symmetry; the best bound on the coupling parameter $\omega(\phi_0)$ from PSR 1913 + 16 is in the hundreds.

However, the recent discovery of the relativistic neutron star/white dwarf binary pulsar J1141-6545, with a 0.19 day orbital period, may ultimately lead to a very strong bound on dipole radiation, and thence on scalar–tensor gravity [62, 63]. The remarkable "double pulsar" J0737-3039 is a binary system with two detected pulsars, in a 0.10 day orbit seen almost edge on, with eccentricity $e = 0.09$, and a periastron advance of 17° per year. A variety of novel tests of relativity, neutron star structure, and pulsar magnetospheric physics will be possible in this system [64–66]. For a review of binary pulsar tests, see [3].

5 Gravitational-Wave Tests of Gravitation Theory

The detection of gravitational radiation by ground-based laser interferometers, such as LIGO in the USA, or VIRGO and GEO in Europe, by the proposed space-based laser interferometer LISA, or by resonant-bar detectors, will usher in a new era of gravitational-wave astronomy [67,68]. Furthermore, it will yield new and interesting tests of general relativity in its radiative regime [69].

5.1 Polarization of Gravitational Waves

A laser-interferometric or resonant-bar gravitational-wave detector measures the local components of a symmetric 3 × 3 tensor which is composed of the "electric" components of the Riemann tensor, R_{0i0j}. These six independent components can be expressed in terms of polarizations (modes with specific transformation properties under null rotations). Three are transverse to the direction of propagation, with two representing quadrupolar deformations and one representing an axisymmetric "breathing" deformation. Three modes are longitudinal, with one an axially symmetric stretching mode in the propagation direction, and one quadrupolar mode in each of the two orthogonal planes containing the propagation direction. General relativity predicts only the first two transverse quadrupolar modes, independently of the source, while scalar–tensor gravitational waves can in addition contain the transverse breathing mode. More general metric theories predict up to the full complement of six modes. A suitable array of gravitational antennas, or a single antenna changing its orientation suitably during the passage of a wave, could delineate or limit the number of modes present. If distinct evidence were found of any mode other than the two transverse quadrupolar modes of general relativity, the result would be disastrous for the theory. On the other hand, the absence of a breathing mode would not necessarily rule out scalar–tensor gravity, because the strength of that mode depends on the nature of the source.

5.2 Speed of Gravitational Waves

According to general relativity, in the limit in which the wavelength of gravitational waves is small compared to the radius of curvature of the background spacetime, the waves propagate along null geodesics of the background spacetime, i.e. they have the same speed, c, as light. In other theories, the speed could differ from c because of coupling of gravitation to "background" gravitational fields. For example, in some theories with a flat background metric $\boldsymbol{\eta}$, gravitational waves follow null geodesics of $\boldsymbol{\eta}$, while light follows null geodesics of $\mathbf{g}$ [1]. In brane-world scenarios, the apparent speed of gravitational waves could differ from that of light if the former can propagate off the brane into the higher dimensional "bulk". Another way in which the speed of gravitational waves could differ from c is if gravitation were propagated by a massive field (a massive graviton), in which case v_g would be given by, in a local inertial frame,

$$\frac{v_g}{c} = \left(1 - \frac{m_g^2 c^4}{E^2}\right)^{1/2} \approx 1 - \frac{1}{2}\frac{c^2}{f^2 \lambda_g^2}, \tag{10}$$

where m_g, E and f are the graviton rest mass, energy and frequency, respectively, and $\lambda_g = h/m_g c$ is the graviton Compton wavelength ($\lambda_g \gg c/f$ assumed).

The most obvious way to test for a massive graviton is to compare the arrival times of a gravitational wave and an electromagnetic wave from the same event, e.g. a supernova. For a source at a distance D, the resulting bound on the difference $|1 - v_g/c|$ or on λ_g is

$$|1 - \frac{v_g}{c}| < 5 \times 10^{-17} \left(\frac{200\,\text{Mpc}}{D}\right) \left(\frac{\Delta t}{1\,\text{s}}\right), \tag{11}$$

$$\lambda_g > 3 \times 10^{12}\,\text{km} \left(\frac{D}{200\,\text{Mpc}} \frac{100\,\text{Hz}}{f}\right)^{1/2} \left(\frac{1}{f \Delta t}\right)^{1/2}, \tag{12}$$

where $\Delta t \equiv \Delta t_a - (1+Z)\Delta t_e$, with Δt_a and Δt_e denoting the differences in arrival time and emission time, respectively, of the two signals, and Z denoting the redshift of the source. In many cases, Δt_e is unknown, so that the best one can do is employ an upper bound on Δt_e based on observation or modelling.

However, there is a situation in which a bound on the graviton mass can be set using gravitational radiation alone [71]. That is the case of the inspiralling compact binary, the final stage of evolution of systems like the binary pulsar, in which the loss of energy to gravitational waves has brought the binary to an inexorable spiral toward a final merger. Because the frequency of the gravitational radiation sweeps from low frequency at the initial moment of observation to higher frequency at the final moment, the speed of the gravitational waves emitted will vary, from lower speeds initially to higher speeds (closer to c) at the end. This will cause a distortion of the observed phasing of the waves and result in a shorter than expected overall time Δt_a of passage of a given number of cycles. Furthermore, through the

Table 3 Potentially achievable bounds on λ_g from gravitational-wave observations of inspiralling compact binaries. Values in parentheses include non-precessing spin–orbit effects

$m_1(M_\odot)$	$m_2(M_\odot)$	Distance (Mpc)	Bound on λ_g (km)
Ground-based (LIGO/VIRGO)			
1.4	1.4	300	4.6×10^{12}
10	10	1,500	6.0×10^{12}
Space-based (LISA)			
10^7	10^7	3,000	$6.9(2.2) \times 10^{16}$
10^5	10^5	3,000	$2.3(0.5) \times 10^{16}$

technique of matched filtering, the parameters of the compact binary can be measured accurately [72], and thereby the effective emission time Δt_e can be determined accurately.

A full noise analysis using proposed noise curves for the advanced LIGO ground-based detectors, and for the proposed space-based LISA antenna yields potentially achievable bounds that are summarized in Table 3 [74, 75]. These potential bounds can be compared with the solid bound $\lambda_g > 2.8 \times 10^{12}$ km, derived from solar system dynamics, which limit the presence of a Yukawa modification of Newtonian gravity of the form $V(r) = (GM/r)\exp(-r/\lambda_g)$ [73], and with the model-dependent bound $\lambda_g > 6 \times 10^{19}$ km from consideration of galactic and cluster dynamics [70].

6 Tests of Gravity in the Strong-Field Regime

John Wheeler emphasized the role of compact objects and strong gravity, notably as embodied in the black hole, as a central testing ground for general relativity.

One of the main difficulties of testing GR in the strong-field regime is the possibility of contamination by uncertain or complex physics. In the solar system, weak-field gravitational effects could in most cases be measured cleanly and separately from non-gravitational effects. The remarkable cleanliness of the binary pulsar permitted precise measurements of gravitational phenomena in a strong-field context. Unfortunately, nature is rarely so kind. Still, under suitable conditions, qualitative and even quantitative strong-field tests of GR can be carried out.

One example is the exploration of the spacetime near black holes and neutron stars via accreting matter. Studies of certain kinds of accretion known as advection-dominated accretion flow (ADAF) in low-luminosity binary X-ray sources may yield the signature of the black hole event horizon [76]. The spectrum of frequencies of quasi-periodic oscillations (QPO) from galactic black hole binaries may permit measurement of the spins of the black holes [77]. Aspects of strong-field gravity and frame-dragging may be revealed in spectral shapes of iron fluorescence lines from the inner regions of accretion disks [78, 79]. Measurements of the detailed shape of the infrared image of the accretion flow around $SgA*$, the massive black hole in the center of our galaxy could also provide tests of the spacetime black hole metric [80]. Because of uncertainties in the detailed models, the results to date of studies like these are suggestive at best, but the combination of higher-resolution observations and better modelling could lead to striking tests of strong-field predictions of GR.

The best tests of GR in the strong field limit may come from gravitational-wave observations. The ground-based interferometers may detect the gravitational waves from the final inspiral and merger of a pair of stellar-mass black holes. Comparison of the observed waveform with the predictions of GR from a combination of analytic and numerical techniques may provide a test of the theory in the most dynamical, strong-field limit. In the future, the proposed space antenna LISA may observe a many as 100 mergers of massive black holes per year, with large signal-to-noise ratio. Such observations could provide precise measurements of black hole masses and spins, and could test the "no hair" theorems of black holes by detecting the spectrum quasi-normal "ringdown" modes emitted by the final black hole [81]. Observations by LISA of the hundreds of thousands of gravitational-wave cycles emitted when a small black hole inspirals onto a massive black hole could test whether the geometry of a black hole actually corresponds to the "hair-free" Kerr metric [82]. As John Wheeler was one of the originators of the concept that "black holes have no hair" (i.e. are described completely by just two parameters, mass and spin), he would surely be delighted if such tests could be made.

7 Conclusions

Einstein's relativistic theories altered the course of science. They were triumphs of the imagination and of theory; experiment played a secondary role. In the past four decades, we have witnessed a second triumph for Einstein, in the systematic, high-precision experimental verification of his theories. Relativity has passed every test with flying colors. But the work is not done. Tests of strong-field gravity in the vicinity of black holes and neutron stars need to be carried out. Gammay-ray, X-ray and gravitational-wave astronomy will play a critical role in probing this largely unexplored aspect of general relativity.

General relativity is now the "standard model" of gravity. But as in particle physics, there may be a world beyond the standard model. Quantum gravity, strings and branes may lead to testable effects beyond standard general relativity. Experimentalists will continue a vigorous search for such effects using laboratory experiments, particle accelerators, space instrumentation and cosmological observations.

John Wheeler's prediction that the most critical tests of Einstein's theory would involve the strongest fields and the realm of the quantum may eventually come true.

Acknowledgments This work was supported in part by the US National Science Foundation, Grant No. PHY 03-53180 and 06-52448 and by the National Aeronautics and Space Administration, Grant No. NNG-06GI60G.

References

1. C. M. Will, *Theory and Experiment in Gravitational Physics*, (Cambridge University Press, Cambridge, 1993).
2. C. M. Will, Living Rev. Relativ. **9**, 3 (2006) [Online article]: cited on 1 June 2007, www.livingreviews.org/lrr-2001-4.
3. I. H. Stairs Living Rev. Relativ. **6**, 5 (2003) [On-line article] Cited on 1 June 2007, www.livingreviews.org/lrr-2003-5.
4. D. Mattingly, Living Rev. Relativ. **8**, 5 (2005) [On-line article] Cited on 1 June 2007, www.livingreviews.org/lrr-2005-5.
5. R. V. Eötvös, V. Pekár and E. Fekete, Ann. d. Physik **68**, 11 (1922).
6. P. G. Roll, R. Krotkov and R. H. Dicke, Ann. Phys. (N.Y.) **26**, 442 (1964).
7. V. B. Braginsky and V. I. Panov, Sov. Phys. JETP **34**, 463 (1972).
8. S. Baessler, B. R. Heckel, E. G. Adelberger, J. H. Gundlach, U. Schmidt and H. E. Swanson, Phys. Rev. Lett. **83**, 3585 (1999).
9. E. G. Adelberger, Class. Quantum Gravit. **18**, 2397 (2001).
10. J. G. Williams, X. X. Newhall and J. O. Dickey, Phys. Rev. D **53**, 6730 (1996).
11. J. G. Williams, S. Turyshev and T. W. Murphy, Int. J. Mod. Phys. D **13**, 567 (2004).
12. T. M. Niebauer, M. P. McHugh, and J. E. Faller, Phys. Rev. Lett. **59**, 609 (1987).
13. J. C. Long, H. W. Chan and J. C. Price, Nucl. Phys. B **539**, 23 (1999).
14. C. D. Hoyle, U. Schmidt, B. R. Heckel, E. G. Adelberger, J. H. Gundlach, D. J. Kapner and H. E. Swanson, Phys. Rev. Lett. **86**, 1418 (2001).
15. C. D. Hoyle, D. J. Kapner, B. R. Heckel, E. G. Adelberger, J. H. Gundlach, U. Schmidt, and H. E. Swanson, Phys. Rev. D **70**, 042004 (2004).
16. J. Chiaverini, S. J. Smullin, A. A. Geraci, D. M. Weld, A. Kapitulnik, Phys. Rev. Lett. **90**, 151101 (2003).
17. J. C. Long, H. W. Chan, A. B. Churnside, E. A. Gulbis, M. C. M. Varney and J. C. Price, Nature **421**, 922 (2003).
18. A. A. Michelson and E. W. Morley, Am. J. Sci. **134**, 333 (1887).
19. R. S. Shankland, S. W. McCuskey, F. C. Leone and G. Kuerti, Rev. Mod. Phys. **27**, 167 (1955).
20. T. S. Jaseja, A. Javan, J. Murray and C. H. Townes, Phys. Rev. **133**, A1221 (1964).
21. A. Brillet and J. L. Hall, Phys. Rev. Lett. **42**, 549 (1979).
22. H. E. Ives and G. R. Stilwell, J. Opt. Soc. Am. **28**, 215 (1938).
23. B. Rossi and D. B. Hall, Phys. Rev. **59**, 223 (1941).
24. F. J. M. Farley, J. Bailey, R. C. A. Brown, M. Giesch, H. Jöstlein, S. van der Meer, E. Picasso and M. Tannenbaum, Nuovo Cimento **45**, 281 (1966).
25. K. Brecher, Phys. Rev. Lett. **39**, 1051 (1977).
26. T. Alväger, F. J. M. Farley, J. Kjellman and I. Wallin, Phys. Lett. **12**, 260 (1977).
27. D. C. Champeney, G. R. Isaak and A. M. Khan, Phys. Lett. **7**, 241 (1963).
28. E. Riis, L.-U. A. Anderson, N. Bjerre, O. Poulson, S. A. Lee and J. L. Hall, Phys. Rev. Lett. **60**, 81 (1988); *ibid* **62**, 842 (1989).
29. T. P. Krisher, L. Maleki, G. F. Lutes, L. E. Primas, R. T. Logan, J. D. Anderson and C. M. Will, Phys. Rev. D **42**, 731 (1990).
30. M. P. Haugan and C. M. Will, Phys. Today **40**, 69 (May) (1987).
31. M. D. Gabriel and M. P. Haugan, Phys. Rev. D **41**, 2943 (1990).
32. A. P. Lightman and D. L. Lee, Phys. Rev. D **8**, 364 (1973).
33. D. Colladay and V. A. Kostelecký, Phys. Rev. D **55**, 6760 (1997).
34. D. Colladay and V. A. Kostelecký, Phys. Rev. D **58**, 116002 (1998).
35. V. A. Kostelecký and M. Mewes, Phys. Rev. D **66**, 056005 (2002).
36. V. W. Hughes, H. G. Robinson and V. Beltran-Lopez, Phys. Rev. Lett. **4**, 342 (1960).
37. R. W. P. Drever, Phil. Mag. **6**, 683 (1961).
38. J. D. Prestage, J. J. Bollinger, W. M. Itano and D. J. Wineland, Phys. Rev. Lett. **54**, 2387 (1985).
39. S. K. Lamoreaux, J. P. Jacobs, B. R. Heckel, F. J. Raab and E. N. Fortson, Phys. Rev. Lett. **57**, 3125 (1986).

40. T. E. Chupp, R. J. Hoare, R. A. Loveman, E. R. Oteiza, J. M. Richardson, M. E. Wagshul and A. K. Thompson, Phys. Rev. Lett. **63**, 1541 (1989).
41. T. Jacobson, S. Liberati, and D. Mattingly, Phys. Rev. D **67**, 124011 (2003).
42. R. F. C. Vessot, M. W. Levine, E. M. Mattison, E. L. Blomberg, T. E. Hoffman, G. U. Nystrom, B. F. Farrell, R. Decher, P. B. Eby, C. R. Baugher, J. W. Watts, D. L. Teuber and F. O. Wills, Phys. Rev. Lett. **45**, 2081 (1980).
43. J.P. Turneaure, C.M. Will, B.F. Farrell, E.M. Mattison, and R.F.C Vessot, Phys. Rev. D **27**, 1705 (1983).
44. A. Godone, C. Novero, P. Tavella, and K. Rahimullah, Phys. Rev. Lett. **71**, 2364 (1993).
45. J. D. Prestage, R. L. Tjoelker, and L. Maleki, Phys. Rev. Lett. **74**, 3511 (1995).
46. A. Bauch and S. Weyers, Phys. Rev. D **65**, 081101 (2002).
47. H. Marion, F. Pereira Dos Santos, M. Abgrall, S. Zhang, Y. Sortais, S. Bize, I. Maksimovic, D. Calonico, J. Grünert, C. Mandache, P. Lemonde, G. Santarelli, Ph. Laurent, A. Clairon and C. Salomon, Phys. Rev. Lett. **90**, 150801 (2003).
48. S. Bize, S. A. Diddams, U. Tanaka, C. E. Tanner, W. H. Oskay, R. E. Drullinger, T. E. Parker, T. P. Heavner, S. R. Jefferts, L. Hollberg, W. M. Itano, and J. C. Bergquist, Phys. Rev. Lett. **90**, 150802 (2003).
49. M. Fischer, N. Kolachevsky, M. Zimmermann, R. Holzwarth, Th. Udem, T. W. Hansch, M. Abgrall, J. Grunert, I. Maksimovic, S. Bize, H. Marion, F. Pereira Dos Santos, P. Lemonde, G. Santarelli, P. Laurent, A. Clairon, C. Salomon, M. Haas, U. D. Jentschura, and C. H. Keitel, Phys. Rev. Lett. **92**, 230802 (2004).
50. E. Peik, B. Lipphardt, H. Schnatz, T. Schneider, Chr. Tamm, S. G. Karshenboim, Phys. Rev. Lett. **93**, 170801 (2004).
51. T. Damour and F. Dyson, Nucl. Phys. B **480**, 37 (1996).
52. C. J. A. P. Martins, editor, *The Cosmology of Extra Dimensions and Varying Fundamental Constants* (Kluwer Academic Publishers, The Netherlands, 2003); also published in Astrophys. Space Sci. **283**, 439 (2003).
53. J.-P. Uzan, Rev. Mod. Phys. **75**, 403 (2003).
54. B. Bertotti, L. Iess and P. Tortora, Nature **425**, 374 (2003).
55. S. S. Shapiro, J. L. Davis, D. E. Lebach, and J. S. Gregory, Phys. Rev. Lett. **92**, 121101 (2004).
56. GAIA: for information about the project, see http://astro.estec.esa.nl/GAIA/.
57. S. G. Turyshev, M. Shao, and K. Nordtvedt, Jr., Class. Quantum Gravit. **21**, 2773 (2004).
58. A. Milani, D. Vokrouhlický, D. Villani, C. Bonanno and A. Rossi, Phys. Rev. D **66**, 082001 (2002).
59. For information about Gravity Probe B see http://einstein.stanford.edu.
60. I. Ciufolini and E. C. Pavlis, Nature **431**, 958 (2004).
61. J. M. Weisberg, J. H. Taylor, in *Binary Radio Pulsars*, edited by F. A. Rasio and I. H. Stairs (Astronomical Society of the Pacific Conference Series, Vol. 328), p. 25 (2005).
62. M. Bailes, S. M. Ord, H. S. Knight and A. W. Hotan, Astrophys. J. Lett. **595**, L52 (2003).
63. G. Esposito-Farèse, in *The Tenth Marcel Grossmann Meeting on Recent Developments in Theoretical and Experimental General Relativity, Gravitation and Relativistic Field Theories*, edited by M. Novello and S. P. Bergliaffa (World Scientific, Singapore, 2006), p. 647.
64. A. G. Lyne, M. Burgay, M. Kramer, A. Possenti, R. N. Manchester, F. Camilo, M. A. McLaughlin, D. R. Lorimer, N. D'Amico, B. C. Joshi, J. Reynolds, P. C. C. Freire, Science **303**, 1153 (2004).
65. M. Kramer, D. R. Lorimer, A. G. Lyne, M. A. McLaughlin, M. Burgay, N. D'Amico, A. Possenti, F. Camilo, P. C. C. Freire, B. C. Joshi, R. N. Manchester, J. Reynolds, J. Sarkissian, I. H. Stairs, and R. D. Ferdman, *Proceedings of The 22nd Texas Symposium on Relativistic Astrophysics, Stanfor 2004*, edited by P. Chen, G. Madejski, E. Bloom and V. Petrosian, eConf C041213, 0038 (2005).
66. M. Kramer, I. H. Stairs, R. N. Manchester, M. A. McLaughlin, A. G. Lyne, R. D. Ferdman, M. Burgay, D. R. Lorimer, A. Possenti, N. D'Amico, J. M. Sarkissian, G. B. Hobbs, J. E. Reynolds, P. C. C. Freire and F. Camilo, Science **314**, 97 (2006).
67. K. S. Thorne, In *300 Years of Gravitation*, edited by S.W. Hawking and W. Israel (Cambridge University Press, Cambridge, 1987), p. 330.

68. B. C. Barish and R. Weiss, Phys. Today **52**, 44 (October) (1999).
69. C. M. Will, Phys. Today **52**, 38 (October) (1999).
70. M. Visser, Gen. Relativ. and Gravit. **30**, 1717 (1998).
71. C. M. Will, Phys. Rev. D **57**, 2061 (1998).
72. C. Cutler, T. A. Apostolatos, L. Bildsten, L. S. Finn, É. E. Flanagan, D. Kennefick, D. M. Marković, A. Ori, E. Poisson, G. J. Sussman, and K. S. Thorne, Phys. Rev. Lett. **70**, 2984 (1993).
73. C. Talmadge, J.-P. Berthias, R. W. Hellings, and E. M. Standish, Phys. Rev. Lett. **61**, 1159 (1988).
74. C. M. Will and N. Yunes, Class. Quantum Gravit. **21**, 4367 (2004).
75. E. Berti, A. Buonanno and C. M. Will, Phys. Rev. D **71**, 084025 (2005).
76. K. Menou, E. Quataert and R. Narayan, in *The Eighth Marcel Grossmann Meeting on Recent Developments in Theoretical and Experimental General Relativity, Gravitation and Relativistic Field Theories*, edited by T. Piran (World Scientific, Singapore, 1999), p. 204.
77. D. Psaltis, in *X-ray Timing 2003: Rossi and Beyond*, edited by P. Kaaret, F. K. Lamb and J. H. Swank, Vol. 714 of AIP Conference Proceedings (American Institute of Physics, Melville, U.S.A., 2004), p. 29.
78. C. S. Reynolds, L. W. Brenneman, J. Wilms and M. E. Kaiser, *Mon. Not. Roy. Astron. Soc.*, **352**, 205 (2004).
79. C. S. Reynolds, L. W. Brenneman and D. Garofalo, *Astrophys. Sp. Sci.*, **300**, 71 (2005).
80. A. E. Broderick and A. Loeb, *Astrophys. J. Lett.* **636**, L109 (2006).
81. E. Berti, V. Cardoso and C. M. Will, *Phys. Rev. D* **73**, 064030 (2006).
82. S. A. Hughes, In *Laser Interferometer Space Antenna: 6th International LISA Symposium*, edited by S. M. Merkowitz and J. C. Livas (AIP Conference Proceedings Volume 873, American Institute of Physics, 2006), p. 233.

Measurements of Space Curvature by Solar Mass

John D. Anderson and Eunice L. Lau

Abstract Unlike Newtonian mechanics, Einstein's General Theory of Relativity predicts that the Sun causes the space around it to curve. As a result, a light ray passing near the solar limb will be deflected by twice the amount predicted by Newtonian theory. As John Archibald Wheeler put it, "space-time geometry tells mass-energy how to move and mass-energy tells space-time geometry how to curve." This chapter reviews the experimental verification of light deflection, from an early eclipse expedition in 1919 to more recent measurements using interplanetary spacecraft and very long baseline interferometry (VLBI). It turns out that the Einstein prediction is correct to within a realistic standard error of about 26 parts per million.

1 Introduction

The measurement of the solar gravitational effect on light propagation has been discussed extensively over the years, for example in [37] and references therein. It has long been recognized that the Parameterized Post-Newtonian (PPN) approximation can be useful for detecting deviations from the predictions of General relativity (GR), but that it is most effective in setting the standard deviation (SD) on solar-system tests. The advantage of this metric is that practically all conceivable metric theories of gravitation are included. There are ten free parameters in a general expansion of the post-Newtonian metric in terms of dimensionless gravitational potentials. For the propagation of light in the weak field and slow motion limit, appropriate for the solar system, the solar potential for the Sun is of primary concern. The Sun's quadrupole term is of order 10^{-7} and its effect on space curvature can be neglected at present. However, it can be included in the equations of motion as an added perturbing acceleration at the Newtonian level. The PPN metric for the study of space curvature by the Sun is a generalization of the Schwarzschild metric at the post-

J.D. Anderson (✉) and E.L. Lau
Jet Propulsion Laboratory, California Institute of Technology, Pasadena, CA 91109
e-mail: jdandy@earthlink.net; eunice.l.lau@jpl.nasa.gov

I. Ciufolini and R.A. Matzner (eds.), *General Relativity and John Archibald Wheeler*, Astrophysics and Space Science Library 367, DOI 10.1007/978-90-481-3735-0_7,

Newtonian level. It is given by [13, 26],

$$ds^2 = \left(1 - 2\frac{m}{r} + 2\beta\frac{m^2}{r^2}\right)c^2dt^2 - \left(1 + 2\gamma\frac{m}{r}\right)\left(dr^2 + r^2d\theta^2 + r^2\sin^2\theta d\phi^2\right), \tag{1}$$

where the coordinates are the time t and the standard spherical coordinates r for radius, θ for the polar angle, and ϕ for the azimuthal angle. These coordinates are arbitrary in GR, but they are in general use for fitting solar-system data in the Barycentric Celestial Reference System, BCRF [35]. The Newtonian potential function in the metric is scaled by the normalized mass of the Sun $m = GM_{\odot}/c^2$, which is equal to 1.476625038 km or about 2.12464×10^{-6} solar radii. The nine figure precision for m in units of km is compatible with its accuracy, which of course is far more precision than needed at present. An equivalent reference system to the BCRF has been in use for many years, dating back to the early experiments with planetary and spacecraft ranging [2, 33]. These were the first tests of GR by means of the Shapiro time delay [30], the time delay introduced into a ranging signal by solar gravitation.

There are subtleties to the application of Eq. (1) with its two free parameters β and γ. For example both the Sun and Earth are in motion about the barycenter, the relativistic center of mass of the solar system [18]. These motion effects are not parameterized, hence testing for them is not straightforward, but they are included in fitting models such as the Orbit Determination Program (ODP) of the Jet Propulsion Laboratory [18, 23]. Another subtlety, especially for ranging experiments, is that all calculations must be done consistently in the same reference coordinate system, whether for calculation of the ephemerides of the Sun and planets [24], the calculation of the geometric ray path of a photon, or the transformation from coordinate time t to the actual proper time associated with the data. The ODP documentation for the fitting model [23] makes a convincing case that this is being done to sufficient precision for solar-system experiments carried out to date, including the Cassini Doppler test, the most accurate determination of γ so far [3, 6]. Without a careful consideration of the dependence on coordinates, it is possible to conclude that both γ and β can be determined by ranging to a spacecraft or a single planet, which is impossible [28, 31, 36]. Only space curvature as represented by γ can be measured. In that sense the ranging experiments are closely related to bending experiments and the Cassini Doppler experiment, as discussed in the following sections.

2 Bending Experiments

According to GR, a light ray propagating in the Sun's gravitational field is effectively refracted, with the vacuum index of refraction n augmented by a refractivity inversely proportional to the distance r from the Sun's center of mass. The effective refractivity follows from a linearization of GR by means of a small gravitational perturbation $h_{\mu\nu}$ to the Lorentz space–time metric of special relativity. For the Sun, considered as a massive point of Schwarzschild radius $R_g = 2m$ at a large distance r ($R_g/r \ll 1$), the isotropic metric of GRT can be expanded in series and parameterized as in Eq. (1). All bending experiments are insensitive to the parameter β at their current level of accuracy. As a result, only the parameter γ is of any consequence in determining whether GR provides an acceptable metric description of gravitation. Essentially, solar conjunction experiments are testing GR at an underlying linear level. The same can be said for the current Earth-orbiting gyroscope experiment, Gravity Probe B (GP-B), which is expected to measure γ to the same order of accuracy as Cassini, or at most a factor of ten worse, and in addition measure the dragging of the Earth's inertial frame by its rotation, the Lense–Thirring effect [14].

According to the metric of Eq. (1), the gravitational field causes ray bending towards the Sun's center, one of the classical GR tests, and in addition, electromagnetic waves are time delayed and frequency shifted [22]. The metric of Eq. (1) is zero for light propagation and it follows that the ratio c/v to order m, with v the phase velocity of the light ray, is the effective refractivity n or,

$$n = 1 + (1 + \gamma)\,\frac{m}{r}. \tag{2}$$

The physical interpretation of γ is that it measures the amount of space curvature produced by solar gravity, and its value is exactly unity in GR. For propagation in a flat Newtonian space, the value of γ is zero and the light deflection is one half the GR value. For a source of electromagnetic radiation at infinity with an angular separation χ from the Sun, a light ray, according to the refractivity of Eq. (2), is geometrically bent through an angle $\Delta\chi$ by an angular amount,

$$\Delta\chi = (1 + \gamma)\,\frac{m}{r}\cot\left(\frac{\chi}{2}\right) \tag{3}$$

The apparent angular separation from the Sun is increased by the amount $\Delta\chi$, where the radius r is the distance of the observer from the Sun. At Earth distance with γ equal to unity, the deflection at the solar limb is 1.75166 s of arc (arcsec), where the limb is defined by a solar radius of 695,508 $\pm$ 26 km [10]. The Earth's orbital eccentricity does not enter into the deflection at five places past the decimal, and r in Eq. (3) can be taken equal to the astronomical unit. We take 1.75166 arcsec as the prediction of GR for light bending at the solar limb. Even at a separation angle (elongation) of 90° from the Sun, Eq. (3) predicts a deflection angle of 4.141 marcsec. This is a deflection large enough to be of importance for astrometric satellites such as Hipparcos (1989–1993) and GAIA (launch 2011–2012). Hipparcos has the potential for a measurement of γ to an accuracy of about 0.004, and it is consistent with GR at that level [38]. A new reduction of the raw data is now available for analysis [20]. GAIA is expected to offer improvements in angular astrometry by a factor of 100 or more over Hipparcos, hence γ might be measured at optical wavelengths to an accuracy approaching 10^{-6} within the next few years.

2.1 Measurements During Solar Eclipse

The earliest tests of ray bending relied on the displacement of stars during a solar eclipse. The relative positions of stars on the sky were measured during the eclipse and then measured again some time later for comparison. The deflection results from the data processing and analysis were extrapolated to the solar limb and compared against the predicted 1.75 arcsec. Weinberg gives a table of results dating from the original eclipse expedition in 1919 to an expedition to the Sudan in 1952 [36]. We add a result from a 1961 eclipse [29] and a 1973 eclipse [17] that represent a reduction and analysis of the photographic plates with the best modern techniques available. All the results can be collected into an array consisting of time, deflection value, and SD, and averaged by linear regression. We ignore two measurements stated as a confidence interval, and discard a measurement of 2.73 arcsec from 1936 as an outlier. We also ignore any differences between the older astronomical use of probable error and SD. The resulting array for data analysis is given in Table 1.

A linear regression without weights yields a value of 1.110 ± 0.076 for γ and an insignificant slope to the deflection measurements of -0.0006 ± 0.0035 arcsec/year. The SD for the data is 0.21 arcsec. When the quoted standard errors are used to weight the data, the result for γ is 1.169 ± 0.060 arcsec and there is a slope in the deflection measurements of -0.0065 ± 0.0032 arcsec/year. The solution without weights is in good agreement with values of γ from other techniques that are far more accurate. We conclude that the accuracy of the eclipse measurements is essentially unchanged over the interval 1919–1976 with a SD of about ± 0.2 arcsec. The PPN parameter γ can be inferred from ten measurements over the interval 1999–1976 with a random error of 8%. Further, a single eclipse expedition can determine γ to an accuracy of about 24%, an upper bound on the realistic error from all the eclipse measurements, including both random and systematic error. The first 1919 expedition did indeed verify Einstein's prediction, at least in the sense that the Newtonian prediction of 0.88 arcsec is no longer viable. The space around the Sun is curved, not flat.

Table 1 Solar eclipse observations of the gravitational deflection of light

Date	Deflection (arcsec)	Standard error (arcsec)
1919.50	1.98	0.16
1919.50	1.61	0.40
1922.81	1.77	0.40
1922.81	1.72	0.15
1922.81	1.82	0.20
1929.44	2.24	0.10
1947.47	2.01	0.27
1952.24	1.70	0.10
1961.21	1.98	0.46
1973.58	1.66	0.19

2.2 Bending Measurements at Radio Frequencies

There is no reason to assume that bending of electromagnetic waves by solar gravitation should be any different at radio wavelengths than at optical wavelengths. In fact it is surprising that radio bending experiments can even compete with optical, given the problems with propagation through ionized plasma (solar corona) at cm wavelengths, and the large apertures required for astrometry at radio frequencies. However, by using the fact that the bending by plasma is dispersive or frequency dependent, while gravitational bending is not, radio interferometry techniques can be used to measure the differential bending of compact radio sources near the Sun. The first experiment with three radio sources and an interferometry baseline of 35 km between antennas was performed in March and April 1974 [15] with the result that the deflection angle extrapolated to the solar limb is 1.775 ± 0.019 arcsec. A repeat of the experiment a year later produced a deflection measurement of 1.765 ± 0.016 arcsec [16].

Under the assumption that the bending of light can be extended to radio frequencies, these two measurements can be added to the 10 eclipse measurements and an improved value of γ can be inferred from the combined data set. We weight the eclipse measurements with a uniform SD of 0.19 arcsec, about 90% of their SD when fitted alone, and weight the two radio measurements at their quoted errors. This produces a normalized chi squared of unity for the linear regression. At an epoch of 1973.42, where the determination of γ and the linear term are uncorrelated, the value of γ is 1.022 ± 0.014 and the linear drift in the deflection measurements is -0.0018 ± 0.0014. There is a trend for the deflection measurements to decrease with time over the interval 1919–1975, although it is barely significant. The value of γ implies a weighted average deflection over this interval at the solar limb of 1.772 ± 0.012 arcsec. This differs by 1.6 SD from the Einstein prediction of 1.75166 arcsec. However, the error from the linear regression represents the random SD for the 12 measurements and it does not include systematic error. The small positive bias in γ from bending measurements, and the implied increase in space curvature near the Sun, is not significant enough to consider it a violation of GR. Alternatively, it indicates that bending measurements tend to be positively biased, with the predicted response of fitting all the data equal to 1.871 ± 0.079 arcsec in 1919 and 1.769 ± 0.012 arcsec in 1975.

Nevertheless, at the end of 1975 this was the best information available on the curvature of space by the Sun. It imposed severe limits on the viability of alternative scalar–tensor theories of gravitation that were taken seriously at the time. In particular the Brans–Dicke theory [7] contains an arbitrary constant ω, which couples the metric tensor to a scalar field, and which can be expressed in terms of γ by [37]

$$\gamma = \frac{1+\omega}{2+\omega} \tag{4}$$

Total coupling implies that ω is unity, while for ω equal to infinity there is no scalar field, the GR case. By Eq. (4) the corresponding interval for γ is from 2/3 to unity. Even with a mean value of γ greater than unity, a normal probability distribution can be defined and the probability of ω being greater than a certain value can be computed. The bending experiments demonstrate that ω is greater than 100 with a confidence level of 95%. Similarly, with γ defined over the interval zero to unity, the interval from flat space at the low end to a space curved by GR at the upper end, there is a 95% probability that γ falls between 0.99 and unity.

Although the Brans–Dicke theory is ruled out by modern measurements of γ, it is still a useful prototype for scalar–tensor theories. For example, it has been suggested that the coupling constant ω might itself be a function of the scalar field, not really a constant, and that as the universe evolves γ asymptotically approaches its GR limit of unity [11, 12]. If so, γ is expected to be less than unity by an amount of order 10^{-7}–10^{-5}. This region is just becoming available for testing by modern space experiments, as we discuss below for the 2002 Cassini experiment at solar conjunction [3, 6].

2.3 *VLBI Measurements at Radio Frequencies*

We make a distinction between the radio bending experiments of the previous section and the more recent VLBI measurements. The measurements in the 1970s relied on a differential bending between three radio sources, 0116 + 08 (a source actually occulted by the Sun), 0119+11, and 0111+02. These sources are located nearly on a straight line in the sky. They are particularly well suited for a differential bending measurement which utilizes the nearly optimum interferometric baseline of 35 km used in the experiments of 1974 and 1975 [16]. The source 0116 + 08 was observed in April 1975 at Sun-ray distances between 7 and 22 solar radii (solar elongation angles from 1.9° to 5.9°).

On the other hand, it was pointed out in the 1960s [32] that VLBI measurements with intercontinental baselines of thousands of km could be used to measure the solar gravitational deflection. This is basically a Hipparcos or GAIA experiment performed at radio rather than optical wavelengths. The idea is to build up an astrometric catalogue of compact radio sources and to include the solar deflection in the reduction of the data. The positions will shift on the sky as the Sun follows its annual path through constellations of the zodiac. The observations are naturally concentrated near elongation angles of 90°, but a few measurements can be made with elongation angles as small as 0.7°, and also a few at 180° where the deflection is zero.

An intensive program of VLBI observations was begun in 1979 and it continued until 1999 for purposes of establishing reference coordinate systems and the orientation of the Earth in inertial space [1, 21]. The data are called Geodetic VLBI Data, and they can be obtained from the NASA Goddard Space Flight Center (GSFC), NASA Crustal Dynamics Data Information System. The data are extensive and they require sophisticated data processing and data analysis techniques developed during the 20-year program. An analysis of 10 years of data [27] at elongation angles between 2.5° and 178° yielded the result $\gamma = 1.0002$ with a random error of ± 0.00096

and an estimated total error, random and systematic, of ±0.002. This corresponds to an error of better than 1 marcsec for each measurement of the deflection.

A more recent thorough and careful analysis of the entire Geodetic VLBI data archive from 1979 to 1999 yields the result $\gamma = 0.99983 \pm 0.00045$ [34]. This result represents about 2,500 experiments, defined by 24-h sessions of VLBI data acquisition. The data set is made up of observations from 87 VLBI sites and 541 radio sources. The data archive contains processed data, with each VLBI pair corrected for ionospheric refraction at each site [9]. The error analysis associated with the determination of γ reveals no systematic sources of error above the quoted uncertainty. The realistic error on this all-sky astrometric determination of γ is ±0.00045 and there is agreement with GR. The scalar–tensor coupling constant ω is greater than 1,000 with a 99% confidence level.

3 Spacecraft Radio Tracking Experiments

Radio tracking signals are routinely sent to interplanetary spacecraft equipped with transponders. When scientific use is made of these coherent signals it is called Space Radio Science [39]. The first tests of GR were performed with the Mariner 6 and 7 spacecraft near solar conjunction in April and May 1970 [2]. The experiment depended on the delay in a ranging signal at only one frequency in the S-Band (~13 cm wavelength). Data noise from solar plasma limited the determination of γ to ±0.02, and noise from the spacecraft's attitude control system further limited the accuracy to ±0.06. Nevertheless, the results from both spacecraft were consistent with GR. With the Viking landers on Mars in 1976–1982, the plasma noise was reduced by dual-band tracking of the spacecraft bus in orbit about Mars at S-Band and X-Band (~3.5 cm wavelength). However, the dual-band measurements from the orbiting bus had to be extrapolated to the single S-Band (13 cm) landers on the surface, with the limitation that the ranging signal to the landers could not be calibrated directly for plasma delay. Nevertheless, the best test of space curvature to date was achieved in 1979 with a determination of γ to an accuracy of ±0.002, again consistent with GR [25]. Results from all spacecraft ranging experiments that measure the Shapiro time delay are summarized by Will in graphical form, and they are compared with the deflection measurements discussed above for eclipses and radio interferometry [38]. The best spacecraft test so far is provided not by ranging experiments, but by multi-band Doppler tracking of the Cassini spacecraft at solar conjunction in 2002 [3, 6].

The two groups that have analyzed the 2002 Cassini data are using the same software, JPL's ODP, and essentially the same data, so there should be no disagreement over results. The two values of γ are 1.000021 ± 0.000023 [6] and 0.999987 ± 0.000052 [3]. They differ by 0.000034, which is good agreement, given the standard errors. The difference in the estimate of error, by a factor of two, can be explained by a difference in editing strategies for the data, and to a lesser extent by the inclusion of estimates of spacecraft trajectory perturbations by gas jetting in Anderson et al. (ALG). Basically, the difference in error estimates represents a difference in the estimate of systematic error. With the Doppler frequency data sampled at an interval of 300 s, Bertotti et al. (BIT) retain 1,094 data points in 18 days of tracking, after deleting outliers. With a less aggressive editing of data, Anderson et al. retain 1,697 data points in 17 days of tracking, and as a result the standard error on the data is about a factor of two larger.

In addition to the favorable multi-link tracking at X-Band and Ka-Band (∼9 mm wavelength), the experiment was carried out on reaction wheels for an interval of 30 days, with no spacecraft thrusting for attitude control. In principle, there should be no need to introduce external forces into the fitting model. However, with the introduction of arbitrary constant forces over an interval of a few days, some apparent systematics in the data can be eliminated and the fit is improved. The mean value for γ is decreased by about one SD and the value shifts from slightly greater than unity with no gas leaks to slightly less than unity when gas leaks are included in the fitting model. Both analysis groups include a single arbitrary constant for an adjustment to the model for solar radiation pressure acting over the 30 days of the experiment.

All spacecraft experiments, whether carried out with ranging data or Doppler data, depend on a model for the excess Shapiro time delay. It can be derived from Eq. (2) for a two-way signal transmitted from Earth (uplink), received by a spacecraft, and transmitted back to Earth (downlink). The result for electromagnetic propagation between point 1 and point 2 in heliocentric coordinates is [23],

$$\Delta t = 2\,(1+\gamma)\,\frac{m}{c}\,\ln\left(\frac{r_1 + r_2 + r_{12} + (1+\gamma)\,m}{r_1 + r_2 - r_{12} + (1+\gamma)\,m}\right) \tag{5}$$

where r_1 is the heliocentric coordinate distance to point 1, r_2 is the distance to point 2, and r_{12} is the distance between point 1 and point 2. The term in m inside the natural logarithm results from an integration of the coordinate speed of light ($v < c$) along the curved path between point 1 and point 2. Equation (5) is the expression coded in the software used for the data analysis, JPL's Orbit Determination Program (ODP). Doppler data is represented in the ODP as differenced range by Eq. (5), the model used for the Cassini Doppler data.

When fitting Doppler data, Eq. (5) is still appropriate, but now the Doppler data represents the change in range over some Doppler integration time T_C. Actually, the excess GR time delay must always be referred to some baseline ranging measurement, hence all Shapiro time-delay tests depend on a range change [8]. In addition, the time-delay equation is coordinate dependent. It is only valid if consistent with the coordinate locations of the receiver and transmitter. On the other hand the Doppler frequency shift is a relativistic invariant, similar to perihelion precession or light deflection. It can be defined without reference to any particular coordinate frame [4].

The coordinate-independent GR fractional frequency change $y \equiv \Delta\nu/\nu$ can be obtained by simply differentiating the coordinate-dependent Eq. (5) with respect to time t. When the ray path is near the solar limb, the Doppler shift is proportional to the ray bending. When thought of as a series of frequency measurements, a Doppler experiment has more in common with light deflection than it does with Shapiro time delay. In fact, it is a cleaner test of GR than the ranging tests, in the sense that the correlation of γ with the spacecraft trajectory is not a major source of error. On the other hand, for a ranging test the limiting systematic error, after calibration for plasma, is the trajectory error. Fortunately for the Viking Lander experiment [25], the orbit of Mars is well known, but for free-flying spacecraft, there is always a difficulty of separating the relativistic ranging delay and the trajectory parameters [2, 19].

3.1 The Cassini 2002 Solar Conjunction Experiment

Ideally, a Doppler experiment provides continuous phase data (cycle count) for several days about conjunction, thereby providing accumulative range-change information to much better accuracy than the ranging data. In practice, the Cassini Doppler experiment relies on Doppler data that are not continuous, and each data point is obtained by integrating the measured phase over an interval $T_C = 300$ s, essentially the difference in counted cycles from the beginning of the count interval to its end T_C seconds later. The Doppler data are effectively a time series of frequency measurements, averaged over each T_C interval and labeled at the center of the counting interval [23]. There are both small gaps of several minutes and large gaps of about 16 h or more in the data record. Only one station of the DSN, DSS-25 at Goldstone California, is equipped with a Ka-Band transmitter, required for full Doppler accuracy. The data consist of a pass of Doppler tracking from DSS-25 for a few hours on the days Cassini was tracked, which was not every day of the experiment. These data are available from the NASA/JPL Planetary Data System PDS and consist of the tracking data file plus files of processed water-vapor radiometry data at DSS-25. The radiometry data can be used to calibrate the Doppler data for refraction in the Earth's troposphere. Both plasma calibration and tropospheric calibration are required in order to achieve the stated accuracy of the Cassini test. These two calibrations were done independently by BIT and ALG using both the dual-band Doppler data and the water vapor radiometry data archived in the PDS. This independent processing and analysis of the raw data provides added assurance that all sources of systematic error have been accounted for.

3.2 Data Processing and Data Analysis for the 2002 Cassini Experiment

The processing and analysis of the data by BIT are available in [6] and in the Supplementary Information to their paper at www.nature.com/nature. The final Doppler residuals from the best fit by BIT are available for 18 tracking passes [6] and are plotted in their Fig. 3, for a total of 1094 Doppler measurements retained in the fit. Similar residuals for the best fit by ALG are given in their paper [3], but in three batches and three figures. The first batch contains seven pre conjunction passes for a total of 677 Doppler measurements with a standard error on the residuals of 3.95

μm/s, the second batch contains seven post conjunction passes for a total of 652 Doppler measurements with a standard error of 3.98 μm/s, and the third batch contains three post conjunction passes closest to conjunction for a total of 246 Doppler measurements with a variable standard error, depending on the reduction in plasma noise achieved by ALG with their method of dual-band X- and Ka-Band plasma calibration [5]. Because BIT and ALG used data from different receivers at DSS 25, there is not a perfect one-to-one correspondence between the passes analyzed by the two groups. However the coverage of the conjunction by the two sets of data is basically the same. ALG took data from the closed-loop receiver used by JPL for space navigation. BIT took data from the open-loop receiver used for Radio Science. Data from both receivers are archived in PDS. A total of 1,575 Doppler measurements in 17 passes is retained in the fit by ALG. The difference between 1,575 points retained by ALG and 1,094 points retained by BIT can be attributed to the more aggressive rejection of outliers by BIT, not to the difference in the receivers used for recording of the raw data. For example, there are only six points retained in the pass closest to the Sun by BIT, while ALG retain 55 points in the same pass. Consequently, the noise in that pass is about 30 times larger in the ALG fit than in the BIT fit. In order to compensate for the larger noise, the pass is downweighted by ALG. They also downweight the second and third pass after conjunction so that the weighted rms residual is equal to unity for all three of the noisy passes nearest conjunction. The other 14 passes farther from conjunction are weighted with a standard error of 3.95 μm/s.

The uncalibrated data used by ALG is plotted in Fig. 1. The data after applying the dual-band plasma calibration is plotted in Fig. 2 on the same scale. The algorithm

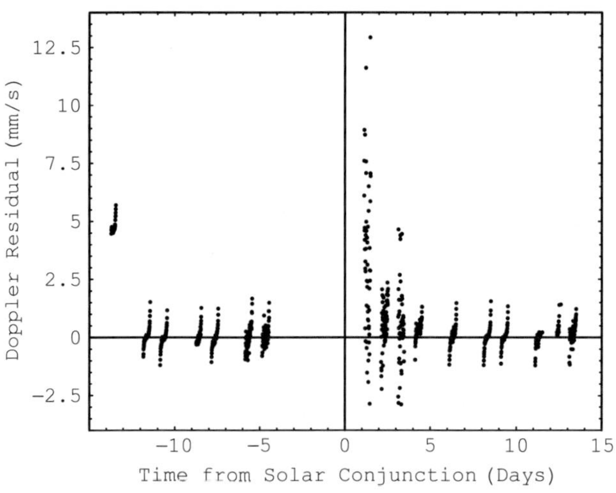

Fig. 1 Uncalibrated Doppler residuals at X-Band sampled at a Doppler integration time of 300 s and converted to units of mm/s by the conversion factor 1/2 c $\Delta\nu/\nu$, with ν the frequency of the radio wave. The residuals are computed for a preliminary spacecraft trajectory obtained before the converged fit to the calibrated data, hence the first pass at approximately 15 days before conjunction is biased by about 5 mm/s

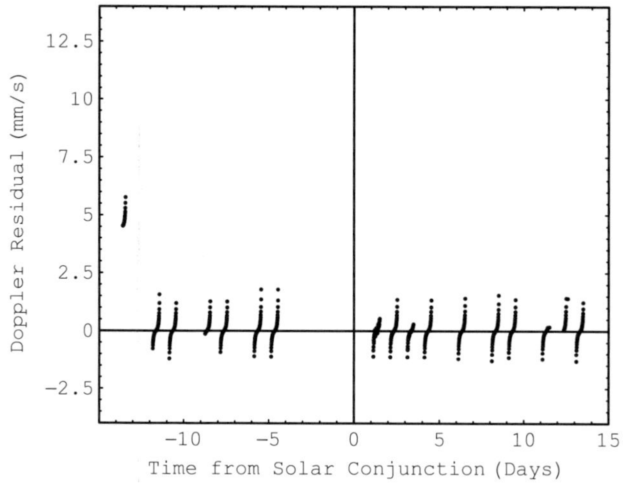

Fig. 2 The same residuals as in Fig. 1 after calibration for solar plasma with simultaneous data at both X-Band and Ka-Band

for the plasma correction in terms of the measured fractional frequency shift y_X at X-Band and y_{Ka} at Ka-Band is [5],

$$y = 1.0575901 y_{Ka} - 0.0575901 y_X. \tag{6}$$

Here y is the plasma-corrected fractional frequency shift, which can be converted to units of Doppler velocity and plotted, as in Fig. 2. The next stage in the data processing is to correct the data of Fig. 2 for refraction in the Earth's troposphere. ALG used the troposphere corrections of the fitting software, the ODP [23]. BIT used the zenith path delays from the water-vapor radiometry data [6]. The results of the ALG approach are shown in Fig. 3. The zenith path delays for each Doppler measurement are computed by the ODP, and the mapping functions to other elevation angles are approximated by continued fractions [23]. The water-vapor radiometry data were used by ALG only to approximate the stochastic variation in the zenith delay over time scales of 30 min. The zenith delays calculated in the ODP are constant over each Doppler pass. The residuals in Fig. 3 do not reflect this final adjustment for the stochastic variation in the zenith delays nor do they reflect a final adjustment of the Cassini orbit that removes some remaining systematic error, including the obvious bias in the first pre conjunction pass of about 5 mm/s. The remaining plasma noise in the first three post conjunction passes is retained in the final fit. There are 14 data points in the first pass after conjunction that are off scale, all near the beginning of the pass when the ray path is closest to the Sun, but these 14 points are in the ALG fit. The largest residual off scale is at 0.224 mm/s. Details of the data processing and analysis can be found in the papers of the two groups [3, 6].

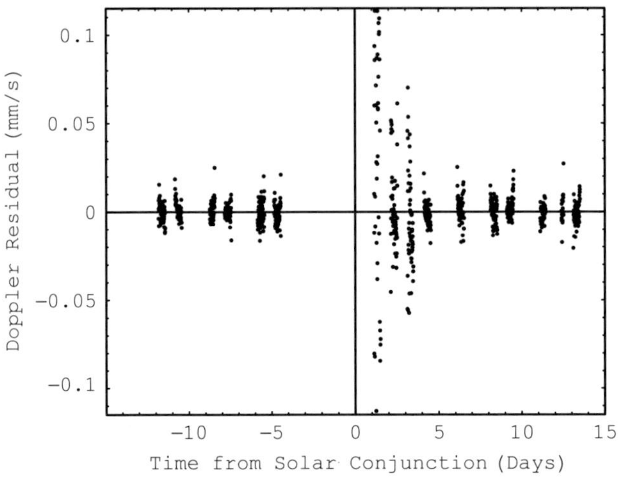

Fig. 3 The same residuals as in Fig. 2 after calibration for refraction in the Earth's troposphere by the method used in the fitting software, JPL's ODP [23]. The scale of the plot is reduced by a factor of 75. There are 14 points off scale near the beginning of the first pass after conjunction. The entire biased pass at 15 days before conjunction is also off scale

The Cassini experiment provides a test of space curvature that agrees with GR with a random error on γ of no more than 2.3×10^{-5}, the BIT estimate of error. The more conservative ALG error of 5.2×10^{-5} represents an upper bound on the realistic error, both random and systematic. Even this more conservative error implies that the scalar–tensor parameter ω is greater than 9,000 at a 95% confidence level.

References

1. Z. Altamimi, P. Sillard, C. Boucher, "ITRF2000: A New Release of the International Terrestrial Reference Frame for Earth Science Applications," Jour. Geophys. Res., **107** B10 (2002).
2. J. D. Anderson, P. B. Esposito, W. Martin, C. L. Thornton, D. O. Muhleman, "Experimental Test of General Relativity Using Time-Delay Data from *Mariner* 6 and *Mariner* 7," Astrophys. Jour. **200**, 221 (1975).
3. J. D. Anderson, E. L. Lau, G. Giampieri, "Measurement of the PPN Parameter γ with Radio Signals from the Cassini Spacecraft at X- and Ka-Bands," in Proceedings of the 22nd Texas Symposium on Relativistic Astrophysics, Stanford University, December 13-17 2004, ed. by P. Chen, E. Bloom, G. Madejski, V. Petrosian, SLAC-R-752, (Stanford Linear Accelerator Center Technical Publications, Menlo Park, CA 2004).
4. B. Bertotti, G. Giampieri, "Relativistic Effects for Doppler Measurements Near Solar Conjunction," Class. Quant. Grav. **9**, 777 (1992).
5. B. Bertotti, G. Giampieri, "Solar Corona Plasma in Doppler Measurements," Solar Phys. **178**, 85–107 (1998).
6. B. Bertotti, L. Iess, P. Tortora, "A Test of General Relativity Using Radio Links with the Cassini Spacecraft," Nature **425**, 374–376 (2003).

7. C. Brans, R. H. Dicke, "Mach's Principle and a Relativistic Theory of Gravitation," Phys. Rev. Lett. **124**, 925–935 (1961).
8. I. Ciufolini, J. A. Wheeler, *Gravitation and Inertia*, (Princeton University Press, Princeton 1995).
9. T. A. Clark et al., "Precision geodesy using the Mark-III very-long-baseline interferometer system," IEEE Trans. Geosci. Remote Sens. **23** 438 (1985).
10. A. N. Cox, *Allen's Astrophysical Quantities Fourth Edition*, Chapter 2, (AIP Press, New York 1999).
11. T. Damour, K. Nordtvedt, "General relativity as a cosmological attractor of tensor-scalar theories," Phys. Rev. Lett. **70**, 2217 (1993a).
12. T. Damour, K. Nordtvedt, "Tensor-scalar cosmological models and their relaxation toward general relativity," Phys. Rev. **D48**, 3436 (1993b).
13. A. S. Eddington, *The Mathematical Theory of Relativity*, (Cambridge University Press, Cambridge, 1922).
14. C. W. F. Everitt, in J. D. Fairbank, B. S. Deaver, Jr., C. W. F. Everitt and P. F. Michelson, *Near Zero: New Frontiers of Physics*, (Freeman, San Francisco, 1988).
15. E. B. Fomalont, R. A. Sramek, "Measurements of the Solar Gravitational Deflection of Radio Waves in Agreement with General Relativity," Phys. Rev. Lett. **36**, 1475–1478 (1976).
16. E. B. Fomalont, R. A. Sramek, "A Confirmation of Einstein's General Theory of Relativity by Measuring the Bending of Microwave Radiation in the Gravitational Field of the Sun," Astrophys. Jour. **199**, 749 (1975).
17. B. F. Jones, "Gravitational Deflection of Light: Solar Eclipse of 30 June 1973 II. Plate Reductions," Astron. Jour. **81**, 455–463 (1976).
18. S. M. Kopeikin, A. G. Polnarev, G. Schaefer, I. Yu. Vlasov, "Gravimagnetic Effect of the Barycentric Motions of the Sun and Determination of the Post-Newtonian Parameter Gamma in the Cassini Experiment," Phys. Lett. A **367** (2007).
19. T. P. Krisher, J. D. Anderson, A. H. Taylor, "Voyager 2 Test of the Radar Time-delay Effect," Astrophys. Jour. **373**, 665 (1999).
20. F. van Leeuwen, *Hipparcos, the New Reduction of the Raw Data*, (Springer, 2007).
21. C. Ma, E. F. Arias, T. M. Eubanks, A. L. Fey, A.-M. Gontier, C. S. Jacobs, O. J. Sovers, B. A. Archinal, P. Charlot, "The International Celestial Reference Frame as Realized by Very Long Baseline Interferometry," Astron. Jour. **116** 516–546 (1998).
22. C. Misner, K. Thorne, J. A. Wheeler, *Gravitation*, (Freeman, 1973).
23. T. D. Moyer, *Formulation for Observed and Computed Values of Deep Space Network Data Types for Navigation*, (Wiley-Interscience, 2003).
24. X X Newhall, E. M. Standish, J. G. Williams, "DE 102 - A numerically integrated ephemeris of the moon and planets spanning forty-four centuries," Astron. Astrophys. **125** 150 (1983).
25. R. D. Reasenberg, et al., "Viking Relativity Experiment: Verification of Signal Retardation by Solar Gravity," Astrophys. Jour. **234**, L219–L221 (1979).
26. H. P. Robertson, In *Space Age Astronomy*, ed. by A. J. Deutsch and W. B. Klemperer, (Academic Press, New York 1962).
27. D. S. Robertson, W. E. Carter, W. H. Dillinger, "New measurement of solar gravitational deflection of radio signals using VLBI," Nature **349** 768–770 (1991).
28. D. K. Ross, L. I. Schiff, "Analysis of the Proposed Planetary Radar Reflection Experiment," Phys. Rev. **141**, 1215 (1966).
29. F. Schmeidler, "Observations of the light deflection during the Solar eclipse on 15th February, 1961," Astron. Nachr. **306**, 71 (1985).
30. I. I. Shapiro, "Fourth Test of General Relativity," Phys. Rev. Lett. **13**, 789–791 (1964).
31. I. I. Shapiro, "Ross-Schiff Analysis of a Proposed Test of General Relativity: A Critique," Phys. Rev. **145** 1005 (1966).
32. I. I. Shapiro, "New Method for the Detection of Light Deflection by Solar Gravity," Science **157** 806 (1967).
33. I. I. Shapiro, M. E. Ash, R. P. Ingalls, W. B. Smith, D. B. Campbell, R. B. Dyce, R. F. Jurgens, G. H. Pettengill, "Fourth Test of General Relativity: New Radar Result," Phys. Rev. Lett. **26** 1132 (1971).

34. S. S. Shapiro, J. L. Davis, D. E. Lebach, J. S. Gregory, "Measurement of the Solar Gravitational Deflection of Radio Waves using Geodetic Very-Long Baselene Interferometry Data, 1979–1999, Phys. Rev. Lett. **92** 121101 (2004).
35. M. Soffel, S. A. Klioner, G. Petit, P. Wolf, S. M. Kopeikin, P. Bretagnon, V. A. Brumberg, N. Capitaine, T. Damour, T. Fukushima, B. Guinot, T.-Y. Huang, L. Lindegren, C. Ma, K. Nordtvedt, J. C. Ries, P. K. Seidelmann, D. Vokrouhlicky, C. M. Will, C. Xu, "The IAU 2000 Resolutions for Astrometry, Celestial Mechanics, and Metrology in the Relativistic Framework: Explanatory Supplement," Astron. Jour. **126**, 2687–2706 (2003).
36. S. Weinberg, *Gravitation and Cosmology: Principles and Applications of the General Theory fo Relativity*, (Wiley, New York 1972).
37. C. M. Will, *Theory and Experiment in Gravitational Physics, Rev. ed.*, (Cambridge University Press, Cambridge 1993).
38. C. M. Will, "was Einstein Right? Testing Relativity at the Centenary," in *100 Years of Relativity*, ed. by Abhay Ashtekar, (World Scientific, Singapore 2005).
39. O. I. Yakolev, *Space Radio Science*, (Taylor and Francis, New York 2002).

Modern Cosmology: Early and Late Universe

Vladimir Nikolaevich Lukash

Abstract Nowadays we are on the verge of discoveries that can change the essence of our world-view. The matter concerns origin of initial conditions and physics of dark matter.

1 "Go There, Don't Know Where. Bring Me That, Don't Know What"

In recent years astronomy has made important advances in getting the observational evidence of dark matter, and now the existence of dark matter can be considered as an established fact. Intriguingly, astronomers *observe* structures made of matter that is unknown to physicists. So the problem of discovering the physical nature of this matter arises.

Contemporary physics does not know particles with dark matter properties. It is necessary to extend the Standard Model. But how and in which direction should we move, what and where should we look for? The words from a Russian fairy tale in the section head describe the current situation very well. Physicists are looking for unknown particles having only general information about the particle properties.

What are these properties?

We only know that dark matter interacts with luminous matter (with baryons) gravitationally, and cosmological density of this cold medium is several times higher than that of the baryons. Thanks to such simple properties dark matter directly affects the development of the Universe gravitational potential. Its density contrast was increasing in time to form gravitationally bound dark matter halos.

It is crucial that this process of gravitational instability could be launched in the Friedmann Universe only if seed density perturbations were present since the beginning. The existence of the latter has nothing to do with dark matter, the perturbations

V.N. Lukash (✉)
Astro Space Center, Lebedev Physical Institute, Russian Academy of Sciences,
Profsoyuznaya ul., 84/32, Moscow, 117997, Russia
e-mail: lukash@asc.rssi.ru

I. Ciufolini and R.A. Matzner (eds.), *General Relativity and John Archibald Wheeler*, Astrophysics and Space Science Library 367, DOI 10.1007/978-90-481-3735-0_8,

result from the Big Bang physics. Thus, another important problem arises, the problem of origin of the seed density perturbations, which later developed in the dark matter structure.

The problem of generating the seed density perturbations will be discussed further. Now let us focus on dark matter.

The baryons were captured into gravitational wells of the dense dark matter regions. That is why light is there where dark matter is, in spite of dark matter particles do not interact with light. Thanks to this remarkable feature of gravitational instability it is possible to study amount, state and distribution of dark matter in observations ranging from radio to x-ray bands.

The conclusions about dark matter properties and other parameters of the Universe are independently confirmed by the data on anisotropy and polarization of cosmic microwave background (CMB), data on light elements abundance and absorbtion line distribution in spectra of distant quasars. Numerical simulations – in cosmological research they replace experiment – play an increasingly important role. A large body of observational data on gravitational lensing of distant sources by closer dark matter structures gives valuable information on the dark matter distribution.

Figure 1 shows a sky region in the direction of one of these dark mass concentrations ($\sim 10^{14} M_{\odot}$). We can see a cluster of galaxies captured by gravitational field of the concentration, hot x-ray gas residing on the bottom of the gravitational well, and a multi-image of one of the background galaxies that happened to be on the line of site of the dark halo and was distorted by its gravitational field.

Table 1 presents average values of the cosmological parameters obtained from astronomical observations (with 10% accuracy). Obviously, the overall energy density of particles in the Universe is not greater than 30% of the critical density (neutrino

Fig. 1 Hubble space telescope photo of a sky region in the direction of cluster of galaxies 0024 + 1654

Table 1 Basic cosmological parameters

Hubble parameter	$h = 0.7$
CMB temperature	$T = 2.725K$
3-space curvature	$\Omega_\kappa = 0$
Cosmological density of baryons	$\Omega_B = 0.05$
Cosmological density of dark matter	$\Omega_{DM} = 0.25$
Cosmological density of dark energy	$\Omega_\Lambda = 0.7$
Power-spectrum index	$n_S = 0.96$

contribution does not exceed a few percent). The rest 70% is in the form that takes no part in gravitational clustering. The only matter with this property is the cosmological constant, or more generally, so-called "dark energy" – medium with negative pressure ($|\epsilon + p| \ll \epsilon$). Discovering the nature of dark energy is a far prospective of the physics development.

This review deals with problems of physical cosmology that are expected to be solved in the nearest future. The most urgent of them are searching for the unknown particles and determining the initial conditions that cause dark matter to form structures.

2 Early Universe and Late Universe

The observed structure of the Universe is a product of start conditions and evolution of the density perturbation field. Up-to-date observational data made possible to determine characteristics of the density perturbation field at different stages of its development. It allowed one to separate information about the initial and development conditions, thus giving rise to independent investigations of the early and late Universe physics.

In modern cosmology the term "Early Universe" stands for the final stage of the accelerated expansion with subsequent transition to the hot phase. Today we do not know the Big Bang parameters, there are only upper bounds (see Eq. 12). However, we have a well-developed theory of gravitational generation of the cosmological perturbations. Using this theory, we can derive the spectra of primordial density perturbations and primordial gravitational waves as functions of cosmological parameters, and constrain the latter if the spectra are known.

The reason why we still have no generally accepted model of the early Universe stems from stable predictions of the Big Bang inflationary paradigm that are obtained in a wide class of model parameters. Namely, the generated spectra are almost flat, the amplitude of the cosmological gravitational waves appears to be relatively small, the 3-space of the visible Universe is Euclidian, etc. Detection of the cosmological gravitational waves would give crucial information about the early Universe. This discovery may come in case the PLANCK experiment succeeds.

Our knowledge of the Late Universe is quite opposite. We have a rather precise model – we know the matter components, structure development laws, cosmological parameters (see Table 1). But there is no generally accepted theory of the origin of the matter components.

The known properties of the visible Universe allow us to describe it in the framework of the perturbation theory. The small parameter (10^{-5}) is the amplitude of the cosmological perturbations.

To zeroth order the Universe is Friedmannian and described with one time function – the scale factor $a(t)$. The first order is a bit more complicated. The metrics perturbations are the sum of three independent modes – the scalar one $S(k)$, the vector one $V(k)$, and the tensor one $T(k)$, each of them being described by its spectrum, the function of the wave number k. The scalar mode describes the cosmological density perturbations, the vector mode is responsible for vortical matter motions and the tensor mode presents the gravitational waves. Therefore, the entire geometry of our Universe is described with four positively defined functions, $a(t)$, $S(k)$, $V(k)$, $T(k)$. Now we only know the first two functions (in some ranges of their definition).

The Big Bang was a catastrophic process of rapid expansion accompanied by intensive fast varying gravitational field. During the cosmological expansion the metrics perturbations were being parametrically born from vacuum fluctuations as it happens to any massless degrees of freedom when they are created under the action of external non-stationary field. The observational data analysis evidences the quantum-gravitational mechanism of the seed perturbation generation. Thus, the large scale structure of the Universe is an example of the solution of measurability problem in quantum field theory.

Let us outline the basic properties of the generated perturbation fields: the Gaussian statistics (random distribution in space), the preferred time phase ("growing" branch of evolution), the absence of characteristic scales in a wide range of wavelengths, a non-zero amplitude of the gravitational waves. The latter is crucial for building-up the model of the early Universe, since the gravitational waves give direct information on the energetic scale of the Big Bang because of the simple coupling to the background metrics.

The development of the scalar perturbations has resulted in formation of galaxies and other astronomical objects. The knowledge of the recent years (the WMAP experiment) is the CMB anisotropy and polarization which emerged long before the birth of galaxies as a result of joint action of all three perturbation modes on the photon distribution.

The combined analysis of the observational data on galaxy distribution and the CMB anisotropy allowed us to relate the S mode and the sum $(V + T)$. Making use of the fact that the sum $S + V + T \simeq 10^{-10}$ is known from the CMB anisotropy, we obtain the upper bound for the vortical and tensor perturbation modes in the Universe (their detection will become possible only with increase of observational precision):

$$\frac{V + T}{S} < 0.1 \tag{1}$$

In case the latter inequality were violated the density perturbation value would not be sufficient to form the observed structure.

3 In the Beginning Was Sound. And the Sound Was of the Big Bang

The effect of the quantum-gravitational generation of massless fields is well-studied. Matter particles can be created with this effect (see [1, 2], etc.) (although the background radiation photons emerged as a result of the proto-matter decay in the early Universe). The gravitational waves [3] and the density perturbations [4] are generated in the same way since they are massless fields and their creation is not suppressed by the threshold energy condition. The problem of the vortical perturbation creation is waiting for its researchers.

The theory of the S and T perturbation modes in the Friedmann Universe reduces to a quantum-mechanical problem of independent oscillators $q_k(\eta)$ in the external parametrical field $\alpha(\eta)$ in Minkovski space with time coordinate $\eta = \int dt/a$. The action and the Lagrangian of the elementary oscillators depend on their spatial frequency $k \in (0, \infty)$:

$$S_k = \int L_k d\eta, \quad L_k = \frac{\alpha^2}{2k^3}\left(q'^2 - \omega^2 q^2\right). \tag{2}$$

A prime denotes derivative with respect to time η, $\omega = \beta k$ is the oscillator frequency, β is the speed of the perturbation propagation in the vacuum-speed-of-light units (henceforth, $c = \hbar = 1$, the sub-index k for q is omitted). In the case of the T mode $q \equiv q_T$ is a transversal and traceless component of the metric tensor,

$$\alpha_T^2 = \frac{a^2}{8\pi G}, \quad \beta = 1 \tag{3}$$

In the case of the S mode $q \equiv q_S$ is a linear superposition of the longitudinal gravitational potential (the scale factor perturbation) and the potential of the 3-velocity of medium times the Hubble parameter [4]:

$$q_S = A + Hv, \quad \alpha_S^2 = \frac{a^2\gamma}{4\pi G\beta^2}, \tag{4}$$

where $A \equiv \delta a/a$, $v \equiv \delta\phi/w$ (see Eqs. 5 and 6), $H = \dot{a}/a$, $\gamma = -\dot{H}/H^2$. A dot denotes derivative with respect to time t.

As it is seen from Eq. (3), the field q_T is fundamental, because it is minimally coupled with background metrics and does not depend on matter properties (in General Relativity the speed of gravitational waves is equal to the speed of light). On the contrary, the relation between q_S and the external field (4) is more complicated: it includes both derivatives of the scale factor and some matter characteristics (e.g., the speed of perturbation propagation in the medium). We do not know anything about the proto-matter in the Early Universe. There are only general suggestions concerning this problem.

Commonly, ideal medium is considered with the energy-momentum tensor depending on the energy density ϵ, the pressure p, and the 4-velocity u^{μ}. For the S mode, the 4-velocity is potential and represented as a gradient of the 4-scalar ϕ:

$$T_{\mu\nu} = (\epsilon + p)u_{\mu}u_{\nu} - p g_{\mu\nu}, \quad u_{\mu} = \phi_{,\mu}/w, \tag{5}$$

where a comma denotes the coordinate derivative, and $w^2 = \phi_{,\nu}\phi_{,\mu}g^{\mu\nu}$ is a normalizing function. The speed of sound is given by "equation of state" and relates comoving perturbations of the pressure and energy density:

$$\delta p_c = \beta^2 \delta\epsilon_c, \tag{6}$$

where $\delta X_c = \delta X - \upsilon \dot{X}$, and $\upsilon \equiv \delta\phi/w$ is the potential of the 3-velocity of medium.

In the linear order of the perturbation theory the ideal medium concept is equivalent to the field concept where the Lagrangian density $L = L(w, \phi)$ is ascribed to the material field ϕ [4–6]. In the field approach the speed of the perturbation propagation is found from equation:

$$\beta^{-2} = \frac{\partial \ln |\partial L/\partial w|}{\partial \ln |w|}, \tag{7}$$

which also corresponds to Eq. (6). To zeroth order, β is a function time. However, in most models of the early Universe one usually assumes $\beta \sim 1$ (e.g., at the radiation-dominated stage $\beta = 1/\sqrt{3}$).

The evolution of the elementary oscillators is given by Klein–Gordon equation:

$$\bar{q}'' + (\omega^2 - U)\bar{q} = 0, \tag{8}$$

where

$$\bar{q} \equiv \alpha q, \quad U \equiv \frac{\alpha''}{\alpha}. \tag{9}$$

The solution of the equation has two asymptotics: an adiabatic one ($\omega^2 > U$) when the oscillator freely oscillates with the excitation amplitude being adiabatically damped ($|q| \sim (\alpha\sqrt{\beta})^{-1}$), and a parametric one ($\omega^2 < U$) when the q field freezes out ($q \to const$). The latter conditions in respect to quantum field theory implies a parametrical generation of a pair of particles from the state with an elementary excitation (see Fig. 2).

Quantitatively, the spectra of the generated perturbations depend on the initial state of the oscillators:

$$T \equiv 2\langle q_T^2 \rangle, \quad S \equiv \langle q_S^2 \rangle, \tag{10}$$

where the field operators are given in the parametrical zone ($q \sim const$). The factor 2 in the tensor mode expression is due to two polarizations of gravitational waves. The state $\langle\rangle$ is considered to be a ground state, i.e., it corresponds to the minimal level of the initial oscillator excitation. This is the basic hypothesis of the Big Bang theory.

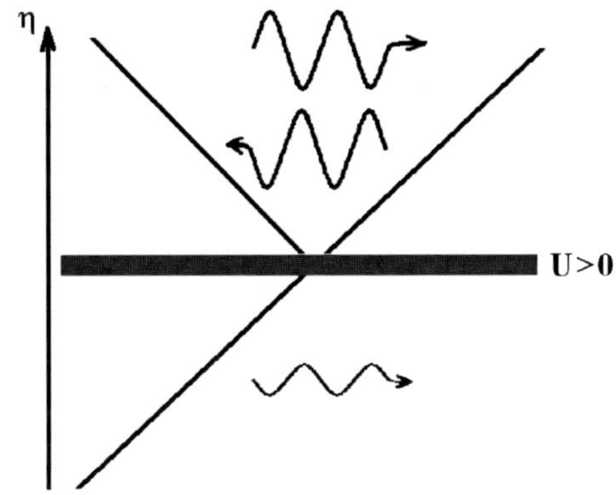

Fig. 2 Illustration of solution of scattering problem for Eq. (8)

In case the adiabatic zone is there, the ground (vacuum) state of the elementary oscillators is unique [7].

Thus, assuming that the function U grows from zero with time (i.e., the initial adiabatic zone is followed by the parametrical one) and $\beta \sim 1$, we obtain a universal and general result for the $T(k)$ and $S(k)$ spectra:

$$T \simeq \frac{|1-\gamma/2|H^2}{M_P^2}, \quad \frac{T}{S} \simeq 4\gamma, \tag{11}$$

where $k = \sqrt{U} \simeq aH$ specifies the moment of creation ($\omega^2 = U$), and $M_P \equiv (8\pi G)^{-1/2}$ is the Planck mass. As it is seen from Eq. (11), the theory does not discriminate the T from S mode. It is the value of the factor γ in the creation period that matters when we relate T and S.

From the observed fact that the T mode is small in our Universe (see Eq. 1) we obtain the upper bound on the energetic scale of the Big Bang and on the parameter γ in the early Universe:

$$H < 10^{13} GeV, \quad \gamma < 0.02. \tag{12}$$

The latter condition implies that the Big Bang took place in the form of inflation ($\gamma < 1$).

We have important information on phases: the fields are generated in certain phase, only the growing evolution branch is parametrically amplified. Let us illustrate it for a scattering problem, with $U = 0$ at the *initial* (adiabatic) and *final* (radiation-dominated, $a \propto \eta$) evolution stages (see Fig. 2).

For either of the two above-mentioned stages the general solution is

$$\bar{q} = C_1 \sin \omega\eta + C_2 \cos \omega\eta, \tag{13}$$

where the operators $C_{1,2}$ yield the amplitudes of the "growing" and "decaying" evolution branches. In the vacuum state the initial time phase is arbitrary:

$\langle|C_1^{(in)}|\rangle = \langle|C_2^{(in)}|\rangle$. However, the solution of the evolution equations yields that only the growing branch of the sound perturbations takes advantage at the radiation-dominated stage[1]: $\langle|C_1^{(fin)}|\rangle >> \langle|C_2^{(fin)}|\rangle$. By the moment of matter-radiation decoupling at the recombination era, the radiation spectrum appears modulated with typical phase scales $k_n = n\pi\sqrt{3}/\eta_{rec}$, where n is a natural number.

It is these acoustic oscillations that are observed in the spectra of the CMB anisotropy (see Fig. 3, the highest peak corresponds to $n = 1$) and the density perturbations, which confirms the quantum-gravitational origin of the S mode. The sound modulation in the density perturbation spectrum is suppressed by the small factor of the baryon fraction in the entire budget of matter density. This allows one to determine this fraction independently on other cosmological tests.

The oscillation scale itself is an example of the standard ruler that is used to determine cosmological parameters of the Universe. The problem of degeneracy of cosmological parameters reconstructed from observational data hindered scientists from building-up the model of the Universe for many years. But now the acuteness of this problem is looser thanks to many independent and complementary observational tests.

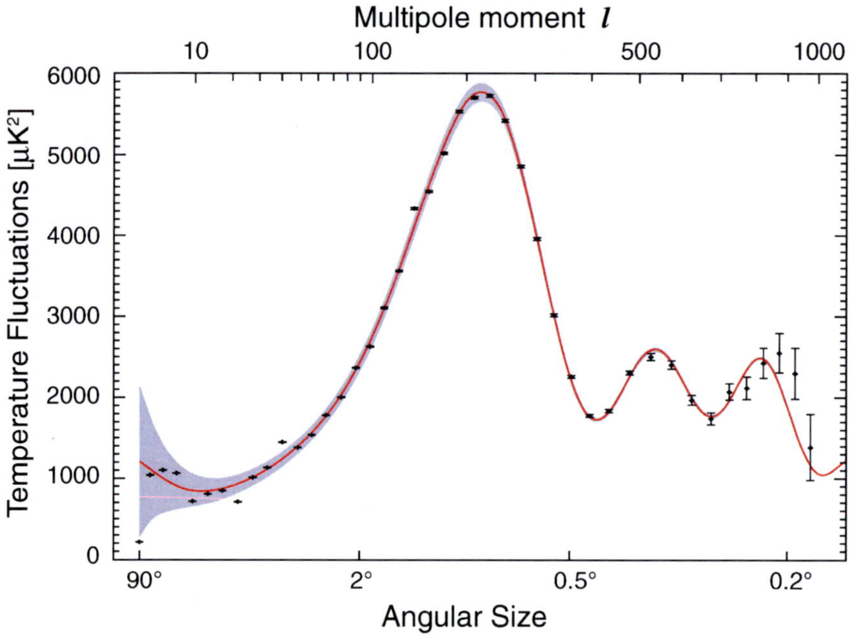

Fig. 3 Manifestation of sound modulation in the CMB anisotropy spectrum (http://map.gsfc.nasa.gov)

[1] This important result can be explained by the fact that only growing solution is consistent with the isotropic Friedmannian expansion at small times ($\omega\eta \ll 1$).

To summarize we can say that in principle the problem of the generation of the primordial cosmological perturbations and the large scale structure of the Universe is solved today. The theory of the quantum-gravitational creation of perturbations in the early Universe will be finally confirmed as soon as the T mode is discovered, which is anticipated in the nearest future. For example, the simplest Big Bang model (the power-law inflation on massive field) predicts the T mode amplitude to be only as much as five times smaller than that of the S mode [8]. Modern devices and technologies are quite able to solve the problem of registering such small signals analyzing observational data on the CMB anisotropy and polarization.

4 Dark Side of Matter

Currently, there are several hypotheses on the origin of the matter, but none of them has been confirmed so far. There exist observational data directly indicating that the dark matter mystery is closely connected with baryon asymmetry in the Universe. However, now there are no generally accepted theories of the dark matter and baryon asymmetry origin.

But where is dark matter?

We know that luminous constituent of matter is observed as stars residing in galaxies of different masses and in the form x-ray gas in clusters of galaxies. However, greater amount of ordinary matter (up to 90%) is contained in rarefied intergalactic gas with temperature of several eV and also in MACHO-objects which are the compact remnants of star evolution and objects of small masses. Since these structures mostly have low luminosity they are traditionally called "dark baryons".

Several scientific groups (MACHO, EROS and others) carried out the investigation of the number and distribution of compact dark objects in the halo of our Galaxy, which was based on micro-lensing events. The combined analysis resulted in an important bound – no more than 20% of the entire halo mass is contained in the MACHO-objects of masses ranging from the Moon to star masses (see Fig. 4). The rest of the halo dark matter consists of unknown particles.

Where else is non-baryonic dark matter hidden?

The development of high technologies in observational astronomy of the twentieth century allowed us to get a clear-cut answer to this question – non-baryonic dark matter is contained in gravitationally bound systems (halos). Dark matter particles are non-relativistic and weakly interacting. Unlike baryons, they do not dissipate whereas baryons are radiationally cooled, settle and gather in the halo centers attaining rotational equilibrium. Dark matter stays distributed around the visible matter of galaxies with characteristic scale of around 200 kpc. For example, in the Local Group, which comprises Andromeda and Milky Way, more than a half of all dark matter belongs to these two large galaxies.

Particles with required properties are absent in the Standard Model of particle physics. An important parameter that cannot be determined from observations due to the Equivalence Principle is the mass of particle. There are several candidates for

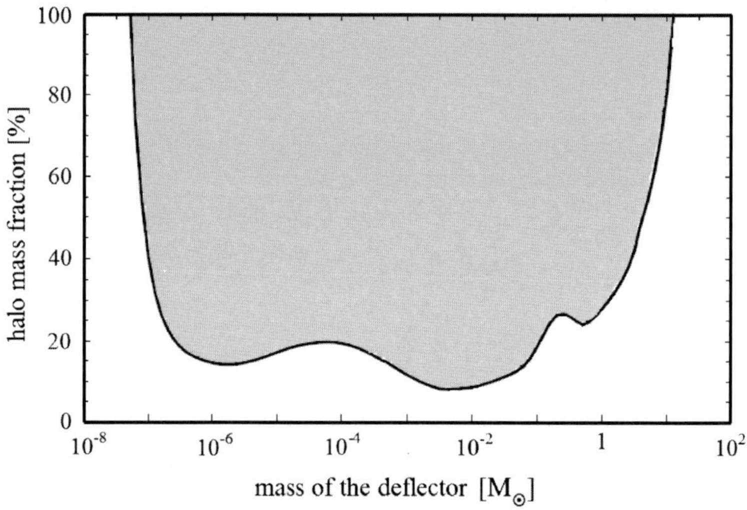

Fig. 4 Upper bound on the fraction of MACHO objects in halo mass of our galaxy vs. lense mass (according to the EROS experiment, see [9])

Table 2 Candidates for non-baryonic dark matter particles

Candidate	Mass
Gravitons	10^{-21} eV
Axions	10^{-5} eV
"Sterile" neutrino	10 keV
Mirror matter	1 GeV
Massive particles	100 GeV
Super-massive particles	10^{13} GeV
Monopoles and defects	10^{19} GeV
Primordial black holes	$10^{-16} - 10^{-7}\ M_\odot$

the dark matter particles in the framework of possible Standard Model extensions. The main candidates are listed in Table 2 in ascending order of their rest masses.

The main version of massive particles nowadays – the neutralino hypothesis – rises from minimal supersymmetry. This hypothesis can be verified in CERN Large Hadron Collider that runs now. The expected mass of these particles is $\sim$100 GeV, and their density in our Galaxy is one particle per cup of coffee.

Dark matter particles are being searched in many experiments all over the world. Interestingly, the neutralino hypothesis can be independently verified both in underground experiments on elastic scattering and by indirect data on neutralino annihilation in Galaxy. So far the positive signal has been found only in one of the underground detectors (DAMA), where a season signal of unknown origin has been observed for several years now. But the range of masses and cross-sections associated with this experiment has not been confirmed in other experiments yet, which makes reliability and meaning of the results quite questionable.

Neutralino give an important possibility of indirect detection by their annihilation gamma-ray flux. During the process of hierarchic clustering these particles could form mini-halos of the Earth masses and characteristic sizes comparable to that of the Solar System. Some of these mini-halos could stay intact till now. With high probability the Earth itself is inside one of these halos where the particle density is as much as tens of times higher than the mean halo density. Hence, the probability of both direct and indirect detection of dark matter in our Galaxy gets higher. Availability of so different search techniques gives a solid hope that the physical nature of dark matter will soon be determined.

5 On the Verge of New Physics

Nowadays it became possible to separately determine properties of the early and late Universe from observational astronomical data. We understand how the primordial cosmological density perturbations that formed the structure of Universe emerged. We know crucial cosmological parameters on which the Standard Model of the Universe is based, and the latter has no viable rivals. However, some fundamental questions of the origin of the Big Bang and of main matter constituents remain unsolved. Observational discovery of the tensor mode of the cosmological perturbations is a key to building-up the model of the early Universe. In this domain of our knowledge we have a clear-cut theory prediction that is already verified in the case of the S mode and can be experimentally verified for the T mode in the nearest future.

Giving a long list of hypothetical possibilities where and how to look for the dark matter particles theoretical physics has exhausted itself. Now it is experiment's turn. The current situation reminds great moments in history of science when quarks, W- and Z-bosons, neutrino oscillations, the CMB anisotropy and polarization were discovered.

One question is beyond the scope of this review – why Nature is so generous and allows us to reveal its secrets?

Acknowledgement This work was partially supported by the Russian Foundation for Basic Research (10-02-00626).

References

1. Grib, A.A., Mamayev, S.G., and Mostepanenko, V.M., Quantum effects in intense external fields (in Russian), Atomizdat (1980)
2. Zel'dovich, Ya.B., and Starobinskiy, A.A., Zh. Eksp. Teor. Fiz. **61**, 161 (1971)
3. Grischuk, L.P., Zh. Eksp. Teor. Fiz. **67** 825 (1974)
4. Lukash, V.N., Zh. Eksp. Teor. Fiz. **79** 1601 (1980)
5. Lukash, V.N., astro-ph/9910009 (1999)
6. Strokov, V.N., Astron. Rep. **51** 431 (2007)
7. Lukash, V.N., Physics-Uspekhi **176** 113 (2006)
8. Lukash, V.N., and Mikheeva, E.V., Int. J. Mod. Phys. A **15** 3783 (2000)
9. Afonso, C., et al., A&A **400** 951 (2003)

Part III
Gravitational Waves

Introduction to Gravitational Waves

Richard Matzner

Abstract We present an introductory description of the production, propagation and detection of gravitational radiation in an astrophysical settings.

1 Introduction

John A. Wheeler was a bridge from the days of quantum theory and Einstein's theory to modern understanding of General Relativity. He provided the impetus to make General Relativity a PHYSICS subject. He was a generous colleague, a wonderful resource of knowledge and insight, and a very good friend. John died 13 April 2008. He is truly missed.

Gravitational radiation is predicted by Einstein's theory of gravity, General Relativity. The Einstein equation is[1]:

$$G_{\mu\nu} \equiv R_{\mu\nu} - \frac{1}{2} g_{\mu\nu} R = \frac{8\pi G}{c^4} T_{\mu\nu}. \tag{1}$$

This very compact form defines the Einstein tensor $G_{\mu\nu}$ in terms of the Ricci curvature $R_{\mu\nu}$ and equates it to constant times the stress–energy tensor $T_{\mu\nu}$. The fundamental quantity is the four-dimensional spacetime metric $g_{\mu\nu}$; in the absence of a gravitational field the spacetime metric is the flat *Minkowski* metric: $g_{\mu\nu} = \eta_{\mu\nu} = diag(-1, 1, 1, 1)$. The Ricci curvature is a nonlinear construction from the metric, quadratic first derivatives of the metric, and second derivatives of the metric. The explicit formulae are given below.

The stress tensor describes the source; in astrophysics the source is large scale mass motion: explosion or very close binary motion of stellar mass objects. The

[1] We use Greek letters ($\mu, \nu, \rho, \cdots = 0, 1, 2, 3 = t, x, y, z$) for spacetime indices, and Latin letters for spatial indices. Henceforth we will use geometrical units, $G = c = 1$.

R. Matzner (✉)
Center for Relativity, University of Texas, Austin, TX 78712
e-mail: matzner2@physics.utexas.edu

I. Ciufolini and R.A. Matzner (eds.), *General Relativity and John Archibald Wheeler*, Astrophysics and Space Science Library 367, DOI 10.1007/978-90-481-3735-0_9,

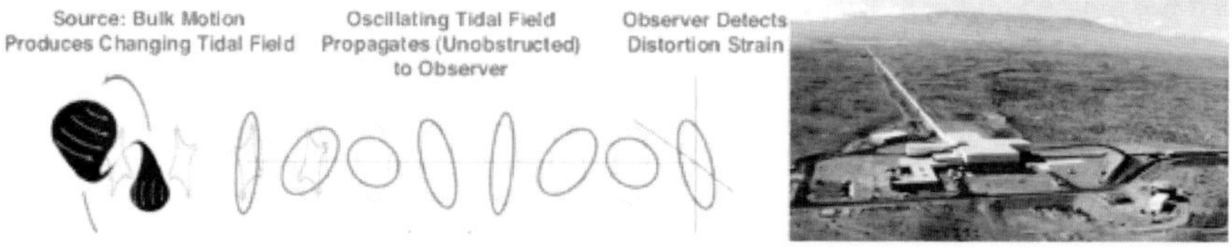

Fig. 1 Motion of large masses produces nonlinear gravitational waves that become weak and effectively linear as they travel to detection in LIGO (LIGO laboratory figures)

strongest field sources are these interacting very compact objects. Neutron star (NS) in orbit around a NS, NS in orbit around a black hole (BH), BH/BH binaries are all possible sources that could in principle be detected in current detectors (LIGO, TAMA, GEO, Virgo). White Dwarf (WD) binaries will also be sources in LISA (a proposed spaceborne detector), and supernovae explosions in our galaxy will be detectable sources for advanced LIGO and for LISA.

Between the source (binary BHs in a distant galaxy, say) and the detector, the Einstein Equation obviously is:

$$G_{\mu\nu} = 0. \tag{2}$$

For interacting black holes, there is no matter involved (black holes are vacuum distortions in the gravitational field), so this describes even the source. At and near the source this leads to a nonlinear, dynamic, in general $3 + 1$-dimensional problem. (This latter means three-dimensional systems evolving in time.) Black holes represent the most concentrated gravitational sources, so interacting black holes would be likely the strongest sources of gravitational radiation. Away from the source the wave amplitude falls off as $\sim 1/r$, where r is the distance from the source. Eventually the wave becomes a weak perturbation on a spacetime with zero gravitational field, the Minkowski spacetime (Fig. 1).

2 Details of Einstein Equations

The details of the geometrical terms in the Einstein equations are given (in a coordinate frame) by the following equations:

$$\Gamma^{\rho}{}_{\mu\nu} = \frac{1}{2} g^{\rho\lambda}(\partial_\mu g_{\lambda\nu} + \partial_\nu g_{\lambda\mu} - \partial_\lambda g_{\mu\nu}); \tag{3}$$

$$\begin{aligned} R_{\alpha\kappa\lambda\mu} = &\frac{1}{2}(g_{\alpha\mu,\lambda\kappa} - g_{\lambda\kappa,\alpha\mu} - g_{\alpha\lambda,\kappa\mu} - g_{\kappa\mu,\alpha\lambda}) \\ &+ g_{\nu\rho}(\Gamma^{\nu}{}_{\kappa\lambda}\Gamma^{\rho}{}_{\alpha\mu} - \Gamma^{\nu}{}_{\kappa\mu}\Gamma^{\rho}{}_{\alpha\lambda}); \end{aligned} \tag{4}$$

and

$$R_{\lambda\nu} = R^{\mu}{}_{\lambda\mu\nu}. \tag{5}$$

The full Einstein equations are difficult to solve analytically, except in cases with special symmetry (e.g., spherical symmetry, axial symmetry, and/or time stationarity), because of their complicated nonlinearity. Astrophysical sources do not have any such symmetry, but are • fully nonlinear, • fully dynamic, and • fully 3 + 1-dimensional. Until recently BH-BH simulations were difficult, with medium-term instabilities. However in 2005, several groups made dramatic improvements in their computational simulation. Pretorius [16] demonstrated binary BH merger using a generalized harmonic formulation, in an explicit second-order in time integration, with excision just inside the apparent horizon of the black holes. Excision removes the region of the spacetime in which the gravitational field. However excision has caused substantial difficulty in implementation. Pretorius [16]'s feat in this effort was to produce stable evolution with excised evolution. A few months later, almost simultaneous with one another, a group at the University of Texas at Brownsville [4], and an independent group at NASA/Goddard [1] produced very successful black hole merger simulations. These latter groups used very similar *puncture* approaches, in which the singular region (the puncture) is not excised, but is kept in the numerical simulation, with only the proviso that in the initial data the puncture is not on a grid point (all these codes are finite difference codes). We will discuss computational relativity below.

3 Linearized Einstein Equations: Weak Fields – Far from the Source

Physics tells us that a region with no gravitational field has a flat space (Minkowski) metric $\eta_{\mu\nu} = \text{diag}\ (-1, 1, 1, 1)$. This assumes that we are using natural rectangular (or, *Cartesian*) coordinates. Far from any source of gravitational radiation, we expect the propagating waves to be weak. Because the wave should be weak we confidently anticipate that we can maintain a similar rectangular coordinate frame, with the wave as a small perturbation on flat space:

$$g_{\mu\nu} = \eta_{\mu\nu} + h_{\mu\nu} + \cdots , \tag{6}$$

where h is the small perturbation which "contains" the waves propagating on a fixed Minkowskian background. The detector reacts to the weak curvature associated with $h_{\mu\nu}$ We shall see in what sense these statements is true, shortly. In what follows we shall assume that $h_{\mu\nu}$ is so small that we can work only to linear order in $h_{\mu\nu}$. In that case the inverse metric is

$$g^{\mu\nu} = \eta^{\mu\nu} - \eta^{\mu\beta}\eta^{\gamma\nu}h_{\beta\gamma} + \cdots . \tag{7}$$

Of interest is the fact that $h_{\mu\nu}$ is a Lorentz tensor. For, consider any Lorentz transformation Λ^{α}_{β} (a constant matrix). Then since $g_{\mu\nu}\Lambda^{\mu}_{\alpha}\Lambda^{\nu}_{\beta} = \eta_{\alpha\beta}$,

$$\bar{g}_{\alpha\beta} = g_{\mu\nu}\Lambda^{\mu}_{\alpha}\Lambda^{\nu}_{\beta} = \eta_{\alpha\beta} + h_{\mu\nu}\Lambda^{\mu}_{\beta}\Lambda^{\nu}_{\beta} + \cdots , \tag{8}$$

which has the same form as $g_{\mu\nu}$, but with Lorentz-transformed $h_{\mu\nu}$.

In linear order, the Riemann tensor is (of course) linear in $h_{\mu\nu}$. Manipulation to construct the Riemann tensor involves taking derivatives, and raising and lowering using $\eta_{\mu\nu}$. (Because the terms are already first order, the additional $h_{\mu\nu}$ terms in the $g_{\alpha\beta}$ and $g^{\alpha\beta}$ are higher order, thus ignored.) Thus $R_{\alpha\beta\gamma\delta}$, $R_{\alpha\beta}$, and $G_{\alpha\beta}$ are Lorentz tensors.

To begin working out the Einstein equations, we note that in the definition of the Riemann tensor, Eq. (4), the terms involving products of Christoffel symbols $\Gamma^{\nu}{}_{\kappa\mu}$ are second order in $h_{\mu\nu}$, and thus ignored. Then the Riemann tensor is explicitly linear, consisting of second derivatives of $h_{\mu\nu}$.

4 Linear Theory Coordinate Transformations More General Than Lorentz; Small Coordinate Transformations

Generally, a coordinate transformation can leave a form of the metric which is strongly different from the usual flat space form; simply imagine transformation to spherical coordinates. Hence to maintain our assumption that the metric is Minkowski plus small perturbation, except for Lorentz transformations, we must have *small* transformations, at most approximately the same size as the metric perturbation $h_{\mu\nu}$.

In this spirit we consider a coordinate transformation of the form

$$x^{\mu'} = x^{\mu}(x^{\alpha}) + \xi^{\mu}(x^{\alpha}), \tag{9}$$

where the primed coordinates are the "new" coordinates, and the notation means that μ' and μ have the same numerical value and ξ is small, $\xi = O(h)$. The standard expression for change of components under a coordinate transformation involves

$$\frac{\partial x^{\mu'}}{\partial x^{\beta}} = \delta^{\mu'}_{\beta} + \frac{\partial \xi^{\mu'}}{\partial x^{\beta}}. \tag{10}$$

For the metric, under this transformation (where ξ is of the same order as $h_{\mu\nu}$) we find

$$g_{\mu\nu} = g_{\alpha'\beta'}\frac{\partial x^{\alpha'}}{\partial x^{\mu}}\frac{\partial x^{\beta'}}{\partial x^{\nu}} = \eta_{\mu\nu} + h'_{\mu'\nu'} + 2\xi_{(\mu,\nu)} + O((\xi)^2). \tag{11}$$

(The parentheses in the subscript represent symmetrization.) This coordinate transform is equivalent to a *gauge* transformation on the metric perturbation:

$$h_{\mu\nu} = h'_{\mu'\nu'} + 2\xi_{(\mu,\nu)}. \tag{12}$$

The name *gauge transformation* is applied because this resembles the change of the electromagnetic potential under an electromagnetic change of gauge; we discuss it more fully below. We have seen that a gauge transformation is identically the same as a small coordinate transformation. We also see that such a transformation changes the metric perturbation. How can we extract physics from a quantity that depends on the coordinates used?

5 Gauge Invariance of the Riemann Tensor

We sketch two proofs that the Riemann Tensor is component by component invariant under a small coordinate transformation (gauge transformation):

$$R_{\alpha\kappa\lambda\mu} = R_{\alpha'\kappa'\lambda'\mu'}. \tag{13}$$

(1)

$$\begin{aligned} R_{\alpha\kappa\lambda\mu} &= R_{\alpha'\kappa'\lambda'\mu'} \frac{\partial x^{\alpha'}}{\partial x^{\alpha}} \frac{\partial x^{\kappa'}}{\partial x^{\kappa}} \frac{\partial x^{\lambda'}}{\partial x^{\lambda}} \frac{\partial x^{\mu}}{\partial x^{\mu'}} \\ &= O(h)(1 + O(h))(1 + O(h))(1 + O(h))(1 + O(h)) \\ &= R_{\alpha'\kappa'\lambda'\mu'}. \end{aligned} \tag{14}$$

(2) Compute $R_{\alpha\kappa\lambda\mu}$ from $h_{\mu\nu} + 2\xi_{(\mu,\nu)}$. Compare to $R_{\alpha\kappa\lambda\mu}$ computed from $h_{\mu\nu}$. Because of the structure of antisymmetry of the Riemann tensor, and because partial derivatives commute, the two results are identical, component by component, in our linear regime.

We have found that under gauge (small coordinate) transformation,

$$h_{\mu\nu} \to h_{\mu\nu} + \xi_{\mu,\nu} + \xi_{\nu,\mu}. \tag{15}$$

but

$$R_{\alpha\kappa\lambda\mu} \to R_{\alpha\kappa\lambda\mu}. \tag{16}$$

This is very similar to the situation in electromagnetism, where a gauge transformation is

$$A_{\mu} \to A_{\mu} + \Lambda_{,\mu}. \tag{17}$$

but

$$F_{\alpha\kappa} \to F_{\alpha\kappa}. \tag{18}$$

Here $F_{\alpha\kappa}$, the electromagnetic field tensor, contains antisymmetric combinations of derivatives of the potential A, which (again with the fact that partial derivatives commute) means that the partial derivatives of Λ do not contribute to the field tensor.

There are therefore remarkable similarities between Electromagnetism and weak field (i.e., linearized) gravity. Both are Lorentz invariant, gauge invariant theories on Minkowski space. Because of the Lorentz invariance we can raise and lower indices on $h_{\alpha\beta}$ using the flat space metric. In particular, we have the trace: $h = h^{\alpha}{}_{\alpha} \equiv \eta^{\alpha\beta} h_{\alpha\beta}$.

Both theories contain wave equations (with propagation at the speed of light), and these equations simplify with the choice of appropriate gauge, as we now show for gravity.

6 Linearized Gravity and the Wave Equation

The first order Christoffel symbol is

$$\Gamma^{\rho}_{(1)\mu\nu} = \frac{1}{2}\eta^{\rho\lambda}\left(\partial_{\mu}h_{\lambda\nu} + \partial_{\nu}h_{\lambda\mu} - \partial_{\lambda}h_{\mu\nu}\right), \tag{19}$$

and the Ricci and scalar curvatures are

$$R_{\mu\nu} = \frac{1}{2}\left(-\Box h_{\mu\nu} + \partial_{\lambda}\partial_{\mu}h^{\lambda}{}_{\nu} + \partial_{\lambda}\partial_{\nu}h^{\lambda}{}_{\mu} - \partial_{\mu}\partial_{\nu}h\right) \tag{20}$$

and

$$R = -\Box h + \partial_{\lambda}\partial_{\nu}h_{\lambda\nu}, \tag{21}$$

where $\Box = \eta^{\rho\lambda}\,\partial_{\rho}\partial_{\lambda}$ is the flat spacetime wave operator. The analysis is simplified if we introduce the *trace reversed* perturbation:

$$\bar{h}_{\alpha\beta} = h_{\alpha\beta} - \frac{1}{2}h\eta_{\alpha\beta}. \tag{22}$$

Then the (linearized) Einstein tensor is

$$G_{\alpha\beta} = \frac{1}{2}\left(\bar{h}^{\sigma}{}_{\beta,\alpha\sigma} + \bar{h}^{\sigma}{}_{\alpha,\beta\sigma} - \bar{h}_{\alpha\beta,\sigma}{}^{\sigma} - \eta_{\alpha\beta}\bar{h}^{\lambda\sigma}{}_{\lambda\sigma}\right). \tag{23}$$

Notice that the third term in Eq. (23) is the wave operator acting on the metric. Thus this would be a simple familiar operator, if we can eliminate the remaining terms. In fact each of these other terms has the form of the (gradient of) a divergence, so *if* we can impose a "gauge condition" similar to the Lorentz gauge in electromagnetism, we have accomplished our goal.

With the gauge transformation we have found (under small coordinate transformations) and using the definition of $\bar{h}$, we find for the effect of such a transformation on $\bar{h}$:

$$\bar{h}^{new}_{\alpha\beta} = \bar{h}^{old}_{\alpha\beta} + 2\xi_{(\alpha,\beta)} - \eta_{\alpha\beta}\xi^{\sigma}{}_{\sigma}. \tag{24}$$

Thus

$$\bar{h}^{new\ \alpha}{}_{\beta,\alpha} = \bar{h}^{old\ \alpha}{}_{\beta,\alpha} + \xi_{\beta,\sigma}{}^{\sigma}. \tag{25}$$

If we now demand that the new divergence vanish, we can read this as a wave equation with source for the gauge (coordinate transformation) function ξ, with solution

$$\xi_\beta(x) = \frac{1}{4\pi}\int\left(\frac{\bar{h}^{old\ \alpha}{}_{\beta,\alpha}(x')d^3V'}{|x-x'|}\right)_{ret} + \zeta_\beta, \tag{26}$$

where ζ_β is a solution of the homogeneous (source free) equation $\Box\zeta_\beta = 0$.

Note that since $G_{\alpha\beta}$ is gauge invariant, even though every component of $h_{\mu\nu}$ changes, and every term in every component of $G_{\mu\nu}$ changes, $G_{\mu\nu}$ itself remains component-by-component unchanged. So, away from the source, we can solve $G_{\mu\nu} = 0$ in a coordinate system so that $\bar{h}^\alpha{}_{\beta,\alpha} = 0$, so

$$G_{\mu\nu} = -\frac{1}{2}\Box\bar{h}^{new}_{\mu\nu}. \tag{27}$$

In principal we can then transform back to any other coordinate system, but why bother? In fact the typical approach is to apply more restrictions on $\bar{h}_{\alpha\beta}$ by additional gauge (coordinate) transforms, where the gauge function ζ_β solves $\Box\zeta_\beta = 0$. With u^α a chosen 4-velocity, usually the observer's time direction, additional requirements typically imposed are

- $u^\alpha\bar{h}_{\alpha\beta} = 0,$ (28)

and

- $\bar{h}^\alpha{}_\alpha = 0,$ (29)

Recall also that

- $\bar{h}^\alpha{}_{\beta,\alpha} = 0.$ (30)

The first two of these can be achieved for *waves* with solutions of $\Box\zeta_\beta = 0$. Thus, if we are in the rest frame of the observer and impose the first two conditions, we have $\bar{h}_{0\beta} = 0$ (the first condition), which means the wave has only spacelike components; and (the second condition), the wave is traceless. Furthermore, the wave satisfies $\Box\bar{h}_{ab} = 0$. Notice the indices are now purely spacelike. Since this is the flat space wave equation, which is linear, with constant coefficients, we consider a Fourier decomposition:

$$\bar{h}_{ab}(x,t) = \int\bar{h}_{ab}(k)e^{ikt+k_ix^i}d^3k. \tag{31}$$

Consider a typical Fourier component, and rotate the coordinates so that it is propagating in the $+x$ direction: $\bar{h}_{ab}(k(x-t)) = \bar{h}_{ab}(k)e^{ik(t-x)}$. Then imposing the vanishing divergence, with the other two conditions gives a perturbation $\bar{h}$ which has purely transverse components (no time and no longitudinal (x-direction) components). Thus these conditions produce a transverse-traceless (*TT*) wave. Such a wave has only two components, $\bar{h}_{yy}(= -\bar{h}_{zz})$, and $\bar{h}_{yz}(= \bar{h}_{zy})$. Finally, because this wave is traceless, there is no distinction between $\bar{h}_{ab}$ and h_{ab}.

7 Effects of Gravitational Radiation

One of the most confusing aspects in understanding the effects of gravitational radiation is the observation that an observer at rest in the transverse-traceless coordinate system remains at rest even as a gravitational wave passes. The argument is simple: if a particle (an observer) is at rest, $\dot{x}^i = 0$, then at that instant the geodesic equation is

$$\frac{d^2x^i}{d\tau^2} = -\Gamma^i{}_{00}. \tag{32}$$

But with the metric in the *TT* form, $(g_{0\alpha} = 0)$, $\Gamma^i{}_{00} = 0$. Hence the acceleration is zero, and the particle stays at rest (at constant coordinate position).

A second adjacent particle (say positioned transverse to the wave direction, separated by a coordinate difference Δy) also remains at constant coordinate position as the wave goes past. Does the gravitational wave have any physical effect as it passes?

The resolution of the paradox is that even though these particles remain at constant coordinate position, the metric does vary as the wave goes past. The metric (proper) separation distance between our two particles is

$$\Delta s_y = \sqrt{g_{yy}}\Delta y = \left(1 + \frac{1}{2}h_{yy}(x-t)\right) \tag{33}$$

The change of the metric leads to a variation of the distance between the particles. If we consider a second pair separated a coordinate distance Δz in the z-direction, then the metric z-distance shrinks when the y-distance expands, so one gets the standard picture of quadrupole variation of the distance between test particles. For this demonstration, one can assume that the mixed y–z components of the perturbed metric vanish. Thus in this case the perturbation in the y–z plane is $h(x-t) \times diag\{1, -1\}$.

Consider now a coordinate system rotated around the x-axis by an arbitrary angle θ. Under this transformation, the same perturbation of the metric is

$$h(x-t) \times \begin{bmatrix} \cos 2\theta & \sin 2\theta \\ \sin 2\theta & -\cos 2\theta \end{bmatrix}. \tag{34}$$

Thus if θ is 45°, there are only mixed y–z components; the diagonal y–y and z–z terms vanish. From this and the linearity of the system, one can see that there are two independent polarizations, which differ in their transverse pattern of action by 45°.

7.1 *Riemann Tensor Depends Only on Transverse Traceless Components*

We can easily compute the Riemann Tensor (to first order in the perturbations) in the frame (gauge) where the metric is transverse-traceless (TT) and the wave is traveling in the x-direction. Then

$$R_{aAbB} = -\frac{1}{2}h^{TT}_{AB,ab} \tag{35}$$

where (here only) $\begin{bmatrix} A, B = y, z \\ a, b \ = x, t \end{bmatrix}$.

Since the Riemann tensor is component-by-component invariant under gauge changes, the oscillating *TT* components of the metric must also be gauge invariant. Consider an algebraic projection transverse to the propagation direction, followed by subtraction of the trace. In the *TT* frame this coincides with what we have called the *TT* components. But since the Riemann tensor components are identically the same under any infinitesimal coordinate transformation, and since these components of the Riemann tensor are precisely the ones that are obtained by such a projection, we conclude that this projection produces the *TT* components from any gauge. So we need not worry about gauge functions, we can simply apply algebraic projection to get the *TT* metric.

Also note that the effect of the Riemann tensor is to produce relative accelerations of separated free particles (tidal accelerations); the oscillating part of the *TT* metric is the second integration of the Riemann tensor, producing the instantaneous separation.

7.2 *Do Gravitational Waves Carry Energy?*

Yes! H. Bondi gave the following thought experiment (Fig. 2): We have seen that a passing wave changes the distance between two free particles, and has a very similar effect to tidal forces. Consider two masses in the form of rings. Instead of free, suppose they are arranged to slide on a stick, and there is very slight friction between the rings and the stick. Because the friction is so slight the separation of the rings changes (there is a relative acceleration). But the stick has internal strength, and its motion (change of length) is slight. Hence the rings move relative

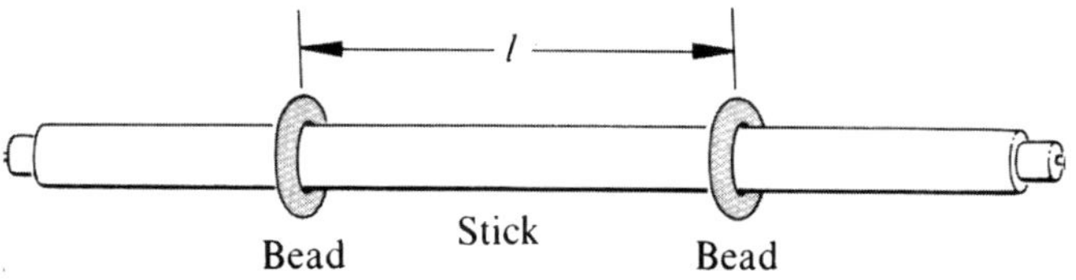

Fig. 2 Bondi's thought experiment [14]

to the stick, and the stick heats up slightly. The wave has heated a material detector (transporting positive energy into the stick). Gravitational waves transport positive energy.

8 Detection of Gravitational Waves

The two basic methods to detect gravitational radiation in terrestrial settings are • by bar detectors (the wave excites one of the longitudinal modes in the detector, and transducers on the bar sample this motion) or • laser interferometry, which measures the instantaneous length of the arms in a Michelson interferometer. LIGO, GEO, TAMA and Virgo are examples of such interferometric detectors.

Figure 3 is a diagram of a Michelson interferometer, similar to a LIGO detector. In LIGO the arms are 4 km in length. The system is set on a dark fringe, so most of the light goes back toward the input laser. Incident gravitational waves change the relative length of the two arms, which results in an imbalance signal into the photodetector.

The first resonant bar detector was created by Joseph Weber in the 1960s, and work has continued with these devices ever since. A resonant bar detector is a massive cylinder of several tons. A passing gravitational wave will excite acoustic waves in the bar. Data extraction is accomplished by measuring the resulting motion of the end of the bar. The response of the detector is narrowly concentrated near its resonant frequencies, in contrast to laser interferometer detectors, which are wideband.

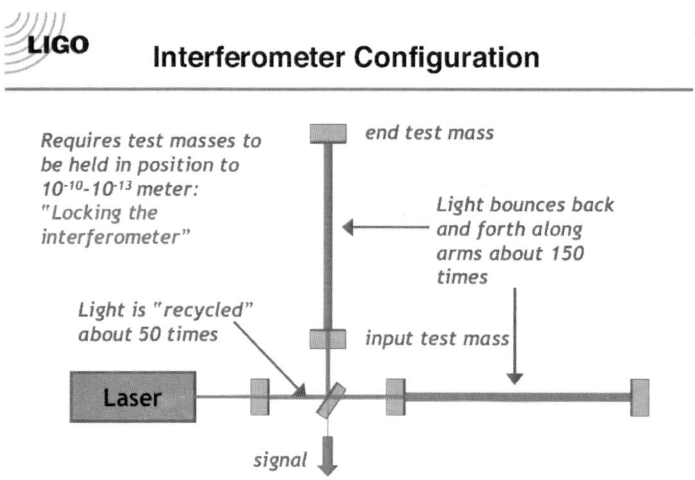

Fig. 3 Michelson interferometer, similar to a LIGO detector. In LIGO the arms are 4 km in length (LIGO laboratory figure)

Fig. 4 The AURIGA bar detector (http://www.auriga.lnl.infn.it/images/auriga.jpg)

Of the currently operating bar antennas, three are Italian efforts:

- AURIGA
  The AURIGA detector (Fig. 4) is run by the Instituto Nazionale di Fisica Nucleare (INFN) in Legnaro, Italy. It is an aluminum bar cooled to 0.1 K. Its center frequency is 900 Hz, with a strain sensitivity of $h \sim 3 \times 10^{-19}$.
- EXPLORER
  The EXPLORER detector is based at CERN and operated by the University of Rome. It is a 2,270 kg aluminum bar cooled to 2°K, with resonance frequencies of 906 and 923 Hz. Its strain sensitivity is 7×10^{-19}.
- NAUTILUS
  The NAUTILUS detector is supported by the Italian National Nuclear Physics Institute (INFN) at the National Scientific Laboratory in Frascati Italy. It is a 2,300 kg aluminum bar operated at cryogenic temperature 0.1 K, with resonant frequencies of 908 and 924 Hz. It has a sensitivity of 6×10^{-19}.

Two other operating bar detectors are:

- ALLEGRO
  ALLEGRO is an experiment operated by the Louisiana State University in Baton Rouge, Louisiana, USA. This experiment employs an aluminum bar cooled to 6°K. It operates at a frequency of 900 Hz, and has a strain sensitivity of 7×10^{-19}.
- NIOBE
  The NIOBE detector is located at Perth, Australia. It is a niobium bar operated by the university of Western Australia at a temperature of 5°K. Its center frequency 700 Hz; its sensitivity is 5×10^{-19}.

9 Strength of Gravitational Waves

This summary is concerned mostly with waves as would be detected in a gravitational wave observatory. Hence we can concentrate on weak (effectively linearized) waves in terms of interaction with the detector. However, the source may involve strong gravitational fields (estimated by $\frac{2m}{R} \sim 1$ in the source).

To correctly describe such strong sources, one needs to resort to computational approaches, since the nonlinearity of the Einstein equations frustrates analytic approaches. Very strong sources include only binary black hole interactions, however. Even a binary neutron star inspiral and merger does not have especially strong gravity, since $\frac{2m}{R} \sim 0.1$ when the neutron stars merge; the potential is less for the evolution prior to the merger. The full nonlinear case requires computational simulation, but an analysis based on the linearized theory provides valuable insight even for the merger of neutron stars.

In this linearized description, the equation to be solved is

$$\bar{h}_{\mu\nu,\alpha}{}^{;\alpha} = -16\pi T_{\mu\nu} \tag{36}$$

where this assumes a weak source, slow motion in the source (which implies that the wavelength is much greater than size of the source).

The solution to this equation (assuming, as usual, outgoing boundary conditions) is:

$$\bar{h}_{\mu\nu} = 4 \int \left(\frac{T_{\mu\nu} dV'}{|x - x'|} \right)_{ret}, \tag{37}$$

so that an approximate solution is

$$\bar{h}_{ij} \approx \frac{4}{r} \int \left(T_{ij} dV' \right)_{ret\ phase} \tag{38}$$

Recall that our previous analysis concentrated on the spatial components of the traveling wave, so the spatial components are of special interest for this analysis. But this appears to require a knowledge of the stress distribution in the source. Such knowledge is not simple to acquire. But since we actually only need spatial integrals of these stresses, a simplification can be done by taking into account the equation of motion (the fact that the our stress–energy tensor has vanishing 4-divergence). In this case, under our assumption of linearity, we should consider the simple partial derivative divergence:

$$T^{\alpha\beta}_{\quad,\beta} = 0 \tag{39}$$

Consider

$$\int x^i x^j \partial_t^2 T_{00} dV, \tag{40}$$

and use Eq. (39). One obtains

$$\int x^i x^j \partial_t^2 T_{00} dV = 2 \int \left(T_{ij} + (W_{ij}^a)_{,a} \right) dV. \tag{41}$$

The term involving W_{ij} is a total derivative and since the matter fields are assumed to vanish at spatial infinity, contributes nothing to an integral over all space. Hence:

$$\int x^i x^j \partial_t^2 T_{00} d V = \frac{d^2}{dt^2} \int x^i x^j T_{00} dV = 2 \int T_{ij} dV. \tag{42}$$

Hence Eq. (38) means

$$\bar{h}_{ij} = \frac{2}{r} \frac{d^2}{dt^2} Q_{ij}|_{ret\ phase}, \tag{43}$$

where the central term in Eq. (42) defines the equadrupole moment Q_{ij}.

9.1 Estimate of Typical Gravitational Radiation

To get an idea for the strength of gravitational radiation in an idealized astrophysical setting, consider a system with two approximately equal masses m_{anis}, moving with speed $\pm v$ in roughly circular orbits.

In this case we have $\frac{d^2 Q}{dt^2} \sim m_{anis} v^2$.

Since the source is a bound system, the Newtonian virial theorem says that $v^2 \sim \phi_{source}$, where ϕ_{source} is the "typical" value of the potential in the source. In the equal mass case, m_{anis} is comparable to the total mass of the system, so the amplitude from Eq. (43) is

$$h_{ij}^{TT} \lesssim \phi_{receiver} \phi_{source}, \tag{44}$$

where $\phi_{receiver}$ is the Newtonian potential of the source as measured at the receiver [18].

As an example, consider a binary neutron star; each of mass $\sim 1.4\ M_{sun} \sim 2$ km. Measurements in LIGO science are usually quoted in Megaparsec; 1 pc $\sim 3 \times 10^{13}$ km (Fig. 5).

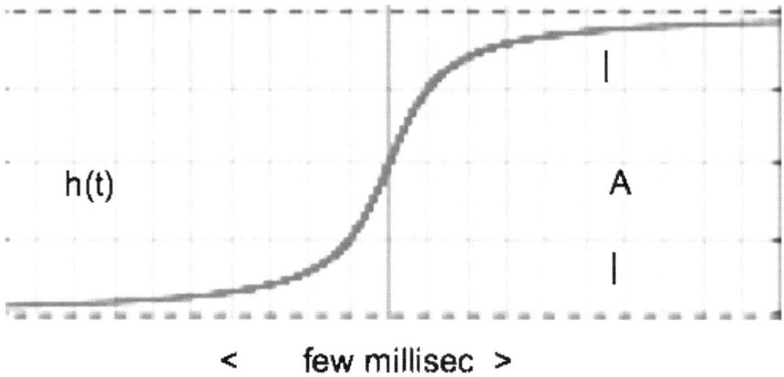

Fig. 5 A rapid change in h in a core bounce supernova (schematic). A is the amplitude. In the text this waveform is approximated as a step function

Thus $\phi_{receiver}|_{1Mpc} \sim 10^{-19}$. This is to be multiplied by the virial potential of the binary system, which never exceeds $\phi_{source=BNS} \sim 0.1$.

It has been determined that to have a sufficient frequency of neutron star mergers to observe several per year, LIGO must survey a sphere about 200 Mpc in radius. At this range, and multiplying the small value of $\phi_{source=BNS}$, we expect amplitude of order 10^{-23} at most for the typical neutron star merger in the final (advanced) LIGO.

We can also describe the effect of a sudden acceleration of mass, for instance in a supernova. Suppose the explosion is very anisotropic, so that half the mass escapes in each direction along the axis, at velocity which exceeds the escape velocity. (Then there will be little deceleration of the motion so we can take the velocity, v, as constant.) In this case the second time derivative of the quadrupole moment after the explosion is $\sim$Mv^2. Before the explosion the second time derivative of the quadrupole moment was effectively zero. There is a very short period of time over which the quadrupole moment changes (at the explosion). Hence if we take h_{ij} to be zero before the explosion, its subsequent evolution is a step function to the value (Eq. 43)

$$h_{ij} \lesssim \frac{M}{r} 10^{-5}, \tag{45}$$

assuming an expansion velocity $\sim$1,000 km/s. A wildly optimistic estimate for the involved mass M is $0.1 M_{sun}$. This gives an estimate like our neutron star estimate above, but with a 10^{-6} coefficient. We thus expect an amplitude $\lesssim 10^{-22}$ at 1 kpc, a typical galactic distance.

The wave amplitude thus can be idealized as an effective *step* function. (The transition in a core bounce supernova is several milliseconds.) Several milliseconds corresponds to a few hundred hertz. This frequency is in the LIGO bandpass, but above the most sensitive frequencies, 100–200 Hz. In fact the LIGO detectors can be characterized as having a center frequency around 100 Hz (slightly higher) and a bandwidth of the same order. The estimated sensitivity of LIGO in this bandpass is a fractional strain somewhat better than 10^{-22}.

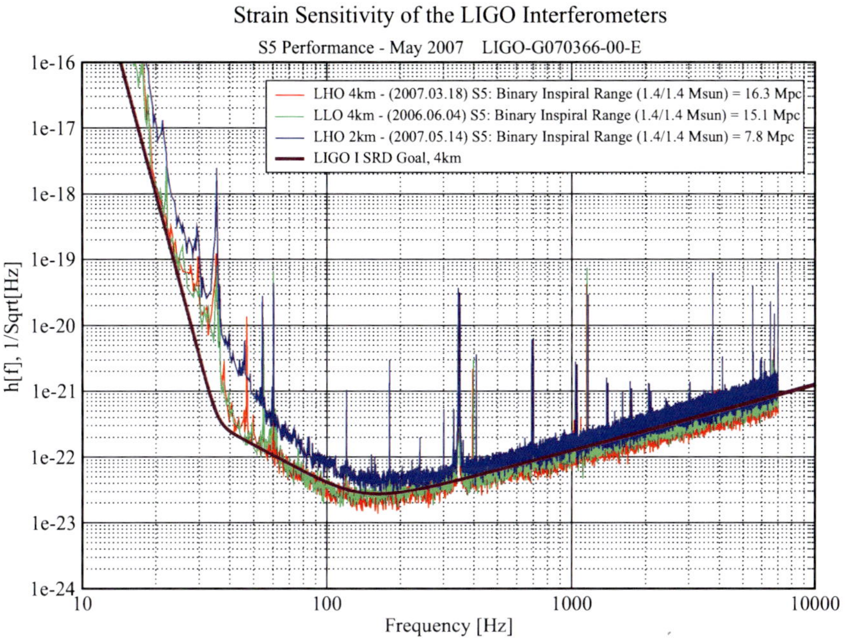

Fig. 6 The LIGO strain sensitivity is maximum around 120 Hz, with a "bucket" of best sensitivity of about 100 Hz width (LIGO laboratory figure)

The center frequency of the proposed LISA detector is of order 10^{-3} Hz, with a bandwidth of the same order. The sensitivity, expressed in terms of h, is proposed to be of the order 10^{-22}, the same order as the LIGO sensitivity (Fig. 6). hence the two instruments have very similar characteristics, but centered at very different frequencies.

To get a rough understanding of such an astrophysical signal, we can suppose that the gravitational wave actually is a step function at $t = 0$.

The response in the detector is given by a convolution of the (Fourier transform of) the signal with the response function of the detector:

$$h_s(t) \sim \int \frac{df \tilde{h}(f)}{S(f)} e^{2\pi i f t}, \tag{46}$$

where $S(f)$ is the noise power spectrum of the detector.

The approximation we consider is a sudden approximation the source jumps in zero time (effectively, the inverse jump time is a frequency greater than can be detected).

For an impulse, $\tilde{h}(f) \sim -A/(i\omega)$, where A is the jump in h. Then

$$h_s(t) \approx \int_{\omega_{min}}^{\omega_{max}} \frac{A}{i\omega} e^{i\omega t} d\omega \sim \frac{A}{i\omega} \int_{\omega_{min}}^{\omega_{max}} e^{i\omega t} \tag{47}$$

define $\omega_0 = (\omega_{min} + \omega_{max})/2$ and $\Delta\omega = \omega_{min} - \omega_{max}$.
Then the response signal is (approximately)

$$h_s(t) \approx \frac{A}{\omega_0 t} sin(\Delta\omega_0 t). \tag{48}$$

Because LIGO and (the proposed) LISA have similar strain sensitivity, and because for each, $\Delta\omega \sim \omega_0$, their responses are just time-stretched versions of each other: a couple of damped cycles centered in the bandpass, with similar signal to noise ratio in the two detectors. The detectability is similar in each detector. Differences in the details (most of which we have glossed over) between the two detectors may lead to important information about short timescale effects in the supernova.

10 Computing Gravitational Wave Signals

It is of great significance to predict the gravitational waveform expected from a given physical configuration of sources. There are several levels of approximation that can describe the behavior of the source and the evolved gravitational radiation:

- The simplest level, used up to this point in this paper, is the assumption of linearized gravity. This implies that gravity is weak. However, within this approximation it allows analytical computation, and is in fact very similar to Electromagnetic theory. If we consider bound (i.e., orbiting) systems, the weak gravity assumption is equivalent to assuming that $v/c \ll 1$.
- Post-Newtonian (PN) expansions are still in principle analytical. They are based on a systematic expansion assuming that the dynamics and the evolved gravitational radiation can be expressed in powers of v/c. As a binary system orbits, initially $v/c \ll 1$ and both the orbit and the radiation can be accurately described by a few (or in the case of systems accurately described by the linear theory, only the first) term in the expansion. As radiation shrinks the orbit the strength of the gravitational fields and the orbital velocities increase (Fig. 7). Then more terms in the series are needed for accurate description. The behavior of the series become problematic for speeds $v/c \sim 0.4$ because there is no clear decrease in size for the higher order terms, at least not for the few terms (up to $(v/c)^7$) that have been explicitly computed. Thus PN methods can provide good prediction of the waveform for inspiraling and merging binary neutron stars, since the stars touch and merge before the fields become too strong for the method to work.

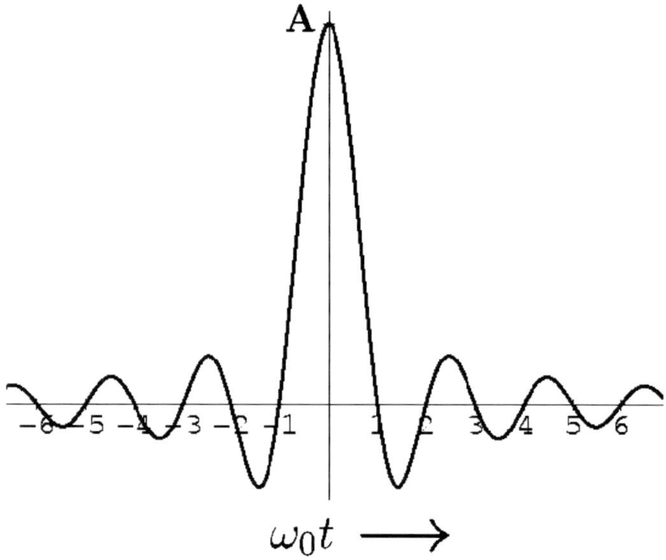

Fig. 7 The time dependent response of the LIGO or LISA detectors to a step function of amplitude A in the strain signal h

But even stronger sources, such as inspiraling black holes, will certainly evolve to conditions that cannot be represented by the PN method. The lectures in this volume by Will describe the PN method in detail; see also [2, 15].

- Such strong sources can only be described by a different approach, a discrete-arithmetic approximation to the full Einstein equations, by a technique called Computational Relativity. In principle this is the most accurate way to compute the dynamics and emitted radiation from a strong field source. Successful computational relativity requires careful analytical and computational formulation. Validation of this method always involves demonstration of convergence of the discrete solution. For convergent solutions, which approach a limiting behavior as the discretization is made finer, the limit for fine discretization is in fact the analytical solution.

Concerted effort on binary black hole computational simulations began in 1993 with an effort called the Binary Black Hole Grand Challenge. This was a five year, eight university effort. By the time the effort ended in 1998, substantial progress had been made, and simulations of several limited binary black hole collisions had been carried out. But progress really took off with the publication of Pretorius [16]. In this paper, Pretorius employed a formulation due to [9], which was novel in computational physics. This formalism leads to a set of coupled second order wave equations with sources. With this approach and employing adaptive mesh refinements with a method of excising the part of spacetime inside the black hole horizons (and thus excising the singularities inside the black hole) Pretorius [16] was able to compute the inspiral and merger of binary black holes and the emitted radiation.

Very shortly after the publication of Pretorius [16], two groups independently and essentially simultaneously (announced at the same conference in November 2005) succeeded in using a *puncture* method to model similar simulations. Baker et al. [1] and Campanelli et al. [4] are references to this fundamental work. Puncture methods work by actually carrying along singular terms (the central singularity of the black holes), and the simplest methods proceed by simply offsetting the singularity in the initial data so that it does not fall on a coordinate point (so that no attempt is made to compute at an actually singular point). Because of this simpler approach, the puncture method has become fairly standard among a number of groups to evolve black hole mergers.

11 Coordinates for Computational Relativity

Numerical relativity is based on the concept of establishing an initial data state, and then evolving it into the future. This introduces the necessity to define the coordinates and the behavior of the coordinate *slicing*.

Four possibilities are:

- Null + Time: (3 + 1): The simple expectation is that a central world line defines the time; the constant-time (usually written $u = const$) slices are the null cones emanating from that world line. However, a more successful method demands smooth (similar to flat space) null "cones" at infinity. A large advantage of this approach is that near infinity, outgoing wave solutions have a phase $e^{i\omega u}$ multiplied by a polynomial in inverse distance. This slow falloff means that the wave all the way to infinity can be accurately modeled in a finite number of integration steps. This *compactification* is a dramatic reduction in the effort needed to achieve a given level of accuracy in the description.
  The difficulty with this approach is that as the null generators are followed inward toward the center there are inevitable *caustics* (rather than a clean point focus) which arise. However this method can work if a different method is used to evolve the central region. Generally this requires some sort of patching together of (at least) two different coordinate systems; one for the inner and one for the outer regions being described.
- Double Null: (2 + 2): The description of flat-space spacetime is based on a combination of outgoing and ingoing null cones. Again, this allows simple accurate description of outgoing waves. Again a problem is that strong gravitational fields can cause the null-ray generators of these null "cones" to cross, forming caustics.
- Space + Time: (3 + 1): This is closest to the Newtonian idea of a universal absolute time, and it handles strong fields well, even within the black hole horizons, if properly set up. The $t = const$ 3-surfaces are spacelike, and can extend to spatial infinity. The infinitesimal normal (timelike) distance between two such spaces is given by $d\tau = \alpha(x^i, t)dt$. The natural choice is that α tends to unity, and that the coordinates become rectangular far from the sources. In this case outgoing waves have the common form $e^{i\omega(r-t)}$, where $r = (\Sigma x^i x^i)^{1/2}$. This form means that

there is variation on the wavelength scale at large distances from the sources, in contrast to the compactification possible with null coordinates. As a consequence this Space + Time method can evolve only a finite portion of spacetime. It appears however to be an ideal "inner filling" for the null systems defined above. However, only limited success at matching such systems has been achieved.

- Hyperboloidal: (3 + 1): This approach attempts an analytic match between the spacelike and the null approaches. The "spaces" are spacelike for all finite radii, but smoothly become null "at" infinity. This may be a very promising approach [8], but has not yet been completely worked out.

12 Space + Time (3 + 1) Coordinates

The most successful approaches to computational relativity and the only approaches to produce waveforms to date are the Space + Time (3 + 1) ones, so we concentrate on them here. Figure 8 gives a sketch of two $t = const$ spaces, $t = t_0$ and $t = t_0 + dt$. The *proper* time separation between these two spaces (normal, i.e., orthogonal to the spaces) is given by $d\tau$. Here "normal" means "as measured by an observer instantaneously at rest in this space", equivalent to "as measured by an observer whose instantaneous local 3-space corresponds to the 3-space $t = const$". Because the coordinate t is arbitrary, there is an arbitrary, specifiable relation between dt and $d\tau$: $d\tau = \alpha(x^i, t)dt$. The function $\alpha(x^i, t)$, which gives this relation is in principle different at different spatial positions x^i in a given $t = const$ space, and this relation can be different at other times. Hence α is a function of four spacetime variables. Figure 8 also points out that the location of a given triplet x^i may not coincide with the normal evolution from one time to the next. The vector $\partial/\partial t$ is the vector that points in a direction where only t varies and the x^i are held constant.

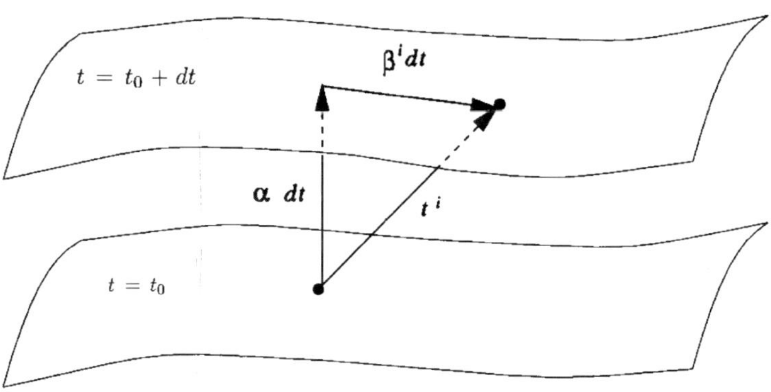

Fig. 8 The *lapse* α and the *shift* β^i can be freely chosen and determine the coordinates into the future

There is thus an offset between the orthogonally evolved point. That offset is the 3-vector $\beta(x^i, t)$, which is a function of the four coordinates for the same reason that α is.

With these definitions for α and β, the full four-dimensional metric can be written:

$$ds^2 = -\alpha^2 dt^2 + g_{ij}(dx^i + \beta^i dt)(dx^j + \beta^j dt). \tag{49}$$

Since α and β^j are completely specifiable (i.e., arbitrary) functions, Eq. (49) shows that the dynamics of spacetime is contained in the 3-metric

$$g_{ij} = g_{ij}\left(x^k, t\right). \tag{50}$$

This is a symmetric 3×3 matrix, hence six independent variables.

Associated with the spatial metric is a momentum K_{ij}: If the derivative is taken orthogonally to the 3-space, then $dg_{ab}/d\tau = -2K_{ab}$. The general definition is:

$$K_{ab} = -\frac{1}{2\alpha}\left(\frac{\partial g_{ab}}{\partial t} + \beta_{a|b} + \beta_{b|a}\right), \tag{51}$$

where the vertical stroke (|) indicates the covariant derivative based on the spatial metric g_{ij}, and the indices on β are raised and lowered using the spatial metric g_{ij}. (We also use the symbol ∇ for the three-dimensional covariant derivative.)

To repeat, there are no equations for α or β; they are completely arbitrary choices. For instance, an apparently simple choice, $\alpha = 1$ and $\beta = 0$ is possible and will work in the short term, but may lead to difficulties in the future evolution.

12.1 Outer Boundary Conditions for Space + Time Computations

In view of the failure to achieve complete matching between 3 + 1 domains and outgoing null representations, various approximations have been employed to provide consistent outer boundary conditions for such computations. The methods that are generally employed in puncture evolutions are variants of Sommerfeld conditions, applied to the difference of the computed solution from a simple flat space solution. The approach of [16] is different; he uses spatial compactification, which gradually absorbs outgoing waves, and eliminates the need for an explicit boundary condition.

13 The Einstein Equations for Space + Time Formulations

We are interested principally in binary black hole interactions, so we need only consider vacuum Einstein equations. These equations naturally divide into two sets. The four equations $G_{0\alpha} = 0$ constitute the constraint equations. They are:

$$H_i \equiv G_{0i} = K^i_{j|i} - K_{,j} = 0. \tag{52}$$

$$2H \equiv 2G_{00} = {}^3R + K^2 - K^{ab}K_{ab}, \tag{53}$$

Equation (52) specifies a kind of transversality for K_{ab}. In Eq. (53), ${}^3R = 0$ is the three-dimensional scalar curvature. Equation (53) states that K_{ij} is algebraically related to the 3-curvature computed from the spatial metric g_{ij}. Notice that these equations contain no time derivatives, so they must hold independently of the value of the time. In particular they must hold at the initial time. They are consistency, or *constraint* equations on the initial data in any physical situations.

The evolution equations are more complicated. They are

$$\frac{\partial g_{ij}}{\partial t} = -2\alpha K_{ij} + \beta^k g_{ij,k} + g_{jk}\beta^k_{\ ,j} + g_{ik}\beta^k_{\ ,j}; \tag{54}$$

$$\frac{\partial K_{ij}}{\partial t} = -\alpha_{|ij} + \alpha({}^3R_{ij} - 2K_{ij}K^k_{\ j} + KK_{ij}) + \beta^k K_{ij,k} + K_{jk}\beta^k_{\ i} + K_{ik}\beta^k_{\ j}. \tag{55}$$

In these equations the vertical stroke indicates the three-dimensional covariant derivative; also K is the trace of K_{ij}, and ${}^3R_{ij}$ is the three-dimensional Ricci tensor. These equations are correct; but they may not be the best form for numerical evolution. To understand these equations, let's have some ...

14 Fun with Exact Solutions

14.1 Schwarzschild in Standard Coordinates

The first exact solution to the Einstein equations is the Schwarzschild [19] solution in standard form, which describes a static spherical black hole:

$$ds^2 = -\left(1 - \frac{2M}{r}\right)dt^2 + \frac{dr^2}{1 - \dfrac{2M}{r}} + r^2(d\theta^2 + \sin^2\theta d\phi^2) \tag{56}$$

Recall: $\alpha^2 = -g_{00} + \beta_i\beta^i$; $\beta_i = g_{0i}$, so here $\alpha^2 = 1 - \frac{2M}{r}$ and $\beta_i = 0$. Because β_i vanishes and the metric has no time dependence, we find from Eq. (54): $\dot{g}_{ij} = 0 = -2\alpha K_{ij}$. Hence for this form of the Schwarzschild metric, we have found $K_{ij} = 0$, and this value persists in time because the whole system is static. With vanishing K_{ij} and β^i, Eq. (55) reads

$$\dot{K}_{ij} = -\alpha_{|ij} + \alpha\ {}^3R_{ij}, \tag{57}$$

which is zero, because of the static nature of the solution. We also find that $H^i = 0$, and $H = {}^3R = 0$.

Note that neither of the terms in Eq. (57) is individually zero. The direct computation of ${}^3R_{ij}$ follows Eqs. (3)–(5), with the indices restricted to the 3-space, Evidently it is rather complicated to evaluate. However, Eq. (57) gives a much easier way to calculate ${}^3R_{ij}$ *in this case*, since taking derivatives of α is simple, and the ${\Gamma^i}_{jk}$ appear only once in the expression.

I emphasize that results hold only for the Schwarzschild spacetime, and only when expressed in these coordinates. (Recall that both β and α are freely specifiable so the *form* of these equations would be different in different coordinate choices.)

14.2 Schwarzschild in Kerr–Schild Coordinates

To emphasize this point, we can repeat the computation for the same Schwarzschild spacetime, expressed in a different set of coordinates, the Eddington–Finkelstein [7], (or, Kerr–Schild [11, 12]) coordinates.

In these coordinate the metric is again static, but there is a nonzero shift vector:

$$ds^2 - -\left(1 - \frac{2M}{r}\right) dt_{KS}^2 + 2\left(\frac{2M}{r}\right) dt_{KS} dr + \left(1 + \frac{2M}{r}\right) dr^2 + r^2(d\theta^2 + \sin^2\theta d\phi^2) \tag{58}$$

This is accomplished by a position-dependent change of the time coordinate (only):

$$t_{KS} = t + 2M\, ln\left(\frac{r}{2M} - 1\right). \tag{59}$$

From the standard form Eq. (49) we see that the shift vector is $\beta_i = g_{0i}$, and that $g_{00} = -\alpha^2 + \beta_i\beta^i$. Thus we find, for this case:

$$\beta_r = \frac{2M}{r}, \tag{60}$$

$$\alpha^2 = \left(1 + \frac{2m}{r}\right)^{-1}. \tag{61}$$

Because the metric is again time independent, but here $\beta \neq 0$, Eq. (54) gives a nonzero K_{ab}.

15 Creating Initial Data

We have noted that the constraint equations $H = 0$ and $H^i = 0$ must be satisfied for all data sets. In particular if we imagine an evolution forward in time, we must begin with a combination of 3-space metric g_{ij} and conjugate momentum K_{ij} that satisfy Eqs. (52) and (53).

There is a standard approach to creating initial data, as developed by [6, 13, 21] and a large number of collaborators. It is called a conformal method.

The process begins with a guess for the initial metric and initial momentum: $\tilde{g}_{ab}(x^i, t_0)$, $\tilde{K}_{ab}(x^i, t_0)$. We then introduce a conformal factor $\phi(x^i)$ to define the physical initial metric g_{ab}:

$$g_{ab}(x^i, t_0) = \phi(x^i)^4 \tilde{g}_{ab}(x^i, t_0). \tag{62}$$

(The same coordinate system is used for the conformal and the physical metric. ϕ is defined only at $t = t_0$, so we write it as only a function of the spatial coordinates.) We also adjust K_{ab} with ϕ, and also introduce a 3-vector $w^i(x^i)$, which is adjusted to soak up the nontransverse parts (in the sense of Eq. 52) of K_{ab}.

$$A^{ij} = \phi^{-10}\left(\tilde{A}^{ij} + (\tilde{lw})^{ij}\right). \tag{63}$$

w^i is a vector potential, and

$$(\tilde{lw})^{ij} \equiv \tilde{\nabla}^i w^j + \tilde{\nabla}^j w^i - \frac{2}{3}\tilde{g}^{ij}\tilde{\nabla}_k w^k. \tag{64}$$

The trace K is taken to be a given function

$$K = \tilde{K}. \tag{65}$$

Writing the Hamiltonian and momentum constraint equations in terms of the quantities in Eqs. (62)–(65), we obtain four coupled elliptic equations for the fields ϕ and w^i [22]:

$$\tilde{\nabla}^2\phi = (1/8)\left(\tilde{R}\phi + \frac{2}{3}\tilde{K}^2\phi^5 - \phi^{-7}(\tilde{A}^{ij} + (\tilde{lw})^{ij})(\tilde{A}_{ij} + (\tilde{lw})_{ij})\right), \tag{66}$$

$$\tilde{\nabla}_j(\tilde{lw})^{ij} = \frac{2}{3}\tilde{g}^{ij}\phi^6\tilde{\nabla}_j K - \tilde{\nabla}_j\tilde{A}^{ij}. \tag{67}$$

Figure 9 shows the result of a solution for data describing spinning orbiting black holes, [20].

16 Computational Evolution of Binary Black Hole Systems

It is of course of interest to see computational evolution of binary black hole systems. We conclude these lectures with a few such examples of current ongoing work. Figure 10 from [17] shows simulations using the hyperbolic method of interactions of two black holes which in one case merge and in the other case orbit but escape. The data are constructed by boosting clumps of scalar field (boson stars) which will subsequently collapse to black holes before approaching one another. The difference in the two simulations is a slight difference in the initial boost.

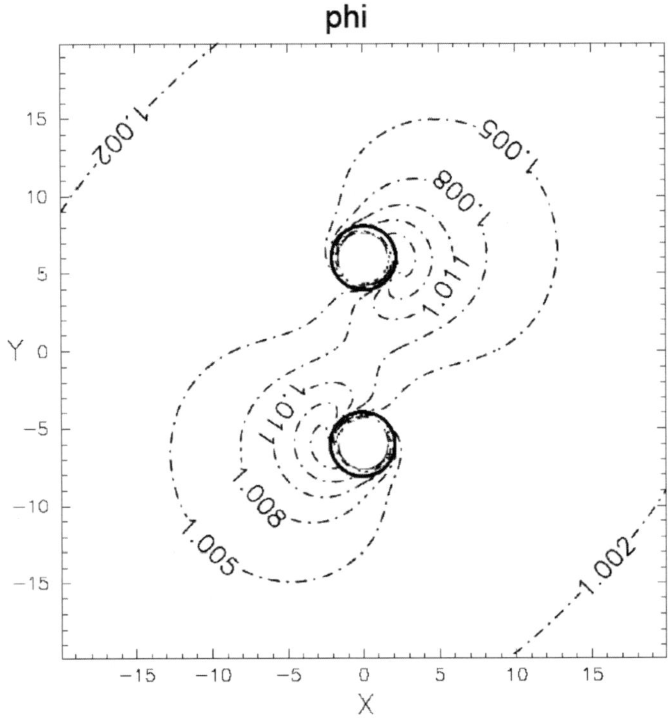

Fig. 9 Initial data: the conformal factor ϕ for two orbiting spinning black holes [20]

Figure 11, also from [17] shows the gravitational waveform (one angular component of the Reimann tensor) for the merging orbit of Fig. 10, together with the quadrupole estimate of radiation from that orbit, and from the orbit in Fig. 10 that did not merge.

The currently most effective method is a *puncture* method. Figure 12 is a result from [1] showing that quasicircular orbits starting from different initial separation all have very similar endpoint orbits as they merge. The orbits are specified by the puncture locations, which can move – and be followed – inside a common horizon, so the orbits continue beyond the asterisk indicating the merger.

Figure 13 show results from [5], giving the out-of-plane motion and precession arising from the spin–orbit interaction of spinning black holes.

The simulation of black holes and the resulting gravitational radiation has astrophysical implications other than possibly-detectable gravitational radiation. For instance, Herrmann et al. [10] discovered substantial recoil velocities from spinning black hole mergers (14). The recoils found, in excess of 400 km/s is sufficient to eject the merged black hole from moderate-sized galaxies. This simulation had equal spin and equal mass black holes with oppositely directed spins perpendicular to the orbit. Other simulations with different configurations (e.g., [5]) even larger kicks. This has stimulated interest in the astrophysical community; for instance

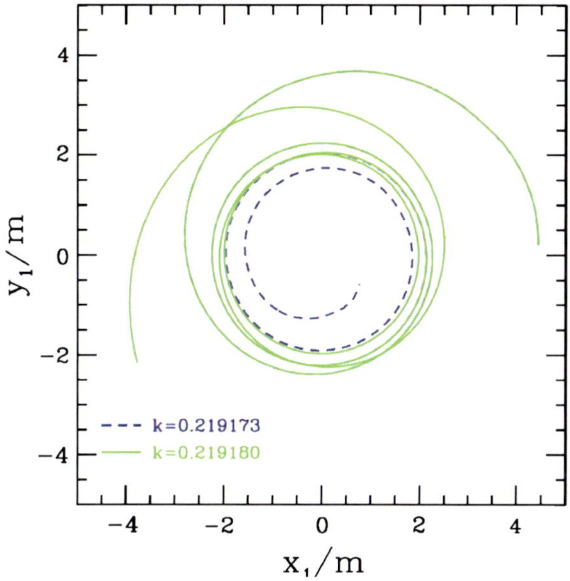

Fig. 10 Two examples of equal-mass orbits (only one of the two black holes is shown in each case). The parameter k is an initial boost parameter perpendicular to the initial line between the (nonspinning) black holes. The value $k = 0.219173$ leads to a merger, but the slightly larger $k = 0.219180$ results in a flyby [17]

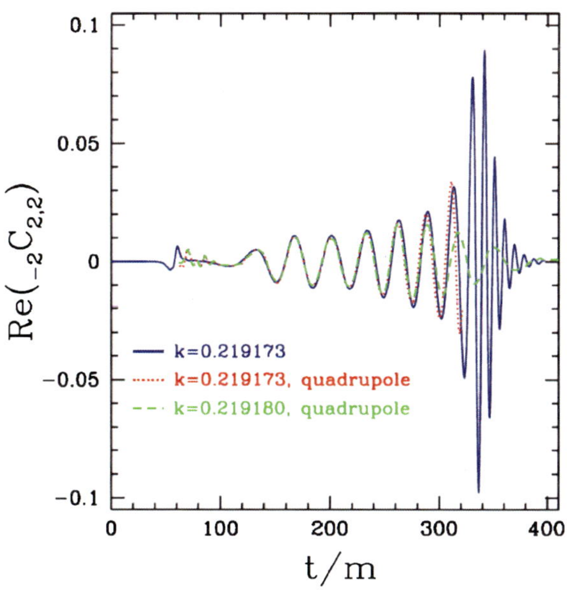

Fig. 11 The waveform from the $k = 0.219173$ merger case, with a quadrupole estimate for this radiation, and a quadrupole estimate for the radiation from the flyby case in Fig. 10 [17]

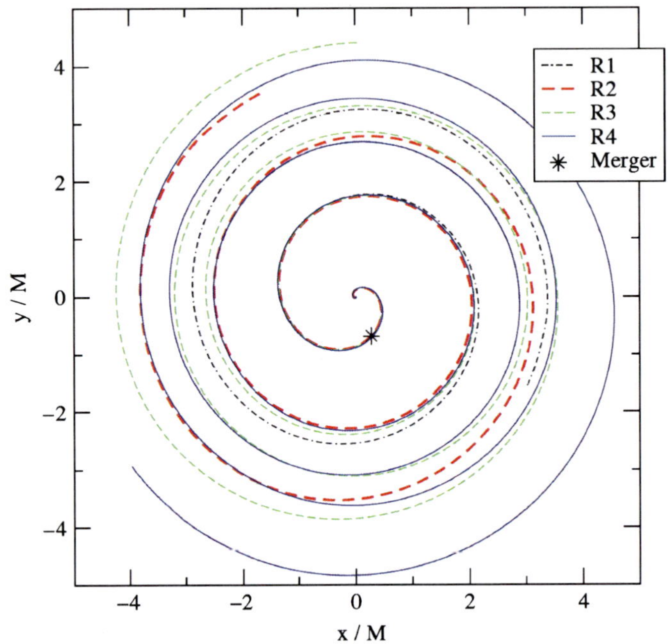

Fig. 12 Quasicircular orbits starting from different initial separation all have similar merger behavior [1]

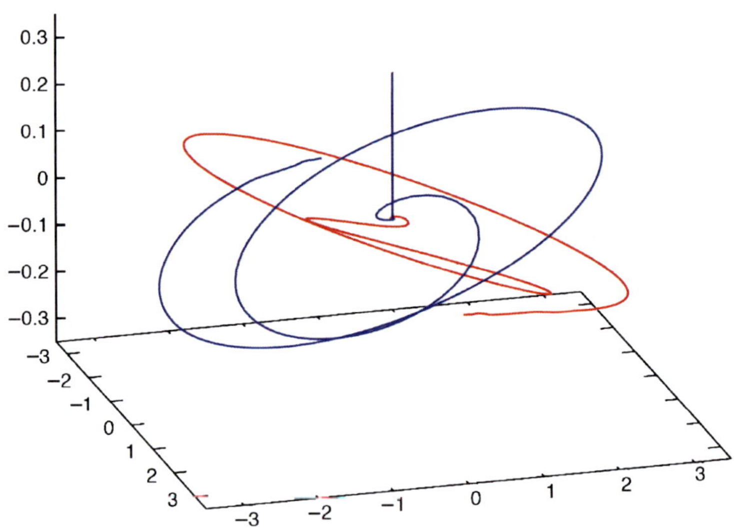

Fig. 13 Out-of-plane motion and precession arising from the spin–orbit interaction of spinning black holes [5]. The Kerr spin parameter a for each hole is $a = 0.5$ m, directed in the plane of the orbit

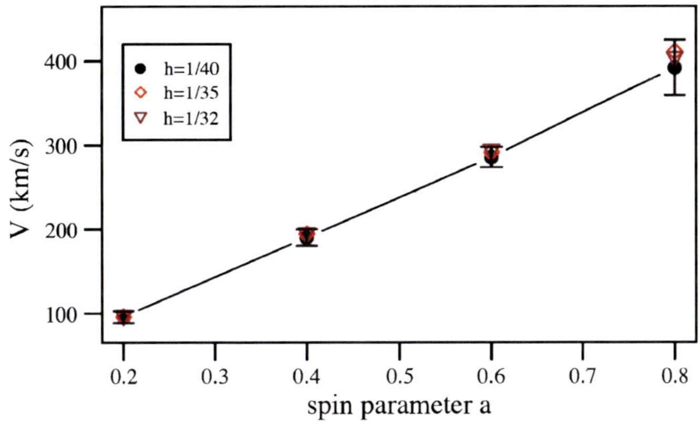

Fig. 14 Recoil velocity for two spinning black holes with equal oppositely directed Kerr spin parameters a perpendicular to the plane of the orbit. The kick is in the plane of the orbit [10]

Bogdanović et al. [3] argues that the Bardeen–Petterson effect (dragging of disk along the rotation of a central black hole) will align the disks and the central black holes in a way that minimizes kicks (Fig. 14).

17 Conclusion

The study of gravitational radiation is in a rapid development, observationally and theoretically. The laser interferometer detectors are on schedule to upgrades which will guarantee detections and the beginning of gravitational radiation astronomy in a few years, given only conventional physics: the validity of Einstein's theory, and the validity of neutron-star astrophysics a seen in our galaxy. In terms of theory we have seen in the last few years an explosion of enterprise in the simulation of black holes and neutron star encounters and the effect of the evolved gravitational radiation in the dynamics of the modelled systems. Much awaits us in the next decade of gravitational wave physics.

References

1. Baker, J. G., Centrella, J., Choi, D.-I., Koppitz, M., & van Meter, J. 2006, Physical Review D, 73, 104002
2. Berti, E., Iyer, S., & Will, C. M. 2007, ArXiv e-prints 0709.2589, 709
3. Bogdanović, T., Reynolds, C. S., & Miller, M. C. 2007, Astrophys. Jour. Lett., 661, L147
4. Campanelli, M., Lousto, C. O., Marronetti, P., & Zlochower, Y. 2006, Physical Review Letters, 96, 111101
5. Campanelli, M., Lousto, C. O., Zlochower, Y., & Merritt, D. 2007, Physical Review Letters, 98, 231102

6. Choquet-Bruhat, Y. & York, J. 1969, General Relativity and Gravitation, A. Held, ed. (Plenum Press, New York), 99
7. Eddington, A. S. 1924, Nature, 113
8. Friedrich, H. 2002, The Conformal Structure of Space-Time, in Lecture Notes in Physics, (Berlin: Springer Verlag), 604
9. Garfinkle, D. 2002, APS Meeting Abstracts, 12004
10. Herrmann, F., Hinder, I., Shoemaker, D., Laguna, P., & Matzner, R. A. 2007, Astrophys. Jour., 661, 430
11. Kerr, R. P. & Schild, A. 1965, "A New Class of Vacuum Solutions of the Einstein Field Equations," in Atti del Convegno Sulla Relativita Generale: Problemi Dell'Energia E Onde Gravitazionale, G. Barbera, ed.
12. —. 1965, "Some Algebraically Degenerate Solutions of Einstein's Gravitational Field Equations, in Applications of Nonlinear Partial Differential Equations", in Mathematical Physics, Proc. of Symposia B Applied Math., Vol. XV11
13. Lichnerowicz, A. 1994, J. Math. Pures Appl., 23, 37
14. Misner, C. W., Thorne, K. S., & Wheeler, J. A. 1973, Gravitation (San Francisco: W.H. Freeman and Co., 1973)
15. Mitchell, T. & Will, C. M. 2007, Phys. Rev. D, 75, 124025
16. Pretorius, F. 2005, Physical Review Letters, 95, 121101
17. Pretorius, F. & Khurana, D. 2007, Classical and Quantum Gravity, 24, 83
18. Schutz, B. 1994, A first course in general relativity, Cambridge University Press
19. Schwarzschild, K. 1916, Sitzber. Deut. Akad. Wiss. Berlin, Kl. Math-Phys. Tech., 189
20. Yo, H.-J., Cook, J. N., Shapiro, S. L., & Baumgarte, T. W. 2004, Phys. Rev. D, 70, 084033
21. York, J. W. 1994, J. Math. Pures Appl., 23, 37
22. York, Jr., J. W. & Piran, T. 1982, in Spacetime and Geometry, The Alfred Schild Lectures; Richard A. Matzner and L. Shepley eds. University of Texas Press, 147

Discovering Relic Gravitational Waves in Cosmic Microwave Background Radiation*

Leonid P. Grishchuk

Abstract The authority of J.A. Wheeler in many areas of gravitational physics is immense, and there is a connection with the study of relic gravitational waves as well. I begin with a brief description of Wheeler's influence on this study. One part of the paper is essentially a detailed justification of the very existence of relic gravitational waves, account of their properties related to the quantum-mechanical origin, derivation of the expected magnitude of their effects, and reasoning why they should be detectable in the relatively near future. This line of argument includes the comparison of relic gravitational waves with density perturbations of quantum-mechanical origin, and the severe criticism of methods and predictions of inflationary theory. Another part of the paper is devoted to active searches for relic gravitational waves in cosmic microwave background radiation (CMB). Here, the emphasis is on the temperature-polarization *TE* cross-correlation function of CMB. The expected numerical level of the correlation, its sign, statistics, and the most appropriate interval of angular scales are identified. Other correlation functions are also considered. The overall conclusion is such that the observational discovery of relic gravitational waves looks like the matter of a few coming years, rather than a few decades.

1 Introduction

It is my honor and pleasure to be a lecturer at the first course of J.A. Wheeler School on Astrophysical Relativity. Wheeler is one of the founding fathers of the field of gravitational physics, and many of us are strongly influenced by his work and his

*Based on the invited lecture at the 1-st J. A. Wheeler School on Astrophysical Relativity, June 2006, Italy.

L.P. Grishchuk (✉)
School of Physics and Astronomy, Cardiff University, Cardiff CF243AA, UK
and
Sternberg Astronomical Institute, Moscow State University, Moscow 119899, Russia
e-mail: grishchuk@astro.cf.ac.uk

I. Ciufolini and R.A. Matzner (eds.), *General Relativity and John Archibald Wheeler*, Astrophysics and Space Science Library 367, DOI 10.1007/978-90-481-3735-0_10,

personality. In particular, we often find inspiration in the great textbook of him and his colleagues [1].

I was fortunate to speak with J.A. Wheeler right before the publication of my first papers on relic gravitational waves [2]. That conversation helped me to shape my views on the subject. Being a young researcher, I was somewhat nervous about evaluation of my work by the towering scientist, but to my relief, Wheeler quickly understood the work and agreed with it. In what follows, we will be discussing relic gravitational waves systematically and in details, but I would like to start from describing my initial doubts and worries, and how Wheeler helped me to see them in a different light.

I showed that the wave-equation for a gravitational wave $h(\eta, \mathbf{x}) = [\mu(\eta)/a(\eta)]e^{i\mathbf{n}\cdot\mathbf{x}}$ in a homogeneous isotropic universe with the scale factor $a(\eta)$ can be written as a Schrodinger-like equation

$$\mu'' + \mu\left[n^2 - \frac{a''}{a}\right] = 0. \tag{1}$$

It follows from this equation that the fate of the wave with the wavenumber n depends on the comparative values of n^2 and the effective potential $U(\eta) = a''/a$. If n^2 is much larger than $|U(\eta)|$, the wave does not feel the potential and propagates with the adiabatically changing amplitude $h \propto 1/a$. In the opposite limit, the interaction with the potential is strong and the wave changes dramatically. The amplitude of the initial wave gets amplified over and above the adiabatic law $h \propto 1/a$, and at the same time a wave propagating in the opposite direction is being created. This process results effectively in the production of standing waves. I called this phenomenon the superadiabatic amplification. Over the years, other cosmological wave equations were also modeled on Eq. (1). In terms of the variable $h(\eta)$, where $h(\eta) = \mu(\eta)/a(\eta)$, Eq. (1) has the form

$$h'' + 2\frac{a'}{a}h' + n^2 h = 0. \tag{2}$$

While analyzing various scale factors $a(\eta)$ and potentials $U(\eta)$, I was imagining them as being 'drawn by a hand'. My first example was the potential $U(\eta)$ architypical for quantum mechanics – a rectangular barrier. This means that $U(\eta) = const$ in some interval of η-time between η_a and η_b, while $U(\eta) = 0$ outside this interval. I have shown that the waves interacting with this barrier are necessarily amplified. However, the postulated $U(\eta)$ caused some concerns.

It is easy to make $U(\eta) = 0$ before η_a and after η_b. Indeed, if $a''/a = 0$, the scale factor $a(\eta)$ is either a constant, like in a flat Minkowski world, or is proportional to η, like in a radiation-dominated universe. It is not difficult to imagine that the Universe was radiation-dominated before and after some crucial interval of evolution. However, if it is assumed that $a''/a = const$ between η_a and η_b, it is not easy to find a justification for this evolution, as the scale factor $a(\eta)$ should depend exponentially on η-time in this interval. In terms of t-time, which is related to η-time by $c\,dt = a\,d\eta$, the scale factor $a(t)$ should be proportional to t. The Einstein equations allow this law of expansion, but they demand that the 'matter' driving this evolution should have the effective equation of state $p = -(1/3)\epsilon$. Some other potentials $U(\eta)$ drawn by a hand do also require strange equations of state.

Having shown the inevitability of superadiabatic amplification, as soon as $a''/a \neq 0$, I was somewhat embarrassed by the fact that in some parts of my study I operated with scale factors drawn by a hand and driven by matter with unusual equations of state. Although equations of state with negative pressure had already been an element of cosmological research, notably in the work of A.D. Sakharov [3], I feared that Wheeler may dislike this idea and may say 'forget it'. To my surprise, he accepted the approach and even suggested a wonderful name: the 'engine-driven cosmology'. The implication was that although we may not know the nature of the 'engine' which drives a particular $a(\eta)$, this knowledge is not, for now, our high priority. Being inspired by Wheeler's attitude, I hurried to include the notion of the engine-driven cosmology, with reference to Wheeler, in the very first paper on the subject [2].

It is interesting to note that E. Schrodinger [4] felt uneasy about wave solutions in an expanding universe. (I became aware of his work much after the time of my first publications.) Schrodinger identified the crucial notion of the "mutual adulteration of positive and negative frequency terms in the course of time". He was thinking about electromagnetic waves and he called the prospect of photon creation in an expanding universe an "alarming phenomenon". From the position of our present knowledge, we can say that Schrodinger was right to be doubtful. Indeed, electromagnetic waves cannot be amplified and photons cannot be created in a nonstationary universe. Even though the wavelengths of electromagnetic and gravitational waves change in exactly the same manner, their interactions with external gravitational field are drastically different. The corresponding effective potential $U(\eta)$ in the Maxwell equations is strictly zero and the "alarming phenomenon" does not take place. All physical fields are tremendously 'stretched' by expansion, but only some of them are being amplified.

In his paper, Schrodinger was operating with a variant of scalar electrodynamics, wherein the coupling of scalar fields to gravity is ambiguous and can be chosen in such a way that the wave equation becomes identical to Eq. (1), making the amplification of scalar waves possible (for more details, see introductory part to Ref. [5]). L. Parker [6] undertook a systematic study of the quantized version of test scalar fields in FLRW (Friedmann–Lemaitre–Robertson–Walker) cosmologies. As for the gravitational waves, there is no ambiguity in their coupling to gravity since the coupling follows directly from the Einstein equations. As we see, the "alarming phenomenon" does indeed take place for gravitational waves [2]. (The authors of publications preceeding Ref. [2] explicitely denied the possibility of graviton creation in FLRW universes. In the end, it was only Ya. B. Zeldovich who wrote to me: "Thank you for your goal in my net".)

It is important to realize that the possibility of generation of relic gravitational waves relies only on the validity of general relativity and quantum mechanics. The governing principles are part of the well-understood and tested physics. The underlying Eq. (1) admits an analogy with the Schrodinger equation (outlined

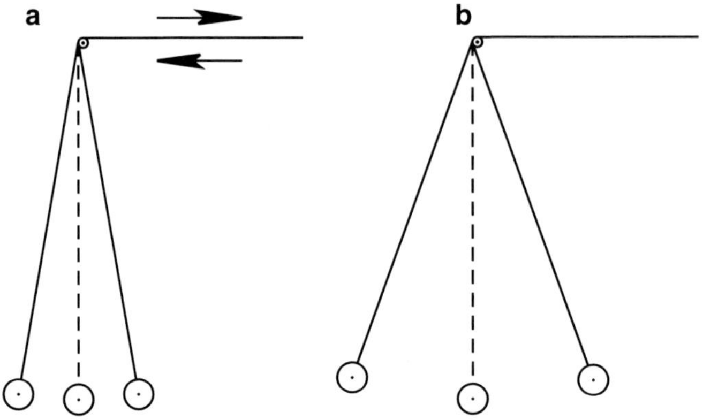

Fig. 1 Parametric amplification. (**a**) Variation of the length of the pendulum. (**b**) Increased amplitude of oscillations

above), but it can also be viewed as an equation for a classical oscillator with variable frequency. The phenomenon of superadiabatic (parametric) amplification of the waves' zero-point quantum oscillations is at the heart of the cosmological generating mechanism. In order to better appreciate this phenomenon, we shall briefly review the closely related laboratory-type problem of parametric amplification in a classical pendulum.

Let us consider an ideal pendulum hanging in a constant gravitational field characterized by the free-fall acceleration g (see Fig. 1). The frequency of the oscillator is given by $\omega_0 = \sqrt{g/l}$, where l is the length of the pendulum,

$$\ddot{x} + \frac{g}{l}x = 0.$$

The amplitude of oscillations can be enhanced either by force acting directly on the pendulum's mass or by an influence which changes a parameter of the oscillator – in this case, its frequency. The equation for small horizontal displacements $x(t)$ of the oscillator's mass takes the form

$$\ddot{x} + \omega^2(t)x = 0. \tag{3}$$

The simplest parametric intervention makes the length l time-dependent, as shown in Fig. 1a. Then, $\omega^2(t)$ in Eq. (3) is $\omega^2(t) = (g - \ddot{l})/l(t)$. Note that even if the gravitational acceleration g remains constant, it still gets modified by the acceleration term $\ddot{l}$ arising due to the variation of length $l(t)$. So, the correct $\omega^2(t)$ differs from the naive expectation $\omega^2(t) = g/l(t)$. (For more details about parametrically excited oscillators, see [7].) In general, both $g(t)$ and $l(t)$ are functions of time, in which case $\omega^2(t) = [g(t) - \ddot{l}]/l(t)$ and Eq. (3) reads

$$\ddot{x} + x\left[\frac{g(t)}{l(t)} - \frac{\ddot{l}}{l}\right] = 0. \tag{4}$$

If the external influence on the oscillator is very slow, i.e. $\omega_0 \gg |\dot{\omega}(t)/\omega(t)|$, the ratio of the slowly changing energy $E(t)$ and frequency $\omega(t)$,

$$\frac{E}{\hbar\omega} = N, \tag{5}$$

will remain constant. This ratio is called an adiabatic invariant [8]. The quantity N is also a 'number of quanta' in a classical oscillator (see paper by Ya. B. Zeldovich, signed by a pseudonym, on how quantum mechanics helps understand classical mechanics [9]).

On the other hand, if the oscillator was subject to some interval of appropriate parametric influence, the amplitude of oscillations and the number of quanta N will significantly increase, as shown in Fig. 1b. To get a significant effect, the function $\omega(t)$ does not have to be periodic, but in its Fourier spectrum there should be enough power at frequencies around ω_0. The final frequency does not need to differ from the initial ω_0.

One can notice the striking analogy between Eqs. (1) and (4). This analogy extends further if one goes over from the displacement $x(t)$ to the dimensionless angle variable $\phi(t)$ related to $x(t)$ by $\phi(t) = x(t)/l(t)$. (Compare with the cosmological relationship $h(\eta) = \mu(\eta)/a(\eta)$.) Then, Eq. (4) takes the form

$$\ddot{\phi} + 2\frac{\dot{l}}{l}\dot{\phi} + \frac{g(t)}{l(t)}\phi = 0, \tag{6}$$

which is strikingly similar to Eq. (2). In cosmological equations, the analog of the ratio $g(t)/l(t)$ is n^2.

There are two lessons to be learned from this discussion. First, the necessary condition for a significant amplification of the wave is the availability of a regime where the characteristic time $|a/a'|$ of variation of the external gravitational field (represented by the scale factor $a(\eta)$) becomes comparable and much shorter than the wave period $2\pi/n$,

$$n \ll \left|\frac{a'(\eta)}{a(\eta)}\right|. \tag{7}$$

Before and after this regime, the wave may be a high-frequency wave, that is, it may satisfy the condition $n \gg |a'(\eta)/a(\eta)|$. But if during some interval of evolution the opposite condition (7) is satisfied, and $a''/a \neq 0$, the wave will be superadiabatically amplified, regardless of whether the model universe is expanding or contracting [2].

Second, a classical pendulum should initially be in a state of oscillations – excited state – in order to have a chance to be amplified by an external 'pump' influence. Otherwise, if it is initially at rest, i.e. hanging straight down, the time-dependent change of l or g will not excite the oscillator, and the energy of oscillations will remain zero. However, in the quantum world, even if the oscillator is in the state of lowest energy (ground, or vacuum, state) it will inevitably possess the zero-point quantum oscillations. One can think of these zero-point quantum oscillations as those that are being amplified by the external pump (assuming, of course, that the oscillator is coupled to the pump). This is where the quantum mechanics enters the picture. The initial ground state of the parametrically excited oscillator evolves into a multi-quantum state, and the mean number of quanta grows, when condition (7) is satisfied [11]. There is no quantum state lower than the ground state, so the engine-driven cosmology will necessarily bring the properly coupled quantum oscillator into an excited state.

The concept of the engine-driven cosmology remains perfectly adequate for the present-day research. We do not know what governed the scale factor of the very early Universe. It could be a lucky version of the scalar field (inflaton) or something much more sophisticated and dictated by the 'theory of everything'. The lack of this knowledge is not important for the time being. By observing relic gravitational waves we may not be able to determine at once the nature of the cosmological 'engine', but we will be able to determine the behaviour of the early Universe's scale factor. Therefore, we will gain unique information about the 'birth' of the Universe and its very early dynamical evolution.

2 Cosmological Oscillators

The physics of laboratory-type oscillators have direct relevance to cosmological oscillators. In cosmology, we normally consider a universe filled with some matter and slightly perturbed in all constituents. It is convenient to write the perturbed metric of a flat FLRW universe in the form

$$\begin{aligned} ds^2 &= -c^2dt^2 + a^2(t)(\delta_{ij} + h_{ij})dx^i dx^j \\ &= a^2(\eta)\left[-d\eta^2 + (\delta_{ij} + h_{ij})dx^i dx^j\right]. \end{aligned} \tag{8}$$

The six functions $h_{ij}(\eta, \mathbf{x})$ can be expanded over spatial Fourier harmonics $e^{\pm i\mathbf{n}\cdot\mathbf{x}}$, where $\mathbf{n}$ is a dimensionless time-independent wave-vector,

$$h_{ij}(\eta, \mathbf{x}) = \frac{\mathcal{C}}{(2\pi)^{3/2}} \int_{-\infty}^{+\infty} d^3\mathbf{n} \frac{1}{\sqrt{2n}} \sum_{s=1,2} \left[\overset{s}{p}_{ij}(\mathbf{n})\, \overset{s}{h}_n(\eta)\, e^{i\mathbf{n}\cdot\mathbf{x}}\, \overset{s}{c}_{\mathbf{n}} + \overset{s}{p}{}^{*}_{ij}(\mathbf{n}) \overset{s}{h}{}^{*}_n(\eta) e^{-i\mathbf{n}\cdot\mathbf{x}}\, \overset{s}{c}{}^{*}_{\mathbf{n}} \right]. \tag{9}$$

This representation requires some explanations. The mode functions $\overset{s}{h}_n(\eta)$ obey differential equations that follow from the perturbed Einstein equations (as an example, look at Eq. (1)). The wavelength of the mode $\mathbf{n}$ is given by $\lambda = 2\pi a/n$, where the wavenumber n is $n = (\delta_{ij}n^i n^j)^{1/2}$. It is convenient to take today's scale factor $a(\eta_R)$ to be equal to the 'size of the Universe', that is, $a(\eta_R) = 2l_H$, where $l_H = c/H_0$ is today's Hubble radius and $H_0 = H(\eta_R)$ is today's Hubble parameter.

Then, for a fixed moment of time, today, we can also write the laboratory-type expression for the same wavelength λ: $\lambda = 2\pi/k$, where k has the dimensionality of inverse length and is related to the dimensionless n by $k = n/2l_H$. The wave whose length is equal to today's 'size of the Universe' has $n = 2\pi$. It is assumed that wavelengths can always be measured by unchangeable laboratory standards.

By systematically writing **n** and distinguishing it from **k** we essentially follow the original motivations of E.M. Lifshitz (see, for example, [10]). In contrast, some modern authors think that Lifshitz needs to be 'simplified' and 'modernized'. They use only one letter **k** and write **k** everywhere where Lifshitz was writing **n**. This created quite a mess in wavenumbers and wavelengths. Among them you will see physical, non-physical, coordinate, comoving, proper, etc.

Moving to complex Fourier coefficients $\overset{s}{c}_{\mathbf{n}}$, $\overset{s}{c}{}^{*}_{\mathbf{n}}$ we note that they are some particular numbers, if the left-hand-side (l.h.s) of Eq. (9) is a deterministic, even if arbitrarily complicated, function. In the rigorous quantum-mechanical version of the theory, these coefficients will be promoted to the status of quantum-mechanical annihilation and creation operators $\overset{s}{c}_{\mathbf{n}}$, $\overset{s}{c}{}^{\dagger}_{\mathbf{n}}$ acting on some quantum states. In the CMB applications, we will treat, for simplicity, $\overset{s}{c}_{\mathbf{n}}$, $\overset{s}{c}{}^{*}_{\mathbf{n}}$ as random numbers taken from some probability distributions. The factor $1/\sqrt{2n}$ in Eq. (9) is a useful insertion inspired by quantum field theories. The normalization constant $\mathcal{C}$ will be discussed later.

The gravitational field polarization tensors $\overset{s}{p}_{ij}(\mathbf{n})$ deserve special attention. As we shall see below, the polarization properties of the CMB radiation – our final destination – are intimately connected with the structure of these metric polarization tensors. Polarization of CMB and polarization of metric perturbations is not simply a coincidence in the usage of the word polarization. The polarization tensors $\overset{s}{p}_{ij}(\mathbf{n})$ have different forms depending on whether the functions $h_{ij}(\eta, \mathbf{x})$ represent gravitational waves, rotational perturbations, or density perturbations. Each class of these perturbations have two polarization states, so $s = 1, 2$ for each of them. In what follows we will be considering gravitational waves and density perturbations.

In the case of gravitational waves, two independent linear polarization states can be described by two real polarization tensors

$$\overset{1}{p}_{ij}(\mathbf{n}) = l_i l_j - m_i m_j, \quad \overset{2}{p}_{ij}(\mathbf{n}) = l_i m_j + m_i l_j, \tag{10}$$

where spatial vectors $(\mathbf{l}, \mathbf{m}, \mathbf{n}/n)$ are unit and mutually orthogonal vectors. The polarization tensors (10) satisfy the conditions

$$\overset{s}{p}_{ij}\,\delta^{ij} = 0, \quad \overset{s}{p}_{ij}\,n^i = 0, \quad \overset{s'}{p}_{ij}\overset{s}{p}{}^{ij} = 2\delta_{s's}. \tag{11}$$

Two circular polarization states are described by

$$\overset{L}{p}_{ij} = \frac{1}{\sqrt{2}}\left(\overset{1}{p}_{ij} + i\,\overset{2}{p}_{ij}\right), \quad \overset{R}{p}_{ij} = \frac{1}{\sqrt{2}}\left(\overset{1}{p}_{ij} - i\,\overset{2}{p}_{ij}\right). \tag{12}$$

The left and right polarizations interchange under a coordinate reflection (altering the sign of l^i or m^i). In other words, gravitational waves can support a chirality, or 'handedness'.

In the case of density perturbations, the polarization tensors are

$$\overset{1}{p}_{ij} = \sqrt{\frac{2}{3}}\delta_{ij}, \quad \overset{2}{p}_{ij} = -\sqrt{3}\frac{n_i n_j}{n^2} + \frac{1}{\sqrt{3}}\delta_{ij}. \tag{13}$$

These polarization tensors satisfy the last of the conditions (11). The polarization tensors (13) remain unchanged under coordinate mirror reflections, so density perturbations cannot support handedness.

It is important to stress that from the position of general relativity, cosmological density perturbations represented by metric perturbations h_{ij} with polarization structure (13), can be viewed as 'scalar', or spin-0, gravitational waves. Indeed, although in the absence of matter, i.e. for $T_{\mu\nu} = 0$, the linearised Einstein equations admit spin-0 wave solutions, i.e. solutions with the structure (13), these solutions do not carry energy, do not affect the relative motion of test masses, and can be nullified by coordinate transformations. It is only spin-2 wave solutions, i.e. solutions with the structure (10), that carry energy, affect the relative motion of test masses, and cannot be removed by coordinate transformations. These spin-2 solutions are called gravitational waves.

However, in cosmology, the spin-0 solutions survive and become non-trivial, as soon as metric perturbations with the structure (13) are supported by non-vanishing matter perturbations, that is, when $\delta T^{\nu}_{\mu} \neq 0$. These solutions, which unite gravitational field and matter perturbations, are called cosmological density perturbations. Therefore, cosmological density perturbations and cosmological gravitational waves, although separate from the point of view of algebraic classification of the tensor h_{ij}, are not entirely disconnected entities that should be treated by different theories. On the contrary, they should be viewed as parts of a common set of gravitational (metric) degrees of freedom. This can be regarded as a physical principle that will later guide our choice of initial conditions for density perturbations.

One more comment is in order. In our presentation, we will be consistently using the class of synchronous coordinate systems (8), that is, we assume that $h_{00} = 0$ and $h_{0i} = 0$. We are not losing anything in terms of physics but we gain significantly in terms of technical simplicity and universality of our approach to gravitational waves and density perturbations. In principle, one can work in arbitrary coordinates, assuming that all components of metric perturbations, including h_{00} and h_{0i}, are non-zero. This will not bring you any real advantages, but will complicate calculations and can mislead you in issues of interpretation. Especially if you attempt to compare, say, gravitational waves described in synchronous coordinates with density perturbations described in the 'Newtonian-gauge' or other 'gauge' coordinates. Nevertheless, since various gauges and gauge-invariant formalisms are popular in contemporary literature, we will indicate, where necessary, how our formulas would be modified had we used arbitrary coordinates.

3 Quantization of Gravitational Waves

Cosmological gravitational waves exist in the absence of matter perturbations, i.e. for $\delta T_{\mu}^{\nu} = 0$. For each wavenumber n and polarization state $s = 1, 2$ the mode functions $\overset{s}{h}_n(\eta)a(\eta) = \overset{s}{\mu}_n(\eta)$ satisfy the familiar equation (1). We assume that each of gravitational-wave oscillators **n** was initially, at some $\eta = \eta_0$, in its ground state. Although certain oscillators could be somewhat excited without violating our perturbative assumptions, we do not see physical justification for such non-vacuum initial states.

There is no such thing as the ground (vacuum) state without explicitly indicating the Hamiltonian for which the state is the ground state. Shortly, we will explicitly write down the Hamiltonian for gravitational waves (and, later, for density perturbations too). Specifying the Hamiltonian will also make more precise the concept of normalization of the initial mode functions to the zero-point quantum oscillations, or in other words the normalization to a 'half of the quantum in each mode'.

It is important to remember, however, that, qualitatively, we already know the answer [2]. The energy of a gravitational wave with wavelength λ_0, contained in a volume $(\lambda_0/2)^3$, is equal to a half of the quantum, i.e. $N = 1/2$ in Eq. (5), if the amplitude of the wave is at the level

$$h_i(n) \approx \sqrt{\frac{G\hbar}{c^3}\frac{n}{a_0}} \approx \frac{l_{Pl}}{\lambda_0}, \tag{14}$$

where $l_{Pl} = \sqrt{G\hbar/c^3}$, $a_0 = a(\eta_0)$ and $\lambda_0 = 2\pi a_0/n$. Equation (14) defines the initial vacuum spectrum of the gravitational wave (g.w.) amplitudes: $h_i(n) \propto n$.

Shifting the initial time η_0 up to the boundary between the adiabatic and superadiabatic regimes at $\eta = \eta_i$, we derive the estimate $h_i \sim l_{Pl}/\lambda_i \sim l_{Pl}H_i/c$. Then, we use the constancy of the metric amplitude h throughout the long-wavelength (superadiabatic) regime $n \ll a'/a$, Eq. (7), that is, the regime in which the wave is 'under the barrier a'/a'. The constancy of h follows from the constancy of the dominant (first) term in the approximate long-wavelength solution [2] to Eqs. (1), (2):

$$\frac{\mu}{a} = C_1 + C_2 \int \frac{d\eta}{a^2}. \tag{15}$$

This allows us to write $h_f \approx h_i$, where h_f is the estimate of h at the end of the superadiabatic regime. After having emerged from 'under the barrier a'/a' the wave will again behave adiabatically.

Using initial conditions (14) and evolving classical mode functions through all the barriers and intervals of adiabatic evolution, one can derive today's metric amplitudes as a function of frequency, i.e. today's amplitude spectrum. For a typical engine-driven expanding cosmology shown in Fig. 2, and for its associated barrier a'/a shown in Fig. 3, one expects to arrive at today's spectrum qualitatively shown in Fig. 4 (for more details, see [12]).

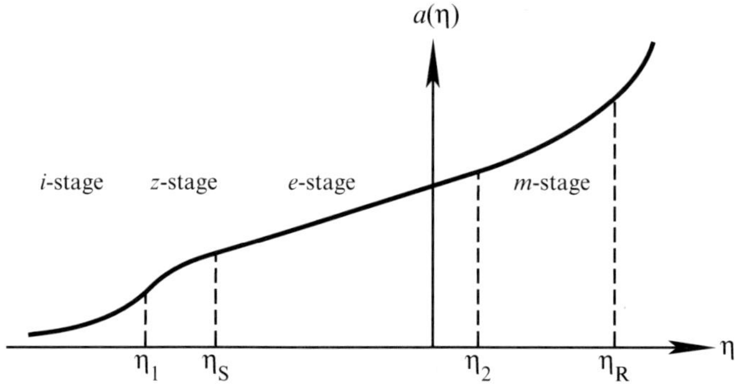

Fig. 2 A typical scale factor $a(\eta)$ as a function of time from the era of imposing the initial conditions, i-stage, and up to the present time

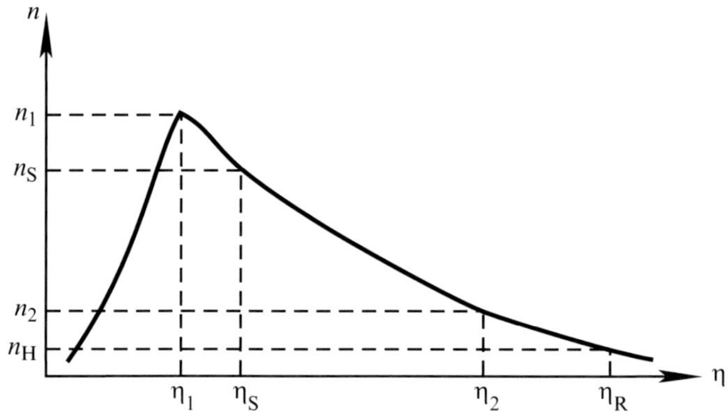

Fig. 3 The barrier a'/a built from the scale factor $a(\eta)$ of Fig. 2 versus crucial wavenumbers n defined by this barrier

The today's amplitudes in different parts of the spectrum are mostly determined by the thickness of the barrier a'/a, that is, by the duration of time that a given mode n spent under the barrier. All modes start with the initial $N_i = 1/2$, but if the mode n enters the barrier (i.e. satisfies the condition $\lambda_i \approx c/H_i$) when the scale factor was $a_i(n)$, and leaves the barrier (i.e. satisfies the condition $\lambda_f \approx c/H_f$) when the scale factor was $a_f(n)$, the final number of quanta in the mode will be $N_f(n) = \left(a_f/a_i\right)^2 = \left(\lambda_f/\lambda_i\right)^2 = \left(H_i/H_f\right)^2$. The very high-frequency modes with n above the tip of the barrier, $n > n_1$, remain in the adiabatic regime throughout their evolution, and their energy is being renormalised to zero. Therefore, the resulting amplitude spectrum quickly drops to zero at the high-frequency end.

The power-law index of today's spectrum $h(n)$ in an arbitrary narrow interval of frequencies, say, between n_a and n_b, as outlined in Fig. 4, is expressible in terms of the initial vacuum spectrum $h_i(n) \propto n$ and the power-law indeces β of the scale factor [2]

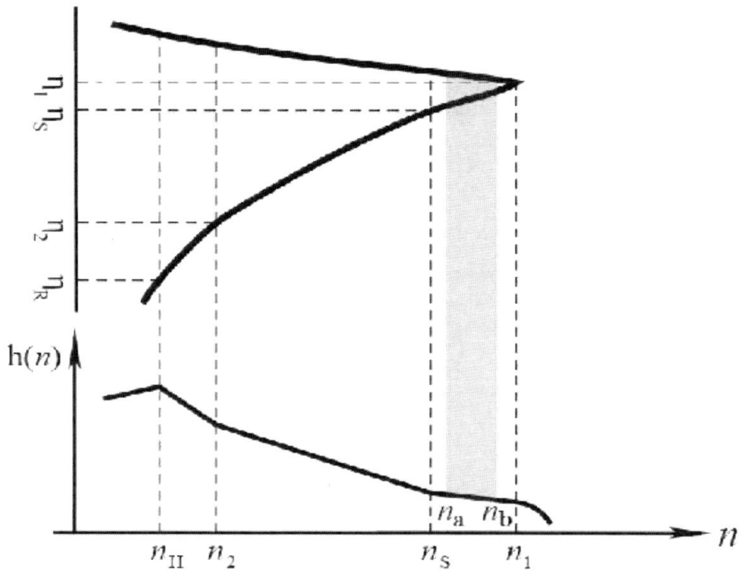

Fig. 4 The characteristic gravitational-wave amplitude $h(n)$ today as a function of frequency n. The size of the amplitude is mostly determined by the thickness of the barrier a'/a in the place, where it is traversed by the wave with a given wavenumber n. The barrier from Fig. 3 is shown at the *top* of the figure

$$a(\eta) \propto |\eta|^{1+\beta}$$

which approximates exact cosmological evolution in short intervals of time when the left and the right sides of the relevant portion of the barrier a'/a were formed. Today's spectral index of $h(n)$ in the discussed narrow interval of frequencies is given by

$$h(n) \propto n\, n^{1+\beta_i}\, n^{-(1+\beta_f)} \propto n^{1+\beta_i-\beta_f}, \tag{16}$$

where β_i and β_f refer to the left and to the right slopes of the barrier, respectively.

The vacuum spectrum processed only on the left slope of the barrier, i.e. considered at times before processing on the right slope of the barrier, is called the primordial spectrum $h_p(n)$:

$$h_p(n) \propto n^{2+\beta_i}. \tag{17}$$

For some historical and notational reasons, one and the same spectral index $2(2+\beta_i)$ in the power spectrum $h^2(n)$ of primordial metric perturbations is called n_t in the case of gravitational waves and $n_s - 1$ in the case of density perturbations (more details below, Eq. (65)). We will use the notation $2(2+\beta_i) = \mathrm{n} - 1$.

If the cosmological evolution $a(\eta)$ is such that the barrier a'/a is 'one-sided' and has only the right slope, the determination of the primordial spectrum requires additional considerations. The under-barrier amplification of the waves is still taking place [2], but the lack of the initial high-frequency regime makes the definition of the initial amplitudes somewhat ambiguous.

To summarise, for a given cosmological evolution $a(\eta)$, one can qualitatively predict today's piecewise amplitudes and slopes of the spectrum $h(n)$ without doing any particularly detailed calculations. A barrier, whose shape is more complicated than the one shown in Fig. 4, would have resulted in a more complicated shape of the generated spectrum. On the other hand, from the measured g.w. spectrum one can, in principle, reconstruct cosmological evolution $a(\eta)$ [13].

We will now turn to rigorous calculations based on quantum theory. The traditional, but not obligatory [14], approach to quantum theory begins with a classical Lagrangian. The Lagrangian for a gravitational-wave oscillator of frequency n, and for each of two polarizations $s = 1, 2$, has the form ([15] and references there):

$$L_{gw} = \frac{1}{2n}\left(\frac{a}{a_0}\right)^2 \left[\left(\bar{h}'\right)^2 - n^2\bar{h}^2\right], \tag{18}$$

where

$$\bar{h} = \left(\frac{\hbar}{32\pi^3}\right)^{1/2} \frac{\lambda_0}{l_{Pl}} h. \tag{19}$$

This Lagrangian can be derived from the total Hilbert–Einstein quadratic action, where both gravity and matter parts of the action are taken into account. The classical equation of motion derivable from the Lagrangian (18) in terms of the variable h is Eq. (2), and the equation of motion derivable in terms of the variable μ, where $h = \mu/a$, is Eq. (1).

The canonical pair of position q and momentum p for the oscillator can be taken as

$$q = \bar{h}, \qquad p = \frac{\partial L_{gw}}{\partial \bar{h}'} = \frac{1}{n}\left(\frac{a}{a_0}\right)^2 \bar{h}'. \tag{20}$$

Then, the classical Hamiltonian $H_{gw} = pq' - L_{gw}$ reads

$$H_{gw} = \frac{n}{2}\left[\left(\frac{a_0}{a}\right)^2 p^2 + \left(\frac{a}{a_0}\right)^2 q^2\right]. \tag{21}$$

We now promote q and p to the status of quantum-mechanical operators and denote them by bold-face letters. Since we are interested in the initial conditions imposed in the early high-frequency regime, i.e. when $n \gg a'/a$, we can write the following asymptotic expressions for the operators:

$$\mathbf{q} = \sqrt{\frac{\hbar}{2}}\frac{a_0}{a}\left[\mathbf{c}e^{-in(\eta-\eta_0)} + \mathbf{c}^\dagger e^{in(\eta-\eta_0)}\right], \tag{22}$$

$$\mathbf{p} = i\sqrt{\frac{\hbar}{2}}\frac{a}{a_0}\left[-\mathbf{c}e^{-in(\eta-\eta_0)} + \mathbf{c}^\dagger e^{in(\eta-\eta_0)}\right]. \tag{23}$$

The commutation relationships for $\mathbf{q}, \mathbf{p}$ operators, and for the annihilation and creation $\mathbf{c}, \mathbf{c}^\dagger$ operators, are given by

$$[\mathbf{q}, \mathbf{p}] = i\hbar, \qquad \left[\mathbf{c}, \mathbf{c}^{\dagger}\right] = 1.$$

The asymptotic expression for the Hamiltonian H_{gw} takes the form

$$\mathbf{H}_{gw} = \hbar\, n\, \mathbf{c}^{\dagger}\mathbf{c}. \tag{24}$$

Obviously, the quantum state $|0\rangle$ satisfying the condition

$$\mathbf{c}|0\rangle = 0$$

is the state of the lowest energy, i.e. the ground (vacuum) state of the Hamiltonian (24). At $\eta = \eta_0$ we get the relationships

$$\langle 0|\mathbf{q}^2|0\rangle = \langle 0|\mathbf{p}^2|0\rangle = \frac{\hbar}{2}, \qquad \Delta\mathbf{q}\Delta\mathbf{p} = \frac{\hbar}{2}.$$

The root-mean-square value of $\mathbf{q}$ in the vacuum state $|0\rangle$ is $q_{rms} = \sqrt{\hbar/2}$. Combining this number with (19) we derive

$$h_{rms} = \left(\langle 0|\mathbf{h}^2|0\rangle\right)^{1/2} = \frac{\sqrt{2}(2\pi)^{3/2} l_{Pl}}{\lambda_0}, \tag{25}$$

which agrees with the qualitative estimate (14). The adopted notations require $\mathcal{C} = \sqrt{16\pi}\, l_{Pl}$ in the general expression (9).

Now we will have to discuss exact Hamiltonians. The Hamiltonian built on the canonical pair μ, $\partial L_{gw}/\partial\mu'$ manifestly illustrates the underlying pair creation process for gravitational waves. Specifically, we introduce

$$Q = \sqrt{\frac{\hbar}{8\pi}}\frac{1}{n\sqrt{n}}\frac{1}{l_{Pl}}\mu, \quad P = \frac{\partial L_{gw}}{\partial Q'} = Q' - \frac{a'}{a}Q$$

and write the classical Hamiltonian in the form

$$H_{gw} = \frac{1}{2}\left[P^2 + n^2 Q^2 + \frac{a'}{a}(PQ + QP)\right].$$

The associated annihilation and creation operators are

$$c = \sqrt{\frac{n}{2}}\left(Q + i\frac{P}{n}\right), \quad c^{\dagger} = \sqrt{\frac{n}{2}}\left(Q - i\frac{P}{n}\right).$$

The full quantum Hamiltonian can be written as [11]

$$\mathbf{H}(\eta) = n\mathbf{c}^{\dagger}\mathbf{c} + \sigma\mathbf{c}^{\dagger 2} + \sigma^{*}\mathbf{c}^2, \tag{26}$$

where coupling to the external field is given by the function $\sigma(\eta) = (i/2)(a'/a)$.

The interaction with the external field can be neglected when $n \gg a'/a$, and then the first term in (26) dominates and represents a free oscillator. The early-time asymptotic expressions for the Heisenberg operators,

$$\mathbf{c}(\eta) = \mathbf{c}e^{-in(\eta-\eta_0)}, \quad \mathbf{c}^\dagger(\eta) = \mathbf{c}^\dagger e^{in(\eta-\eta_0)},$$

enter into formulas (22), (23).

One can note that the notion of the barrier a'/a is convenient when one thinks of the problem in terms of the interaction Hamiltonian (26) and the first-order Heisenberg equations of motion, whereas the barrier a''/a is more convenient when one thinks of the problem in terms of the second-order Schrodinger-like equation (1).

The emerging correlation between the traveling modes $\mathbf{n}$ and $-\mathbf{n}$, which leads to the production of standing waves referred to in the Introduction, is described by the 2-mode Hamiltonian (see [11] and references there):

$$\mathbf{H}(\eta) = n\mathbf{c}_\mathbf{n}^\dagger\mathbf{c}_\mathbf{n} + n\mathbf{c}_{-\mathbf{n}}^\dagger\mathbf{c}_{-\mathbf{n}} + 2\sigma\mathbf{c}_\mathbf{n}^\dagger\mathbf{c}_{-\mathbf{n}}^\dagger + 2\sigma^*\mathbf{c}_\mathbf{n}\mathbf{c}_{-\mathbf{n}}. \tag{27}$$

This Hamiltonian can be viewed as the sum of two Hamiltonians (26). The Hamiltonian (26) is called a 1-mode Hamiltonian.

The defined quantum-mechanical operators and Hamiltonians, plus the assumption about a particular initial state of the field, fully determine dynamical evolution of the field, its statistical properties, correlation functions, and eventually the observational predictions for later times.

4 Squeezing and Power Spectrum

The quantum-mechanical Schrodinger evolution transforms the initial vacuum state $|0_\mathbf{n}\rangle|0_{-\mathbf{n}}\rangle$ into a 2-mode squeezed vacuum state, which is equivalent to a pair of 1-mode squeezed vacuum states (see [11] and references there). In the Heisenberg picture, the initial state of the field does not evolve, but the operators do, and their evolution is ultimately described by the mode functions $\overset{s}{h}_n(\eta)$.

It is the variance of the oscillator's phase that is being strongly diminished (squeezed), whereas the variance of the amplitude is being strongly increased. The Gaussian nature of the initial vacuum state is maintained in the course of the Schrodinger evolution, but the variances of phase and amplitude in the resulting squeezed vacuum quantum state are dramatically different. The parameter of squeezing and the mean number of quanta are growing all the way up in the amplifying regime, and stop growing only at its end [11]. (The multi-quantum nature of the developing squeezed vacuum quantum state is behind the continuing debate over the 'quantum-to-classical' transition, 'decoherence', etc.) The phenomenon of squeezing allows us to treat the resulting quantum states as a stochastic collection of standing waves. The squeezing and the associated picture of standing waves is very important observationally [11]. It leads to oscillatory features in the metric power spectrum and, as a consequence, to oscillatory features in the angular power spectrum of CMB temperature and polarization anisotropies. We will discuss these oscillations later.

Having accepted the initial vacuum state $|0\rangle$ of the gravitational (metric) field (9), we can calculate the variance of the field:

$$\langle 0|\, h_{ij}(\eta, \mathbf{x}) h^{ij}(\eta, \mathbf{x})\, |0\rangle = \frac{C^2}{2\pi^2} \int_0^\infty n^2 \sum_{s=1,2} \left| \overset{s}{h}_n(\eta) \right|^2 \frac{dn}{n}. \tag{28}$$

The quantity

$$h^2(n, \eta) = \frac{C^2}{2\pi^2} n^2 \sum_{s=1,2} \left| \overset{s}{h}_n(\eta) \right|^2 \tag{29}$$

gives the mean-square value of the gravitational field perturbations in a logarithmic interval of n and is called the metric power spectrum. The spectrum of the root-mean-square (*rms*) amplitude $h(n, \eta)$ is determined by the square root of Eq. (29).

Having evolved the classical mode functions $\overset{s}{h}_n(\eta)$ up to some arbitrary instant of time η one can find the spectrum $h(n, \eta)$ at that instant of time. As mentioned before, the spectrum calculated at times when the waves of today's interest were in their long-wavelength regime (that is, longer than the Hubble radius at that times) is called the primordial spectrum. The today's spectrum, i.e. spectrum calculated at $\eta = \eta_R$, is normally expressed in terms of frequency ν measured in Hz, $\nu = nH_0/4\pi = n\nu_H/4\pi$. The spectral *rms*-amplitude is denoted by $h_{rms}(\nu)$, or simply $h(\nu)$. The mean-square value of the gravitational-wave field in some interval of frequencies between ν_1 and ν_2 is given by the integral:

$$\langle h^2 \rangle = \int_{\nu_1}^{\nu_2} h^2(\nu) \frac{d\nu}{\nu}. \tag{30}$$

The spectral function $h^2(\nu)$ depends on frequency ν, but is dimensionless. The dimensionality of Hz^{-1} is carried by the function $h^2(\nu)/\nu$.

For gravitational waves which are comfortably shorter than the Hubble radius, one can also calculate the spectral gravitational-wave energy density $\rho_{gw}(\nu)c^2$ and the total g.w. energy density in some interval of frequencies. These quantities are expressible in terms of $h^2(\nu)$:

$$\rho_{gw}(\nu) = \frac{\pi}{8G} h^2(\nu)\nu^2, \qquad \rho_{gw}(\nu_1, \nu_2) = \int_{\nu_1}^{\nu_2} \rho_{gw}(\nu) \frac{d\nu}{\nu}. \tag{31}$$

For the purpose of comparing a g.w. background with other sorts of matter, it is convenient to introduce the cosmological Ω_{gw}-parameter and its spectral value $\Omega_{gw}(\nu)$.

As with all other sorts of radiation, this parameter is defined by

$$\Omega_{gw}(\nu) = \frac{\rho_{gw}(\nu)}{\rho_{crit}}, \tag{32}$$

where $\rho_{crit} = 3H_0^2/8\pi G$. Using Eq. (31) one can also write [12]

$$\Omega_{gw}(\nu) = \frac{\pi^2}{3} h^2(\nu) \left(\frac{\nu}{\nu_H} \right)^2 . \tag{33}$$

The total Ω_{gw} between frequencies ν_1, ν_2 is given by

$$\Omega_{gw}(\nu_1, \nu_2) = \int_{\nu_1}^{\nu_2} \Omega_{gw}(\nu) \frac{d\nu}{\nu}.$$

One should be weary of the confusing definition

$$\Omega_{gw}(\nu) = \frac{1}{\rho_{crit}} \frac{d\rho_{gw}(\nu)}{d \ln \nu}$$

often floating in the literature. As it stands, this relationship is incorrect. It can be made consistent with the correct definition (32) only if one assumes that what is being differentiated in this formula is not the spectral density $\rho_{gw}(\nu)$, but a logarithmic integral of this quantity in the limits between some fixed ν_1 and a running ν.

To make more precise the expected qualitative graph for $h_{rms}(n)$ in Fig. 4, as well as previous theoretical graphs in Fig. 4 of Ref. [16], we need to use some available observational data. We make the fundamental assumption that the observed CMB anisotropies are indeed caused by cosmological perturbations of quantum-mechanical origin. If so, the contribution of relic gravitational waves to the large-scale anisotropies should be of the same order of magnitude as the contribution of density perturbations (we will show this in more detail below). This allows us to determine the position and the slope of the function $h_{rms}(\nu)$ at frequencies near the Hubble frequency $\nu_H \approx 2 \times 10^{-18}$ Hz and then extrapolate the spectrum to higher frequencies.

We choose $h_{rms}(\nu_H)$ and primordial spectral index n in such a way that the lower-order CMB multipoles produced by relic gravitational waves are at the level of the actually observed values. Then, today's spectra for $h_{rms}(\nu)$ with spectral indeces n $= 1$ ($\beta_i = -2$) and n $= 1.2$ ($\beta_i = -1.9$) are shown in Fig. 5 (for more details, see [12, 17]). At frequencies around ν_H we have $h_{rms}(\nu_H) \approx 10^{-5}$, so that the present-day mean number of quanta N_f exceeds 10^{100}.

When extrapolating the spectrum to higher frequencies we assume that the entire left slope of the barrier in Fig. 3 was formed by a single power-law scale factor with one and the same $\beta_i = \beta$:

$$a(\eta) = l_o |\eta|^{1+\beta}. \tag{34}$$

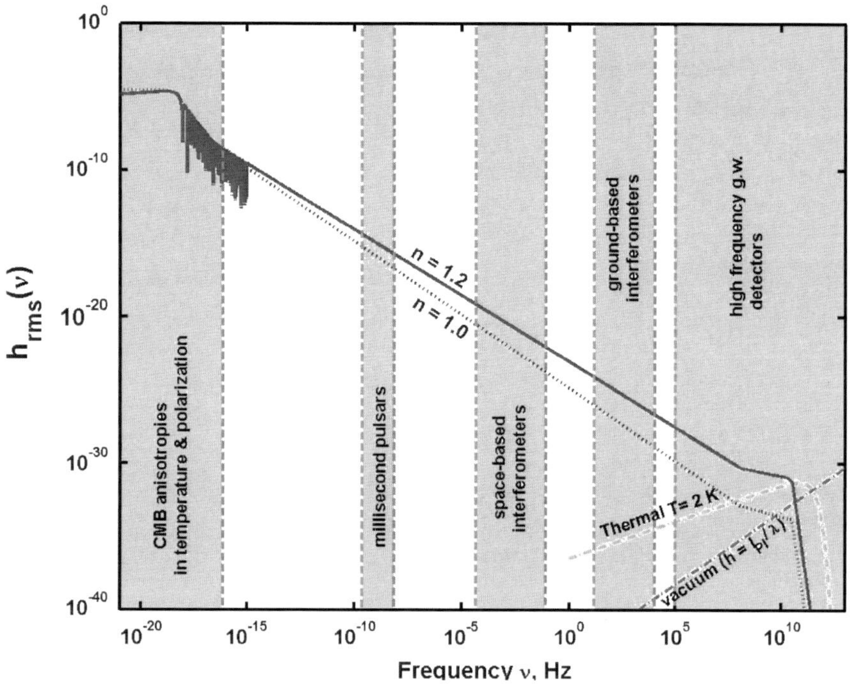

Fig. 5 The present-day spectrum for $h_{rms}(\nu)$. The *solid line* corresponds to the primordial spectral index $\beta = -1.9$, i.e. n = 1.2, while the *dashed line* is for $\beta = -2$, i.e. n = 1

This seems to be a reasonable assumption given the relative featurelessness of the initial stage – the overall energy density was 10 orders of magnitude lower than the Planckian density and was barely changing. Specifically, at the i-stage, we consider two examples: $\beta = -2$ and $\beta = -1.9$. The power-law indeces β_f at the matter-dominated and radiation-dominated stages are well known: $\beta_f = 1$ and $\beta_f = 0$, respectively. The effective pressure p and the energy density ϵ of matter driving the general power-law evolution (34) with a given constant β are related by the effective equation of state $p = [(1-\beta)/3(1+\beta)]\epsilon$.

As for the high-frequency part of the spectrum, it was calculated under the assumption that the z-stage of evolution shown in Fig. 2 was governed by matter with a stiff equation of state $p = \epsilon$ $(\beta_f = -(1/2))$ advocated long ago by Zeldovich. The issue of the back reaction of the created gravitons on the "pump" field $a(\eta)$ becomes important for this sort of values of β_f [2, 18]. Of course, the existence of such an interval in the past evolution of the very early Universe cannot be guaranteed. In any case, the waves with frequencies above 10^{10} Hz have never been in superadiabatic regime, they remain in the vacuum state. The renormalization (subtraction of a "half of the quantum" from each mode) cuts off the spectrum at these high frequencies. At lower frequencies, the renormalization has practically no effect on the spectrum.

The spectra of $\Omega_{gw}(\nu)$, shown in Fig. 6, are derived from $h_{rms}(\nu)$ according to Eq. (33). As was already mentioned, the substantial rise of the spectrum at very high

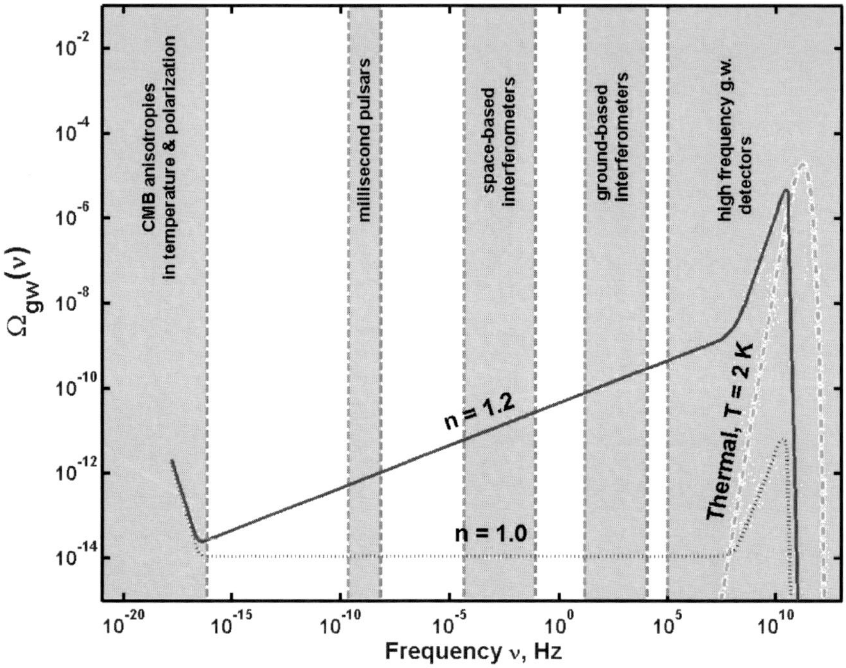

Fig. 6 The present-day spectrum for $\Omega_{gw}(\nu)$. The *solid line* corresponds to the primordial spectral index $\beta = -1.9$, i.e. n = 1.2, while the *dashed line* is for $\beta = -2$, i.e. n = 1

frequencies cannot be guaranteed, and the shown graphs in this area of frequencies should be regarded as the upper allowed limits (especially for the model with the primordial spectral index n = 1.2).

There are two comments to be made about Figs. 5 and 6. First, the phenomenon of squeezing and standing waves production is reflected in the oscillations of the metric power spectrum and Ω-spectrum as functions of frequency ν. The first few cycles of these oscillations are shown in the graph of $h_{rms}(\nu)$. In the CMB sections of the paper, we will show how these oscillations that existed in the recombination era translate into peaks and dips of the angular power spectra for CMB temperature and polarization anisotropies observed today.

Second, since the primordial spectrum of quantum-mechanically generated perturbations has the form (see Eq. (17)):

$$h^2(n) \propto n^{2(\beta+2)}, \tag{35}$$

theoretical considerations suggest that the preferred range of values for the primordial spectral index n is n > 1, i.e. preferred spectra are 'blue'. Indeed, for the opposite range n < 1, the primordial spectra would be 'red', and the integral (28) would be power-law divergent at the lower limit. To avoid the infra-red divergency one would need to make extra assumptions about bending the spectrum down in the region of very small wavenumbers n. In its turn, a primordial spectral index $\mathrm{n} > 1$ requires that the power-law index β of the initial scale factor $a(\eta)$ should be $\beta > -2$. This is because β and n are related by $\mathrm{n} = 2\beta + 5$. The value $\mathrm{n} = 1$ ($\beta = -2$) describes the spectrum known as a flat, or Harrison–Zeldovich–Peebles, or scale-invariant spectrum. It was originally proposed in the context of density perturbations. The spectral index $\mathrm{n} = 1$ marks the beginning of the trouble – the metric variance (28) is logarithmically divergent at the lower limit.

5 Density Perturbations

To discover relic gravitational waves in CMB anisotropies we have to distinguish their effects from other possible sources of CMB anisotropies, and first of all from density perturbations. In comparison with gravitational waves, the possibility of quantum-mechanical generation of density perturbations requires an extra hypothesis, namely, the appropriate parametric coupling of the matter field to external gravity. But we can assume that this hypothesis was satisfied by, for example, a lucky version of the scalar field. Can the amplitudes of quantum-mechanically generated density perturbations be many orders of magnitude larger than the amplitudes of relic gravitational waves? This would make the search for relic gravitational waves practically hopeless. We will show that this can never happen. Density perturbations can compete with relic gravitational waves in producing the large-scale CMB anisotropies, but can never be overwhelming.

We are mostly interested in the gravitational field (metric) sector of density perturbations. Indeed, it is perturbations in the gravitational field that survived numerous transformations of the matter content of the Universe. Primordial matter (for example, scalar field) has decayed, together with its own perturbations, long ago. But it is primordial metric perturbations that have been inherited by density perturbations at radiation-dominated and matter-dominated stages.

As a driving 'engine' for the very early Universe, it is common to consider the so-called minimally-coupled scalar field $\varphi(\eta, \mathbf{x})$, with the energy-momentum tensor

$$T_{\mu\nu} = \varphi_{,\mu}\varphi_{,\nu} - g_{\mu\nu}\left[\frac{1}{2}g^{\alpha\beta}\varphi_{,\alpha}\varphi_{,\beta} + V(\varphi)\right]. \tag{36}$$

The polarization structure (13) of density perturbations allows us to write the $\mathbf{n}$-mode metric perturbation:

$$h_{ij} = h(\eta)Q\delta_{ij} + h_l(\eta)n^{-2}Q_{,ij}, \tag{37}$$

where $Q = e^{\pm i\mathbf{n}\cdot\mathbf{x}}$. Two polarization amplitudes $h(\eta)$ and $h_l(\eta)$ are accompanied by the third unknown function – the amplitude $\varphi_1(\eta)$ of the scalar field perturbation:

$$\varphi(\eta, \mathbf{x}) = \varphi_0(\eta) + \varphi_1(\eta)Q.$$

The perturbed Einstein equations allow us to find all three unknown functions. The important fact is that, for any $V(\varphi)$, there exists only one second-order differential equation, of the same structure as Eq. (1), that needs to be solved [19]:

$$\mu'' + \mu\left[n^2 - \frac{(a\sqrt{\gamma})''}{a\sqrt{\gamma}}\right] = 0. \tag{38}$$

This equation coincides with the gravitational wave equation (1) if one makes there the replacement

$$a(\eta) \to a(\eta)\sqrt{\gamma(\eta)}. \tag{39}$$

We will call function μ satisfying Eq. (38) μ_S, and that satisfying Eq. (1) μ_T.

The function $\gamma(\eta)$ is defined by

$$\gamma(\eta) = 1 + \left(\frac{a}{a'}\right)' = -\frac{c}{a}\frac{H'}{H^2}.$$

For power-law scale factors (34), this function reduces to a set of constants,

$$\gamma = \frac{2+\beta}{1+\beta}.$$

The constant γ degenerates to zero in the limit of the expansion law with $\beta = -2$, that is, in the limit of the gravitational pump field (an interval of deSitter evolution) which is responsible for the generation of cosmological perturbations with flat primordial spectrum $\mathrm{n} = 1$.

In terms of t-time the function γ is

$$\gamma(t) = -\frac{\dot{H}}{H^2}.$$

It is also related to the cosmological deceleration parameter $q(t) = -a\ddot{a}/\dot{a}^2$: $\gamma(t) = 1 + q(t)$. The unperturbed Einstein equations allow us to write

$$\kappa\left(\varphi_0{}'\right)^2 = 2\left(\frac{a'}{a}\right)^2\gamma(\eta), \tag{40}$$

or, equivalently,

$$\frac{\dot{\varphi}_0}{H} = \sqrt{\frac{2}{\kappa}}\sqrt{\gamma(t)}, \tag{41}$$

where $\kappa = 8\pi G/c^4$. The function $\gamma(t)$ is sometimes denoted in the literature by $\epsilon(t)$.

As soon at the appropriate solution to Eq. (38) is found, all unknown functions are easily calculable:

$$h(\eta) = \frac{1}{c}H(\eta)\left[\int_{\eta_0}^{\eta}\mu\sqrt{\gamma}\,d\eta + C_i\right], \tag{42}$$

$$h_l{}'(\eta) = \frac{a}{a'}\left[h'' - \frac{H''}{H'}h' + n^2 h\right], \tag{43}$$

$$\varphi_1(\eta) = \frac{\sqrt{\gamma}}{\sqrt{2\kappa}}\left[\frac{\mu}{a\sqrt{\gamma}} - h\right]. \tag{44}$$

The arbitrary constant C_i in Eq. (42) reflects the remaining coordinate freedom within the class of synchronous coordinate systems (8). Indeed, a small coordinate transformation [19]

$$\overline{\eta} = \eta - \frac{C}{2a}Q \tag{45}$$

generates a 'gauge transformation'

$$\overline{h}(\eta) = h(\eta) + C\frac{a'}{a^2} = h(\eta) + C\frac{H(\eta)}{c}. \tag{46}$$

The term with constant C is automatically present in the properly written general solution (42) for $h(\eta)$.

By combining the transformation (46) with its time-derivative, one can build a quantity which does not contain the constant C and therefore is a 'gauge-invariant' quantity with respect to transformations preserving synchronous coordinates. Specifically, one can check that

$$h - h'\frac{H}{H'} = \overline{h} - \overline{h}'\frac{H}{H'}.$$

Denoting this quantity by ζ, and taking into account Eq. (42), we write

$$\zeta \equiv h - h'\frac{H}{H'} = \frac{\mu_S}{a\sqrt{\gamma}}. \tag{47}$$

In terms of the variable $\zeta(\eta)$ the fundamental equation (38) takes the form

$$\zeta'' + 2\frac{(a\sqrt{\gamma})'}{a\sqrt{\gamma}}\zeta' + n^2\zeta = 0. \tag{48}$$

Not surprisingly, this equation coincides with the gravitational-wave equation (2), if one replaces the gravitational-wave function h with ζ, and a with $a\sqrt{\gamma}$. For density perturbations, the metric variable ζ is the physically relevant quantity that plays the same role as the metric variable h for gravitational-wave perturbations.

The fact that we are working with scalar gravitational waves supported by scalar field fluctuations, instead of normal tensor gravitational waves, has boiled down to the necessity of a single modification: the substitution (39) in the gravitational-wave equations.

It is appropriate to say a few words about cosmological gauge transformations in general. Their origin is related to the notion of Lie transport performed on a manifold covered by some coordinates x^α [20]. Lie transport is being carried out along a given vector field $\xi^\mu(x^\alpha)$. This is a quite formal construction which respects only the transformation properties of fields defined on the manifold. These fields are not required to satisfy any physical equations. An infinitesimal Lie transport changes the field by an amount equal to its Lie derivative. This change can be viewed as a rule by which new values of the field are assigned to the same point x^α. It is only with some reservation that we borrow the name gauge transformation from physical field theories and apply it to this, always valid, mathematical procedure.

In gravitational applications, the vector field $\xi^\mu(x^\alpha)$ is usually associated with an infinitesimal coordinate transformation $\overline{x}^\mu = x^\mu - \xi^\mu(x^\alpha)$ [10]. Then, for example, an infinitesimal increment of the metric tensor $g_{\mu\nu}(x^\alpha)$ is given by the Lie derivative of $g_{\mu\nu}$. It can be written as

$$\delta g_{\mu\nu}(x^\alpha) = \xi_{\mu;\nu} + \xi_{\nu;\mu}.$$

A particular case of these transformations is represented by our Eqs. (45), (46).

Since we operate with only four components of an arbitrary vector ξ^μ and, potentially, with many fields transforming with the same ξ^μ, one can build some combinations of these transformed fields, in which the functions ξ^μ cancel out. Therefore, these combinations become gauge-invariant quantities. An example is given by our Eq. (47).

Certainly, it is an exaggeration to claim that "only gauge-invariant quantities have any inherent physical meaning". It is as if you were told that where specifically you are going in your car has no physical meaning, because the components of velocity are coordinate-dependent, and it is only the readings of your speedometer that have an inherent physical meaning, because they are coordinate-independent. In any case, there exists an infinite number of gauge-invariant quantities. For example, a product of a gauge-invariant quantity with any 'background' function produces a new gauge-invariant quantity. The fact that a quantity is gauge-invariant does not answer the question of its physical interpretation. Nevertheless, it is useful to know gauge-invariant quantities.

Had we started from arbitrary coordinates, the perturbed metric (8) would have contained two more unknown functions, namely $A(\eta)$ and $B(\eta)$, [21]

$$g_{\eta\eta} = -a^2(\eta)(1 + 2A(\eta)Q), \qquad g_{\eta i} = a^2(\eta)B(\eta)\frac{1}{n}Q_{,i}.$$

The η-transformation generalizing our Eq. (45) would read

$$\overline{\eta} = \eta + T(\eta)Q, \tag{49}$$

where $T(\eta)$ is an arbitrary function of η. The perturbed metric components h and A would transform as follows (any possible x^i-transformations do not participate in these relationships):

$$\overline{h} = h - 2\frac{a'}{a}T, \quad \overline{A} = A - T' - \frac{a'}{a}T. \tag{50}$$

The functions T and T' cancel out in the gauge-invariant metric combination

$$\zeta_g = h - \frac{H}{H'}h' - \frac{2A}{\gamma}.$$

Obviously, ζ_g reduces to ζ for transformations (45) preserving synchronous conditions, that is, when $A = \overline{A} = 0$. The previously introduced [22] quantity ζ_{BST}, where *BST* stands for Bardeen, Steinhardt, Turner, can also be reduced, after some work, to our ζ, up to a coefficient $-(1/2)$.

Since we assume that in addition to the tensor metric field $g_{\mu\nu}$, a scalar field

$$\varphi(\eta, \mathbf{x}) = \varphi_0(\eta) + \delta\varphi(\eta)Q$$

is also defined on the manifold, we can write down its gauge transformation under the action of Eq. (49):

$$\overline{\delta\varphi} = \delta\varphi + \varphi_0' T. \tag{51}$$

The arbitrary function $T(\eta)$ cancels out in the gauge-invariant combination

$$V(\eta) = \delta\varphi + \frac{1}{2}\frac{a}{a'}\varphi_0' h. \tag{52}$$

Although the combination (52) is often quoted in the literature, this object is something like a 'half of a horse plus half of a cow'. It combines physically separate quantities – one from metric another from matter – and its physical interpretation is obscure. However, using the unperturbed Einstein equation (40) and solution (44) for $\delta\varphi = \varphi_1$ one can show that V reduces to the gauge-invariant metric perturbation ζ times a background-dependent factor:

$$V = \frac{1}{\sqrt{2\kappa}}\sqrt{\gamma}\zeta.$$

The quantity $\sqrt{2\kappa}aV$ is called Chibisov, Lukash, Mukhanov, Sasaki [23–25] variable u_{CLMS}:

$$u_{CLMS} = a\sqrt{\gamma}\zeta. \tag{53}$$

In our approach, this variable is simply the function μ_S satisfying Eq. (38), modeled on Eq. (1). The factor $\sqrt{\gamma}$ in Eq. (53) will be a matter of great attention when we come to the discussion of quantization procedures.

Returning to the function $\zeta(\eta)$, one can notice that this metric amplitude is practically constant in the regime when n^2 is much smaller than the effective potential $(a\sqrt{\gamma})''/a\sqrt{\gamma}$. This behaviour is similar to the constancy of the gravitational wave amplitude $h = \mu_T/a$ throughout the long-wavelength regime. Indeed, in full analogy with the long-wavelength solution (15), the general solution to Eq. (38) in this regime reads

$$\frac{\mu_S}{a\sqrt{\gamma}} = C_1 + C_2 \int \frac{d\eta}{(a\sqrt{\gamma})^2}.$$

For usually considered expanding cosmologies, the term with constant C_2 is decreasing, and therefore the dominant solution is $h \approx C_1$ for gravitational waves and $\zeta \approx C_1$ for density perturbations. The constancy of ζ allows one to easily estimate the value of metric perturbations at much later times, at the radiation-dominated and matter-dominated stages, as soon as one knows the initial value of ζ.

The constancy of ζ (when one can neglect the term with C_2) is sometimes called a conservation law. This association is incorrect. Genuine conservation laws reflect symmetries of the system, and conserved quantities are constants independently of the initial conditions. For example, the energy of a free oscillator is a constant independently of initial positions and velocities. In our problem, the genuine conservation law for ζ would look like an empty statement $0 = 0$.

6 Quantization of Density Perturbations

A quantum system is defined by its Hamiltonian. Whether or not the Hamiltonian follows from some assumed classical Lagrangian should not be a question of major concern [14]. From this point of view, the quantization of density perturbations, similarly to the quantization of gravitational waves (26), is defined by the Hamiltonian [19]

$$\mathbf{H}(\eta) = n\mathbf{c}^{\dagger}\mathbf{c} + \sigma\mathbf{c}^{\dagger 2} + \sigma^*\mathbf{c}^2. \tag{54}$$

The pair creation of scalar gravitational waves ζ by the external pump field is regulated by the coupling function $\sigma(\eta) = (i/2)[(a\sqrt{\gamma})'/(a\sqrt{\gamma})]$. However, a more traditional approach begins with a Lagrangian, and here too, the Hamiltonian (54) can be derived from quadratic perturbation terms in the total Hilbert–Einstein action, where both gravity and matter (scalar field) are taken into account.

The Lagrangian for an n-mode of density perturbations, after some transformations of the total Hilbert–Einstein quadratic action, can be written in the form [15]

$$L_{dp} = \frac{1}{2n}\left(\frac{a\sqrt{\gamma}}{a_0\sqrt{\gamma_0}}\right)^2 \left[\left(\bar{\zeta}'\right)^2 - n^2\bar{\zeta}^2\right], \tag{55}$$

where

$$\bar{\zeta} = \left(\frac{\hbar}{32\pi^3}\right)^{1/2} \frac{\lambda_0}{l_{Pl}}\zeta, \tag{56}$$

and a_0, γ_0 are values of the functions $a(\eta)$, $\gamma(\eta)$ at $\eta = \eta_0$ where the initial conditions are being set. The Euler–Lagrange equations derivable from this Lagrangian in terms of ζ are Eq. (48), and in terms of μ_S – Eq. (38).

Obviously, the Lagrangian (55) for ζ coincides with the gravitational wave Lagrangian (18) for h after the replacement of the factor a/a_0 with the factor $a\sqrt{\gamma}/a_0\sqrt{\gamma_0}$. Starting from the Lagrangian (55) and building on the canonical pair μ_S, $\partial L_{dp}/\partial\mu_S'$ one can derive the Hamiltonian (54) by doing exactly the same steps that have led us from the Lagrangian (18) to the Hamiltonian (26).

It should be noted that the total classical Lagrangian (55) admits, as always, some freedom of modifications without affecting the Euler–Lagrange equations. In particular, the derived equations of motion (48), (38) will remain exactly the same, if one changes the Lagrangian (55) to a new one by multiplying (55) with a constant, for example with a constant γ_0:

$$
\begin{aligned}
L_{dp(new)} &= \frac{1}{2n}\left(\frac{a\sqrt{\gamma}}{a_0\sqrt{\gamma_0}}\right)^2 \gamma_0 \left[\left(\bar{\zeta}'\right)^2 - n^2\bar{\zeta}^2\right] \\
&= \frac{1}{2n}\left(\frac{a}{a_0}\right)^2 \gamma \left[\left(\bar{\zeta}'\right)^2 - n^2\bar{\zeta}^2\right]. \qquad (57)
\end{aligned}
$$

This new Lagrangian degenerates to zero in the limit $\gamma \to 0$. We will discuss later the subtleties in quantum theory that arise after such a modification of the Lagrangian.

Similarly to gravitational waves, we impose initial conditions in the early regime of a free oscillator, that is, when $n \gg (a\sqrt{\gamma})'/(a\sqrt{\gamma})$. We choose the canonical pair

$$
q = \bar{\zeta}, \qquad p = \frac{\partial L_{dp}}{\partial \bar{\zeta}'} = \frac{1}{n}\left(\frac{a\sqrt{\gamma}}{a_0\sqrt{\gamma_0}}\right)^2 \bar{\zeta}', \qquad (58)
$$

and write the asymptotic expressions for the operators:

$$
\mathbf{q} = \sqrt{\frac{\hbar}{2}}\frac{a_0\sqrt{\gamma_0}}{a\sqrt{\gamma}}\left[\mathbf{c}e^{-in(\eta-\eta_0)} + \mathbf{c}^\dagger e^{in(\eta-\eta_0)}\right], \qquad (59)
$$

$$
\mathbf{p} = i\sqrt{\frac{\hbar}{2}}\frac{a\sqrt{\gamma}}{a_0\sqrt{\gamma_0}}\left[-\mathbf{c}e^{-in(\eta-\eta_0)} + \mathbf{c}^\dagger e^{in(\eta-\eta_0)}\right], \qquad (60)
$$

with

$$
[\mathbf{q}, \mathbf{p}] = i\hbar, \qquad \left[\mathbf{c}, \mathbf{c}^\dagger\right] = 1. \qquad (61)
$$

Obviously, a quantum state satisfying the condition

$$
\mathbf{c}|0\rangle = 0 \qquad (62)
$$

is the ground (vacuum) state of the Hamiltonian (54). Calculating the mean square values of $\mathbf{q}$ and its canonically conjugate momentum $\mathbf{p}$, we find

$$
\langle 0|\mathbf{q}^2|0\rangle = \langle 0|\mathbf{p}^2|0\rangle = \frac{\hbar}{2}, \qquad \Delta\mathbf{q}\Delta\mathbf{p} = \frac{\hbar}{2}.
$$

Returning to ζ from $\bar{\zeta}$ according to Eq. (56), we find

$$
\zeta_{rms} = \left(\langle 0|\zeta^2|0\rangle\right)^{1/2} = \frac{\sqrt{2}(2\pi)^{3/2} l_{Pl}}{\lambda_0}, \qquad (63)
$$

that is, exactly the same value as the initial amplitudes (25) for each of two polarization components of gravitational waves.

Extrapolating the initial time η_0 up to the boundary between the adiabatic and superadiabatic regimes at $\eta = \eta_i$, we derive the evaluation $\zeta_{rms} \sim l_{Pl}/\lambda_i$. This evaluation, plus the constancy of the quantity ζ throughout the long-wavelength regime, is the foundation of the result according to which the final (at the end of the long-wavelength regime) metric amplitudes of gravitational waves and density perturbations should be roughly equal to each other [19].

The primordial ζ-spectrum has the same form as the primordial gravitational-wave spectrum in Eq. (35):

$$\zeta^2(n) \propto n^{2(\beta+2)}. \tag{64}$$

In this approximation, the spectral indeces are equal:

$$n_s - 1 = n_t = 2(\beta + 2) \equiv \mathrm{n} - 1. \tag{65}$$

Scalar fields (36) can support cosmological scale factors with β in the interval $-1 \leq \beta, \beta \leq -2$. The ratio of (35) to (64) is approximately 1 for all spectral indeces near and including $\beta = -2$ ($\mathrm{n} = 1$).

Having strictly defined the dynamical equations and the initial (quantum ground state) conditions, one can calculate from formula (29) the exact power spectrum of metric perturbations associated with density perturbations. In this formula, two polarization components $\overset{s}{h}_n(\eta)$ are now determined by the conventions (13), (37), and the constant $\mathcal{C}$ is $\mathcal{C} = \sqrt{24\pi} l_{Pl}$.

The employed approximations cannot guarantee that in the real Universe the coefficients in Eqs. (35) and (64) should be exactly equal to each other. But there is no reason for them to be different by more than a numerical factor of order 1. Therefore, the lower order CMB multipoles, induced primarily by metric perturbations which are still in the long-wavelength regime, should be approximately at the equal numerical levels for, both, density perturbations and gravitational waves.

As we shall now see the inflationary theory differs from the described quantum theory of density perturbations by many orders of magnitude, let alone numerical coefficients of order 1. The discrepancy becomes infinitely large in the limit of the observationally preferred spectral index $\mathrm{n} = 1$ ($\beta = -2$, $\gamma = 0$).

7 What Inflationary Theory Says About Density Perturbations, and What Should be said About Inflationary Theory

For many years, inflationists keep insisting that in contrast to the generation of gravitational waves [2], which begins with the amplitude $h \approx l_{Pl}/\lambda$ ("half of the quantum in the mode") and finishes with the amplitude $h \approx l_{Pl}/(c/H) \approx H/H_{Pl}$ and power $h^2 \approx (H/H_{Pl})^2$ (see [2] and Sects. 3, 4), the generation of primordial density perturbations (i.e. scalar metric perturbations) is more efficient by many orders of magnitude. Contrary to the calculations reviewed in Sec. 6, inflationary theory claims that in a cosmological model with the same H, the resulting amplitude and power of the quantity ζ should contain a huge extra factor, tending to infinity in the limit of the standard de Sitter inflation.

Traditionally, the claimed inflationary derivation of density perturbations goes along the following lines. One starts from the ground state quantum fluctuations taken from a different theory, namely from a theory of a free test scalar field, where metric perturbations are ignored altogether, and writes

$$\delta\varphi \approx \frac{H}{2\pi}.$$

Then, from the gauge transformation (51), assuming that the l.h.s. is equal to zero, one finds the characteristic time interval δt "to the end of inflation" and puts into δt the estimate of $\delta\varphi$ from the above-mentioned 'quantum' evaluation:

$$\delta t \approx \frac{\delta\varphi}{\dot{\varphi}_0} \approx \frac{H}{2\pi\dot{\varphi}_0}.$$

Then, one declares that the dimensionless ratio $\delta t/H^{-1}$ is what determines the dimensionless metric and density variation amplitudes in the post-inflationary universe. And this is being presented as the "famous result" of inflationary theory:

$$\delta_H \approx \frac{\delta\rho}{\rho} \approx \frac{H^2}{2\pi\dot{\varphi}_0} = \frac{H/2\pi}{\dot{\varphi}_0/H}.$$

This "inflationary mechanism" of generation of density perturbations is claimed to have been confirmed in numerous papers and books[1]. The most dramatic feature of the claimed inflationary result is the factor $\dot{\varphi}_0/H$ in the denominator of the final expression. I call it a 'zero in the denominator' factor [15]. According to Eq. (41) this factor is $\sqrt{\gamma}$; in the literature, this factor appears in several equivalent incarnations including such combinations as $V_{,\varphi}/V$, $1 + p/\rho c^2 \equiv 1 + w$, ϵ, where $\epsilon \equiv \gamma \equiv -\dot{H}/H^2$, and so on.

The early Universe Hubble parameter H featuring in the numerator of the final expression cannot be too small because $H(t)$ in scalar field driven cosmologies is a decreasing, or at most constant, function of time. The parameter H in the very early Universe cannot be smaller than, say, the H in the era of primordial nucleosynthesis. So, the numerator of the "famous result" cannot be zero, but the denominator can, at $\gamma = 0$. Therefore, the "famous result" prescribes the arbitrarily large numerical values to the amplitudes of density perturbations in the limit of the de Sitter evolution $\gamma = 0$ ($\beta = -2$). One has to be reminded that it is this gravitational pump field that generates gravitational waves and other perturbations with primordial spectrum n = 1, advocated long ago on theoretical grounds by Harrison, Zeldovich, Peebles,

[1] All inflationary quotations in the present article are taken from real publications. However, I do not see much sense in giving precise references. I am confronting here something like a culture, distributed over hundreds of publications, rather than a persistent confusion of a few authors. This situation was once qualified [26] as "the controversy between Grishchuk and the rest of the community".

and which is in the vicinity of the primordial spectral shape currently believed to be preferred observationally.

The inflation theory claims to have predicted a "nearly" scale-invariant spectrum of density perturbations because on the strictly scale-invariant spectrum the predicted amplitudes blow up to infinity.

More recent literature operates with the 'curvature perturbation $\mathcal{R}$', equivalent to our ζ from Eq. (47). A typical quotation states: "The amplitude of the resulting scalar curvature perturbation is given by

$$\langle \mathcal{R}^2 \rangle^{1/2} = \left(\frac{H}{\dot{\phi}} \right) \langle \delta\phi^2 \rangle^{1/2},$$

where H is the Hubble parameter, $\dot{\phi}$ is the time derivative of the inflation ϕ, and $\delta\phi$ is the inflation fluctuation on a spatially flat hypersurface. The quantum expectation value of the inflaton fluctuations on super-horizon scales in the de Sitter space–time is

$$\langle \delta\phi^2 \rangle = \left(\frac{H}{2\pi} \right)^2 ."$$

Here again one is invited to believe that the ground state quantum fluctuations in one theory (without metric perturbations) are responsible for the appearance of arbitrarily large (factor γ in the denominator of the $\mathcal{R}$ power spectrum) gravitational field fluctuations in another theory. Inflationists are keen to make statements about the resulting gauge-invariant curvature perturbations without having included metric perturbations in the initial conditions. The infinitely large curvature perturbation $\mathcal{R}$ is supposed to occur at $\gamma = 0$, that is, exactly in the de Sitter model from which all the 'quantum' reasoning about scalar field fluctuations has started.

According to inflationary views on the generating process, the amount of the created scalar particles-perturbations is regulated not by the strength of the external gravitational 'pump' field (basically, space–time curvature in the early Universe) but by the closeness of the metric to a de Sitter one. In a sequence of space–times with very modest and approximately equal values of H you are supposed to be capable of generating arbitrarily large amplitudes of the scalar metric perturbation by simply going to smaller and smaller values of $\dot{H}$.

Since the quantity $\mathcal{R}$ (as well as the quantity ζ and the gravitational wave function h) oscillates with a slowly decreasing amplitude in the initial short-wavelength regime (see Eq. (48)) and remains constant during the subsequent long-wavelength regime, the arbitrarily large amplitude of the resulting "inflation-predicted" scalar curvature perturbation $\mathcal{R}$ must have been implanted from the very beginning, i.e. from the times in the high-frequency regime of evolution, when the initial quantum state for density perturbations was defined. It is important to remember that for many years inflationists claimed that the reason for the huge difference between the resulting scalar and tensor perturbations was the "big amplification during reheating" experienced by the long-wavelength scalar metric perturbations. These days, the explanation via the "big amplification during reheating" is not even mentioned. These days, the most sophisticated inflationary texts put forward, as the foundation for their belief in arbitrarily large resulting scalar metric perturbations $\mathcal{R}$ and ζ, the "Bunch–Davies vacuum", i.e. a concept from the theory of a test scalar field, where the gravitational field (metric) perturbations are absent altogether. (Moreover, the Bunch–Davies vacuum was originally introduced as a de Sitter invariant state prohibiting the particle production by definition). The absurd proposition of inflationists with regard to the density perturbations is sometimes called a "classic result", and it is widely used for derivation of further incorrect conclusions in theory and data analysis.

The most common, and most damaging, inflationary claim is that the amplitudes of relic gravitational waves must be "suppressed", "sub-dominant", "negligibly small" in comparison with primordial density perturbations, especially for models with $\gamma \to 0$. Having arrived at a divergent formula for density perturbations, i.e. with a 'zero in the denominator' factor, inflationists compose the ratio of the gravitational-wave power spectrum $\mathcal{P}_T$ to the derived divergent scalar metric power spectrum $\mathcal{P}_\zeta$ (the so-called 'tensor-to-scalar ratio' $r \equiv \mathcal{P}_T/\mathcal{P}_\zeta$) and write it as

$$r = 16\epsilon = -8n_t. \tag{66}$$

The inflationary theory binds the amplitude of scalar perturbations with the spectral index, and makes the absurd prediction of arbitrarily large amplitudes of density perturbations in the limit of models with $\epsilon \equiv \gamma = 0$ ($n_s = 1$, $n_t = 0$). But the inflationary "consistency relation" (66) encourages and misleads one to believe that everything is perfect – as if it were the amount of gravitational waves that must go to zero in this limit. To make the wrong theory look acceptable, inflationary model builders keep $\mathcal{P}_\zeta$ fixed at the observationally required level of 10^{-9} or so, and move H/H_{Pl} down whenever ϵ goes to zero in the inflationary divergent formula

$$\mathcal{P}_\zeta \approx \frac{1}{\epsilon}\left(\frac{H}{H_{Pl}}\right)^2,$$

thus making the amount of relic gravitational waves arbitrarily small.

It is instructive to see how the problem of initial conditions is dealt with by S. Weinberg [27]. The author operates with equations for ζ (equivalent to our Eq. (48)), gravitational waves (equivalent to our Eq. (2) for h), and a test scalar field σ (which is known [2] to be similar to the equation for gravitational waves). When it comes to the initial conditions, the author says that they are "designed to make" $\zeta\dot{\varphi}_0/H$, h, and σ behave like conventionally normalized free fields in the remote past. In other words, the initial conditions for h and σ are not "designed to make" a potentially vanishing factor to enter the normalization, but the initial conditions for ζ are. The normalization of ζ becomes now proportional to $1/\sqrt{\epsilon}$, and the power spectrum $\mathcal{P}_\zeta$ of ζ acquires the factor ϵ in the denominator. The author calls this divergent power spectrum of scalar metric perturbations a "classic" result.

One can imagine why this divergent formula is called "classic". Something repeated in the literature so many times could become "classic" more or less automatically, regardless of its true value. However, Weinberg has not explicitly stated that the "classic" result is a correct result. On the contrary, the recent paper [28], which deals with gravitational waves, makes assumptions diametrically opposite to the prescriptions of the "classic" result (but the paper does not explicitly say that the "classic" result is an incorrect result). That paper chooses for the analysis a model with $n_t = 0$ and 'tensor/scalar ratio' $r = 1$. This choice is in agreement with conclusions of the quantum theory that I advocated and reviewed in Section 6, but it is in conflict with what is demanded by inflationary Eq. (66) (quoted also in [29]). Indeed, the "classic" result for the case $n_t = 0$ implies the non-existence of the very subject of discussion, namely, relic gravitational waves. Although the choice $n_t = 0$, $r = 1$ has been made [28] purely for numerical convenience, it seems to me that the cautious formulation "designed to make" [27], together with the earlier [29] and more recent [28] treatments, testify to a certain evolution of views on the subject of initial conditions.

It appears that some attempts of technical derivation of the inflationary ζ-normalization suffer from serious inaccuracies in dealing with quantum operators and quantum states. Certainly, it is incorrect to think that by "demanding that $a^\dagger$ and a obey the standard creation and annihilation commutation relations we get a normalization condition for ζ". To put it in the context of a medical analogy: the rules for a surgical operation do not identify the patient on whom you want to operate. Let us discuss this point in more detail.

Suppose, being (mis)guided by various inflationary prejudices, you decided to write, instead of Eqs. (59), (60), the following asymptotic expressions for the operators $\mathbf{q}$, $\mathbf{p}$:

$$\mathbf{q} = \sqrt{\frac{\hbar}{2}\frac{a_0}{a}}\frac{1}{\sqrt{\gamma}}\left[\mathbf{b}e^{-in(\eta-\eta_0)} + \mathbf{b}^\dagger e^{in(\eta-\eta_0)}\right], \tag{67}$$

$$\mathbf{p} = i\sqrt{\frac{\hbar}{2}\frac{a}{a_0}}\sqrt{\gamma}\left[-\mathbf{b}e^{-in(\eta-\eta_0)} + \mathbf{b}^\dagger e^{in(\eta-\eta_0)}\right]. \tag{68}$$

The commutation relationships $[\mathbf{q}, \mathbf{p}] = i\hbar$ dictate exactly the same commutation relationships for the operators $\mathbf{b}^\dagger$, $\mathbf{b}$ as they did for the operators $\mathbf{c}^\dagger$, $\mathbf{c}$. Namely, $\left[\mathbf{b}, \mathbf{b}^\dagger\right] = 1$.

The commutation relationships for the annihilation and creation operators are exactly the same, but the quantum state $|0_s\rangle$ annihilated by $\mathbf{b}$,

$$\mathbf{b}|0_s\rangle = 0,$$

is totally different, it is not the ground state of the Hamiltonian (54). Calculation of the mean square value of the variable $\bar{\zeta}$ and its canonically-conjugate momentum gives at $\eta = \eta_0$:

$$\langle 0_s|\mathbf{q}^2|0_s\rangle = \frac{\hbar}{2}\frac{1}{\gamma_0}, \quad \langle 0_s|\mathbf{p}^2|0_s\rangle = \frac{\hbar}{2}\gamma_0,$$

and the factor $\sqrt{\gamma_0}$ cancels out in the uncertainty relation:

$$\Delta\mathbf{q}\Delta\mathbf{p} = \frac{\hbar}{2}.$$

The initial *rms* value of ζ is proportional to $1/\sqrt{\gamma_0}$ and therefore contains the 'zero in the denominator' factor, but this happens only because the quantum state $|0_s\rangle$ is an excited (multi-quantum) squeezed vacuum state (for more details, see [15] and references there). The choice of this state as an initial state for ζ-perturbations would make them arbitrarily large, in the limit of $\gamma_0 \to 0$, right from the very beginning, i.e. from the time of imposing the initial conditions at $\eta = \eta_0$. The multi-quantum state $|0_s\rangle$ is not a choice of the initial state that is regarded physically motivated.

At the level of classical equations, inflationists have reacted to the visual analogy between Eqs. (1), (38), rather than, say, to the visual analogy between Eqs. (2), (48). They have modeled initial conditions for the function μ_S on the initial conditions for the function μ_T, instead of modeling initial conditions for ζ on the initial conditions for the g.w. function h. The inflationary initial conditions are usually written in the form

$$\mu_{(T,S)} = \frac{1}{\sqrt{2k}} e^{-ik\eta} \quad \text{for } k|\eta| \gg 1.$$

These initial conditions are correct for gravitational waves, but are incorrect for density perturbations. As is seen from Eq. (47), these initial conditions for μ_S would require the gauge-invariant metric perturbation ζ (as well as curvature perturbation $\mathcal{R}$) to be arbitrarily large, in the limit of models with $\gamma \to 0$, right from the very beginning, i.e. from the time of imposing initial conditions in the short-wavelength regime. In other words, the inflationary initial conditions are "designed to make" the quantity ζ to be divergent as $1/\sqrt{\gamma}$ from the start. These incorrect initial conditions are used in all inflationary calculations and conclusions, including the latest claims of inflationists on what the observations "really" tell them about inflation.

Finally, let us assume that the 'correct' Lagrangian is given by Eq. (57). (Surely, you may assume correct whichever Lagrangian you wish, because it is taxpayers [30] who will be paying the price at the end of the day.) In contrast to the Lagrangians (18), (55), this new Lagrangian vanishes in the most interesting limit of models with $\gamma \to 0$.

For this new Lagrangian, the canonical quantization would require us to write (instead of Eqs. (59), (60)):

$$\mathbf{q} = \sqrt{\frac{\hbar}{2}} \frac{a_0\sqrt{\gamma_0}}{a\sqrt{\gamma}} \frac{1}{\sqrt{\gamma_0}} \left[\mathbf{d}e^{-in(\eta-\eta_0)} + \mathbf{d}^\dagger e^{in(\eta-\eta_0)}\right], \tag{69}$$

$$\mathbf{p} = i\sqrt{\frac{\hbar}{2}} \frac{a\sqrt{\gamma}}{a_0\sqrt{\gamma_0}} \sqrt{\gamma_0} \left[-\mathbf{d}e^{-in(\eta-\eta_0)} + \mathbf{d}^\dagger e^{in(\eta-\eta_0)}\right], \tag{70}$$

where $[\mathbf{q}, \mathbf{p}] = i\hbar$ and $\left[\mathbf{d}, \mathbf{d}^\dagger\right] = 1$.

The ground state $|0_{new}\rangle$ of the new Hamiltonian associated with this new Lagrangian obeys the condition

$$\mathbf{d}|0_{new}\rangle = 0.$$

The mean square value of the variable $\bar{\zeta}$ in this state at $\eta = \eta_0$ is given by

$$\langle 0_{new}|\mathbf{q}^2|0_{new}\rangle = \frac{\hbar}{2}\frac{1}{\gamma_0}.$$

Technically speaking, one could argue that although the initial *rms* value of ζ is divergent as $1/\sqrt{\gamma_0}$, the divergency takes place for the ground state (of this new Hamiltonian), not for an excited state (of this new Hamiltonian). But this technical subtlety requires a vanishing Lagrangian, and in any case it does not change much from the physical point of view.

Indeed, independently of technical arguments, it is important to realize that a proposal for a divergent scalar metric power spectrum in the limit of n = 1, whatever the reasons for this proposal might be, is in conflict with available observations. The currently derived 'best fit' value for n is slightly lower than 1, but the data allow n = 1, even if with a smaller likelihood. This means that for the tested spectral indeces in the vicinity and including n = 1 the data are consistent with finite and small amplitudes at the level of the best fit amplitude (as implied by the quantum theory discussed in Section 6). But there is absolutely nothing in the data that would suggest the need for a catastrophic growth of the amplitude (demanded by inflationary theory) when the data are fitted against spectra with indeces approaching and crossing n = 1. (Surely, the same comparison of inflationary predictions with available observations is addressed by inflationists as "one of the most remarkable successes… confirmed by observations".) What would we conclude about the assumed Lagrangian (57) whose associated Hamiltonian leads to predictions contradicting observations? We "would conclude that it was the wrong Lagrangian" [14].

The inflationary 'zero in the denominator' factor (if you decided to commit suicide and include it in the scalar metric power spectrum) should be taken at the moment of time, for a given mode n, when the mode begins its superadiabatic evolution. In general, this factor is n-dependent, and hence it affects not only the overall normalization of the spectrum, but also its shape. It enables one to 'generate' a flat, or even a blue, power spectrum of scalar perturbations by gravitational pump fields which in reality can never do this. It enables one to derive all sorts of wrong conclusions in various subjects of study, ranging from the formation of primordial black holes and up to perturbations in cyclic and brane-world cosmologies. In particular, recent claims stating that a given ultra-modern theory predicts an "extremely small $r \lesssim 10^{-24}$" or such an r that "a tensor component… is far below the detection limit of any future experiment", mean only that the predictions were based on the incorrect (inflationary) formula for scalar metric perturbations, with the 'zero in the denominator' factor. Whatever the incorrectly derived r may be, small or large, the use of inflationary Eq. (66) in data analysis (what, unfortunately, is regularly being done in the CMB data analysis [31–33]) can only spoil the extraction of physical information on relic gravitational waves from the data.

The self-contradictory nature of conclusions based on the inflationary formula (66) is dramatically illustrated by the claimed derivation of limits on the amount of gravitational waves from the WMAP data [34]. The authors use Eq. (66)

and explicitly quote, in the form of their Eqs. (17) and (18), the power-spectrum amplitude $\Delta^2_{\mathcal{R}}$ of inflationary scalar perturbations and the 'tensor-to-scalar' ratio r:

$$\Delta^2_{\mathcal{R}} = \frac{V/M^4_{Pl}}{24\pi^2\epsilon}, \qquad r = 16\epsilon. \qquad \text{(I)}$$

The conclusion of this (and many other subsequent papers) is such that according to the likelihood analysis of the data the value of r must be small or zero. In other words, the maximum of the likelihood function is either at $r = 0$ or, in any case, the value of the observed r is perfectly well "consistent with zero". Comparing this conclusion with the inflationary formulas (I), quoted in the same paper, one has to decide whether this conclusion means that the WMAP data are also perfectly well consistent with arbitrarily large scalar amplitudes $\Delta^2_{\mathcal{R}}$, and hence with arbitrarily large CMB anisotropies caused by density perturbations, or that the cited inflationary formulas are wrong and have been rejected by observations (see also a discussion below, in Section 12). The CMB data and the inflationary theory of density perturbations are in deep conflict with each other for long time, but inflationists and their followers keep claiming that they are in "almost perfect agreement".

It seems to me that the situation in this area of physics and cosmology remains unhealthy for more than 25+ years. (This is my mini-version of 'an obligation to inform the public' – from [35].) It appears that more than 2+ generations of researchers have been 'successfully' misled by the "standard inflationary results". One can only hope that the present generation of young researchers will be smarter and more insightful.

8 Why Relic Gravitational Waves should be Detectable

The detailed analysis in previous sections is crucial for proper understanding of the very status of relic gravitational waves. Are we undertaking difficult investigations because we want to find an optional "bonus", "smoking gun", "limitation" on dubious theories, or we search for something fundamental that 'must be there', and at a measurable level?

Strictly speaking, it is still possible that the observed CMB anisotropies have nothing to do with cosmological perturbations of quantum-mechanical origin, that is, with the superadiabatic evolution of the ground state of quantized perturbations. The first worry is that, even if the superadiabatic (parametric) mechanism is correct by itself, the 'engine' that drove the cosmological scale factor was unfortunate and the pump field was too weak. In this case, a relic gravitational-wave signal (as well as scalar perturbations of quantum-mechanical origin) could be too small for discovery. I think this possibility is unlikely. First, it is difficult to imagine an equally unavoidable mechanism – basic laws of quantum mechanics and general relativity – for the generation of the presently existing, as well as the processed in the past, long-wavelength cosmological perturbations. Second, the inevitable quantum-mechanical squeezing and the standing-wave character of the generated perturbations, which, among other things, should have resulted in the scalar metric power spectrum oscillations (see Section 4), seems to have already revealed itself [36] in the observed CMB power spectrum oscillations. (These oscillations are often being associated – in my opinion, incorrectly – with baryonic acoustic waves and are called "acoustic" peaks).

A second worry is that we may be wrong in extrapolating the laws of general relativity and quantum mechanics to extreme conditions of the very early Universe. Although we apply these laws in environments that are still far away from any Planckian or 'trans-Planckian' ambiguities, it is nevertheless an extremely early and unfamiliar Universe (which we want to explore). In principle, it is possible that something has intervened and invalidated the equations and rules that we have used for derivation of relic gravitational waves. This would also invalidate the equations and rules used in the derivation of density perturbations, but the generation of scalar perturbations requires an extra hypothesis in any case. In other words, even if the driving cosmological 'engine' was right, the employed quantum theory of arising perturbations could be wrong. Hopefully, this complication also did not take place.

Assuming that the dangers did not materialize, we come to the conclusion that we have done our job properly. The theoretical calculations were adequate, the normalization of h_{rms} on the CMB lowest-order multipoles was justified, and therefore the relic gravitational wave background must exist, and probably at the level somewhere between the dotted and solid lines in Fig. 5. The other side of the same argument is that if we do not detect relic gravitational waves at this level, something really nasty, like the above-mentioned dangers, had indeed happened. From not seeing relic gravitational waves, we would at least learn something striking about the limits of applicability of our main theories.

It is important to stress again that what is called here the relic gravitational waves is not the same thing which is sometimes called the inflationary gravitational waves. As the name suggests, statements about the inflationary gravitational waves are based on the inflation theory, as applied to density perturbations and gravitational waves. Although, conceptually, the inflationary derivation of density perturbations is an attempt of using the mechanism of superadiabatic (parametric) amplification (see Section 1), the actual implementation of this approach has led inflationists to the divergency in the scalar metric spectrum (see Section 7), which has been converted, for the purpose of "consistency", into a statement about small r. In particular, for the "standard inflation" ($\gamma \equiv \epsilon = 0$), the inflationary theory predicts (see Eq. (66)) the "smoking gun of inflation" in the form of a zero amount of inflationary gravitational waves ($r = 0$). This is in sharp contrast with the superadibatic (parametric) prediction of a finite and considerable amount of relic gravitational waves, as shown by a dotted line in Fig. 5.

Some recent papers, still based on the incorrect theory with Eq. (66), start making additional artificial assumptions about the "natural" inflationary conditions, which amounts to postulating that ϵ should not be too small, but should, instead, be at the level of, say, 0.02. This kind of argument gives some authors "more reason for optimism" with regard to the detection of primordial gravitational waves. Surely, an incorrect theory can lead to predictions not very much different from predictions of a correct theory in some narrow range of parameters that are supposed to be deduced from observations. One should remember, however, that theories are tested not only by what they predict but also by what they do not predict. Any observation consistent with the parameter value other than you postulated is against your theory.

Since the spectrum of relic gravitational-wave amplitudes is decreasing towards the higher frequencies, see Fig. 5, it is the lowest frequencies, or better to say the longest wavelengths, that provide the most of opportunities for the (indirect) detection. It is known for long time [37–40] that gravitational waves affect the CMB temperature and polarization. The low-order CMB multipoles ($\ell \lesssim 100$) are mostly induced by cosmological perturbations with wavelengths ranging from 10 times longer and up to 10 times shorter than the present-day Hubble radius l_H. And this is the range of scales that will be in the center of our further analysis.

9 Intensity and Polarization of the CMB Radiation

A radiation field, in our case CMB, is usually characterized by four Stokes parameters (I, Q, U, V) [10,41]. I is the total intensity of radiation (or its temperature T), Q and U describe the magnitude and direction of linear polarization, and V is the circular polarization. The Stokes parameters of the radiation field arriving from a particular direction in the sky are functions of a point θ, ϕ on a unit sphere centered on the observer. The metric tensor g_{ab} on the sphere can be written as:

$$d\sigma^2 = g_{ab}dx^a dx^b = d\theta^2 + \sin^2\theta d\phi^2. \tag{71}$$

The Stokes parameters are also functions of the frequency of radiation ν, but angular dependence is more important for our present discussion.

The Stokes parameters are components of the polarization tensor P_{ab} [10] which can be written

$$P_{ab}(\theta, \phi) = \frac{1}{2}\begin{pmatrix} I + Q & -(U - iV)\sin\theta \\ -(U + iV)\sin\theta & (I - Q)\sin^2\theta \end{pmatrix}. \tag{72}$$

As every tensor, the polarization tensor P_{ab} transforms under arbitrary coordinate transformations on the sphere, but some quantities remain invariant.

We can build invariants, linear in P_{ab} and its derivatives, with the help of the tensor g_{ab} and the unit antisymmetric pseudo-tensor ϵ^{ab}. First two invariants are easy to build:

$$I(\theta, \phi) = g^{ab}(\theta, \phi)P_{ab}(\theta, \phi), \quad V(\theta, \phi) = i\epsilon^{ab}(\theta, \phi)P_{ab}(\theta, \phi). \tag{73}$$

Two other invariants involve second covariant derivatives of P_{ab} [17]:

$$E(\theta, \phi) = -2\left(P_{ab}(\theta, \phi)\right)^{;a;b}, \quad B(\theta, \phi) = -2\left(P_{ab}(\theta, \phi)\right)^{;b;d}\epsilon^{a}{}_{d}, \tag{74}$$

Being invariant, quantities E, B do not mix with each other. To calculate E and B in a given point θ, ϕ we do not need to know the polarization pattern all over the sky, but we do need to know the derivatives of P_{ab} at that point. Whatever the numerical values of E or B in a given point are, E and B will retain these values under arbitrary coordinate transformations (smoothly reducible to an ordinary rotation). The invariant B is built with the help of a pseudo-tensor ϵ_{ab}, and therefore B is a pseudo-scalar rather than an ordinary scalar. While E does not change sign under a coordinate reflection, B does. With B one can associate the notion of chirality, or handedness (compare with polarization tensors (12), (13)). Clearly, if given cosmological perturbations are such that they themselves are incapable of supporting the handedness, it will not arise in the CMB polarization which these perturbations are responsible for.

The invariant quantities I, V, E, B can be expanded over ordinary spherical harmonics $Y_{\ell m}(\theta, \phi)$, $Y^*_{\ell m} = (-1)^m Y_{\ell, -m}$:

$$I(\theta, \phi) = \sum_{\ell=0}^{\infty} \sum_{m=-\ell}^{\ell} a^T_{\ell m} Y_{\ell m}(\theta, \phi), \tag{75a}$$

$$V(\theta, \phi) = \sum_{\ell=0}^{\infty} \sum_{m=-\ell}^{\ell} a^V_{\ell m} Y_{\ell m}(\theta, \phi), \tag{75b}$$

$$E(\theta, \phi) = \sum_{\ell=2}^{\infty} \sum_{m=-\ell}^{\ell} \left[\frac{(\ell+2)!}{(\ell-2)!}\right]^{\frac{1}{2}} a^E_{\ell m} Y_{\ell m}(\theta, \phi), \tag{75c}$$

$$B(\theta, \phi) = \sum_{\ell=2}^{\infty} \sum_{m=-\ell}^{\ell} \left[\frac{(\ell+2)!}{(\ell-2)!}\right]^{\frac{1}{2}} a^B_{\ell m} Y_{\ell m}(\theta, \phi). \tag{75d}$$

The set of (complex) multipole coefficients $(a^T_{\ell m}, a^V_{\ell m}, a^E_{\ell m}, a^B_{\ell m})$ completely characterizes the intensity and polarization of the CMB.

The same multipole coefficients participate in the expansion of the tensor P_{ab} itself, not only in the expansion of its invariants, but the expansion of P_{ab} requires the use of generalized spherical functions, the so-called spin-weighted or tensor spherical harmonics [42, 43]. For example, the tensor P_{ab} can be written as

$$P_{ab} = \frac{1}{2} \sum_{\ell=0}^{\infty} \sum_{m=-\ell}^{\ell} \left(g_{ab} a^T_{\ell m} - i \epsilon_{ab} a^V_{\ell m} \right) Y_{\ell m}(\theta, \phi)$$
$$+ \frac{1}{\sqrt{2}} \sum_{\ell=2}^{\infty} \sum_{m=-l}^{l} \left(-a^E_{\ell m} Y^G_{(\ell m)ab}(\theta, \phi) + a^B_{\ell m} Y^C_{(\ell m)ab}(\theta, \phi) \right),$$

where $Y^G_{(\ell m)ab}(\theta, \phi)$ and $Y^C_{(\ell m)ab}(\theta, \phi)$ are the 'gradient' and 'curl' tensor spherical harmonics [43].

In what follows, we will not be considering the circular polarization V, and we will sometimes denote the multipole coefficients collectively by $a^X_{\ell m}$, where $X = I, E, B$. These coefficients are to be found from solutions to the radiative transfer equations in a slightly perturbed universe.

10 Radiative Transfer in a Perturbed Universe

Polarization of CMB arises as a result of Thompson scattering of the initially unpolarized light on free electrons residing in a slightly perturbed universe, Eq. (8). Following [40, 41, 44], it is convenient to describe Stokes parameters in terms of a 3-component symbolic vector $\hat{\mathbf{n}}$:

$$\hat{\mathbf{n}} = \begin{pmatrix} \hat{n}_1 \\ \hat{n}_2 \\ \hat{n}_3 \end{pmatrix} = \frac{c^2}{4\pi\hbar\nu^3} \begin{pmatrix} I + Q \\ I - Q \\ -2U \end{pmatrix}. \tag{76}$$

In the zero-order approximation, we assume that all $h_{ij} = 0$ and the CMB radiation field is fully homogeneous, isotropic, and unpolarized. Then,

$$\hat{\mathbf{n}}^{(0)} = n_0(\nu a(\eta))\hat{\mathbf{u}}, \tag{77}$$

where

$$\hat{\mathbf{u}} = \begin{pmatrix} 1 \\ 1 \\ 0 \end{pmatrix}.$$

In the presence of metric perturbations, we write

$$\hat{\mathbf{n}} = \hat{\mathbf{n}}^{(0)} + \hat{\mathbf{n}}^{(1)}, \tag{78}$$

where $\hat{\mathbf{n}}^{(1)}$ is the first order correction. The functions $\hat{\mathbf{n}}^{(1)}$ depend on $(\eta, x^i, \tilde{\nu}, e^i)$, where $\tilde{\nu} = \nu a(\eta)$ and e^i is a unit spatial vector along the photon's path. Our final goal is to predict, with as much completeness as possible, the values of the Stokes parameters at the time of observation $\eta = \eta_R$.

The functions $\hat{\mathbf{n}}^{(1)}$ satisfy the linear version of the radiation transfer equations. These equations can be written [17]

$$\left[\frac{\partial}{\partial\eta} + q(\eta) + e^i\frac{\partial}{\partial x^i}\right]\hat{\mathbf{n}}^{(1)}(\eta, x^i, \tilde{\nu}, e^i)$$
$$= \frac{f(\tilde{\nu})n_0(\tilde{\nu})}{2}e^i e^j \frac{\partial h_{ij}}{\partial\eta}\hat{\mathbf{u}} + q(\eta)\frac{1}{4\pi}\int d\Omega'\, \hat{\mathbf{P}}(e^i; e'^j)\hat{\mathbf{n}}^{(1)}(\eta, x^i, \tilde{\nu}, e'^j). \tag{79}$$

The astrophysical inputs from unpolarized radiation and free electrons are described, respectively, by $f(\tilde{\nu})n_0(\tilde{\nu})$ and $q(\eta)$, while the scattering process is described by the Chandrasekhar matrix $\hat{\mathbf{P}}(e^i; e'^j)$. The input from the gravitational field (metric) perturbations is given by the combination $e^i e^j \partial h_{ij}/\partial\eta$. Certainly, when all $h_{ij} = 0$, all $\hat{\mathbf{n}}^{(1)}$ vanish if they were not present initially, which we always assume.

The combination $e^i e^j \partial h_{ij}/\partial\eta$ gives rise to disparate angular structures for gravitational waves and density perturbations. Let us consider a particular Fourier mode with the wavevector $\mathbf{n} = (0, 0, n)$. The polarization tensors (12) of gravitational waves generate the structure [17]

$$e^i e^j \overset{s}{p}_{ij}(\mathbf{n}) = (1 - \mu^2)e^{\pm 2i\phi}, \tag{80}$$

where $\mu = \cos\theta$ and the $\pm$ signs correspond to the left $s = 1 = L$ and right $s = 2 = R$ polarization states, respectively. Solving Eq. (79) for $\hat{\mathbf{n}}^{(1)}$ in terms of a series over $e^{im\phi}$, one finds that the terms proportional to $e^{\pm 2i\phi}$ are retained while other components $e^{im\phi}$ are not arising. This is because the Chandrasekhar matrix does not create any new $m\phi$ dependence.

In contrast to gravitational waves, the same metric combination with polarization tensors (13) of density perturbations produces the structure which is ϕ-independent. Although in the case of density perturbations, Eqs. (79) contain one extra term, proportional to the electron fluid velocity, $e^i v_i = -i\mu v_b$ (expressible in terms of metric perturbations via perturbed Einstein equations), this term is also ϕ-independent. Since the invariant B depends on the derivative of $\hat{\mathbf{n}}^{(1)}$ over ϕ, one arrives at the conclusion that $B = 0$ for density perturbations and $B \neq 0$ for gravitational waves. These results for one special Fourier mode $\mathbf{n} = (0, 0, n)$ can then be generalized to any arbitrary $\mathbf{n}$. As we anticipated on the grounds of handedness, gravitational waves can generate the B-mode of CMB polarization, but density perturbations cannot [42, 43, 45, 46].

11 Statistics and Angular Correlation Functions

The linear character of the radiation transfer equations (79) makes the randomness of h_{ij} being inherited by $\hat{\mathbf{n}}^{(1)}$. A consistent handling of Eq. (79) requires us to use the spatial Fourier expansion

$$\hat{\mathbf{n}}^{(1)}(\eta, x^i, \tilde{\nu}, e^i) = \frac{\mathcal{C}}{(2\pi)^{3/2}} \int_{-\infty}^{+\infty} \frac{d^3\mathbf{n}}{\sqrt{2n}} \sum_{s=1,2} \Big[\hat{\mathbf{n}}^{(1)}_{\mathbf{n},s}(\eta, \tilde{\nu}, e^i) e^{i\mathbf{n}\cdot\mathbf{x}} \overset{s}{c}_{\mathbf{n}} + \hat{\mathbf{n}}^{(1)*}_{\mathbf{n},s}(\eta, \tilde{\nu}, e^i) e^{-i\mathbf{n}\cdot\mathbf{x}} \overset{s}{c}{}^{*}_{\mathbf{n}} \Big], \tag{81}$$

where random coefficients $\overset{s}{c}_{\mathbf{n}}$ are the same entities which enter the metric field Eq. (9). The CMB anisotropies are random because the underlying metric perturbations are random.

In the strict quantum-mechanical version of the theory, where $\overset{s}{c}_{\mathbf{n}}$ are quantum-mechanical operators, the CMB field $\hat{\mathbf{n}}^{(1)}$ itself becomes a quantum-mechanical operator. If the initial quantum state of the system is chosen to be the ground state $|0\rangle$, $\overset{s}{c}_{\mathbf{n}}\,|0\rangle = 0$, all statistical properties of the system are determined by this choice.

Note the extra degree of uncertainty that we will have to deal with. Usually, the signal is deterministic, even if totally unknown, and the randomness of the observed outcomes arises at the level of the measurement process, as a consequence of the uncontrollable noises. In our problem, if cosmological perturbations do indeed have quantum-mechanical origin, the signal itself is inherently random and is characterized by a quantum state, or a wave-function, or a probability distribution function. At best, we can predict only a probability distribution function for possible CMB maps (outcomes). This is true even if the dynamics and cosmological parameters are strictly fixed and the measurement process is strictly noiseless.

Each mode of cosmological perturbations has started its life in the initial vacuum state. This state is Gaussian. In course of time it evolved into a squeezed vacuum state. Squeezed vacuum states retain the Gaussianity, even though developing the strongly unequal variances in amplitudes and phases. Therefore, the actually observed coefficients $a^X_{\ell m}$ (our own realisation of the CMB map belonging to the theoretical ensemble of CMB maps) are supposed to be drawn from the zero-mean Gaussian distributions for $a^X_{\ell m}$. If the observed CMB map looks a bit strange to you, due to the presence, for example, of some hints on 'axes' and 'voids', this is not necessarily an indication of non-Gaussianity of the underlying cosmological perturbations. Even if the observed CMB map consists entirely of your own images, this is not a proof of non-Gaussianity. It is a legitimate procedure to postulate some sort of a non-Gaussian distribution, by introducing a new parameter f_{NL}, and try to find f_{NL} from the data of a single actually observed map. But it is even more a legitimate procedure to insist on Gaussianity of perturbations, because it follows from the very deep foundations explained above, and therefore regard $f_{NL} \equiv 0$ by definition. From this position, if the set of the observed coefficients $a^X_{\ell m}$ looks a bit strange to you, this is simply because our Universe is not as dull and 'typical' as you might expect.

In this paper, we simplify the problem and treat cosmological fields classically rather than quantum-mechanically. We also make mild statistical assumptions. Specifically, we assume that classical random complex numbers $\overset{s}{c}_{\mathbf{n}}$ satisfy a limited set of statistical requirements:

$$\left\langle \overset{s}{c}_{\mathbf{n}} \right\rangle = \left\langle \overset{s'*}{c}_{\mathbf{n}'} \right\rangle = 0, \ \left\langle \overset{s}{c}_{\mathbf{n}} \overset{s'*}{c}_{\mathbf{n}'} \right\rangle = \left\langle \overset{s*}{c}_{\mathbf{n}} \overset{s'}{c}_{\mathbf{n}'} \right\rangle = \delta_{ss'}\delta^{(3)}(\mathbf{n}-\mathbf{n}'), \ \left\langle \overset{s}{c}_{\mathbf{n}} \overset{s'}{c}_{\mathbf{n}'} \right\rangle = \left\langle \overset{s*}{c}_{\mathbf{n}} \overset{s'*}{c}_{\mathbf{n}'} \right\rangle = 0, \tag{82}$$

where the averaging is performed over some probability distributions. We do not even assume outright that these distributions are Gaussian. The rules (82) are sufficient for the most of our further calculations.

We want to know the value of quantities $a^X_{\ell m}$ at the time of observation $\eta = \eta_R$. To find these quantities we have to integrate Eq. (79) over time, with all the astrophysical and gravitational inputs taken into account. The derived coefficients $a^X_{\ell m}$ are random, because the participating coefficients $\overset{s}{c}_{\mathbf{n}}$ are random. We can calculate various correlation functions of the CMB by calculating the quantities $\langle a^{X*}_{\ell m} a^{X'}_{\ell' m'} \rangle$. Using the rules (82), one can show that these quantities take the form

$$\langle a^{X*}_{\ell m} a^{X'}_{\ell' m'} \rangle = C^{XX'}_{\ell} \delta_{\ell\ell'} \delta_{mm'}, \tag{83}$$

where $C^{XX'}_{\ell}$ depend on the gravitational mode functions and astrophysical input. $C^{XX'}_{\ell}$ are calculable as general theoretical expressions.

The quantities $C^{XX'}_{\ell}$ are called the multipoles of the corresponding CMB power spectrum XX'. As usual, the power spectrum of a field contains less information than the field itself, but we will mostly ignore this loss of information. Also, it is worth noting that practically any feature in the CMB power spectrum can be "predicted" and "explained" solely by the properly adjusted primordial spectrum of cosmological perturbations. But we will not go along this line and will stick to simple power-law primordial spectra.

In the case of relic gravitational waves, the general shape of today's CMB power spectra and their features in the ℓ-space are almost in one-to-one correspondence with the processed metric power spectra and their features in the wavenumber n-space. The processed metric power spectra should be taken at the time of decoupling of CMB. The *TT* power spectrum is determined by the power spectrum of the metric itself, *hh*. The *EE* and *BB* power spectra are largely determined by the power spectrum of the metric's first time-derivative, $h'h'$. And *TE* power spectrum is determined by the cross power spectrum of the metric and its first time-derivative, hh'. For the gravitational-wave background with parameters indicated by dotted line in Fig. 5, the corresponding metric and CMB power spectra are shown in Figs. 7 and 8. For the gravitational-wave background with parameters indicated by a solid line in Fig. 5, the summary of CMB spectra is shown in Fig. 9. The summary also includes the reionization 'bump' at $\ell \lesssim 12$. (More details about these figures can be found in [17].)

All CMB power spectra have been calculated as averages over a theoretical ensemble of all possible realizations of the CMB field. In its turn, this randomness of CMB anisotropies ensues from the randomness of the gravitational field coefficients $\overset{s}{c}_{\mathbf{n}}$, $\overset{s}{c}{}^*_{\mathbf{n}}$. The characteristic parameters of a stochastic field, such as its mean values and variances, are, by definition, averages over the ensemble of realizations. However, in CMB observations, we have access to only one realization of this ensemble, which can be thought of as a single observed set of coefficients $a^X_{\ell m}$. Is it possible to find the parameters of a stochastic process by studying only one realization of this process?

The answer is yes, if the correlations of the stochastic process decay sufficiently quickly at large separations in time or space where the process is defined [47]. The process allowing the derivation of its true parameters, with probability arbitrarily

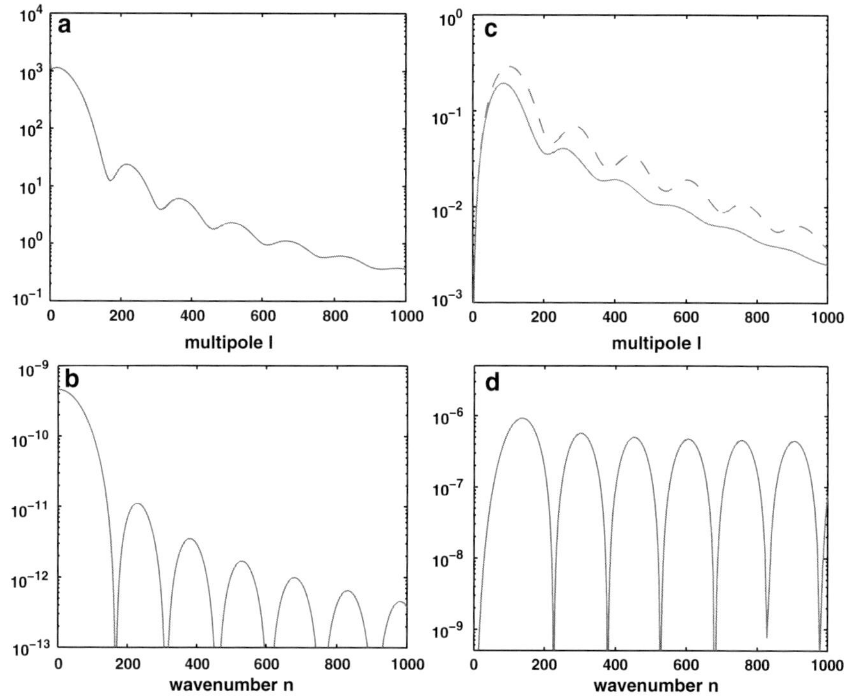

Fig. 7 The *left panel* shows (**a**) the power spectrum of temperature anisotropies $\ell(\ell+1)C_\ell^{TT}$ (in μK^2) generated by (**b**) the power spectrum of g.w. metric perturbations hh (29), $\beta = -2$. The *right panel* shows (**c**) the power spectra of polarization anisotropies $\ell(\ell+1)C_\ell^{BB}$ (*solid line*) and $\ell(\ell+1)C_\ell^{EE}$ (*dashed line*), panel (**d**) shows the power spectrum of the first time derivative of the same g.w. field, $h'h'$

close to 1, from a single realization is called ergodic. For example, the distribution of galaxies, or a stochastic density field, in an infinite 3 - space (our Universe) may be ergodic, and then by studying a single realization of this distribution we could extract the true parameters of the underlying random process. In the theory that we are discussing here, the randomness of the linear density field is also described by the random coefficients $\overset{s}{c}_{\mathbf{n}}$, $\overset{s}{c}{}^{*}_{\mathbf{n}}$ appearing in the metric perturbations.

If the process is non-ergodic, there will be an inevitable uncertainty surrounding the parameter's estimation derived from a single realization. This uncertainty should not be mistaken for the 'cosmic variance', often quoted and plotted on the observational graphs. The cosmic variance is a mathematically correct statement about the size of the variance of the χ^2-distribution for $2\ell+1$ independent Gaussian variables. There is nothing particularly cosmic in the cosmic variance. The elementary theorem, called cosmic variance, has nothing to say about the ergodicity or non-ergodicity of a given stochastic process. In particular, the universal validity of cosmic variance cannot prevent the extraction of exact parameters from the observation of a single realization of the stochastic density field in our Universe, if the density field is ergodic, – and it might be ergodic.

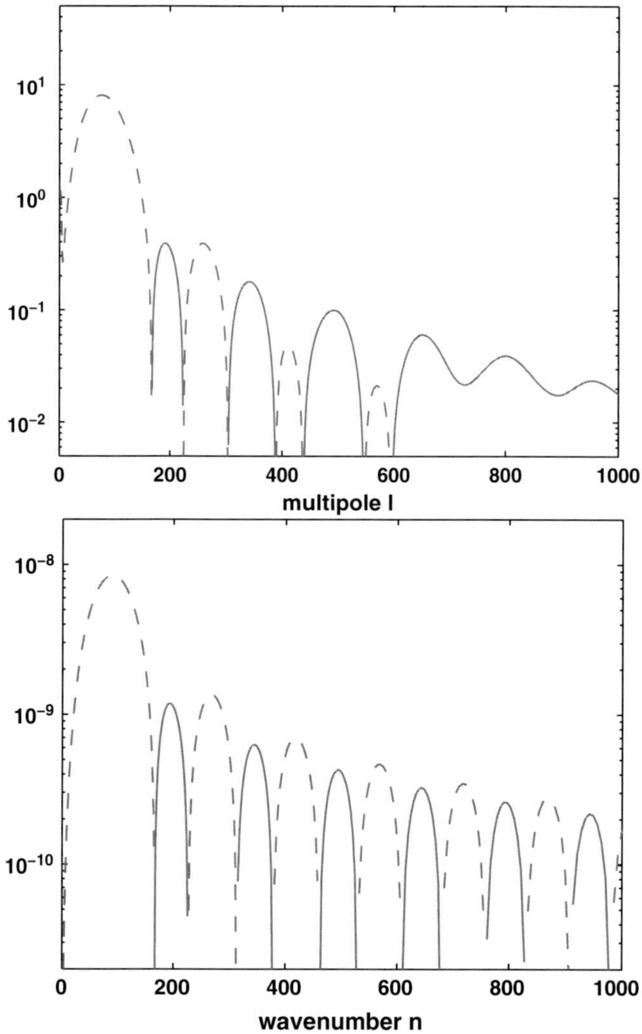

Fig. 8 The *bottom panel* shows the cross-power spectrum hh' of gravitational waves, whereas the *top panel* shows the angular power spectrum $\ell(\ell+1)C_\ell^{TE}$ caused by these waves. The negative values of these functions are depicted by *broken lines*

The problem is, however, that on a compact 2-sphere, where the random CMB is defined, ergodic processes do not exist. We will always be facing some uncertainly related to non-ergodicity. The size of this uncertainty about the derived parameter depends on the statistics and employed estimator. Under some conditions, this uncertainty is close, numerically, to the size of the usually quoted cosmic variance [48]. This discussion is important, because we are now approaching the observational predictions and the ways of discovering relic gravitational waves in the CMB data.

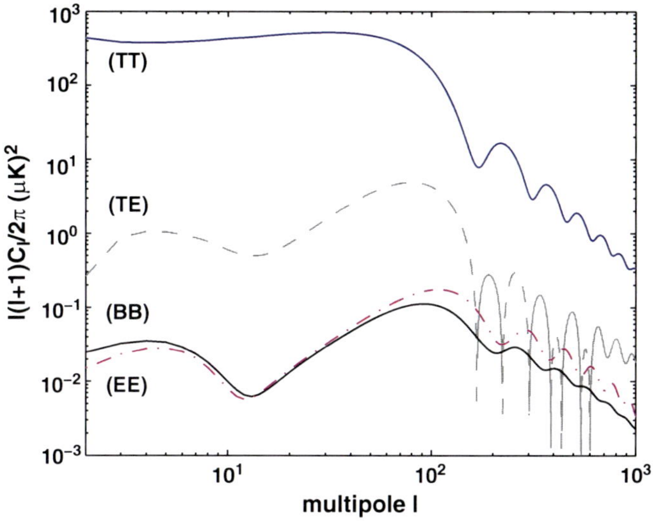

Fig. 9 The summary of CMB temperature and polarization anisotropies due to relic gravitational waves with $n = 1.2$ and $R = 1$

12 Temperature-Polarization Cross-Correlation Function

The numerical levels of primordial spectra for tensor and scalar metric perturbations are approximately equal. This means that the amplitudes of those Fourier modes n which have not started yet their short-wavelength evolution are numerically comparable, and the metric mode functions are practically constant in time. In contrast, in the short-wavelength regime, the amplitudes of gravitational waves adiabatically decrease, while the amplitudes of gravitational field perturbations associated with density perturbations may grow. Specifically, at the matter-dominated stage, the function $h_l(\eta)$ grows and overtakes $h(\eta)$, which remains constant.

In the context of CMB anisotropies, the crucial time is the epoch of decoupling. The wavelengths of modes with $n \lesssim 100$ were comfortably longer than the Hubble radius at the decoupling. The influence of these modes, both in gravitational waves (g.w.) and density perturbations (d.p.), have been projected into today's XX' anisotropies at $\ell \lesssim 100$. In this interval of ℓ, the g.w. contribution is not a small effect in comparison with the d.p. contribution. The g.w. contribution to the CMB power spectra is illustrated in Figs. 7, 8, and 9.

One way of detecting relic gravitational waves is based on measuring the BB auto-correlation. This method is clean, in the sense that density perturbations do not intervene, but the expected signal is very weak, see Figs. 7, 9.

We propose [17] to concentrate on the TE cross-correlation (without, of course, neglecting the BB searches). The TE signal is about two orders of magnitude stronger than the BB signal, and the use of a cross-correlation is always better than an auto-correlation, in the sense of fighting the noises. The special feature allowing to distinguish the g.w. part of *TE* from the d.p. part of *TE* is the difference in their signs. In the interval $\ell \lesssim 100$ the $(TE)_{gw}$ must be negative (see Fig. 8), while the $(TE)_{dp}$ must be positive at lowest $\ell's$ and up to, at least, $\ell \approx 50$. This difference in sign of *TE* correlation functions is the consequence of the difference in sign of the g.w. and d.p. metric cross-power spectra hh' in the interval $n \lesssim 70$ [17]. (The difference in sign of *QT* correlation functions is discussed in the earlier paper [49].)

An example of expected g.w. and d.p. contributions to the *TE* correlation function is shown in Fig. 10. To include reionization and enhance the lowest $\ell's$, we plot the function $(\ell+1)/2\pi\ C_\ell^{TE}$ rather than the usual $\ell(\ell+1)/2\pi\ C_\ell^{TE}$. The g.w. and d.p. metric power spectra are normalized in such a way that they give $R = 1$, where

$$R \equiv \frac{C_{\ell=2}^{TT}(gw)}{C_{\ell=2}^{TT}(dp)}.$$

More specifically, they give equal contributions to the total 'best fit' [32] temperature quadrupole $C_{\ell=2}^{TT}$. The dashed line in Fig. 10 shows the effect of the g.w. background marked by dotted line in Fig. 5. The dotted line in Fig. 10 shows the effect of d.p. metric perturbations, with the same primordial spectral index, n = 1. The sum of the two contributions is shown by a solid line. The fact that the two contributions may almost cancel each other does not mean that the g.w. signature is weak and hard to measure. The g.w. signal is in the strong deviation of the total *TE* spectrum from the expectation of the d.p. model.

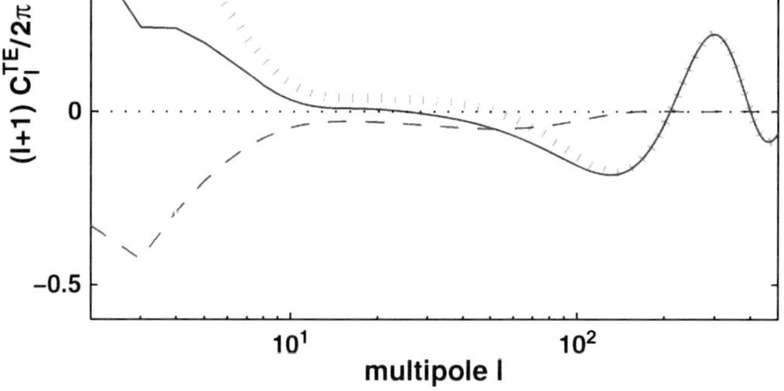

Fig. 10 The *dotted line* shows the contribution of density perturbations alone, while the *dashed line* shows the contribution of gravitational waves alone. The *solid line* is the sum of these contributions. It is seen from the graph that the inclusion of g.w. makes the total curve to be below the d.p. curve

It is important to remember that the widely publicized "tight limits" on the 'tensor-to-scalar ratio' r which allegedly rule out $R = 1$ and any $R \neq 0$, unless R is small, were derived by making use of the inflationary "consistency relation" (see Section 7). In these derivations, the g.w. spectral index n_t is being taken from the relationship $n_t = -r/8$, which automatically sends r to zero when n_t approaches zero. If this relationship is regarded as an artificial extra condition on g.w. parameters, then the results of such a data analysis may be of some value to those who are interested in this ad hoc condition, but not to those who are interested in determination of the true amount of relic gravitational waves. If, on the other hand, this relationship is regarded as part of inflationary theory, then such a data analysis is deeply self-contradictory. Indeed, the invariably derived conclusion, according to which the maximum of the likelihood function for r is at $r = 0$, or at least the value of the observed r is "consistent with zero", means that the most likely values of the inflationary density perturbation amplitudes, together with the CMB anisotropies induced by them, are infinitely large, or at least the data are "consistent" with such an infinity (see end of Section 7). It goes without saying that we are not using this relationship. But we do use the relationship $n_t = n_s - 1$ which is a consequence of the superadiabatic generating mechanism, see Eq. (65).

The WMAP community seems to be satisfied with the data analysis which states that the CMB data can be described by a small number of parameters which include density perturbations, but with no necessity for gravitational waves. The usual logic in these derivations is first to find the best-fit parameters assuming that only the density perturbations are present, and then to claim that there is no much room left for inclusion of gravitational waves. This looks like being satisfied with the statement that most of what is known about the human race can be described by a small number of parameters which includes one leg of individuals, but with no need for another. And when you propose to the data analysts that it is better to treat the data under the assumption that humans have two legs, they reply that this would be one extra parameter, and the proposer should be penalized for that. Anyway, in our analysis, we include the (inevitable) gravitational waves from the very beginning.

The total *TE* signal is the sum of g.w. and d.p. contributions. Even if this sum is positive in the interval $\ell \lesssim 70$, the effect of gravitational waves can (and expected to) be considerable. In this case, the amount of gravitational waves can be estimated through the analysis of all correlation functions together. However, if the total *TE* signal is negative in this interval of ℓ, there is little doubt that a significant g.w. component is present and is responsible for the negative signal, because the mean value of the *TE* signal cannot be negative without gravitational waves (the issues of statistics are discussed below).

It is intriguing that the WMAP team [32] explicitly emphasizes the detection of a negative correlation (i.e. anticorrelation) at the multipoles near $\ell \approx 30$: "The detection of the *TE* anticorrelation near $\ell \approx 30$ is a fundamental measurement of the physics of the formation of cosmological perturbations...". The motivation for this statement is the continuing concern of CMB observers about the so-called 'defect' models of structure formation. These 'causal' models cannot produce any correlations, positive or negative, at $\ell \lesssim 100$. So, a detected correlation near $\ell \approx 30$ is an evidence against them. At the same time, the better detected *TE* anticorrelation in the region of higher $\ell \approx 150$ could still be accommodated by the causal models, and therefore this more visible feature is not a direct argument against these models. However, the motivation for the WMAP team's statement is not essential for the present discussion. Even if the available data are not sufficient to conclude with confidence that the excessive *TE* anticorrelation at lower ℓ's has been actually detected, it seems that the WMAP's published data (together with the published statement, quoted above) can serve at least as an indication that this is likely to be true. In the framework of the theory that we are discussing here, such an *TE* anticorrelation is a natural and expected feature due to relic gravitational waves, but of course this needs to be thoroughly investigated.

The ensemble-averaged correlation functions do not answer all the questions. It is necessary to know what one would get with individual realizations of $a_{\ell m}^X$ caused by individual realizations of random coefficients $\overset{s}{c}_{\mathbf{n}}, \overset{s}{c}{}_{\mathbf{n}}^{*}$. The general unbiased estimator $D_\ell^{XX'}$ of $C_\ell^{XX'}$ is given by

$$D_\ell^{XX'} = \frac{1}{2(2\ell+1)} \sum_{m=-\ell}^{\ell} \left(a_{\ell m}^X a_{\ell m}^{X'*} + a_{\ell m}^{X*} a_{\ell m}^{X'} \right), \tag{84}$$

where $a_{\ell m}^X$ depend in a complicated, but calculable, manner on mode functions, astrophysical input, and, in general, on all random coefficients $\overset{s}{c}_{\mathbf{n}}, \overset{s}{c}{}_{\mathbf{n}}^{*}$. (It was explicitly shown [48] that the estimator D_ℓ^{TT} is not only unbiased, but also the best, i.e. the minimum-variance, estimator among quadratic estimators.) One can write, symbolically,

$$a_{\ell m}^X = \int_{-\infty}^{+\infty} d^3\mathbf{n} \sum_{s=1,2} \left[f_{\ell m}^X(n,s)\, \overset{s}{c}_{\mathbf{n}} + f_{\ell m}^{X*}(n,s)\, \overset{s}{c}{}_{\mathbf{n}}^{*} \right]. \tag{85}$$

Some features of the averaged XX' functions remain the same in any realization, i.e. for any choice of random coefficients $\overset{s}{c}_{\mathbf{n}}, \overset{s}{c}{}_{\mathbf{n}}^{*}$. One example is the absence of BB correlations for density perturbations. In the case of density perturbations, the BB correlation function vanishes not just on average, but in every realization. This happens because all the functions $f_{\ell m}^B(n,s)$ in Eq. (85) are zeros. Another example is the retention, in every realization, of positive sign of the auto-correlation functions XX', with $X' = X$. Indeed, expression (85) may be arbitrarily complicated, but the estimator (84) for $X' = X$ is always positive, as it is the sum of strictly positive terms. In different realizations the estimates will be different, but they will always be positive. That is, in any realization, including the actually observed one, the quantity D_ℓ^{XX} has the same sign as the quantity C_ℓ^{XX}.

The important question is how often the *TE* estimations (84) retain the same sign as the already calculated ensemble-averaged quantity C_ℓ^{TE}. In other words, suppose a negative *TE* is actually observed in the interval $\ell \lesssim 50$. Can it be a statistical fluke of density perturbations alone, rather than a signature of gravitational waves? In general, the answer is 'yes', but we will present arguments why in the problem under discussion the answer may be 'no', or 'very likely no'.

To quantify the situation, it is instructive to start from two zero-mean Gaussian variables, say, a^T and a^E:

$$\langle a^T a^T \rangle = \sigma^2, \quad \langle a^E a^E \rangle = \sigma^2, \quad \langle a^T a^E \rangle = \rho\sigma^2, \quad 0 \leq \rho \leq 1. \tag{86}$$

The probability density function for the product variable $a^T a^E$ involves the modified Bessel function K_0 and is known as an exact expression [50]. In the situation, like ours, where ρ is close to 1, the probability of finding negative products $a^T a^E$ is small. A crude evaluation shows that, with probability 68% and higher, the values of $a^T a^E$ lie in the interval (0.3–1.7) around the positive mean value $\langle a^T a^E \rangle$, whereas for $\rho = 1$ the probability of finding a negative $\langle a^T a^E \rangle$ is strictly zero. For $2\ell + 1$ degrees of freedom the scatter around the mean value is expected to be much narrower. Nevertheless, if $\rho \neq 1$, infrequent realizations with negative values of the product $a^T a^E$ are still possible.

We are probably quite fortunate in our concrete case of *TE* cross-correlation functions. It is true that $a^T_{\ell m}$ and $a^E_{\ell m}$ contain a large number of independent random coefficients $\overset{s}{c}_{\mathbf{n}}$, multiplied by different deterministic functions $f^T_{\ell m}(n,s)$, $f^E_{\ell m}(n,s)$. However, in practice, these deterministic functions are such that the integrands in Eq. (85) are quite sharply peaked at $n \approx \ell$. Ideally, for every ℓ, the random variables $a^T_{\ell m}$, $a^E_{\ell m}$ become proportional to one and the same linear combination of random coefficients with $|\mathbf{n}| = n \approx \ell$. The square of this combination is always positive, so the sign of the estimator D^{TE}_ℓ would be the same as the sign of the mean value C^{TE}_ℓ. This sign is determined by the known deterministic functions, not by statistics. Although this practical retention of the sign is not a strictly proven theorem, I think it is a very plausible conjecture.

If the above-mentioned conjecture is correct, the negative sign of any observed *TE* correlation in the interval $\ell \lesssim 50$ is unlikely to be a statistical fluke of density perturbations alone, it should be a signature of presence of gravitational waves. One especially interesting interval of *TE* searches is the interval between $\ell \approx 30$ and $\ell \approx 50$. In this interval of ℓ, the total signal C^{TE}_ℓ should be negative, whereas the contribution from density perturbations is still positive (see Fig. 10 for a realistic example). The total *TE*-signal is a factor 50, or so, larger than the expected *BB*-signal. All the logic described above suggests that if a negative *TE* is detected in this interval of ℓ's it must have occurred due to relic gravitational waves.

13 Prospects of the Current and Forthcoming Observations

It is difficult to predict the future, but it seems to me that the Planck mission [51], as well as the ground-based experiments, such as BICEP [52, 53], Clover [54], QUIET [55], have a very good chance of detecting relic gravitational waves through the *TE* anticorrelation described above. A level of the expected *BB* correlation is shown in Fig. 9. This level is natural for the discussed theory, but may be a little bit optimistic, as it is shown for a relatively high primordial spectral index n = 1.2. Observations with the help of space and ground facilities, and the measurement of all relevant correlation functions, should bring positive results given the expected sensitivities of those observations. It seems to me that the detection of relic gravitational waves in CMB is the matter of a few coming years, rather than a few decades. The analysis of the latest WMAP data provides serious indications of the presence of relic gravitational waves in the CMB anisotropies, see [56, 57].

Acknowledgments I am grateful to D. Baskaran and A.G. Polnarev for collaboration on the joint paper [17] that was extensively quoted in this presentation, and to S. Weinberg for helpful comments on this manuscript.

References

1. C. W. Misner, K. S. Thorne, and J. A. Wheeler. *Gravitation*, Freeman and Co., San Francisco, 1973
2. L. P. Grishchuk. *Zh. Eksp. Teor. Fiz.* **67**, 825 (1974) [Sov. Phys. JETP **40**, 409 (1975)]; *Ann. NY Acad. Sci.* **302**, 439 (1977); *Pisma Zh. Eksp. Teor. Fiz.* **23**, 326 (1976) [JETP Lett. **23**, 293 (1976)]; *Uspekhi Fiz. Nauk* **121**, 629 (1977) [Sov. Phys. Usp. **20**, 319 1977]
3. A. D. Sakharov. *Zh. Eksp. Teor. Fiz.* **49**, 345 (1965) [Sov. Phys. JETP **22**, 241 (1966)]
4. E. Schrodinger. *Physica* **6**, 899 (1939)
5. L. P. Grishchuk. *Phys. Rev.* D**48**, 5581 (1993)
6. L. Parker. *Phys. Rev. Lett.* **21**, 562 (1968); *Phys. Rev.* **183**, 1057 (1969)
7. B. Ya. Zeldovich. *Impedance and parametric excitation of coupled oscillators*, to be published in *Physics-Uspekhi* (2008)
8. L. D. Landau and E. M. Lifshitz. *Mechanics*. Oxford, Pergamon Press, 1975
9. P. Paradoksov. *Uspekhi Fiz. Nauk* **89**, 707 (1966) [Sov. Phys. Usp. **9**, 618 (1967)]
10. L. D. Landau and E. M. Lifshitz. *The Classical Theory of Fields*. Oxford, Pergamon Press, 1975
11. L. P. Grishchuk. *Class. Quant. Gravity* **10**, 2449 (1993)
12. L. P. Grishchuk. *Lecture Notes in Physics*, Vol. 562: Gyros, Clocks, Interferometers: Testing Relativistic Gravity in Space, 167 (2001)
13. L. P. Grishchuk and M. Solokhin. *Phys. Rev.* D**43**, 2566 (1991)
14. S. Weinberg. *The Quantum Theory of Fields*. Cambridge U. Press, 1995
15. L. P. Grishchuk. *Uspekhi Fiz. Nauk* **175(12)** 1289–1303 (2005) [Physics-Uspekhi, **48**(12) 1235–1247 (2005], arXiv:gr-qc/0504018
16. L. P. Grishchuk and A. G. Polnarev. In: *General Relativity and Gravitation, 100 Years After the Birth of A. Einstein*, ed. A. Held (Plenum Press, NY, 1980) Vol. 2, p. 393
17. D. Baskaran, L. P. Grishchuk, and A. G. Polnarev. *Phys. Rev.* D**74**, 083008 (2006)
18. Ya. B. Zeldovich and I. D. Novikov. *The Structure and Evolution of the Universe*. Chicago U. Press, 1983
19. L. P. Grishchuk. *Phys. Rev.* D**50**, 7154 (1994)
20. L. P. Eisenhart. *Continuous Groups of Transformations*. Dover, 1933
21. J. M. Bardeen. *Phys. Rev.* D**22**, 1882 (1980)
22. J. M. Bardeen, P. J. Steinhardt, and M. S. Turner. *Phys. Rev.* D**28**, 679 (1983)
23. V. N. Lukash. *Pis'ma Zh. Eksp. Teor. Fiz.* **31**, 631 (1980) [JETP Lett. **6**, 596 (1980)]
24. G. Chibisov and V. Mukhanov. *Mon. Not. R. Astr. Soc.* **200**, 535 (1982)
25. M. Sasaki. *Prog. Theor. Phys.* **76**, 1036 (1986)
26. W. Unruh. *Cosmological long wavelength perturbations*, arXiv:astro-ph/9802323

27. S. Weinberg. *Phys. Rev.* D**72**, 043514 (2005)
28. R. Flauger and S. Weinberg. *Phys. Rev.* D**75**, 123505 (2007) (arXiv:astro-ph/0703179)
29. S. Weinberg. *Phys. Rev.* D**69**, 023503 (2004)
30. V. N. Lukash. *On the relation between tensor and scalar perturbation modes in Friedmann cosmology,* arXiv:astro-ph/0610312
31. D. N. Spergel *et al. WMAP Collaboration, Astrophys. J. Suppl.* **170**, 377 (2007) (arXiv:astro-ph/0603449)
32. L. Page *et al. WMAP Collaboration, Astrophys. J. Suppl.* **170**, 335 (2007) (arXiv:astro-ph0603450)
33. L. Pagano, A. Cooray, A. Melchiori, and M. Kamionkowski. *Red Density Perturbations and Inflationary Gravitational Waves*, arXiv:0707.2560 (In this fresh paper one can also find some previous history of derivation and application of the inflationary 'consistency relation')
34. H. V. Peiris *et al. Astrophys. J. Suppl.* **148**, 213 (2003)
35. S. Weinberg. website: www.amacad.org/publications/bulletin/summer2003/weinberg.pdf
36. S. Bose and L.P. Grishchuk. *Phys. Rev.* D**66**, 043529 (2002)
37. R. K. Sachs and A. M. Wolfe. *Astrophys. J.* **147**, 73 (1967)
38. G. Dautcourt. *Month. Not. R. Astr. Soc.* **144**, 255 (1969)
39. L. P. Grishchuk and Ya. B. Zeldovich. *Astron. Zh.* **55**, 209 (1978) [Sov. Astron. **22**, 125 (1978)]
40. A. G. Polnarev. Sov. Astron. **29**, 607 (1985)
41. S. Chandrasekhar. *Radiative Transfer*. New York, Dover Publications Inc., 1960
42. M. Zaldarriaga and U. Seljak. *Phys. Rev. D* **55**, 1830 (1997)
43. M. Kamionkowski, A. Kosowsky, and A. Stebbins. *Phys. Rev. D* **55**, 7368 (1997)
44. M. M. Basko and A. G. Polnarev. *Sov. Astron.* **24**, 268 (1980)
45. U. Seljak. *Astrophys. J.* **482**, 6 (1997)
46. W. Hu and M. White. *New Astron.* **2**, 323 (1997)
47. A. M. Yaglom. *An Introduction to the Theory of Stationary Random Functions*. Prentice-Hall, Englewood Cliffs, NJ, 1962
48. L. P. Grishchuk and J. Martin. *Phys. Rev.* D**56**, 1924 (1997)
49. R. Crittenden, D. Coulson, and N. Turok. *Phys. Rev.* D**52**, R5402 (1995)
50. L. P. Grishchuk. *Phys. Rev.* D**53**, 6784 (1996)
51. The Planck Collaboration. *arXiv:astro-ph/0604069*
52. B. Keating and N. Miller. *New Astron. Review,* **50**, 184 (2006)
53. J. M. Kovac and D. Barkats. *CMB from the South Pole: Past, Prsent, and Future*, arXiv:0707.1075
54. A. Taylor *et al.* In: *Proceedings of 39th Recontres de Moriond* (Frontieres, 2004)
55. C. Lawrence. *The QUIET Experiment*, (http://hea.iki.rssi.ru/Z-90/). Invited talk at the conference "Zeldovich-90", Moscow, 2004
56. W. Zhao, D. Baskaran, and L. P. Grishchuk. *Phys. Rev. D* **79**, 023002 (2009)
57. W. Zhao, D. Baskaran, and L. P. Grishchuk. *Phys. Rev. D* **80**, 083005 (2009)

Status of Gravitational Wave Detection

Adalberto Giazotto

Abstract We review the status of gravitational wave detection.

1 INFN – Pisa and European Gravitational Observatory

Taylor and Hulse [1] discovery of Gravitational Wave [2] (GW) emission by the compact binary system composed by two Neutron Stars PSR1913 + 16 has been, for the experimental physicists working in this field, the definitive thrust allowing to reach the extremely sophisticated technology needed for investigating in this field of research.

J. Taylor

R. Hulse

In Fig. 1 the famous histogram showing the changes of periaster passage time produced by the associated emission of GW together with the prediction evaluated through General Relativity (GR), is shown.

A. Giazotto (✉)
INFN, Edificio C - Polo Fibonacci, Largo B. Pontecorvo, 3, 56127 Pisa
e-mail: adalberto.giazotto@pi.infn.it

I. Ciufolini and R.A. Matzner (eds.), *General Relativity and John Archibald Wheeler*, Astrophysics and Space Science Library 367, DOI 10.1007/978-90-481-3735-0_11,

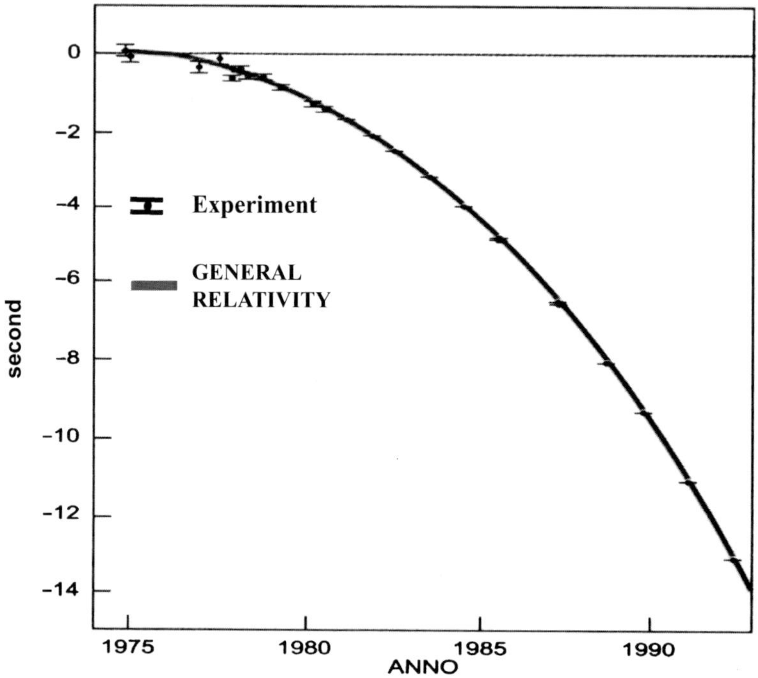

Fig. 1 Orbital period decreasing changes periaster passage time in total agreement with GR

Today at least other six similar systems are investigated and energy loss remarkably agrees with the prediction of GR.

But, after this extremely short presentation of one of the most important physics measurements of all times I think it is interesting trying to shortly explore the steps that experimentalists followed for the understanding of the very GW emission mechanism and the principles on which a GW detector is working, to the technology of today's detectors.

2 The Generation of Gravitational Waves [3]

Gravitational Waves are a consequence of Einstein's GR; they are shown to be ripples in the space–time curvature traveling at light speed. In curved space–time the invariant interval ds between two neighboring events is described by Metric Tensor $g_{\mu\nu}$:

$$\begin{cases} ds^2 = g_{\mu\nu} dx^\mu dx^\nu \\ g_{\mu\nu} g^{\nu\lambda} = \delta_\mu^\lambda \end{cases} \quad (1)$$

In Flat Space it is possible to find coordinates such that:

$$g_{\mu\nu} \Rightarrow \eta_{\mu\nu} = \begin{pmatrix} 1 & 0 & 0 & 0 \\ 0 & -1 & 0 & 0 \\ 0 & 0 & -1 & 0 \\ 0 & 0 & 0 & -1 \end{pmatrix} \tag{2}$$

This is not true when the Momentum–Energy tensor $T_{\mu\nu} \neq 0$ because $g_{\mu\nu}$ must satisfy the Einstein's equations:

$$\begin{cases} R^k_{k\mu\nu} = R_{\mu\nu} = \dfrac{8\pi G}{c^4}\left(T_{\mu\nu} - \dfrac{1}{2}g_{\mu\nu}T^l_l\right) & \text{Einstein Equations} \\ R^i_{klm} = \dfrac{\partial\Gamma^i_{km}}{\partial x^l} - \dfrac{\partial\Gamma^i_{kl}}{\partial x^m} + \Gamma^i_{nl}\Gamma^n_{km} - \Gamma^i_{nm}\Gamma^n_{kl} & \text{Riemann Tensor} \\ \Gamma^i_{kl} = \dfrac{1}{2}g^{im}\left(\dfrac{\partial g_{mk}}{\partial x^l} + \dfrac{\partial g_{ml}}{\partial x^k} - \dfrac{\partial g_{kl}}{\partial x^m}\right) & \text{Christoffel Symbols} \end{cases} \tag{3}$$

When gravity is "weak" we may write $g_{\mu\nu} = \eta_{\mu\nu} + h_{\mu\nu}$, where $|h_{\mu\nu}| \ll 1$ $h\mu\nu$ $|h_{\mu\nu}| \ll 1$; to first order in $h_{\mu\nu}$ Riemann tensor and Einstein's equations become:

$$\begin{aligned} R_{iklm} &= \frac{1}{2}\left(\frac{\partial^2 h_{im}}{\partial x_l \partial x_k} + \frac{\partial^2 h_{lk}}{\partial x_i \partial x_m} - \frac{\partial^2 h_{km}}{\partial x_i \partial x_l} - \frac{\partial^2 h_{il}}{\partial x_k \partial x_m}\right) \\ \psi_{\mu\nu} &= \frac{8\pi G}{c^4}\tau_{\mu\nu} \end{aligned} \tag{4}$$

$$\Psi_{\mu\nu} = h_{\mu\nu} - \frac{1}{2}\delta_{\mu\nu}h^\lambda_\lambda \tag{5}$$

where

and $\tau_{\mu\nu}$ is $T_{\mu\nu}$ expanded to first order in $h_{\mu\nu}$.

Considering that:

$$\tau_{00} = \rho c^2 \tag{6}$$

where ρ is the matter density, it follows:

$$\Psi_{\alpha\beta} = -\frac{2G}{c^4 R_o}\left(\frac{\partial^2}{\partial t^2}\int \rho x_\alpha x_\beta dV\right)_{t-R_0/c} \tag{7}$$

Where R_0 is the source distance.

3 The $\Psi_{\mu\nu}$ Reduction

1. $\Psi\mu\nu$ has 16 elements but is symmetrical, hence it reduces to 10.
2. The Gauge trasf. x' = x + ε transforms h_{ik}:

$$h'_{ik} = h_{ik} - \frac{\partial\varepsilon_i}{\partial x_k} - \frac{\partial\varepsilon_k}{\partial x_i} \tag{8}$$

with the condition

$$\varepsilon_{\mu} = o$$

it leaves Riemann tensor unchanged.

This four new equations reduce $\Psi\mu\nu$ to six independent components.

3. From momentum–energy conservation

$$\partial^{\mu}\tau_{\mu\nu} = o \tag{9}$$

it follows the Lorentz condition on $\Psi\mu\nu$

$$\partial_{\mu}\Psi_{\mu\nu} = \frac{8\pi G}{c^4}\partial_{\mu}\tau_{\mu\nu} \rightarrow \partial_{\mu}\Psi_{\mu\nu} = 0 \tag{10}$$

These three conditions can be chosen to give $\Psi\mu\nu$ Transversality and Tracelessness (TT) and reduce it to two independent components:

$$\begin{cases} h^{TT}_{\mu\nu} = h^{TT}_{11}\begin{pmatrix} 0 & 0 & 0 & 0 \\ 0 & 1 & 0 & 0 \\ 0 & 0 & -1 & 0 \\ 0 & 0 & 0 & 0 \end{pmatrix} + h^{TT}_{12}\begin{pmatrix} 0 & 0 & 0 & 0 \\ 0 & 0 & 1 & 0 \\ 0 & 1 & 0 & 0 \\ 0 & 0 & 0 & 0 \end{pmatrix} = h^{+}e^{+} + h^{X}e^{x} \\ GW\ Along\ X_3 \end{cases} \tag{11}$$

The polarizations e^{+}_{ik} and e^{x}_{ik} are exchanged with a $\pi/4$ rotation around x_3 axis i.e. GW are spin two massless fields:

$$R(\pi/4) = \frac{1}{\sqrt{2}}\begin{pmatrix} 0 & 0 & 0 & 0 \\ 0 & 1 & 1 & 0 \\ 0 & -1 & 1 & 0 \\ 0 & 0 & 0 & 0 \end{pmatrix} \tag{12}$$

$$R(\pi/4)\, e^{x} R^{-1}(\pi/4) = -e^{+} \tag{13}$$

$$R(\pi/4)\, e^{+} R^{-1}(\pi/4) = -e^{x} \tag{14}$$

The Riemann tensor in the TT gauge reaches its simplest form:

$$R_{iklm} \Rightarrow R_{0\alpha 0\beta} = -\frac{1}{2}\ddot{h}^{TT}_{\alpha\beta} \tag{15}$$

$$\ddot{h}^{TT}_{\alpha\beta} = -\frac{2G}{c^4 R_0}\frac{d^2}{dt^2}\int \rho\left(x_{\alpha}x_{\beta} - \frac{1}{3}\eta_{\alpha\beta}x^2\right)^{TT} dV \tag{16}$$

Equation (16) shows that only asymmetric sources can emit GW; most efficient are coalescing binaries sources, while pulsar can rely only on ellipticities, usually very small. Supernovae could have relevant asymmetries but this is very unlikely.

4 The Detection of GW [3]

In his fundamental papers on Gravitational Waves, A. Einstein shows how a mechanical system is creating GW. For the detection of GW he shows that GW couple to matter and increase the energy of a suited mechanical system but it was F.A.E. Pirani [4] in 1956 who proposed the geodesic deviation equation as a tool for designing a practical GW receiver. The method suggested by E.A. Pirani is described in the following.

A particle moving freely under the action of a gravitational field has its x_μ coordinate satisfying the geodesic equation:

$$\frac{d^2x^\mu}{d\tau^2} = \Gamma^\mu_{\nu\lambda}\frac{dx^\nu}{d\tau}\frac{dx^\lambda}{d\tau} \tag{17}$$

where τ is the proper time.

It is always possible to find a space–time trajectory in which $\Gamma^\mu{}_{\nu\lambda} = 0$ at any time; along this trajectory a particle is freely falling. Let's consider two freely falling particles A and B (see Fig. 2), their separation $\xi_\alpha = (X_A - X_B)_\alpha$ satisfies the geodesic deviation equation:

$$\frac{D^2\xi^\alpha}{d\tau^2} + R^\alpha_{\beta\gamma\delta}\xi^\gamma\frac{dx^\beta}{d\tau}\frac{dx^\delta}{d\tau} = 0 \tag{18}$$

where D^2 is the second covariant derivative:

$$\frac{D^2\xi^\alpha}{d\tau^2} = \frac{d^2\xi^\alpha}{d\tau^2} + \frac{d\Gamma^\alpha_{\beta\mu}}{d\tau}\xi^\beta\frac{dx^\mu}{d\tau} + \Gamma^\alpha_{\beta\mu}\frac{d}{d\tau}\left(\xi^\beta\frac{dx^\mu}{d\tau}\right)$$
$$+\Gamma^\alpha_{\beta\mu}\left(\frac{d\xi^\beta}{d\tau} + \Gamma^\beta_{\varepsilon\nu}\xi^\varepsilon\frac{dx^\nu}{d\tau}\right)\frac{dx^\mu}{d\tau} \tag{19}$$

With the purpose of evaluating ξ_α let's put $(X_A)_{1,2,3} = 0\,XA = 0\ (X_A)_{1,2,3} = 0$ in the A particle CMS, $(X_A)_0$ being equal to proper time τ and the coordinate axis connected to gyroscopes carried by A. At $(X_A)_{1,2,3} = 0\ (X_A)_{1,2,3} = 0$ since A is freely falling along the geodesic line, we can obtain $(\Gamma\mu\nu\lambda)_{x=0} = 0$,

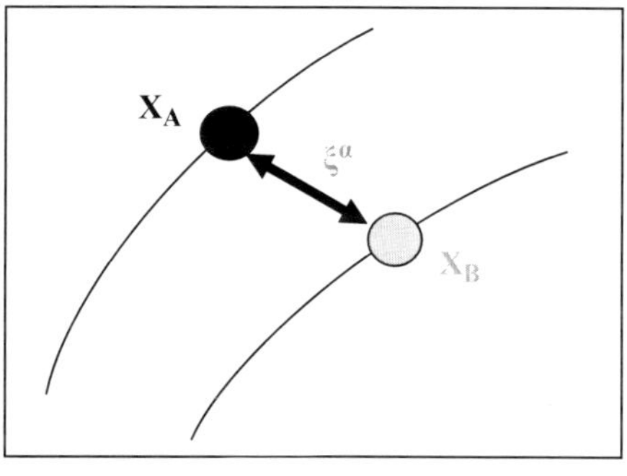

Fig. 2 The two masses in X_A and X_B follow their geodesic line; their separation ξ_α fluctuate due to GW presence. This elementary GW detector is the scheme of principle on which existing GW detectors works

$(d\Gamma\mu\nu\lambda/d\tau)_{x=0} = 0$, consequently:

$$\frac{D^2\xi^\alpha}{d\tau^2} = \frac{d^2\xi^\alpha}{d\tau^2} = -R^\alpha_{\beta\gamma\delta}\xi^\gamma \frac{dx^\beta}{d\tau}\frac{dx^\delta}{d\tau} \tag{20}$$

To first order in hTTμν, since the only surviving terms are for $\beta = \delta = 0$ and $\tau = t$, and remembering that

$$R_{0\alpha 0\beta} = -\frac{1}{2}\ddot{h}^{TT}_{\alpha\beta} \tag{21}$$

it follows that the relative acceleration of B with respect A is

$$\frac{d^2\xi_\alpha}{d\tau^2} = \frac{1}{2}\ddot{h}^{TT}_{\alpha\beta}\xi^\beta \tag{22}$$

and the Riemann force is:

$$\begin{cases} F_\alpha = \frac{1}{2}M\ddot{h}^{TT}_{\alpha\beta}\xi^\beta \\ \xi < \lambda_{GW} \end{cases} \tag{23}$$

Hence the GW receiver is a device measuring space–time curvature, i.e. the relative acceleration of two freely falling masses or, equivalently, their relative displacement. Equation (23) hold for separation ξ_α smaller than GW wavelength.

5 Short Outline About GW Sources

Coalescing Binaries [5] These systems are formed by two compact stars, neutron stars and/or black-holes, rotating one around the other. In Fig. 3 the reference system for a coalescing binaries system having with respect to the equatorial system (X_0, Y_0, Z_0), direction (α, δ) and angles (i, ϑ) of the orbit plane normal with respect to the direction (α, δ).

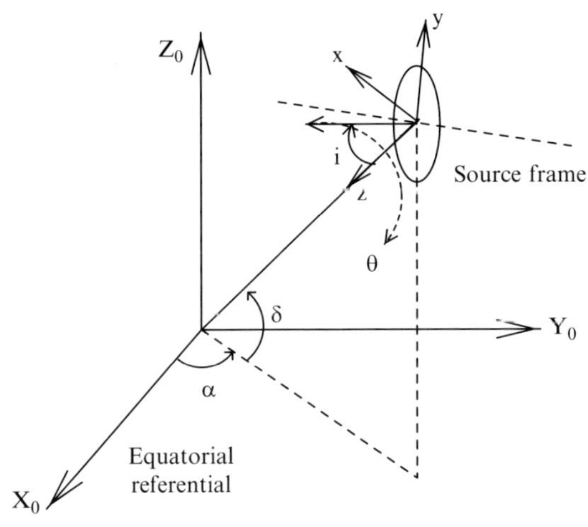

Fig. 3 The reference system for a coalescing binaries system having with respect to the equatorial system (X_0, Y_0, Z_0), direction (α, δ) and angles (i, ϑ) of the orbit plane normal with respect to the direction (α, δ)

The detector responds to the GW action in the following way:

$$h(t) = h_+(t)\, F_+(\alpha, \delta) + h_\times(t)\, F_\times(\alpha, \delta)$$

Where F_+ and F_x depend from detector orientation and h_+ and h_x are given by

$$\begin{pmatrix} h_+ \\ h_\times \end{pmatrix} = 2.6 \times 10^{-21} \left[\frac{1Mpc}{D}\right] \left[\frac{\mu^{3/5} M^{2/5}}{M_\oplus}\right]^{5/3} \left[\frac{\nu(t)}{100\,Hz}\right]^{2/3} \cdot$$
$$\cdot \begin{pmatrix} \cos\Phi(t)\dfrac{\cos^2 i + 1}{2}\cos(2\theta) + \sin\Phi(t)\cos i \sin 2\theta \\ \sin\Phi(t)\cos i \cos(2\theta) - \cos\Phi(t)\dfrac{\cos^2 i + 1}{2}\sin(2\theta) \end{pmatrix}$$

Where $\mu = M_1 M_2/(M_1 + M_2)$ is the reduced mass and $M = M_1 + M_2$ is the total mass, $M_\oplus$ the Sun mass and Φ is

$$\Phi(t) = \Phi_0 + \int_{t_0}^{t} \nu(\tau) d\tau$$

The frequency as a time function is given, at the first order in a post Newtonian expansion, by

$$\nu(t) = \frac{1}{\pi}\left(\frac{5}{256}\frac{1}{K^{5/3}}\frac{1}{(t - t_0)}\right)^{3/8}$$

Where $K = (\mu M^{2/3})^{3/5}$ is the Chirp Mass and t_0 the coalescence time.

The typical received signal will have a shape similar to the one shown in Fig. 4; on the last ms strong GRT effects will occur, so a post Newtonian expansion is no more accurate.

The signal Fourier transform $h(\nu)$ is given by

$$h(\nu) \propto \nu^{-7/6} e^{i\Phi(\nu)}$$

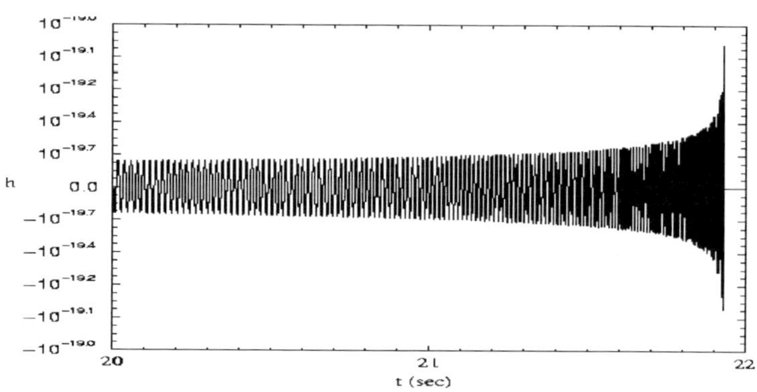

Fig. 4 The chirp wave form typical of a coalescing binary system

The signal analysis could consists in filtering the detector signal with a series of templates in which the unknown parameters M, μ, the initial phase. . .are cyclically explored by means of Wiener integrals:

$$\frac{\mathrm{s}}{\mathrm{n}} = \mathrm{Re} \int \frac{(\mathrm{n}+\mathrm{h}) \, . \mathrm{V}^{*}_{\mathrm{M},\mu,\ldots,\tau}}{\mathrm{S}^{2}} \mathrm{d}\nu$$

Where S is the spectral noise and $n + h$ is the detector signal + noise, $V \ldots$ is a template which corresponds to a given choice of the parameters (τ is the arrival time).

The expected rate is some event per year in a 50 Mpc sphere.

Supernovae [6] (SN) The theoretical evaluation of GW emission from SN explosions has a large uncertainty since the fraction of energy converted in GW is uncertain too; it is assumed a fraction between 10^{-6} and 10^{-3}. The expected rate is about 1/30 years in our Galaxy with $h \sim 10^{-20}$ and about 1/year in the Virgo cluster with $h \sim 10^{-23}$.

Periodic Sources [7] In our Galaxy are about $10^9 - 10^{10}$ Neutron Stars (NS) which may emit because of some mass distribution asymmetry both at the rotation and twice the rotation frequency ν.

Since Earth rotates both around the Sun and around his polar axis, the detector receives a non-monochromatic signal because of Doppler shift:

$$e^{i\omega t} \Rightarrow e^{i\omega(t - \vec{n} \bullet \vec{R}/c)}$$

where $\bar{n}$ is the NS direction, $\omega = 2\pi\nu$ and $\bar{R}$ is the vector connecting the detector to the cms of the Sun–Earth system. The Doppler shift widens the relative natural frequency width $\Delta\nu = 1/\mathrm{T}$, where T is the observation time, to $\Delta\nu/\nu \sim 10^{-4}$. For eliminating this effect we need to know ω and $\bar{n}$ and this happens only for known Pulsars.

The known Pulsars are about 1,400 and those in the Virgo bandwidth, above 4 Hz, about 700.

If ω and $\bar{n}$ are unknown the optimal analysis method it the "blind search" consisting in dividing the space $(\omega, \ \bar{n})$ in cells and evaluate $(\omega_i, \ \bar{n}_i)$ in each cell and compute the Fourier Transform (FF) of the data multiplied by; the $e^{+i\omega_i \vec{n}_i \bullet \vec{R}/c}$FF will be repeated in every cell and the sky will be explored till a high S/N ratio is found.

The payed price is enormous because for obtaining the maximal frequency resolution $\Delta\nu = 1/\mathrm{T}$ we need about 10^{22} sky cells; consequently the computing power rquired becomes impossibly large hence hierarchical methods are under study.

Cosmological BKG [8] Perhaps the most important signal to search for; it gives informations on the Big-Bang at time close to Planck time. For being detected it requires coincidence between two close high sensitivity interferometers.

Joseph Weber

- Born on 17 May 1919 in Paterson, New Jersey.
- In the early 1960s, J.Weber [9] turned his attention to testing the general relativistic prediction of gravitational waves from strong gravity collisions.
- Then, in 1959, he studied a detector that might be able to measure displacements smaller than the size of the nucleus.
- He was alone in charting these unknown waters: His first paper on how to build a gravity wave detector was published in 1959.
- Weber developed an experiment using a large suspended bar of aluminum, with a high resonant Q at a frequency of about 1 kHz.
- In 1969, he was claiming evidence for observation of gravitational waves (*Physical Review Letters* (1969)) based on coincident signals from two bars separated by 1,000 km. Those events, later, have been demonstrated to be consistent with cosmic rays.
- Weber also proposed the idea of doing an experiment to detect gravitational waves using laser interferometric techniques.
- Robert Forward, Weber's student and postdoctoral associate, started in 1970 the first laser interferometric gravitational wave experiments at Hughes Research Labs.
- Joseph Weber, professor emeritus of physics at the University of Maryland, College Park, died on 30 September 2000 in Pittsburgh, Pennsylvania.

JOSEPH WEBER WITH GRAVITATIONAL WAVE DETECTOR

The coupling of GW to the metal cylinder, typically composed by an Al alloy having high mechanical Q, produces forces according to Eq. (23) (see Fig. 5). We schematize the Al cylinder as two masses separated by a spring (see Fig. 6):

The Amplitude of GW Riemann force acting on the harmonic oscillator masses is expressed in Eq. (24), while the amplitude of Thermal Stochastic force acting on

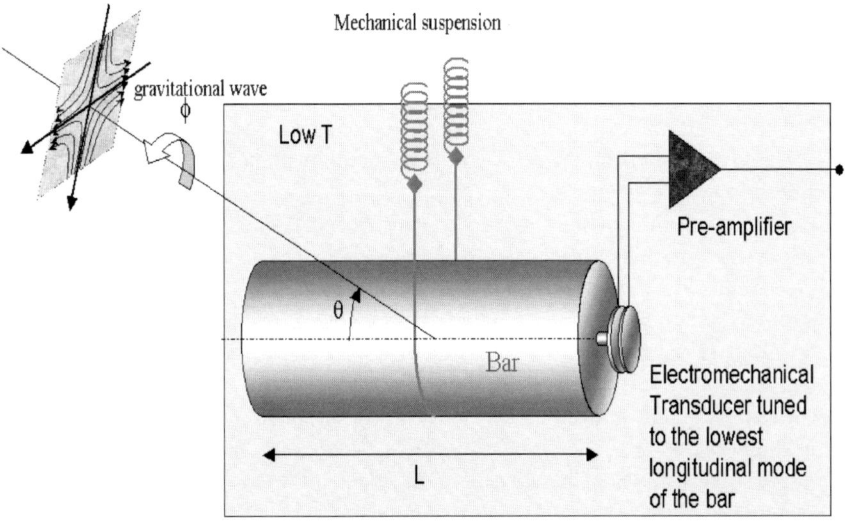

Fig. 5 A Schematic diagram showing a bar detector with the transducer. The GW Riemann force excite longitudinal modes of the Bar. Bar elongation is detected by the transducer and transformed to an electrical signal. The length L is tipically 2–3 m (Picture courtesy M. Cerdonio)

Fig. 6 A metallic cylinder of mass 2 M is schematized as a simple harmonic oscillator having resonance frequency $\nu_0 = \Omega_0/2\pi$

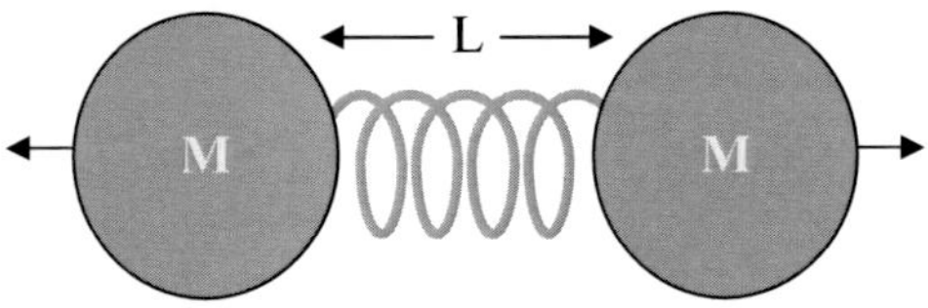

the harmonic oscillator is expressed in Eq. (25).

$$F_{Riemann} \approx \frac{1}{2} M \Omega^2 L \tilde{h}(\Omega) \tag{24}$$

$$F_{Thermal} \approx \sqrt{\frac{4KTM\Omega_0}{Q}} \tag{25}$$

$$\Delta x(\Omega) \approx \frac{\sqrt{\left(\Omega^2 L \tilde{h}(\Omega)/2\right)^2 + \frac{4KT\Omega_0}{QM}}}{-\Omega^2 + \Omega_0^2 + iM\frac{\Omega\Omega_0}{Q}} \tag{26}$$

The spectral displacement $\Delta\tilde{x}(\Omega)$ resulting from GW and Thermal Stochastic force on the harmonic oscillator is expressed in Eq. (26).

For transforming the cryogenic bar oscillations in electrical signals (see Fig. 7) a resonant transducer acting as a capacitor, is attached to the bar end; the capacitor variable current, produced by bar oscillations, is sent to an extremely sensitive SQUID amplifier.

It is evident that Transducer electronic noise determines the cryogenic bar bandwidth (see Fig. 8)

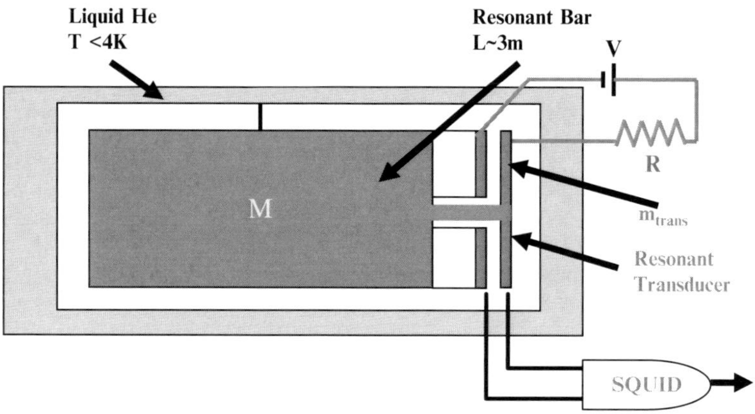

Fig. 7 The cryogenic Al bar is kept at a temperature of 4 K; bar oscillations are measured by the Resonant Transducer

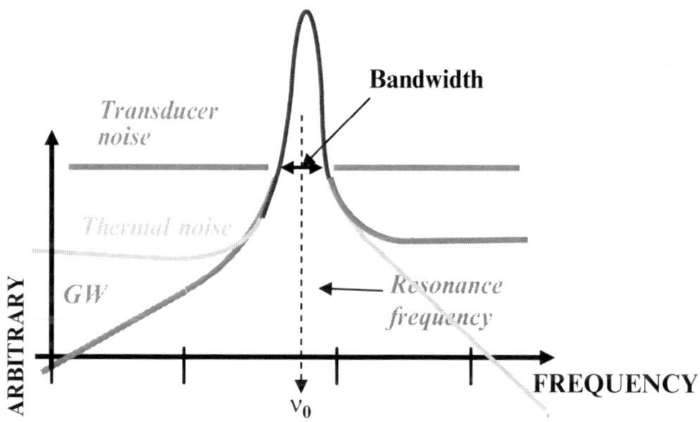

Fig. 8 A conceptual diagram showing how the transducer noise affects bandwidth

6 Modern Bar Detectors: Cryogenic Bars [10]

All these modern detectors were originating from early Weber's detectors.

In the world five cryogenic bar detectors, as shown in Fig. 10, have been built for working at temperatures <4 K:

Explorer at CERN (see Fig. 11), Nautilus at Frascati INFN National Laboratory (see Fig. 12), Auriga at Legnaro INFN National Laboratory (see Fig. 13), Allegro at Luisiana State University (see Fig. 14) and Niobe in Perth.

Instrumental details can be found in an extensive literature. I report the sensitivity and stability of performances of only four of them since Niobe stopped operation in 1997.

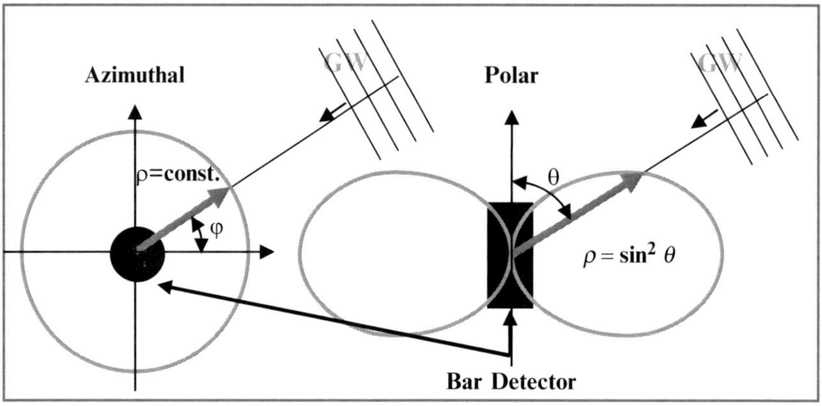

Fig. 9 Bar detector Antenna Pattern summed on polarizations

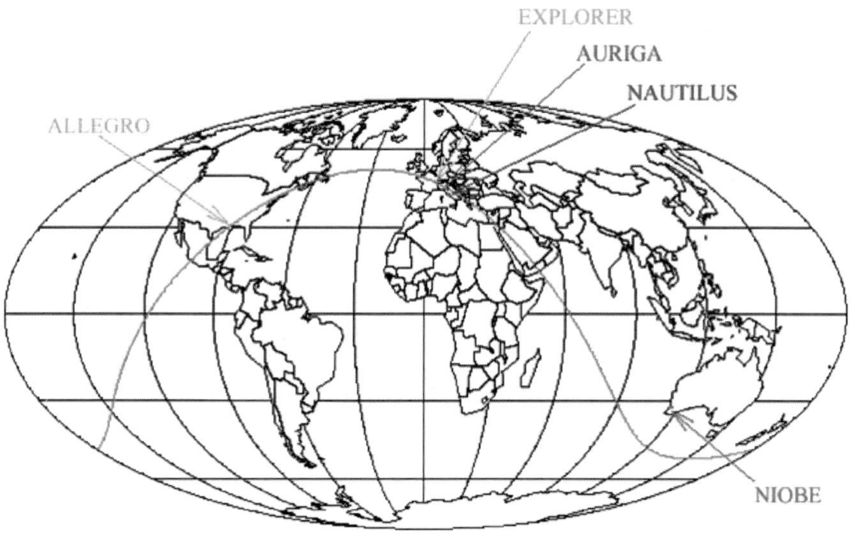

Fig. 10 World location of the five cryogenic Bar detectors

IGEC (Resonant Bar Detectors International Gravitational Event Collaboration) was established. in 1997 in Perth. IGEC has been the first GW Detector network. During IGEC-1 (1997–2000) the following coincidences where registered:

29 days of fourfold coinc.

178 days of threefold coinc.

713 days of twofold coinc.

Sensitivity and bandwidth are shown in Fig. 15.

Fig. 11 The INFN Explorer Cryogenic Detector at CERN

Fig. 12 The INFN Nautilus Cryogenic Detector at INFN Frascati National Laboratory

Fig. 13 The INFN Auriga Cryogenic Detector at INFN Legnaro Laboratory

Followed by a series of upgrades, resumed operations EXPLORER in 2000, AURIGA and NAUTILUS in 2003, ALLEGRO in 2004 the Collaboration started to take data.

With IGEC-2 (2005–) first data analyzed covered May–November 2005 when no other observatory was operating. Sensitivity and operation stability are shown in Figs. 16 and 17 respectively.

Bar Detectors situation at present (Oct. 2007) is the following: NIOBE (Perth) stopped operation and did not join IGEC-2, ALLEGRO (LSU) stopped operation in 2007. In 2006 INFN stopped R&D on Spherical Detectors and left running Auriga, Nautilus and Explorer on an annual evaluation. These three detectors are currently running with very good sensitivity and duty cycle. Their role is particularly relevant when Virgo and LIGO will be shut down for upgradings to Virgo.

Plus and Henanced LIGO; their running is aimed to a coincidence with GEO 600 and Tama 300 interferometers for a kind of observation which has been named "Astro watch".

Fig. 14 The LSU Allegro Cryogenic Detector at Baton Rouge (LU)

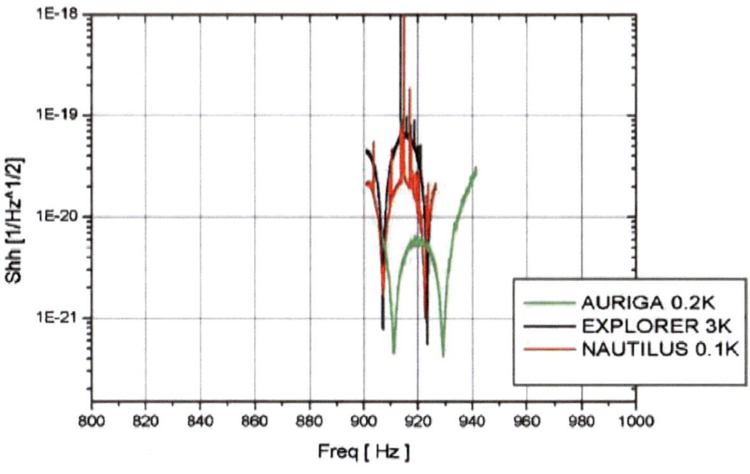

Fig. 15 Sensitivity of Auriga, Explorer and Nautilus during IGEC1 runs. Note the very narrow bandwidth

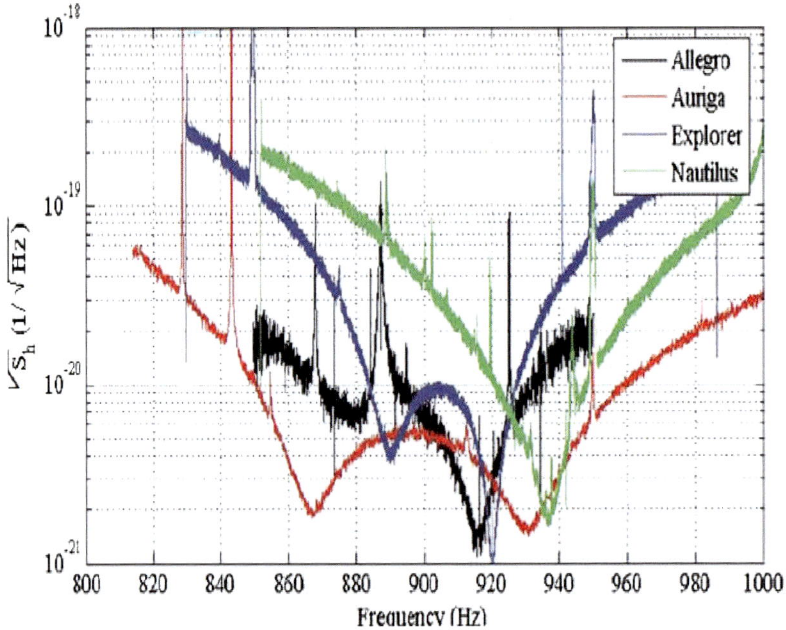

Fig. 16 Sensitivity of Allegro (*black*), Auriga (*red*), Explorer (*blue*) and Nautilus (*green*) during IGEC2 runs. Note the very improved bandwidth

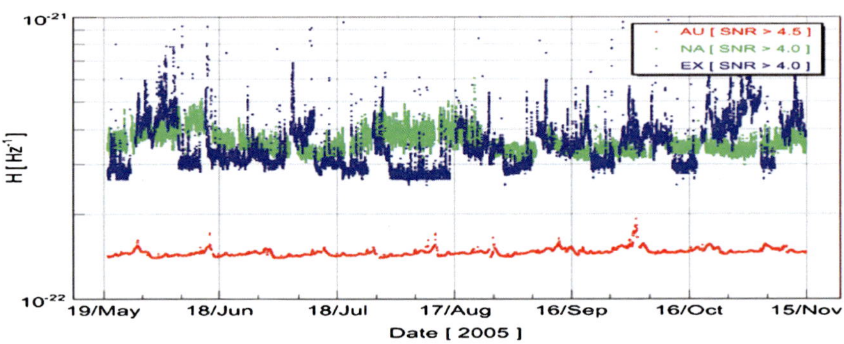

Fig. 17 Operation stability of Auriga (*red*), Explorer (*blue*) and Nautilus (*green*) during IGEC2 runs

7 Spherical Detectors [11]

A spherical antenna will be able to determine the wave polarization and localize its astrophysical source on the sky and is never 'blind' to any particular direction or polarization of the arriving wave. Both these advantages of a spherical antenna are due to its omnidirectionality achieved by the use of at least six transducers

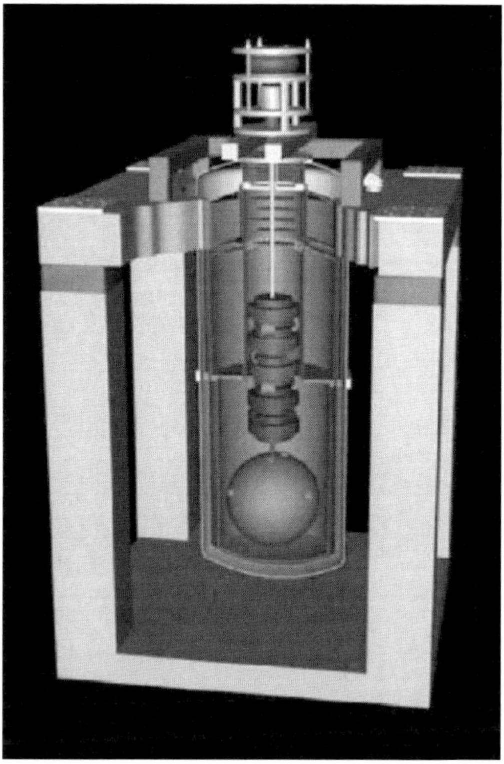

Fig. 18 Lay out of the spherical detector Mario Schenger

placed according to the truncated icosahedron configuration. Another very important advantage of this kind of detector comes exactly from the fact that it has many transducers monitoring many quadrupolar modes.

Mario Schenberg: this detector (see Fig. 18) is built in San Paolo (Brazil) and has a 65 cm-diameter CuAl6% sphere weighing about 1.15 ton; the test mass is only a factor of 2 less massive than existing bar antennas, but by cooling it to much lower temperatures (15–20 mK) and by making it spherical (omnidirectional), we expect a considerable improvement in sensitivity (see Fig. 19).

In this detector each one of the five quadrupole modes of the sphere has an effective mass for oscillation of 287 kg. These five independent modes are coupled to the modes of the six (or nine) transducers, which have oscillating masses of 53.6 g and 10 mg, and form a geometric series with the masses of the sphere modes. All these modes are tuned to the 3.2 kHz resonant frequency and the energy flowing from the sphere modes to the 10 mg transducer masses due to this coupling produces an amplitude gain of about $(287\,\mathrm{kg}/10\,\mathrm{mg})^{1/2} = 5\,k$.

The MiniGRAIL detector is a cryogenic 68 cm diameter spherical gravitational wave antenna (see Fig. 20) built at Kamerlingh Onnes Laboratory of Leiden University, made of CuAl(6%) alloy with a mass of 1,400 kg, a resonance frequency

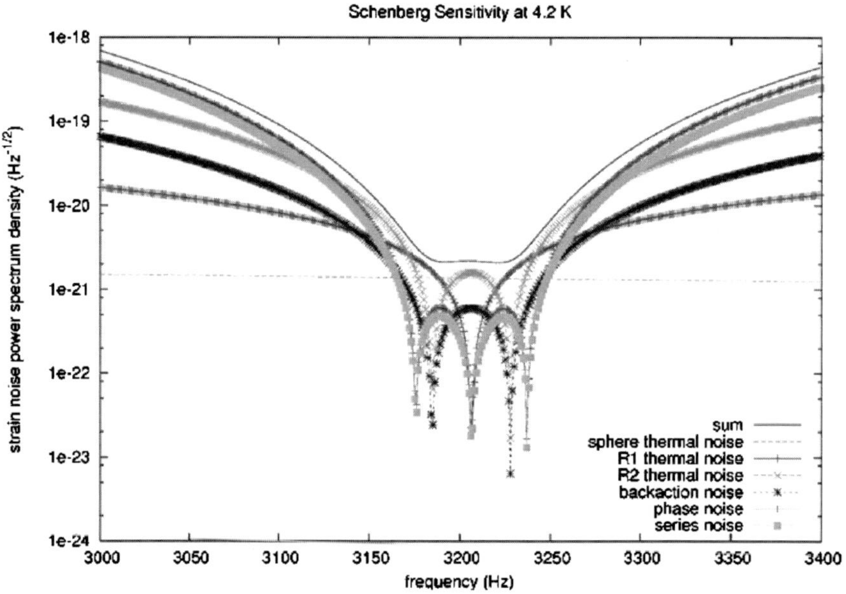

Fig. 19 The Schenberg sensitivity curves at 4.2 K (*upper line*) for a single quadrupole mode and the individual contributions of the noise sources

Fig. 20 Minigrail spherical detector

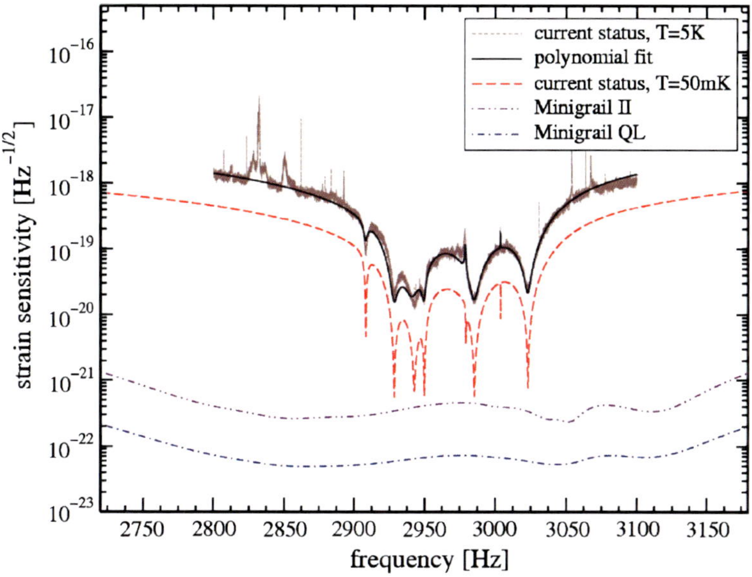

Fig. 21 Sensitivity of the spherical gravitational wave detector MiniGRAIL operating at 5 K. L. Gottardi, A. de Waard, O. Usenko, G. Frossati, M. Podt, J. Flokstra, M. Bassan, V. Fafone, Y. Minenkov, and A. Rocchi. Phys. Rev. D 76, 102005 (2007)

of 2.9 kHz and a bandwidth around 230 Hz, possibly higher. The quantum-limited strain sensitivity dL/L would be $\sim 4 \times 10^{-21}$ (see Fig. 21). The antenna will operate at a temperature of 20 mK.

8 Interferometric Detectors [12]

In this case free falling masses are interferometer mirrors (see Fig. 22) which can be separated km far apart (3 km for Virgo), consequently GW Riemann force can be several order of magnitude larger than in bar detectors. Accompanying large sensitivity is the very large Bandwidth (10–10,000 Hz) due to the fact that mirrors are suspended to pendula having resonance in the Hz region; above resonance frequency mirrors are freely falling masses in the horizontal plane.

With reference to Fig. 23 the two polarizations h_x and h_+ gives contribution to total mirror displacement according to Eq. (27).

From Eq. 27 it is easy to evaluate the sensitivity pattern summed on GW polarizations (see Fig. 24) for an

$$\frac{\Delta L_1 - \Delta L_2}{L} = \frac{1 + (n.\zeta)^2}{2} \frac{(n.\xi)^2 - (n.\eta)^2}{1 - (n.\zeta)^2} h_+ + 2 \frac{(n.\xi)(n.\eta)(n.\zeta)}{1 - (n.\zeta)^2} h_x \tag{27}$$

interferometer having arms at 90°; the antenna pattern is shown in Fig. 25.

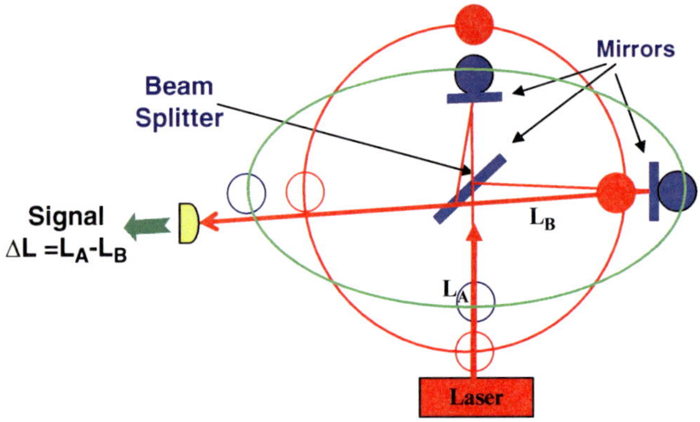

Fig. 22 GW deform a test masses circular ring (*red*) in an elliptical one (*green*) and the *red* masses displace in the *blue* position. If two mirrors are attached to the test masses the arm unbalance $\Delta L = L_A - L_B$ can be measured interferometrically. Interferomer's displacement sensitivity can reach $\sim 10^{-19} - 10^{-20}$ m, then, for measuring $h = \Delta L/L \sim 10^{-22}$ L_A and L_B should be km long

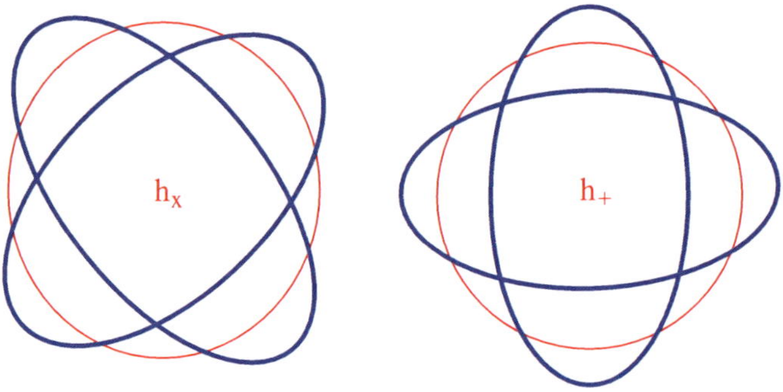

Fig. 23 Effect of the two GW polarizations on a *circular ring* (*red*) of masses for a wave travelling normally to the *circle plane*

Fig. 24 Coordinate system having ξ and η axis coinciding with interferometer arms; n is GW versor and θ and φ are his polar and azimuthal angles

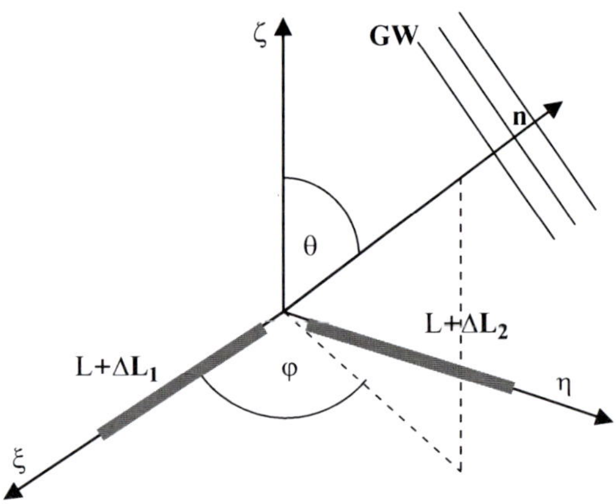

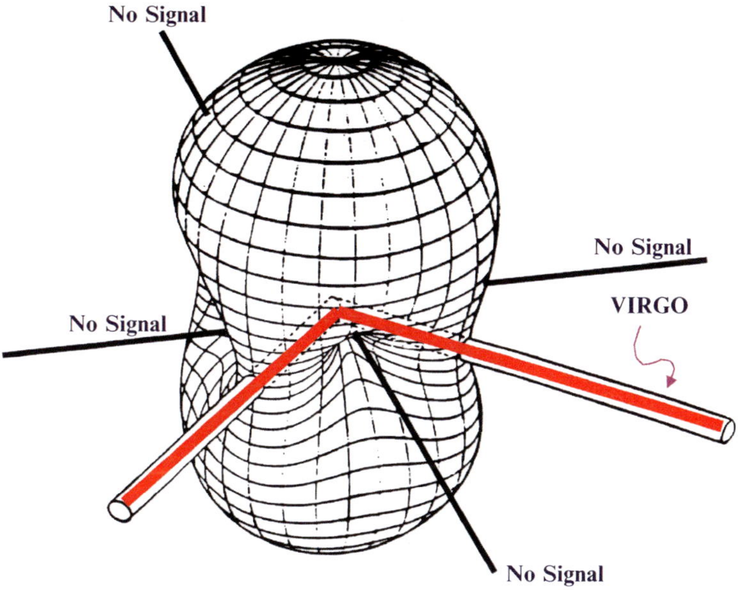

Fig. 25 Antenna Pattern for an interferometer (*in red*) having arms at 90°; the pattern is evaluated by summing on GW polarizations. The zero sensitivity regions are bisector lines in *black*, laying on the interferometer plane

9 Interferometer Noises

It is easy to imagine that a plethora of noises is affecting interferometers for GW detection; infact this kind of interferometers has mirrors suspended and free to move in the horizontal plane. We should imagine how small is the displacement to be measured by these machines i.e. $<10^{-19}$ m/sqrt(Hz) in the frequency band 4–10,000 Hz.

Just for making a mere list we have (see Figs. 26, 27):

Optical Noises Shot Noise, Laser frequency fluctuations, Laser power fluctuations, Diffused and Scattered light hitting vacuum vessels, where collects vibrations, and partially reflected back in the main optical path, Acoustic noise....

Force Noises Radiation pressure fluctuations, Thermal noise in the mirrors and in the mirror suspensions, Electromagnetic noises producing forces on the mirrors....

In Fig. 28 I have kept only two kinds of fundamental noises: Thermal noise and Optical noises, i.e. Shot noise and Radiation pressure fluctuation.

In the following I will try to give some ideas on how Quantum Non-Demolition (QND) techniques can reduce optical noises in GW interferometers.

If we define W the laser power, F and L the cavity finesse and length respectively, ν_0 the laser frequency, Ω the observation frequency and M the mirror mass, the

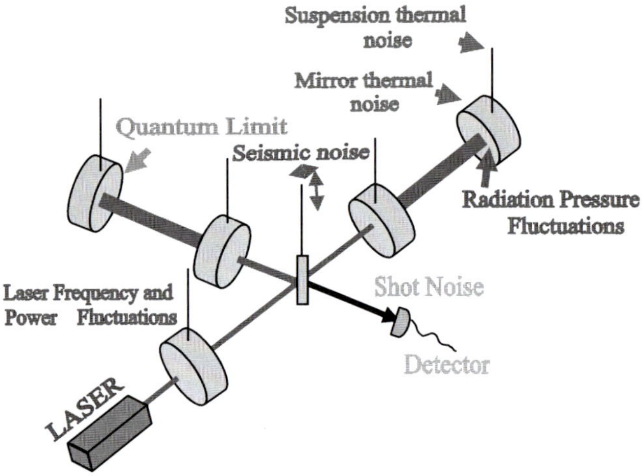

Fig. 26 Pictorial view of noises affecting a suspended mirror interferometer

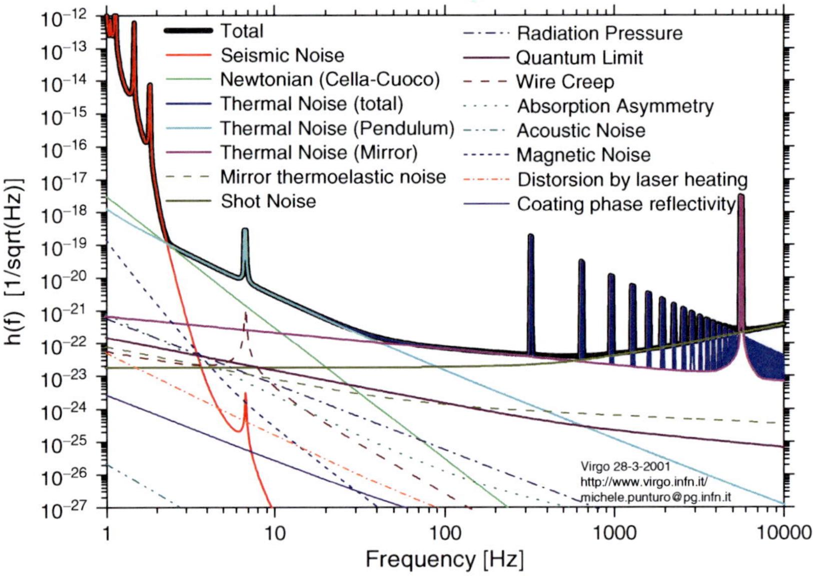

Fig. 27 Some noises affecting Virgo

optical noises will give the following limit to $\tilde{h}$:

Radiation pressure fluctuations Shotnoise

$$\tilde{h}^2 > \left(\frac{2}{LM\Omega^2 c}\right)^2 h\upsilon_0 WF^2 + \left(\frac{\lambda}{4\pi}\right)^2 \frac{h\upsilon_0}{WF^2} \frac{1 + \left(\frac{\Omega FL}{c}\right)^2}{L^2} \tag{28}$$

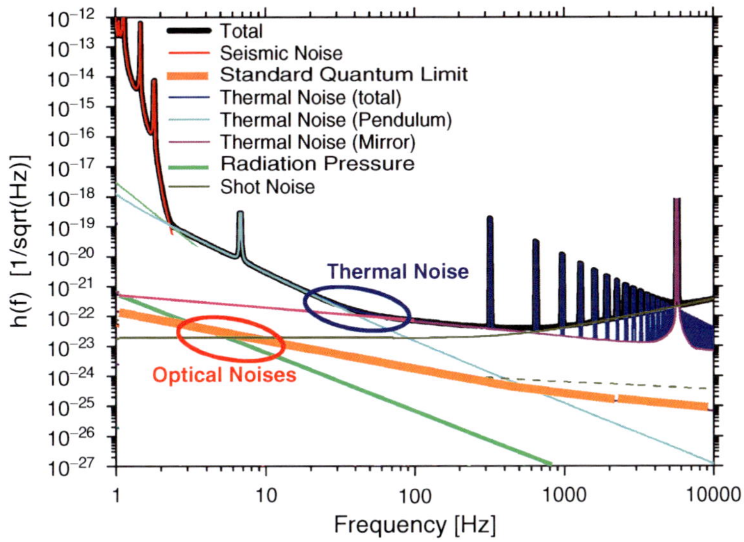

Fig. 28 In this plot the accent is given on two kinds of fundamental noises: Thermal noise and Optical Noises, i.e., shot noise and radiation pressure fluctuation noise. Thermal noise is likely to be reduced by keeping the Interferometer mirrors close to 4 K

In Eq. (28) the first term on the right side is due to to radiation pressure fluctuation while the second is due to shot noise. By minimizing $\tilde{h+}$ with respect WF^2 and specializing Eq. (28) to a Michelson interferometer configuration, we obtain the Standard Quantum Limit (SQL) for both the GW amplitude h and the laser power on the beam splitter, W_{SQL}:

$$\tilde{h}_{SQL} > \tilde{h}_{SQL} > \sqrt{\frac{8\hbar}{M\Omega^2L^2}\left(1+\left(\frac{\Omega FL}{c}\right)^2\right)^{1/2}}$$

$$W_{SQL} = \frac{\lambda M\Omega^2 c}{F^2 8\pi}\sqrt{1+\left(\frac{\Omega FL}{c}\right)^2} \tag{29}$$

10 Modern Interferometers with QND Signal Readout [13]

We can describe the evolution of the total electromagnetic field (emf) as in Eq. (30):

$$E_{Total} = C\left[(\alpha_0 + a_0(t))\cos\omega_0 t + \left(\alpha_{\pi/2} + a_{\pi/2}(t)\right)\sin\omega_0 t\right] \tag{30}$$

Where $(\alpha_0)^2 + (\alpha_{\pi/2})^2 = 2W/(h\nu_0)$ is the amplitude of the carrier in a coherent state.

The carrier and the vacuum fluctuations amplitudes are described by the doublets:

$$\begin{array}{ll} \textbf{Carrier} & \textbf{Quadratures} \\ \text{ClassicalFields} & \text{QuantumFields} \\ \begin{pmatrix} \alpha_0 \\ \alpha_{\pi/2} \end{pmatrix} & \begin{pmatrix} a_0(\Omega) \\ a_{\pi/2}(\Omega) \end{pmatrix} \end{array}$$

$$\begin{cases} a_0(\Omega) = \dfrac{a_+ + a_-^+}{\sqrt{2}}, a_{\pi/2}(\Omega) = \dfrac{a_+ - a_-^+}{i\sqrt{2}} \\ \left[a_+, a_+^{'+}\right] = \left[a_-, a_-^{'+}\right] = 2\pi\delta(\Omega - \Omega') \end{cases} \tag{31}$$

the quadratures satisfy the usual commutation relation for emf creation and annihilation operators. The quantum field parallel to classical field can be interpreted as intensity fluctuations/operator, while the orthogonal can be interpreted as phase fluctuations. We can describe the evolution of the emf vacuum fluctuations propagating in a optical system by means of the quadrature state Eq. (31). The classical and fluctuating fields α and a are entering the optical cavity while the reflected fields are Q and $F_{0,\pi/2}$ respectively; $F_{0,\pi/2}$ is also including contributions from inner vacuum fields created by mirror losses B (for the sake of simplicity I have attributed the same loss coefficient to the two mirrors) (see Fig. 29).

In Eq. (32) the general equations, valid for any frequency ω_0 and Ω, for F, Q, P and G, are presented.

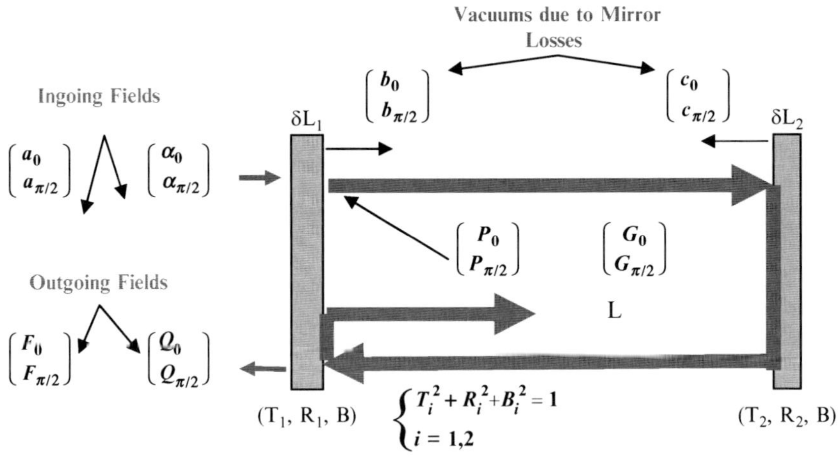

Fig. 29 The in fields α and a, and the out fields Q and F after interacting with a Fabry Perot cavity having length L, mirror transmittance T and reflectivity R and losses B. δL_1 and δL_2 are the infinitesimal mirror displacements produced by radiation pressure, GW Riemann force and by any other external force. P and G are the intracavity fields

$$
\left\{
\begin{aligned}
&\begin{pmatrix} F_0 \\ F_{\pi/2} \end{pmatrix} = \frac{1 - R_1 e^{-\frac{2i\Omega L}{c}} R\left(-\omega_0 \frac{2L}{c}\right)}{1 - 2R_1 e^{-\frac{2i\Omega L}{c}} \cos \omega_0 \frac{2L}{c} + R_1^2 e^{-\frac{4i\Omega L}{c}}} \left[\left(e^{-\frac{2i\Omega L}{c}} R\left(\omega_0 \frac{2L}{c}\right) - R_1\right) \begin{pmatrix} a_0 \\ a_{\pi/2} \end{pmatrix} \right. \\
&\quad + \sqrt{2} B T_1 e^{-\frac{2i\Omega L}{c}} R\left(\omega_0 \frac{2L}{c}\right) \begin{pmatrix} w_0 \\ w_{\pi/2} \end{pmatrix} \\
&\quad \left. + T_1^2 2 \frac{\omega_0 \left(\delta \tilde{L}_2 e^{-\frac{i\Omega L}{c}} - \delta \tilde{L}_1\right)}{c} \frac{1 - R_1 R\left(-\omega_0 \frac{2L}{c}\right)}{1 - 2R_1 \cos \omega_0 \frac{2L}{c} + R_1^2} R\left(\omega_0 \frac{2L}{c} + \frac{\pi}{2}\right) \begin{pmatrix} \alpha_0 \\ \alpha_{\pi/2} \end{pmatrix} \right] \\
&\begin{pmatrix} Q_0 \\ Q_{\pi/2} \end{pmatrix} = \frac{\left(R\left(\omega_0 \frac{2L}{c}\right) - R_1\right)\left(1 - R_1 R\left(-\omega_0 \frac{2L}{c}\right)\right)}{1 - 2R_1 \cos \omega_0 \frac{2L}{c} + R_1^2} \begin{pmatrix} \alpha_0 \\ \alpha_{\pi/2} \end{pmatrix} \\
&\begin{pmatrix} P_0 \\ P_{\pi/2} \end{pmatrix} = T_1 \frac{1 - R_1 e^{-\frac{2i\Omega L}{c}} R\left(-\omega_0 \frac{2L}{c}\right)}{1 - 2R_1 e^{-\frac{2i\Omega L}{c}} \cos \omega_0 \frac{2L}{c} + R_1^2 e^{-\frac{4i\Omega L}{c}}} \left\{ \left[\begin{pmatrix} a_0 \\ a_{\pi/2} \end{pmatrix} + \frac{\sqrt{2} B}{T_1} \begin{pmatrix} w_0 \\ w_{\pi/2} \end{pmatrix} \right] \right. \\
&\quad \left. + \frac{2\omega_0 \frac{\delta L_2 e^{-\frac{i\Omega L}{c}} - \delta L_1}{c} R_1 \left(1 - R_1 R\left(-\omega_0 \frac{2L}{c}\right)\right) R\left(\omega_0 \frac{2L}{c} + \frac{\pi}{2}\right) \begin{pmatrix} \alpha_0 \\ \alpha_{\pi/2} \end{pmatrix}}{1 - 2R_1 \cos \omega_0 \frac{2L}{c} + R_1^2} \right\} \\
&\begin{pmatrix} G_0 \\ G_{\pi/2} \end{pmatrix} = T_1 \frac{1 - R_1 R\left(-\omega_0 \frac{2L}{c}\right)}{1 - 2R_1 \cos \omega_0 \frac{2L}{c} + R_1^2} \begin{pmatrix} \alpha_0 \\ \alpha_{\pi/2} \end{pmatrix} \\
&\begin{pmatrix} G_0 & G_{\pi/2} \end{pmatrix} = T_1 \begin{pmatrix} \alpha_0 & \alpha_{\pi/2} \end{pmatrix} \frac{1 - R_1 R\left(\omega_0 \frac{2L}{c}\right)}{1 - 2R_1 \cos \omega_0 \frac{2L}{c} + R_1^2} \\
&R(\vartheta) = \begin{pmatrix} \cos\vartheta & -\sin\vartheta \\ \sin\vartheta & \cos\vartheta \end{pmatrix}
\end{aligned}
\right.
\tag{32}
$$

The field w is resulting from combination of the two vacuum fields b and c (see Fig. 29). The mirror 2 is supposed to have $R_2 \sim 1$, typical value for mirror configuration in GW detectors, for avoiding loosing power from cavity backs.

Where $F = 1/(1 - \mathrm{R}_1)$ is cavity finesse.

With reference to Eq. (32), radiation pressure fluctuations couple to intracavity power W_{ic} and, by keeping only first order terms in P and in resonance, give the power fluctuations δW_{ic}:

$$
W_{ic}(t) = \frac{h\nu_0}{2} \left((P_0(t) + G_0) \cos \omega_0 t + \left(P_{\pi/2}(t) + G_{\pi/2}\right) \sin \omega_0 t \right)^2
$$

$$
\delta W_{ic}(t) \cong \frac{h\nu_0}{2} \left((P_0(t)\, G_0) + \left(P_{\pi/2}(t)\, G_{\pi/2}\right) \right) \tag{33}
$$

By means of δW_{ic} we may, after straightforward calculations, evaluate $(\delta L_2 - \delta L_1)$ due to Radiation Pressure fluctuations:

$$\delta\tilde{L}_2 e^{-\frac{i\Omega L}{c}} - \delta\tilde{L}_1 \approx 4\frac{\delta W_{ic}(\Omega)}{Mc\left(-\Omega^2+\Omega_0^2\right)}$$
$$= 4\frac{\frac{h\nu_0}{2}}{Mc\left(-\Omega^2+\Omega_0^2\right)}\left(\left(P_0(t)\,G_0\right)+\left(P_{\pi/2}(t)\,G_{\pi/2}\right)\right) = 4\frac{\frac{h\nu_0}{2}T_1^2}{Mc\left(-\Omega^2+\Omega_0^2\right)}\begin{pmatrix}\alpha_0 & \alpha_{\pi/2}\end{pmatrix}$$
$$* \frac{\left(1-R_1R\left(\omega_0\frac{2L}{c}\right)\right).\left(1-R_1e^{-\frac{2i\Omega L}{c}}R\left(-\omega_0\frac{2L}{c}\right)\right)}{\left(1-2R_1\cos\omega_0\frac{2L}{c}+R_1^2\right)\left(1-2R_1e^{-\frac{2i\Omega L}{c}}\cos\omega_0\frac{2L}{c}+R_1^2e^{-\frac{4i\Omega L}{c}}\right)}$$
$$* \left[\left[\begin{pmatrix}a_0\\a_{\pi/2}\end{pmatrix}+\frac{\sqrt{2}B}{T_1}\begin{pmatrix}w_0\\w_{\pi/2}\end{pmatrix}\right] + \frac{2\omega_0\frac{\delta\tilde{L}_2e^{-\frac{i\Omega L}{c}}-\delta\tilde{L}_1}{c}R_1\left(1-R_1R\left(-\omega_0\frac{2L}{c}\right)\right)R\left(\omega_0\frac{2L}{c}\right)\begin{pmatrix}-\alpha_{\pi/2}\\\alpha_0\end{pmatrix}}{1-2R_1\cos\omega_0\frac{2L}{c}+R_1^2}\right] \quad (34)$$

Where Ω_0 is the mirror pendum suspension frequency, usually in the Hz region, and M the mirror mass. Substituting $\delta L_2 e^{-i\Omega L/c} - \delta L_1$ in Eq. (32), at resonance $(2\omega_0 L/c = 2n\pi)$ and putting $\alpha_{\pi/2} = 0$, we obtain:

$$\begin{pmatrix}F_0\\F_{\pi/2}\end{pmatrix} = \left[\frac{\left(e^{-\frac{2i\Omega L}{c}}-R_1\right)}{\left(1-R_1e^{-\frac{2i\Omega L}{c}}\right)}\begin{pmatrix}a_0\\a_{\pi/2}\end{pmatrix}+\frac{\sqrt{2}BT_1e^{-\frac{2i\Omega L}{c}}}{\left(1-R_1e^{-\frac{2i\Omega L}{c}}\right)}\begin{pmatrix}w_0\\w_{\pi/2}\end{pmatrix}\right]$$
$$+8\frac{\omega_0 W}{Mc^2\left(-\Omega^2+\Omega_0^2\right)}\frac{(1+R_1)^2}{\left(1-R_1e^{-\frac{2i\Omega L}{c}}\right)^2}$$
$$\times\left[a_0+\frac{\sqrt{2}B}{\sqrt{1-R_1}\sqrt{1+R_1}}w_0\right]\begin{pmatrix}0\\1\end{pmatrix} \quad (35)$$
$$+2\frac{\omega_0\sqrt{\frac{2W}{h\nu_0}}\left(\delta L_2e^{-\frac{i\Omega L}{c}}-\delta L_1\right)(1+R_1)}{c\left(1-R_1e^{-\frac{2i\Omega L}{c}}\right)}\begin{pmatrix}0\\1\end{pmatrix}$$

Equation (35) has the following structure: the first term in rhs is the contribution from external and internal vacuums, the second therm is the contribution from radiation pressure fluctuations acting on the mirrors itself and consequently creating a phase shift, the third term creates a phase shift when GW or other forces act on the mirrors.

We can put Eq. (35) in a simpler form since $R_1 \sim 1$, if $\Omega L/c \ll 1$:

$$\begin{cases} \tilde{\Phi}_{Sig} = \sqrt{|K|}\sqrt{\dfrac{2\pi Mc^2\left(-\Omega^2+\Omega_0^2\right)}{h}}\left(\delta\tilde{L}_2 e^{-\dfrac{i\Omega L}{c}} - \delta\tilde{L}_1\right)_{Sig} \\ K = -\dfrac{32\omega_0 W}{Mc^2\left(-\Omega^2+\Omega_0^2\right)}\dfrac{F^2}{\left(1+4\left(F\sin\dfrac{\Omega L}{c}\right)^2\right)} \\ \Gamma = \begin{pmatrix} a_0 \\ a_{\pi/2} - Ka_0 + \tilde{\Phi}_{Sig} \end{pmatrix} \\ \begin{pmatrix} F_0 \\ F_{\pi/2} \end{pmatrix} = \Gamma + 2B\sqrt{F}\begin{pmatrix} w_0 \\ w_{\pi/2} + \dfrac{K}{2}w_0 \end{pmatrix} \end{cases} \tag{36}$$

Where $F = 1/(1-R_1)$ is cavity finesse and K is the Squeezing factor; evaluation of the covariance matrix of Γ allows to see the effect of squeezing on SNR:

$$C = \begin{pmatrix} 1 & -K \\ -K & 1+K^2 \end{pmatrix} \tag{37}$$

This matrix shows that the initial coherent state, due to the fact that mirrors are moved by radiation pressure, is transformed in a squeezed state:

From Eq. (36) and Fig. 30 it follows signal to noise ratio to be

$$\frac{S}{N} = \frac{\sqrt{\tilde{\Phi}_{Sig}^2}}{\sqrt{1+K^2}} \tag{38}$$

From Eqs. (27), (36) and (38) it follows the spectral function for the GW amplitude:

$$\tilde{h}_\Omega = \sqrt{\tilde{h}^2} > \sqrt{\frac{\hbar}{Mc^2L^2\left(-\Omega^2+\Omega_0^2\right)}\left(\frac{1}{K}+K\right)} \tag{39}$$

Eq. (39) shows how ponderomotive effects limit interferometer sensitivity to Standard Quantum Limit Eq. (29).

In GW detection interferometer the interesting signal is hidden in the phase φ measurement and not in the number of photon N measurement; hence, since, by uncertainty principle $\Delta N\ \Delta\varphi > 1$, we may completely disregard ΔN by trying to minimize $\Delta\varphi$. This can be done by rotating by an angle Φ the ellips of Fig. 23, in

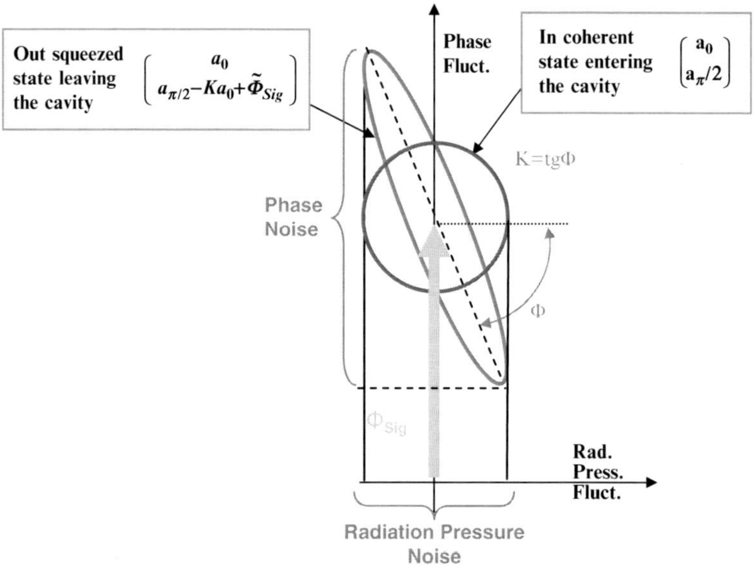

Fig. 30 The state in is in a coherent state while ponderomotive forces created by radiation pressure fluctuations displace mirrors, then creating a phase shift Ka_0

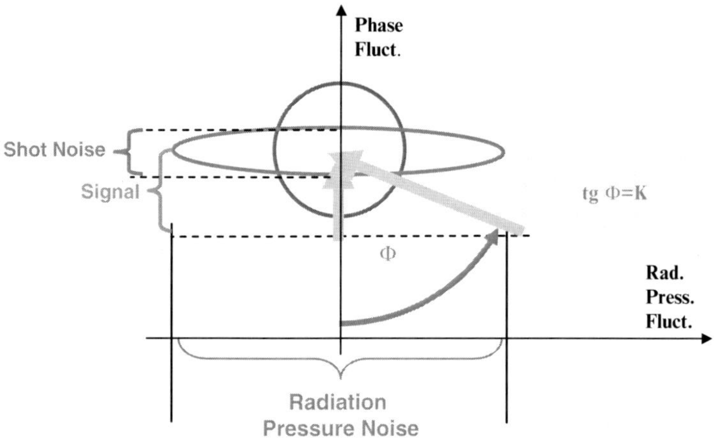

Fig. 31 Rotations in the quadrature space allow to optimize S/N; in this case since we only measure phase shifts we can improve phase sensitivity at expenses of of number of photons uncertainty

fact S/N is optimized if we rotate the quadrature spinor by an angle $\Phi = \text{arctgK}$ (see Fig. 31). The rotation operator being:

$$\begin{cases} \begin{pmatrix} \cos\Phi & -\sin\Phi \\ \sin\Phi & \cos\Phi \end{pmatrix} = \frac{1}{\sqrt{1+K^2}} \begin{pmatrix} 1 & -K \\ K & 1 \end{pmatrix} \approx \begin{pmatrix} 1/K & -1 \\ 1 & 1/K \end{pmatrix} \\ K \gg 1 \end{cases} \tag{40}$$

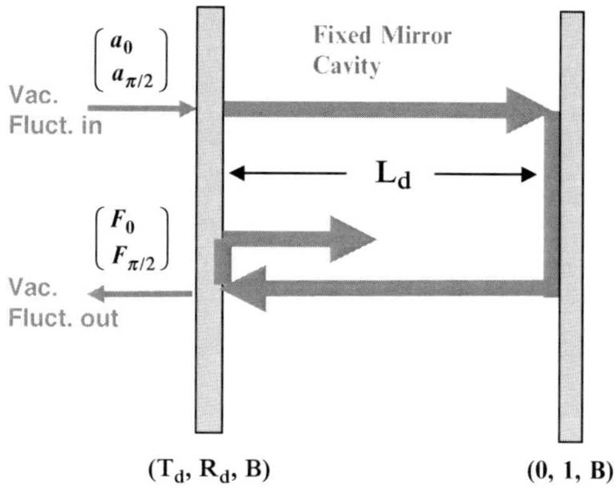

Fig. 32 Rotations in the quadrature space are performed by a fix mirror detuned FP cavity

Operating on Γ we obtain:

$$\begin{pmatrix} 1/K & -1 \\ 1 & 1/K \end{pmatrix} \begin{pmatrix} a_0 \\ a_{\pi/2} - K a_0 + \tilde{\Phi}_{Sig} \end{pmatrix} = \begin{pmatrix} a_0 \left(\frac{1}{K} + K \right) + a_{\pi/2} + \tilde{\Phi}_{Sig} \\ \frac{a_{\pi/2} + \tilde{\Phi}_{Sig}}{K} \end{pmatrix} \tag{41}$$

And S/N becomes:

$$\frac{S}{N} = \frac{\sqrt{\tilde{\Phi}^2_{Sig}}}{1} >> \frac{\sqrt{\tilde{\Phi}^2_{Sig}}}{\sqrt{1+K^2}} \tag{42}$$

The unitary matrix performing this rotation can be found by analyzing the rotation properties of a fix mirror Fabry–Perot optical cavity; let's consider the cavity shown in Fig. 32:

The input–output matrix is

$$\frac{\begin{bmatrix} -R_d\left(1+e^{-\frac{4i\Omega L_d}{c}}\right)+e^{-\frac{2i\Omega L_d}{c}}\left(1+R_d^2\right)\cos\omega_0\frac{2L_d}{c} & -e^{-\frac{2i\Omega L_d}{c}}\left(1-R_d^2\right)\sin\omega_0\frac{2L_d}{c} \\ e^{-\frac{2i\Omega L_d}{c}}\left(1-R_d^2\right)\sin\omega_0\frac{2L_d}{c} & -R_d\left(1+e^{-\frac{4i\Omega L_d}{c}}\right)+e^{-\frac{2i\Omega L_d}{c}}\left(1+R_d^2\right)\cos\omega_0\frac{2L_d}{c} \end{bmatrix}}{1-2R_d e^{-\frac{2i\Omega L_d}{c}}\cos\omega_0\frac{2L_d}{c}+R_d^2 e^{-\frac{4i\Omega L_d}{c}}} \tag{43}$$

With reference to Eq. (40), since the rotation should be such that tgΦ = K, and in Eq. (36) it is always $\Omega^2 \gg {\Omega_0}^2$, the following equation should be satisfied:

$$\begin{cases} tg\Phi = \dfrac{\left(1-R_d^2\right)\sin\omega_0\dfrac{2L_d}{c}}{-2R_d\cos\dfrac{2\Omega L_d}{c}+\left(1+R_d^2\right)\cos\omega_0\dfrac{2L_d}{c}} = K \\ K \approx \dfrac{32\omega_0 W}{Mc^2\Omega^2}\dfrac{F^2}{\left(1+4\left(F\sin\dfrac{\Omega L}{c}\right)^2\right)} \\ F_d = \dfrac{1}{1-R_d} \end{cases} \tag{44}$$

For giving tgΦ an approximate $1/\Omega^2$ behavior (as K does), we should have

$$\begin{cases} -2R_d + \left(1 + R_d^2\right)\cos\omega_0 \dfrac{2L_d}{c} = 0 \\ \sin\omega_0 \dfrac{2L_d}{c} = \dfrac{1}{F_d}\sqrt{\dfrac{2}{\left(1 + R_d^2\right)}} \end{cases} \tag{45}$$

Since $\sin(2\omega_0 L_d/c) \sim 1/\mathrm{F_d}$, the Intra Cavity Intensity becomes half max.

Eq. (44) becomes:

$$F_d = \frac{c^2}{8L_d F}\sqrt{\frac{Mc^3\lambda}{4\pi W}\left(1 + 4\left(F\sin\frac{\Omega L}{c}\right)^2\right)\sqrt{\frac{2\left(1 + R_d\right)^2}{R_d^2\left(1 + R_d^2\right)}}} \tag{46}$$

Since $R_d \sim 1$, by putting $M = 100$ kg, laser wavelenght $\lambda = 10^{-6}$ m, input power $W = 1{,}000$ w, with the condition $\sin(2\Omega L/c) \ll 1$, it follows:

$$F_d \approx \frac{10^8}{3,2L_d F} \tag{47}$$

This finesse value is very high and can be reduced by making F and L_d large.

(a) How Radiation Pressure and Shot noise affects noise in a Michelson Interferometer

It is very interesting now to understand how Radiation Pressure and Shot affects noise in a resonating Michelson Interferometer; we start by studying the effects of the vacuum fluctuations entering both Laser and Beam Splitter ports of a FP Michelson interferometer shown in Fig. 33.

We define:

$W_{1,2}$ Power entering FP cavities
$L_{1,2}$ FP cavity length
$D_{1,2}$ FP-BS Length
$R_{1,2}$ FP entrance mir. Transmittance of far away mirrors is tuned on zero.
$B_{1,2}$ FP mirror losses
$\Delta l_{1,2}$ FP mirrors displacement
$b_{0,\pi/2}$ Laser Port Vac. fluct.
$v_{0,\pi/2}$ Beam Splitter Port Vac. Fluct.
$w1_{0,\pi/2}$, $w2_{0,\pi/2}$ Internal Vac. fluct.
$\alpha_{0,\pi/2}$, $\beta_{0,\pi/2}$ Classical Fields
$F_{1,2}$ Finesse

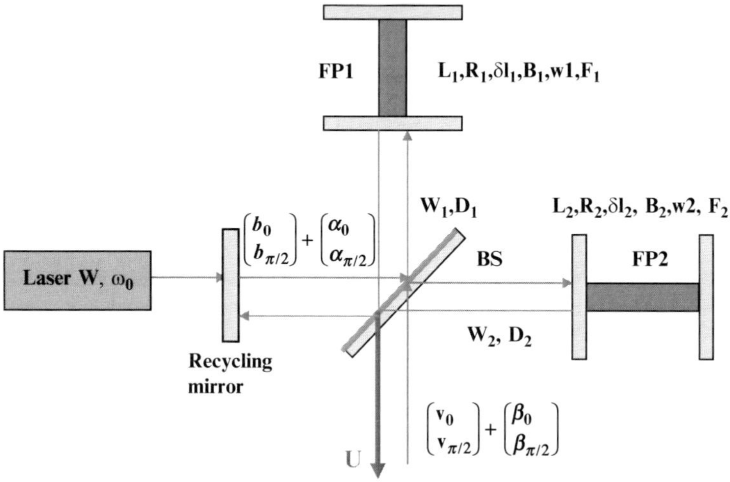

Fig. 33 Vacuums and classical fields entering a Michelson Fabry–Perot interferometer for GW detection

With reference to Fig. 33 we obtain for the output field U:

$$U = \frac{1}{2}\begin{pmatrix} \cos 2w_0 D_1/c & -\sin 2w_0 D_1/c \\ \sin 2w_0 D_1/c & \cos 2\omega_0 D_1/c \end{pmatrix} \Bigg[\underbrace{\begin{pmatrix} b_0 + v_0 \\ b_{p/2} + v_{p/2} - (b_0+v_0)K_1 \end{pmatrix} e^{2i\Lambda_1}}_{\text{FP Ponderomotive}} + \underbrace{2B_1\sqrt{F_1}\begin{pmatrix} w_{1,0} \\ w_{1,p/2} - \frac{K_1}{2} w_{1,0} \end{pmatrix} e^{2i\Lambda_1}}_{\text{FP cavities losses}} + \underbrace{\Gamma_1 L_1 \tilde{h}_1 \begin{pmatrix} 0 \\ 1 \end{pmatrix}}_{\text{GW signal}} \Bigg]$$
$$+ \frac{1}{2}\begin{pmatrix} \cos 2w_0 D_2/c & -\sin 2w_0 D_2/c \\ \sin 2w_0 D_2/c & \cos 2w_0 D_2/c \end{pmatrix} \Bigg[\begin{pmatrix} -b_0 + v_0 \\ -b_{p/2} + v_{p/2} - (-b_0+v_0)K_2 \end{pmatrix} e^{2i\Lambda_2} + 2B_2\sqrt{F_2}\begin{pmatrix} w_{2,0} \\ w_{2,p/2} - \frac{K_2}{2} w_{2,0} \end{pmatrix} e^{2i\Lambda_2} - \Gamma_2 L_2 \tilde{h}_2 \begin{pmatrix} 0 \\ 1 \end{pmatrix} \Bigg] \tag{48}$$

Where:

$$K_J \approx \frac{32 W_J \omega_0 F_J^2}{M c^2 \Omega^2} \quad \Gamma_J = \frac{2\pi}{\lambda}\sqrt{\frac{32 W_J}{h\nu_0}} \frac{e^{i\Omega L_J/c}}{1 - R_J e^{-2i\Omega L_J/c}} \quad \begin{matrix} \Lambda_J = 2\Omega L_J F_J/c \\ F_J = 1/(1-R_J) \end{matrix}$$
$$J = 1, 2 \tag{49}$$

When the two FP cavities are equal i.e:

$D = D1 = D2$
$L = L1 = L2$
$\Gamma = \Gamma 1 = \Gamma 2$
$B = B_1 = B_2$
$\Lambda = \Lambda_1 = \Lambda_2$
$F = F1 = F2$
$K = K1 = K2$

Since w_1 and w_2 are incoherent vacuums we can sum them to form a new vacuum w multiplied with amplitude multiplied by $2^{1/2}$:

$$U = \frac{1}{2}\begin{pmatrix} \cos 2\omega_0 D/c & -\sin 2\omega_0 D/c \\ \sin 2\omega_0 D/c & \cos 2\omega_0 D/c \end{pmatrix}\left[\begin{pmatrix} v_0 \\ v_{\pi/2} - v_0 K \end{pmatrix} e^{2i\Lambda} + 2\sqrt{2}B\sqrt{F}\begin{pmatrix} w_0 \\ w_{\pi/2} + w_0 K/2 \end{pmatrix} e^{2i\Lambda} + \Gamma L\left(\tilde{h}_1 - \tilde{h}_2\right)\begin{pmatrix} 0 \\ 1 \end{pmatrix}\right] \quad (50)$$

Eq. (50) clearly shows that, in a resonant symmetrical configuration, the only contribution to the noise is given by the vacuum quadrature operators $v_{0,\pi/2}$ entering the BS from the dark fringe port; in this configuration there is no noise contribution from $b_{0,\pi/2}$. *This is a fundamental result showing that we may overcome SQL by injecting in the Michelson's BS Black Port a squeezed vacuum.*

(b) Signal Recycling

This scheme is very important because allows to reduce radiation pressure noise; GEO 600 has already implemented succesfully this technique. This technique will be extensively used in next generation interferometers for GW detection.

By adding a further mirror at the output port of the optical configuration shown in Fig. 33 we have a Signal recycling configuration (see Fig. 34).

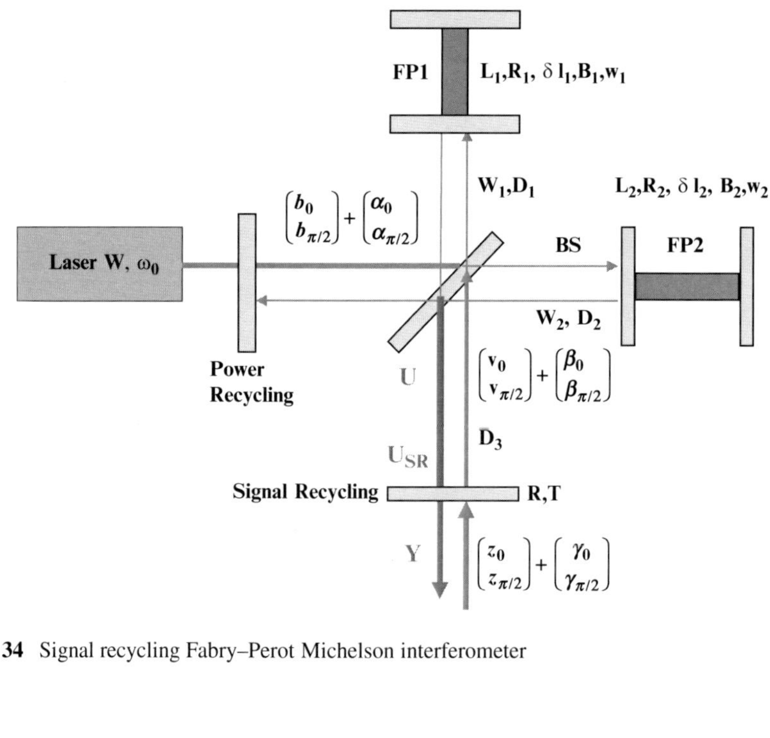

Fig. 34 Signal recycling Fabry–Perot Michelson interferometer

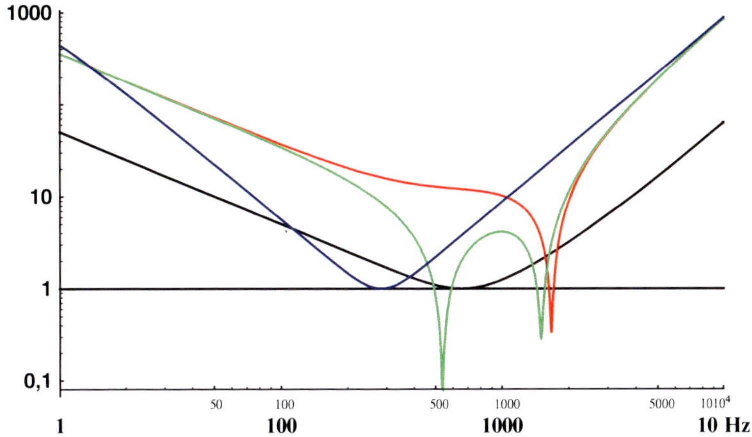

Fig. 35 Strain sensitivity, espressed in units of $\hat{h}_{SQL}$ of Eq. (29), for a signal recycled interferometer. The input power is fixed to $W_{SQL}/10$ (see Eq. (29)), where W_{SQL} is the intensity needed to obtain the standard quantum limit at the Fabry–Perot cavity cut off frequency, in absence of recycling. The *black curve* is the case without recycling, plotted for reference. The *red, green* and *blue curves* correspond to a detuning $2\omega_0 D_3/c = \pi/4, -\pi/4, 0$. The reflectivity R of the recycling mirror is set to 0.99 (Courtesy G. Cella)

Under the hypothesis of totally symmetric resonant interferometer, we obtain:

$$
\begin{aligned}
U_{SR} &= R\left(\omega_0 \frac{D_3}{c}\right)\frac{1}{2}\left[\begin{pmatrix} v_0 \\ v_{\pi/2} - v_0 K \end{pmatrix} e^{2i\Lambda} + 2\sqrt{2}B\sqrt{F}\begin{pmatrix} w_0 \\ w_{\pi/2} + w_0 K/2 \end{pmatrix} e^{2i\Lambda} + \Gamma L\left(\tilde{h}_1 - \tilde{h}_2\right)\begin{pmatrix} 0 \\ 1 \end{pmatrix}\right] \\
V &= TR\left(\omega_0 \frac{D_3}{c}\right) Z - RR\left(\omega_0 \frac{D_3}{c}\right) U_{SR} \\
Y &= TU_{SR} - RZ \\
Y &= T\frac{\begin{pmatrix} 1 + R\left(\cos 2\omega_0 \frac{D_3}{c} + \frac{K}{2}\sin 2\omega_0 \frac{D_3}{c}\right)\frac{e^{2i\Lambda}}{2} & R\left(\sin 2\omega_0 \frac{D_3}{c} + K\sin^2\omega_0 \frac{D_3}{c}\right)\frac{e^{2i\Lambda}}{2} \\ R\left(-\sin 2\omega_0 \frac{D_3}{c} + \mathrm{K}\cos^2\omega_0 \frac{D_3}{c}\right)\frac{e^{2i\Lambda}}{2} & 1 + R\left(\cos 2\omega_0 \frac{D_3}{c} - \frac{K}{2}\sin 2\omega_0 \frac{D_3}{c}\right)\frac{e^{2i\Lambda}}{2} \end{pmatrix}}{1 + R\left(2\cos 2\omega_0 \frac{D_3}{c}\right)\frac{e^{2i\Lambda}}{2} + \frac{e^{4i\Lambda}}{4}R^2 - \frac{e^{4i\Lambda}}{4}R^2\frac{K}{2}\sin 4\omega_0 \frac{D_3}{c} - R^2\frac{e^{4i\Lambda}}{4}\frac{K^2}{2}\sin^2 2\omega_0 \frac{D_3}{c}} \\
&\quad \cdot R\left(\omega_0 \frac{D_3}{c}\right)\frac{1}{2}\left[\begin{pmatrix} 1 & 0 \\ -K & 1 \end{pmatrix} TR\left(\omega_0 \frac{D_3}{c}\right) Z e^{2i\Lambda} + 2\sqrt{2}B\sqrt{F}\begin{pmatrix} w_0 \\ w_{\pi/2} + w_0 K/2 \end{pmatrix}\right. \\
&\quad \left. e^{2i\Lambda} + \Gamma L\left(\tilde{h}_1 - \tilde{h}_2\right)\begin{pmatrix} 0 \\ 1 \end{pmatrix}\right] - RZ
\end{aligned}
\tag{51}
$$

Where Y is the signal. The effect of recycling mirror is shown in Figs. 35–39 where different conditions of reflectivity R, delays D_3 and power W on the beam splitter are presented. It is clear the radiation pressure fluctuation cancellation and the possibility to overcome SQL.

$$
h_{SQL} = \sqrt{\frac{8\hbar}{m\Omega^2 L^2}} \tag{52}
$$

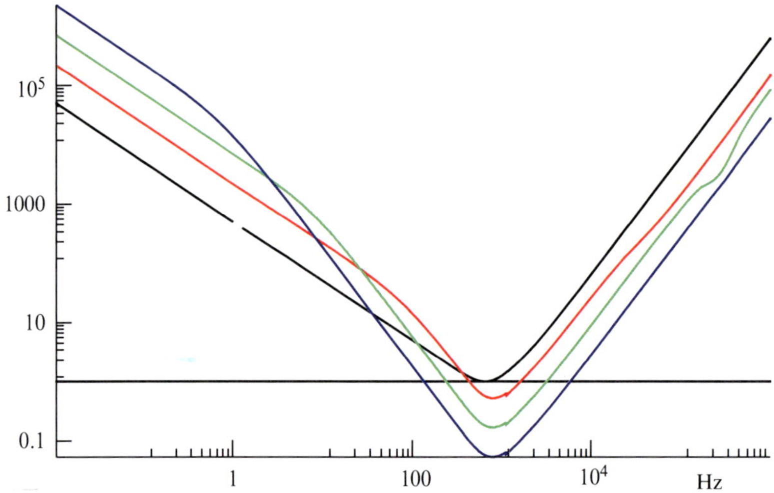

Fig. 36 The strain sensitivity in h_{SQL} units minimized at each frequency over the detuning of the recycling cavity. The input power is the same of the previous plot, and the reflectivity R of the recycling mirror is set to 0.9, 0.99 and 0.999 (*red, green* and *blue plot*). This gives the best sensitivity that can be obtained using signal recycling at a given frequency, for a fixed value of mirror reflectivity and input power. In the real case this best sensitivity can be obtained only at a single frequency. The *black curve* is the sensitivity without recycling, plotted as a reference (Courtesy G. Cella)

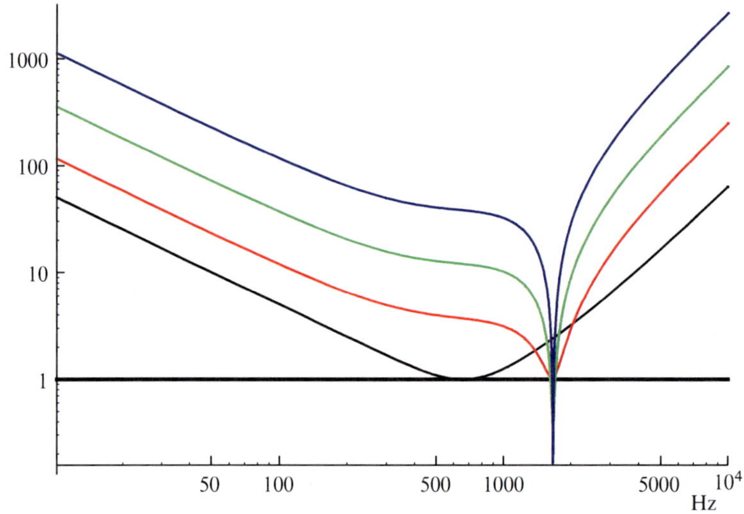

Fig. 37 The influence of recycling mirror sensitivity. Input power is fixed to the same value of the previous plot. The detuning of the recycling cavity is set to $-1/4/4$. The recycling mirror reflectivity is set to 0.9 (*red*) 0.99 (*green*) 0.999 (*blue*). *Black curve* is the reference no recycling case (Courtesy G. Cella)

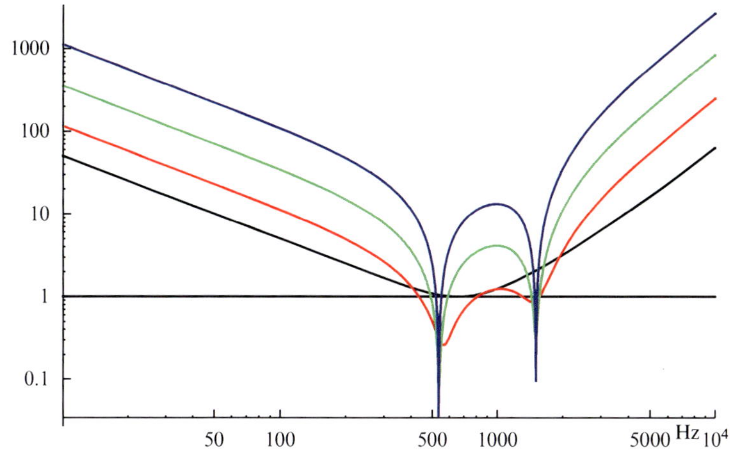

Fig. 38 Same as the previous plot, but with detuning of recycling cavity set to $\pi/4$ (Courtesy G. Cella)

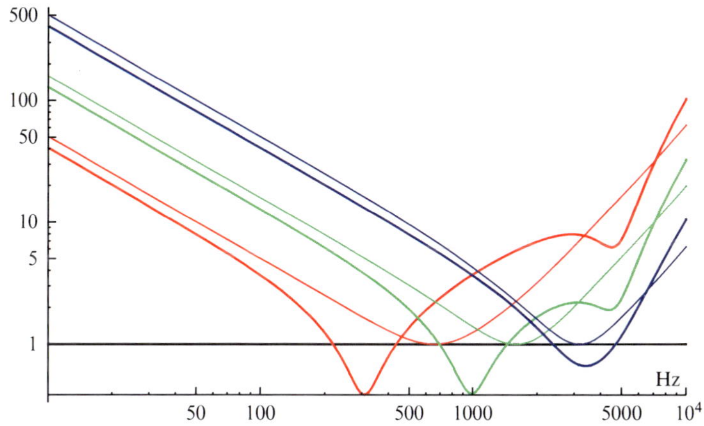

Fig. 39 Effect of the input power. Reflectivity R of recycling mirror is fixed to 0.9, and the detuning angle is 1.256. The input power is set to $I_{SQL}/10$ (*red*), I_{SQL} (*green*) and 10 I_{SQL} (*blue*) *Thin curves* correspond to the no recycling case (same input power of the thick one) (Courtesy G. Cella)

(c) The Photodiode Problem

Signal extraction is affecting S/N ratio since photodiodes are very lossy devices. Let's suppose the Michelson Interferometer to be in resonance and to extract *Y* after a detuned cavity (see Fig. 40):

$$\begin{aligned} U = & \begin{pmatrix} v_0 \\ v_{\pi/2} - v_0 K + \Gamma e^{-2i\Lambda} L \left(\tilde{h}_1 - \tilde{h}_2\right) \end{pmatrix} e^{2i\Lambda} \\ & + 2\sqrt{2} B \sqrt{F} \begin{pmatrix} w_0 \\ w_{\pi/2} + w_0 K/2 \end{pmatrix} e^{2i\Lambda} \end{aligned} \tag{53}$$

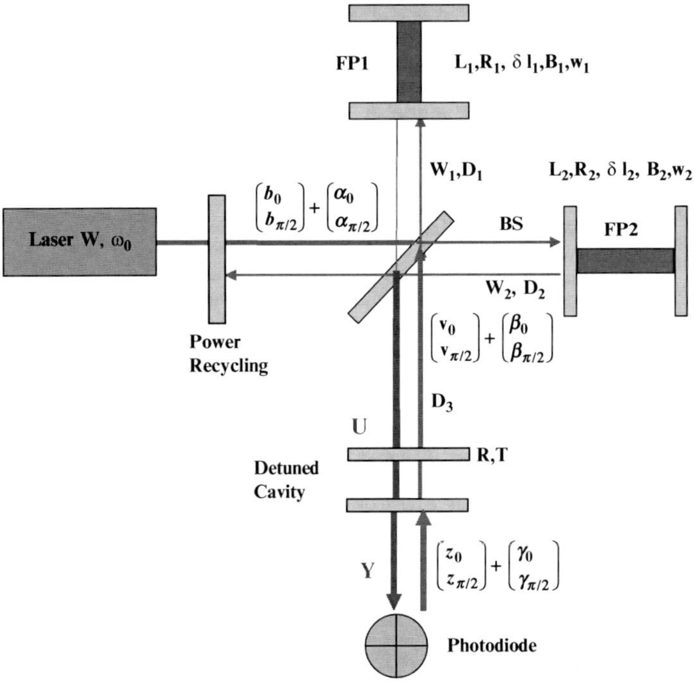

Fig. 40 Squeezed signal Y is sent to a photodiode; squeezing is spoiled since this device is very lossy

$$Y = \begin{pmatrix} \dfrac{v_0}{K} - v_{\pi/2} + v_0 K - \Gamma e^{-2i\Lambda} L\left(\tilde{h}_1 - \tilde{h}_2\right) \\ \dfrac{v_{\pi/2} + \Gamma e^{-2i\Lambda} L\left(\tilde{h}_1 - \tilde{h}_2\right)}{K} \end{pmatrix} e^{2i\Lambda} + 2\sqrt{2}B\sqrt{F} \begin{pmatrix} w_0/K - w_{\pi/2} - w_0\dfrac{K}{2} \\ \dfrac{w_0}{2} + \dfrac{w_{\pi/2}}{K} \end{pmatrix} e^{2i}\Lambda \tag{54}$$

A photodiode (PD) with efficiency $\mathbf{1-\varepsilon}$ can be represented by an efficiency 1 PD, with in front a beam splitter having reflectance $\mathbf{R} = (\mathbf{1-\varepsilon})^{1/2}$ and transmittance $\mathbf{T} = \boldsymbol{\varepsilon}^{1/2}$ (see Fig. 41). An external vacuum η, due to BS transmittance, can reach the PD and then spoil squeezing.

The overall signal entering PD is:

$$Q = \begin{pmatrix} \sqrt{1-\varepsilon}\left(\dfrac{v_0}{K} - v_{\pi/2} + v_0 K - \Gamma e^{-2i\Lambda} L\left(\tilde{h}_1 - \tilde{h}_2\right)\right) + \sqrt{\varepsilon}\eta_0 e^{-2i\Lambda} \\ \sqrt{1-\varepsilon}\dfrac{v_{\pi/2} + \Gamma e^{-2i\Lambda} L\left(\tilde{h}_1 - \tilde{h}_2\right)}{K} + \sqrt{\varepsilon}\eta_{\pi/2} e^{-2i\Lambda} \end{pmatrix} e^{2i\Lambda} + 2\sqrt{2}B\sqrt{F}\sqrt{1-\varepsilon} \begin{pmatrix} w_0/K - w_{\pi/2} - w_0\dfrac{K}{2} \\ \dfrac{w_0}{2} + \dfrac{w_{\pi/2}}{K} \end{pmatrix} e^{2i\Lambda} \tag{55}$$

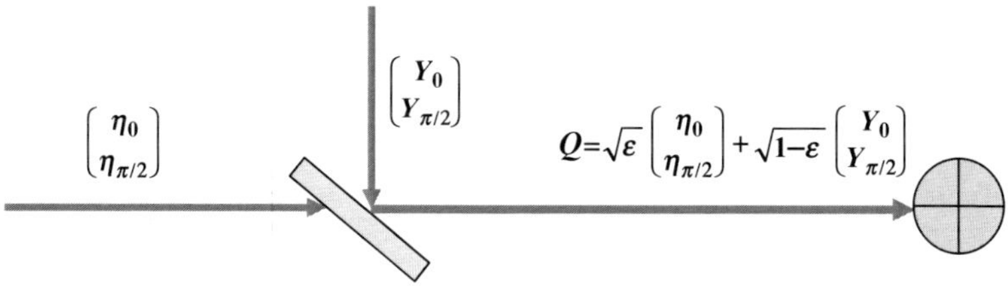

Fig. 41 Equivalent optical diagram of a photodiode having efficiency 1-ε. η is a vacuum entering the photodiode due to its inefficiency

By considering the lower line, signal is detectable if the following inequalicy holds:

$$\Gamma \mathrm{L}\left(\tilde{\mathrm{h}}_1-\tilde{\mathrm{h}}_2\right) \geq \sqrt{1+K^2\frac{\varepsilon}{1-\varepsilon}+8K^2\mathrm{B}^2\mathrm{F}\left(\frac{1}{4}+\frac{1}{K^2}\right)} \tag{56}$$

For the sake of simplicity in Eq. (56), we can put $\Lambda = 0$ and neglect W terms, then signal can be detected if

$$\Gamma \mathrm{L}\left(\tilde{\mathrm{h}}_1-\tilde{\mathrm{h}}_2\right) \geq \sqrt{1+K^2\frac{\varepsilon}{1-\varepsilon}} \tag{57}$$

If we require that the photodiode inefficiency should not spoil S/N, the following inequalicy should hold:

$$K < \sqrt{\frac{1-\varepsilon}{\varepsilon}} \tag{58}$$

Per example if $\varepsilon = 4.10^{-2}$ then $K < 5$ i.e. PD inefficiency limits maximum obtainable squeezing. This shows how important is the that PD should have very high efficiency.

A way of overcoming this problem could be to find a way of amplifying signal and vacuum fluctuations; it seems that for doing this we could inject U (see Eq. (53)) in the following matrix product (for the sake of simplicity we omit terms due to losses):

$$\begin{pmatrix}0&-1\\1&0\end{pmatrix}\begin{pmatrix}1&0\\-\mathrm{K}'&1\end{pmatrix}\begin{pmatrix}\frac{1}{\mathrm{K}'}&-1\\1&\frac{1}{\mathrm{K}'}\end{pmatrix}\begin{pmatrix}\frac{1}{\mathrm{K}}&-1\\1&\frac{1}{\mathrm{K}}\end{pmatrix}\begin{pmatrix}\mathrm{v}_0\\ \mathrm{v}_{\pi/2}+\Gamma \mathrm{e}^{-2\mathrm{i}\Lambda}\mathrm{L}\left(\tilde{\mathrm{h}}_1-\tilde{\mathrm{h}}_2\right)-\mathrm{K}\mathrm{v}_0\end{pmatrix}$$
$$=\begin{pmatrix}-\left(\mathrm{K}'+\frac{1}{\mathrm{K}'}\right)\frac{\mathrm{v}_{\pi/2}+\Gamma \mathrm{e}^{-2\mathrm{i}\Lambda}\mathrm{L}\left(\tilde{\mathrm{h}}_1-\tilde{\mathrm{h}}_2\right)}{\mathrm{K}}\\ \frac{1}{\mathrm{K}'}\left(\frac{\mathrm{v}_0}{\mathrm{K}}-\mathrm{v}_{\pi/2}+\mathrm{v}_0\mathrm{K}-\Gamma \mathrm{e}^{-2\mathrm{i}\Lambda}\mathrm{L}\left(\tilde{\mathrm{h}}_1-\tilde{\mathrm{h}}_2\right)\right)-\frac{\mathrm{v}_{\pi/2}+\Gamma \mathrm{e}^{-2\mathrm{i}\Lambda}\mathrm{L}\left(\tilde{\mathrm{h}}_1-\tilde{\mathrm{h}}_2\right)}{\mathrm{K}}\end{pmatrix} \tag{59}$$

By keeping terms up to $1/K'$, Q becomes:

$$Q \approx \left(\frac{\sqrt{1-\varepsilon}K' \dfrac{v_{\pi/2} + \Gamma e^{-2i\Lambda} L\left(\tilde{h}_1 - \tilde{h}_2\right)}{K} + \sqrt{\varepsilon}\eta_0 e^{-2i\Lambda}}{\sqrt{1-\varepsilon}\dfrac{1}{K'}\left(\dfrac{1}{K}v_0 - 2v_{\pi/2} - 2\Gamma e^{-2i\Lambda} L\left(\tilde{h}_1 - \tilde{h}_2\right) + Kv_0\right) + \sqrt{\varepsilon}\eta_{\pi/2} e^{-2i\Lambda}} \right) e^{2i\Lambda} \tag{60}$$

The detection condition, neglecting W terms and putting $\Lambda = 0$, becomes:

$$\Gamma L \left(\tilde{h}_1 - \tilde{h}_2\right) \geqslant \sqrt{1 + \frac{\varepsilon}{1-\varepsilon}\left(\frac{K}{K'}\right)^2} \tag{61}$$

It is now clear from Eq. (61) that signal detection, unlike in Eq. (57), is not limited by squeezing.

The four matrices of Eq. (59) could be obtained in the following way:

$\begin{pmatrix} 0 & -1 \\ 1 & 0 \end{pmatrix}$ can be obtained by delaying optical signal with respect the classical one by $\pi/2$.

$\begin{pmatrix} 1 & 0 \\ -K' & 1 \end{pmatrix}$ can be obtained by filtering light through a ponderomotive cavity having squeezing K'.

$\begin{pmatrix} -\frac{1}{K} & 1 \\ -1 & -\frac{1}{K} \end{pmatrix}$ can be obtained by filtering optical signal through a detuned cavity whose working point is at half height of Airy peak.

11 The Network of Interferometric GW Detectors

Some relevant reasons for starting network coherent analysis are

- Sensitivity increase
- Source direction determination from time of flight differences
- Sources polarizations measurement
- Test of GW Theory and GW physical properties

Some relevant astrophysical targets are

- Far Universe expansion rate measurement
- GW energy density in the Universe
- Knowledge of Universe at times close to Planck's time

In Fig. 42 the optical diagram of Virgo, a typical large base GW detector, is shown.

With reference to Fig. 42, the laser beam is prestabilized up to 2 Hz, at a frequency stability $\mathbf{\Delta \upsilon = 10^{-4} Hz^{1/2}}$ by means of a Ultra Low Expansion cavity and

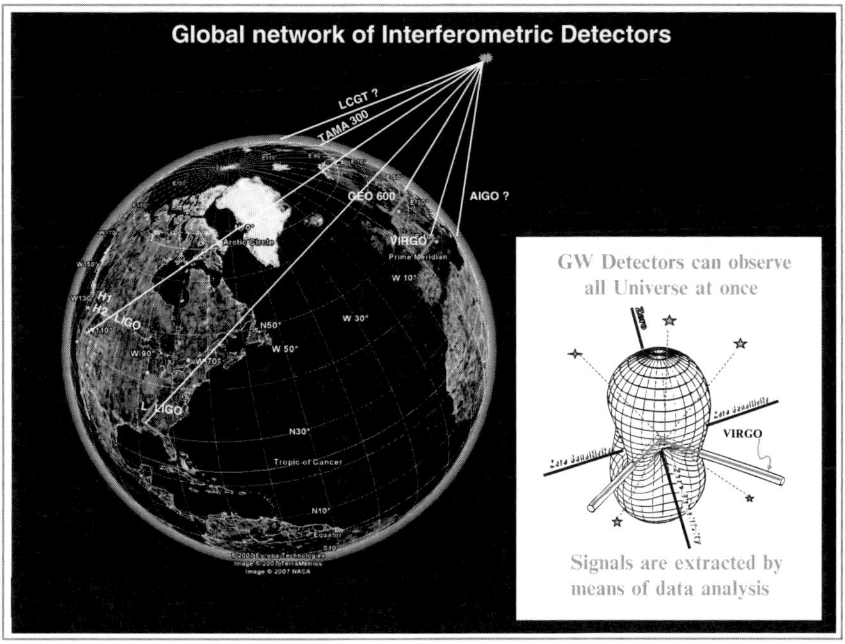

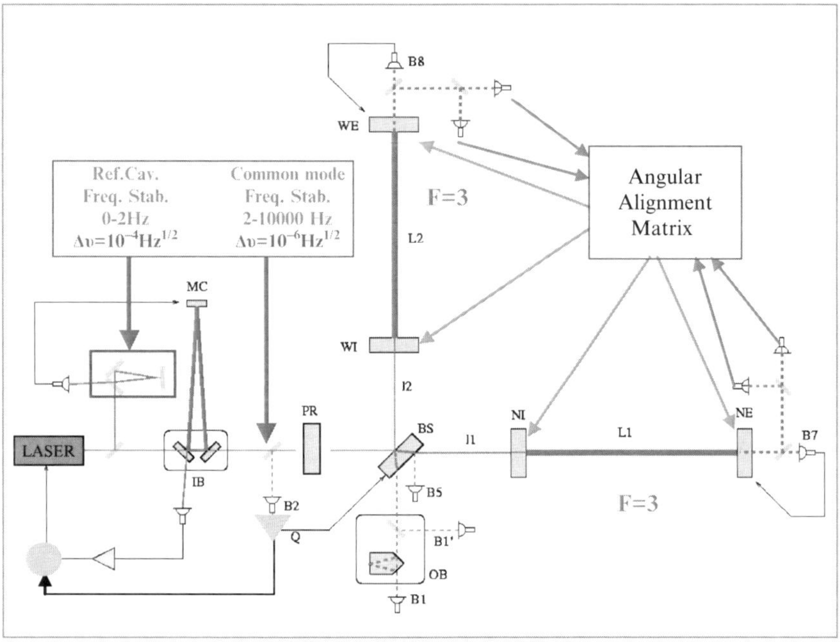

Fig. 42 The Virgo optical diagram

then, by using the 3 + 3 km cavities operated in common mode, for frequencies in the range 2–10,000 Hz, to a frequency stability level of $\boldsymbol{\Delta\upsilon = 10^{-6}\mathrm{Hz}^{1/2}}$.

The power recycling mirror PR allows to increase by a factor 20 the power on the beam splitter mirror BS; the two 3 km Fabry–Perot cavities have a Finesse of 30 and cavities mirror are aligned by using Anderson's technique. In the next a brief description of GW interferometers performances is presented.

LIGO [14]

LIGO project, run by Caltech and MIT, is the largest existing GW detector, composed by 3 interferometers: a 4 + 2 km interferometers inside the same vacuum pipe in Hanford (Washington state) (see Fig. 43) and a 4 km detector in Livingston (Louisiana state) (see Fig. 44).

The distance between Hanford and Livingston is about 3,000 km.

The sensitivity progress of Hanford machine, really outstanding, is shown in Fig. 45; the graph shows only the sensitivity of Hanford machine, but a graph showing sensitivities of all three LIGO machines will be shown later; operation stability is shown in Figs. 46 and 47. The science run time progression is shown in Fig. 48.

Virgo [15]

Virgo is collaboration, initially between INFN (Italy), CNRS (France) at which later added NIKHEF (Holland), has built a 3 km interferometer (see Fig. 48) located in Cascina, about 20 km from Pisa.

Fig. 43 The Hanford machines: one vacuum tube with two interferometers 4 and 2 km long, respectively

Fig. 44 The Livingston 4 km interferometer

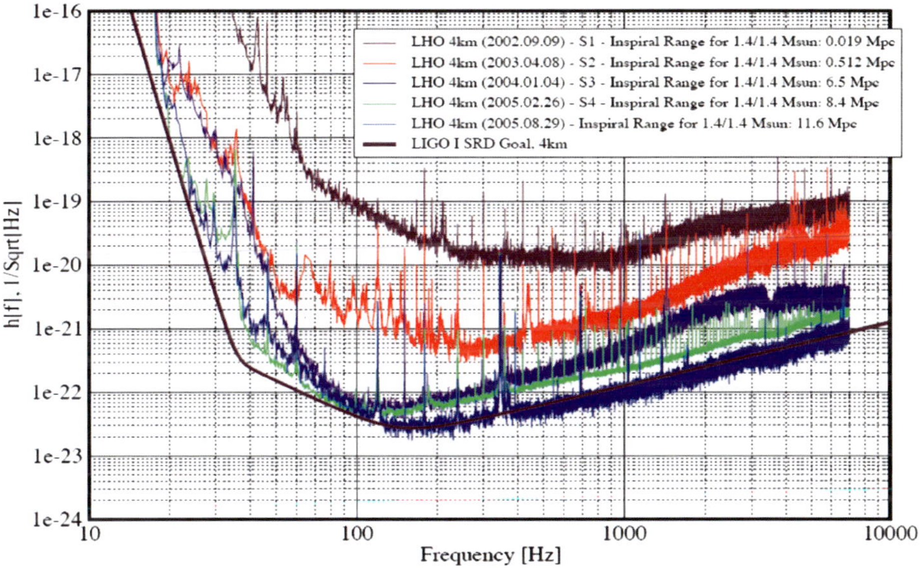

Fig. 45 Progress of LIGO 4 km Hanford sensitivity

Virgo has a very sophisticate antiseismic system whose basic elements are called Superattenuators (SA) (see Fig. 49). These devices, thoroughly described elsewere [16, 17], are capable to attenuate seismic noise below expected thermal noise down to 4 Hz. In Fig. 50 the measured SA attenuation transfer function, measured stage by stage, while the point at 4 Hz has been otained by shaking SA suspension point. In Fig. 51 the Virgo sensitivity progress is presented; it is noticeable the impressive low frequency sensitivity improvements spanning about 7 orders of magnitude.

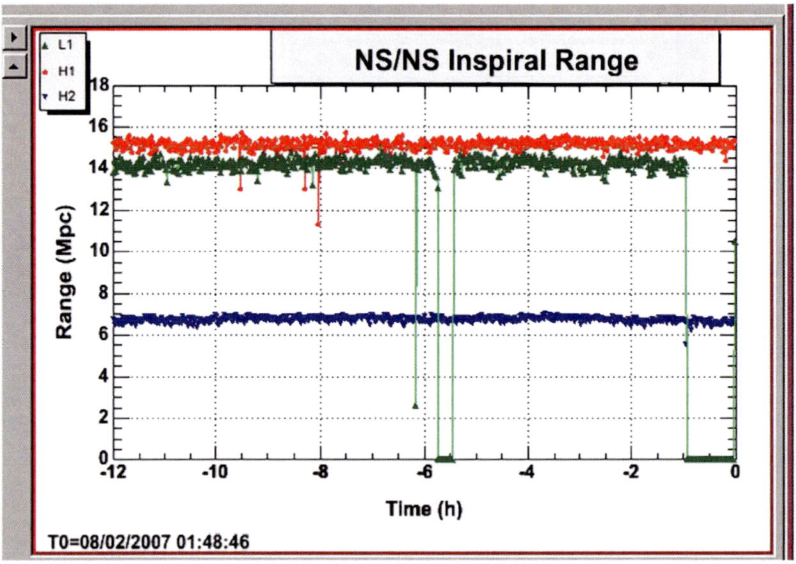

Fig. 46 LIGO operation stability; in the ordinate axis of this plot is the distance in Mpc at which a 1.4–1.4 M_0 NS-NS coalescing binary system, averaged on orbital plane directions, is detectable. Sensitivity uniformity and operation stability are remarkable

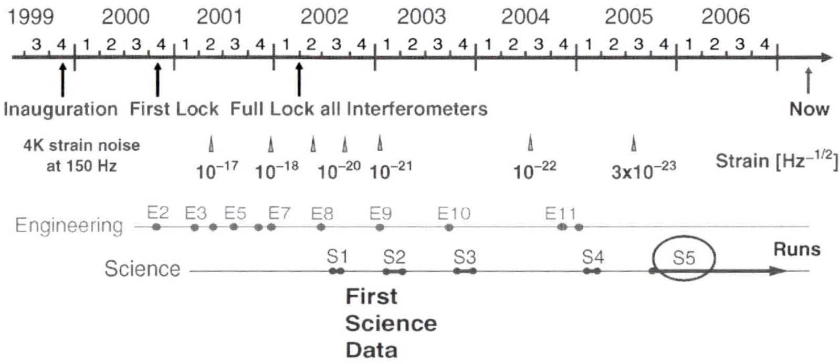

Fig. 47 The LIGO sensitivity milestones and Science runs progression

Figure 52 shows the horizon for 1.4–1.4 solar masses coalescing binaries, while Fig. 53 shows Virgo duty cycle.

GEO600 [18]

GEO600, a collaboration between Great Britain and Germany, is a 600 m long arms signal and power recycled interferometer located near Hannover (see Fig. 54).

Fig. 48 The Virgo 3 km interferometer

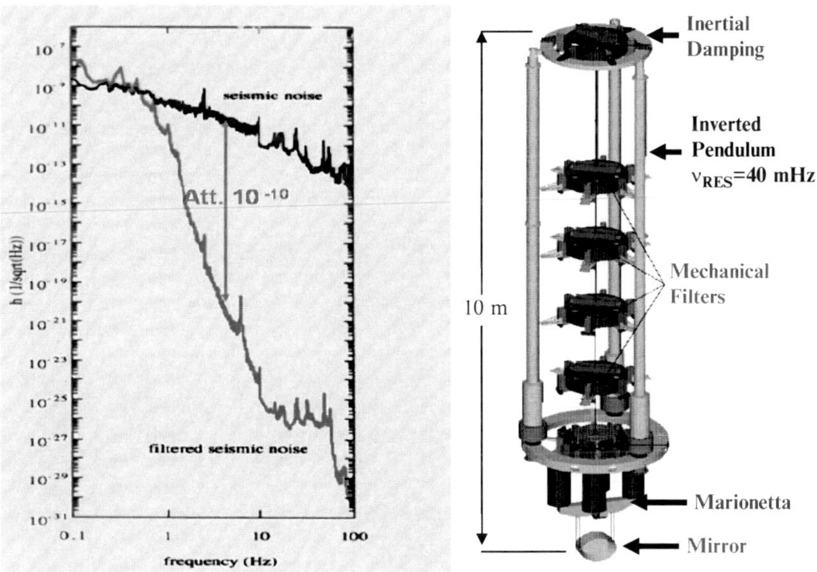

Fig. 49 The Virgo Superattenuator

The optical diagram is shown in Fig. 55; the interferometer has two single pass arms and is equipped with both signal and power recycling mirrors. Mirrors are suspended by means of Silica fibers for reducing thermal noise.

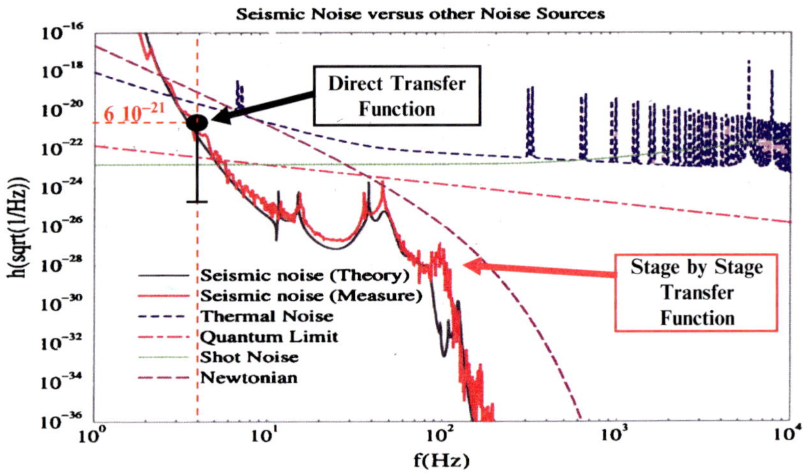

Fig. 50 The seismic noise remnant amplitude having assumed a seismic spectral displacement $10^{-7}/f^2$ m/sqrt (Hz). The Superattenuator transfer function measured stage by stage (*red curve*) and directly by shaking SA suspension point (*black point*)

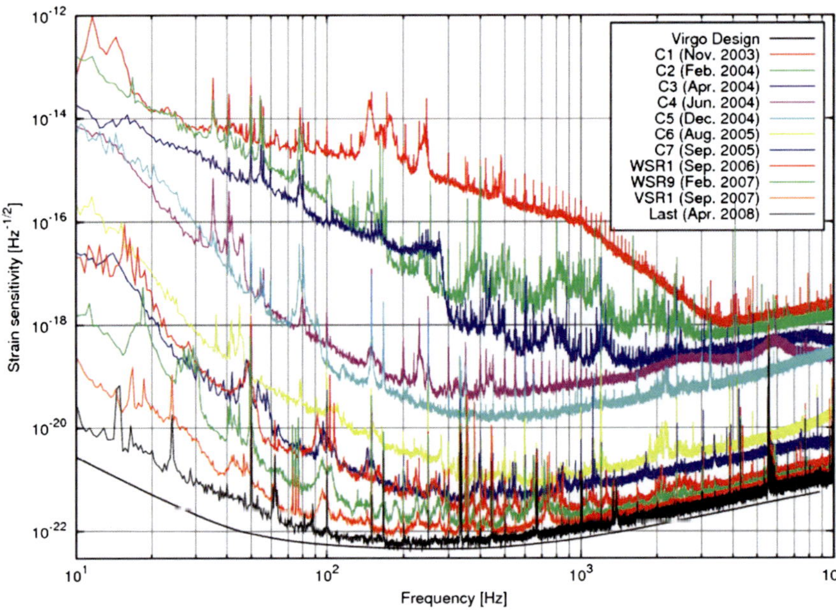

Fig. 51 Virgo sensitivity progress is shown; it is impressive the low frequency improvement. At high frequency design sensitivity has not yet been reached since Virgo is running at low power due to mirror lensing effects. Lensing corrective devices are being tested

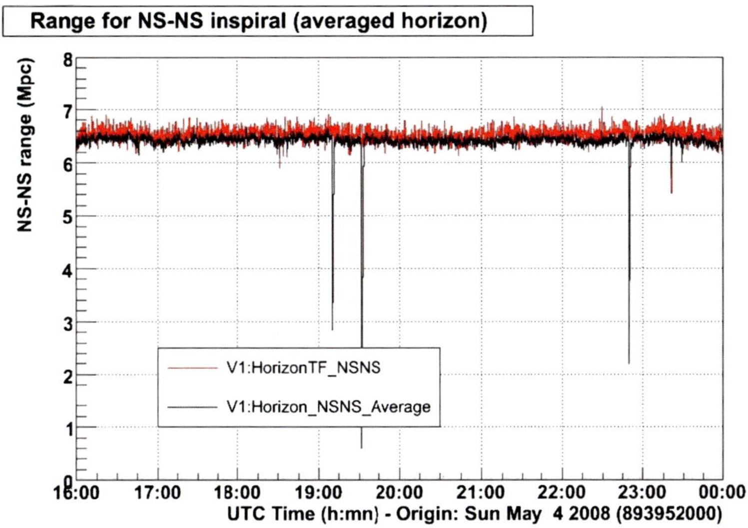

Fig. 52 Virgo operation stability over 8 hours of running; in the ordinate axis of this plot is the distance in Mpc at which a 1.4–1.4 M_0 NS-NS coalescing binary system, averaged on orbital plane directions, is detectable

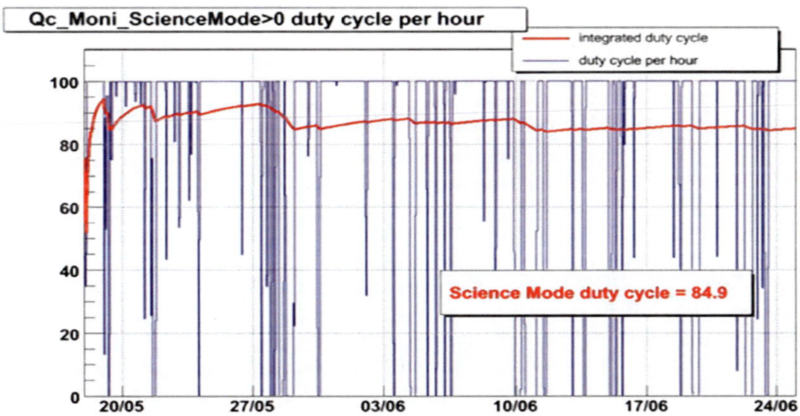

Fig. 53 Virgo operation duty cycle over 1 month of running

Signal recycling is an important addition since performs radiation pressure fluctuation noise reductions in some frequency range. In Fig. 56 sensitivity progression is shown, while in Fig. 57 the effect of quadrature rotation, produced by signal recycling mirror is shown.

Fig. 54 GEO 600 located closed to Hannover

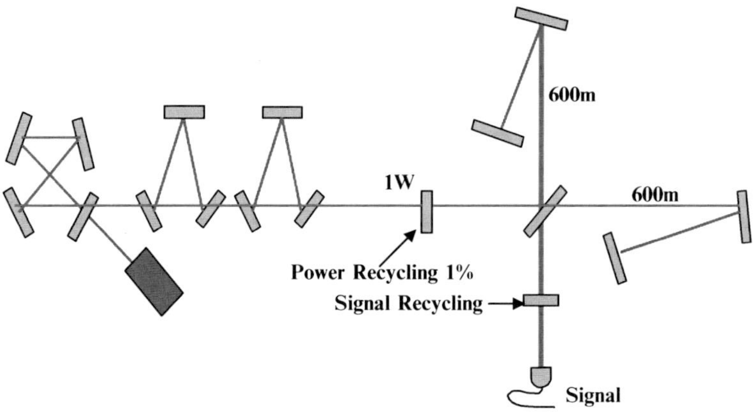

Fig. 55 Optical diagram of GEO600

TAMA 300 [19]

TAMA 300 m interferometer (see Fig. 58), located at National Astronomical Observatory of Japan (Tokyo) has been the first interferometric detector reaching sensitivity good enough for starting a realistic gravitational event search. TAMA construction started in 1995, and, even if the project was run by a relatively small

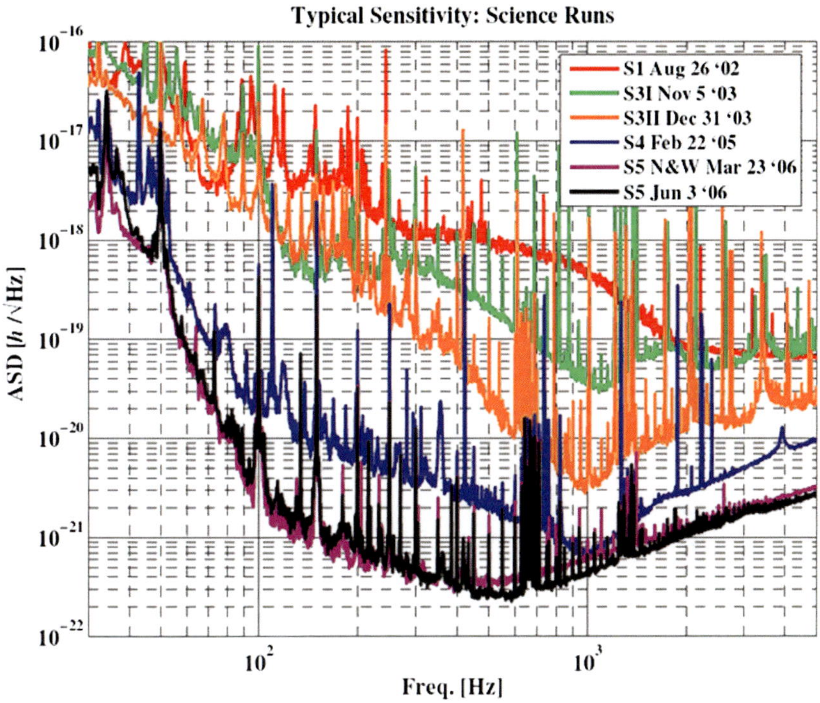

Fig. 56 GEO600 sensitivity progress

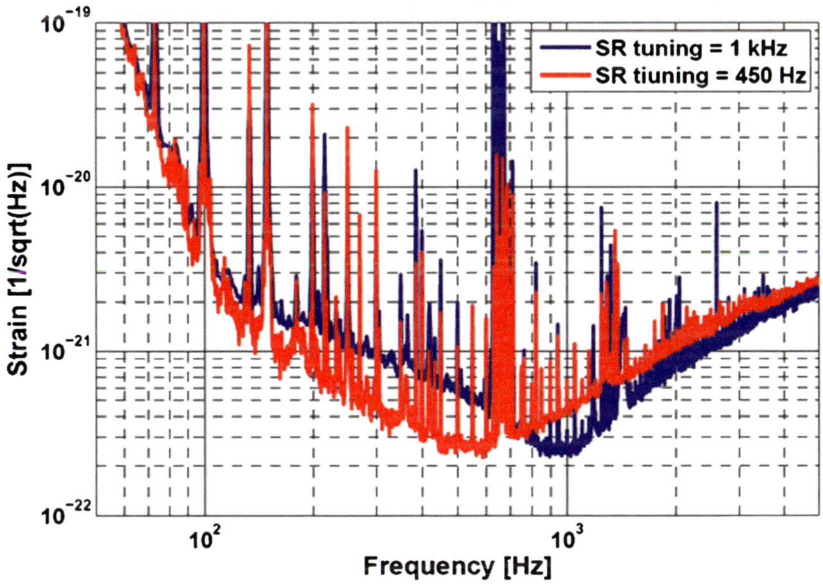

Fig. 57 Signal recycling mirror parameters tuning for producing sensitivity increase at 450 Hz (*red curve*) and 1 KHz (*blue curve*)

Fig. 58 TAMA 300 is located in Mitaka campus, National Astronomical Observatory, in Tokyo

Table 1 TAMA data taking runs progress and running times (Courtesy Kazuaki Kuroda for the TAMA/CLIO/LCGT Collaboration)

TAMA data-taking runs including long-term observations			
Run	Term	Year	Live Time (h)
DT1	6-Aug→7-Aug	1999	7
DT2	17-Sept→20-Sept	1999	31
DT3	20-Apr→23-Apr	2000	13
DT4	21-Aug→4-Sept	2000	161
DT5	2-Mar→8-Mar	2001	111
DT6	15-Aug→20-Sept	2001	1,038
DT7	31-Aug→2-Sept	2002	25
DT8	14-Feb→14-Apr	2003	1,158
DT9	28-Nov→10-Jan	2004	558

number of scientists, it was immediately supplying the world best performances; during the period 2000–2002 attained the world best sensitivity.

In Table 1 TAMA milestone progress is presented: with DT1 In 1999 TAMA started to make observations, in DT4 attained the world best sensitivity, in DT6 performed 1,000 h of data taking and with DT7 started operating with power recycling. Joint observation with LIGO was performed during DT7 to DT9.

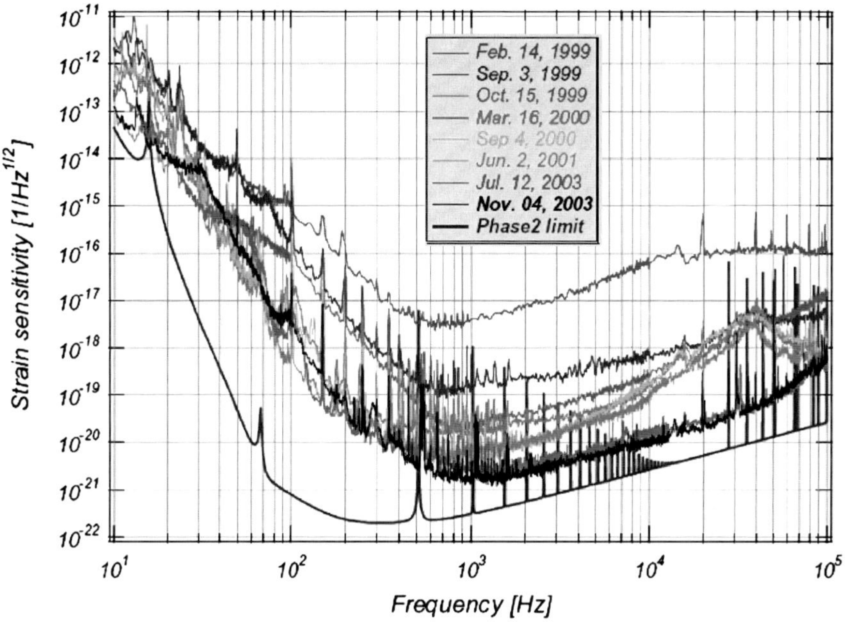

Fig. 59 Sensitivity progress of TAM 300

In Fig. 59 the sensitivity progress is presented; best sensitivity attained was

$$\tilde{h} = 1.710^{-21}\ 1/\sqrt{Hz}@1KHz$$

with a recycling factor of 4.5.

In Fig. 60 sensitivities of all kinds of operating GW detectors are displayed; original sensitivity graphs for bar detectors can be seen in Fig. 16 and for TAMA in Fig. 59.

12 Brief Status of GW Detection

IGEC: Bar Detector Network (Auriga, Explorer, Nautilus, Allegro) started 1997 for detecting impulsive GW [20]. **No evidence of any significant signal**

Periodic Sources [21]

LIGO-GEO600: OG da Pulsar (28 known) – $\varepsilon < 10-5-10-6$ (no mountains $>$ 10 cm) – Upper Limits h: 2.10^{-24}@200 Hz, 5.10^{-24}@400 Hz, 10^{-23}@1 KHz. See Figs. 61 and 62.

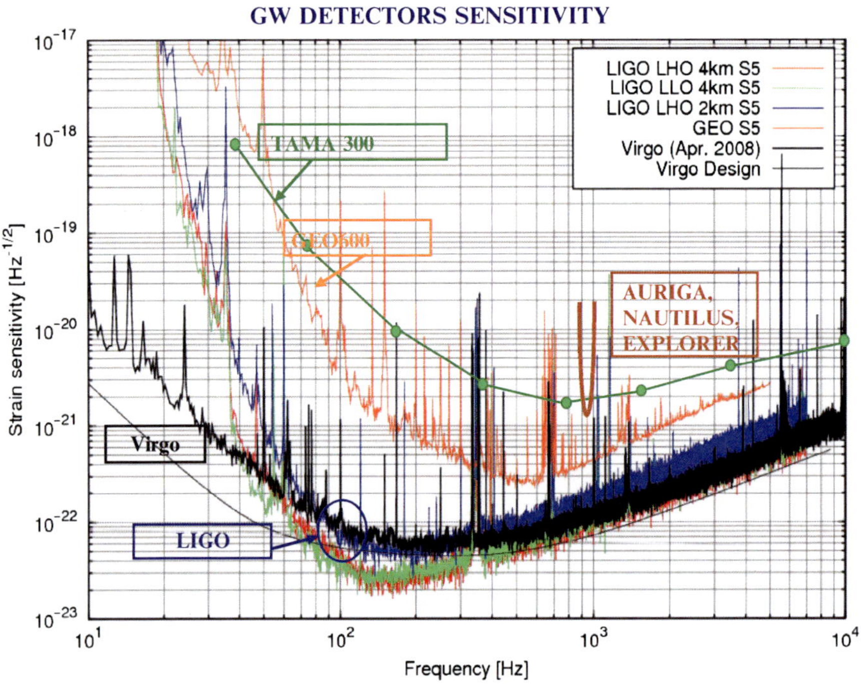

Fig. 60 Sensitivity of all operating detectors, more details on cryogenic bar detectors sensitivity can be seen in Fig.16

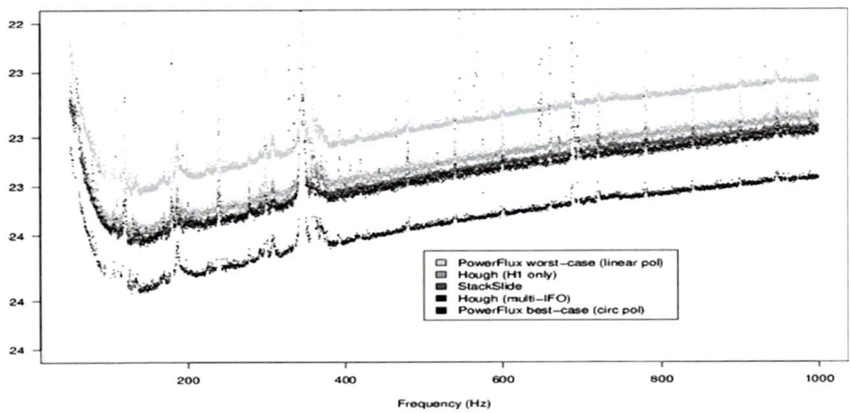

Fig. 61 Periodic sources upper limits: this is the Hanford all sky upper limit for periodic sources strain (95% confidence level), obtained for the Hanford observatory. The plot compares several search method, documented in the S4 paper LIGO-P060010-05-Z

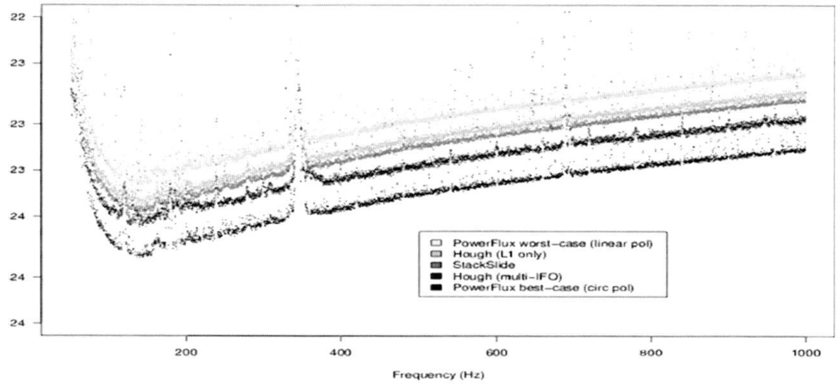

Fig. 62 Periodic sources upper limits: the same of the previous figure, for the Livingston observatory

Fig. 63 Upper limit inspirals: upper limits on the binary inspiral coalescence rate per year and per L10 as a function of total mass of the binary, for Primordial Black Hole binaries. The *darker area* shows the excluded region after accounting for marginalization over estimated systematic errors. The *lighter area* shows the additional excluded region if systematic errors are ignored

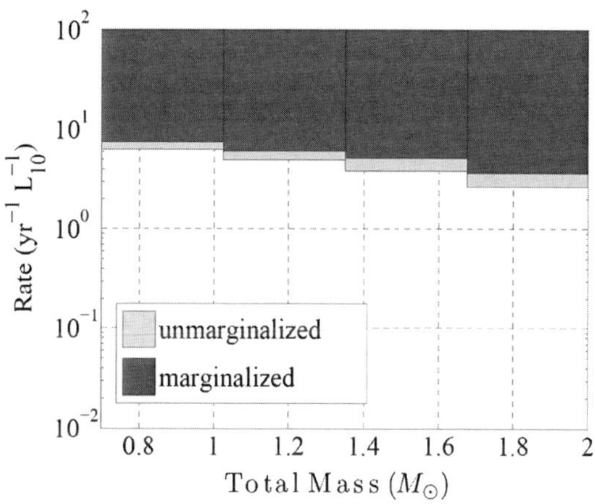

Coalescence [22]

LIGO,GEO600,TAMA (see LIGO paper arXiv:0704.3368) upper limits (see Figs. 63–65): Upp. lim. Coalescing NS-NS < 1 event/(galaxy.year) $2 < M_0 < 6$ Coalescing BH-BH < 1 event/(galaxy.year) $10 < M_0 < 80$

Impulsive [18]

LIGO and LIGO-Tama upper limits for S1, S2, S4 scientific runs are shown in Figs. 66 and 67.

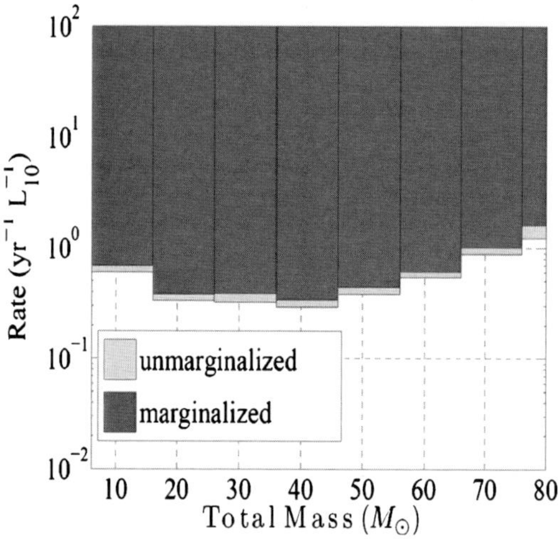

Fig. 64 Upper limits inspirals: same as the previous figure for Binary Black Holes

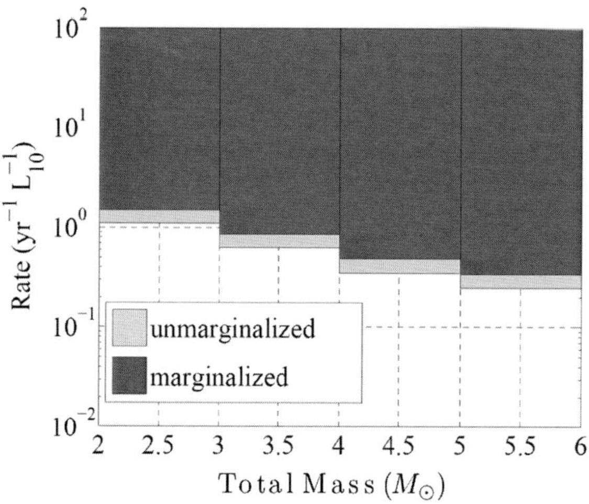

Fig. 65 Upper limits inspirals: same as the previous figure for Binary Neutron Stars

Upper bounds on stochastic background: 90% Upper Limit on GW spectrum at 100 Hz (see the model on the right) as a function of α for S3 H1L1 and S4 H1L1+H2L1 combined, and expected final sensitivities of LIGO H1L1 and H1H2 pairs, assuming LIGO design sensitivity and one year of exposure, are shown in Fig. 68.

Virgo, LIGO, GEO 600 on May 18, 2007 started common and coherent data taking analysis; main target impulsive events.

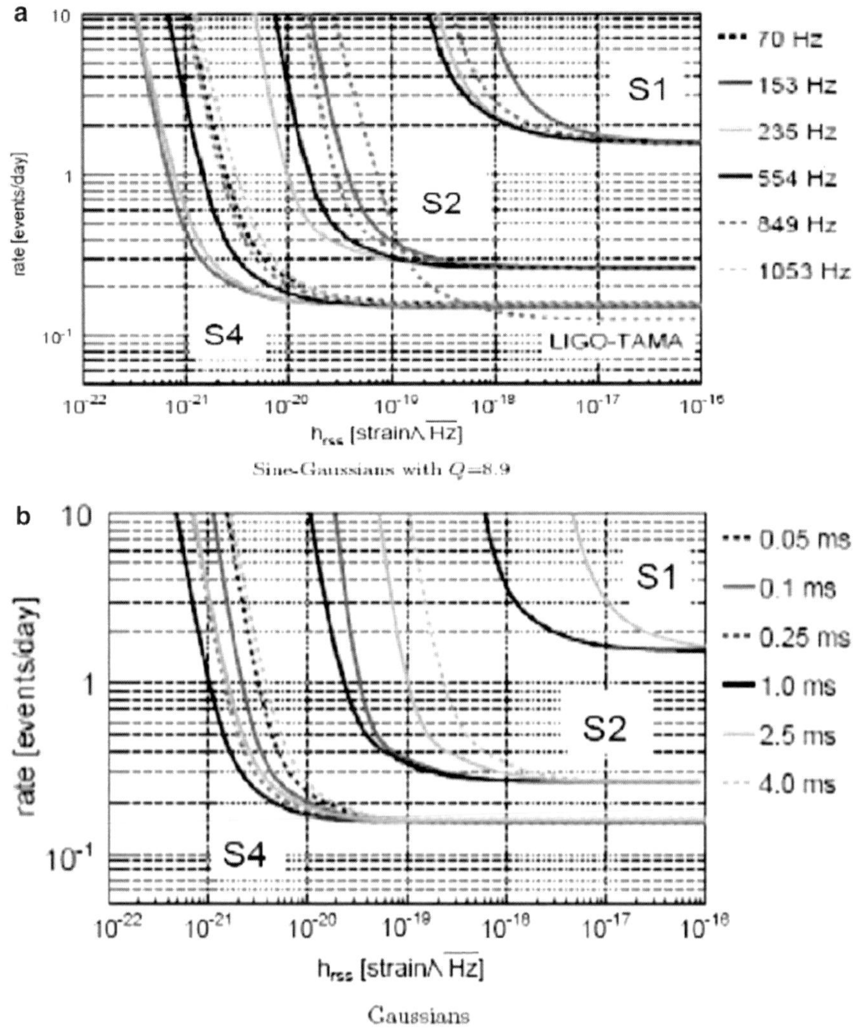

Figs. 66 and 67 Exclusion diagrams (rate limit at 90% confidence level, as a function of signal amplitude) for (**a**) sine-Gaussian and (**b**) Gaussian simulated waveforms for this S4 analysis compared to the S1 and S2 analyses (the S3 analysis did not state a rate limit). These curves incorporate conservative systematic uncertainties from the fits to the efficiency curves and from the interferometer response calibration. The 849 Hz curve labeled "LIGO-TAMA" is from the joint burst search using LIGO S2 with TAMA DT8 data [8], which included data subsets with different combinations of operating detectors with a total observation time of 19.7 days and thereby achieved a lower rate limit. The hrss sensitivity of the LIGO-TAMA search was nearly constant for sine-Gaussians over the frequency range 700–1,600 Hz

$$\Omega_{GW}(f) = \frac{f}{\rho_c}\frac{d\rho GW}{df}$$

$$\Omega_{GW}(f) = \Omega_\alpha \left(\frac{f}{100Hz}\right)^\alpha$$

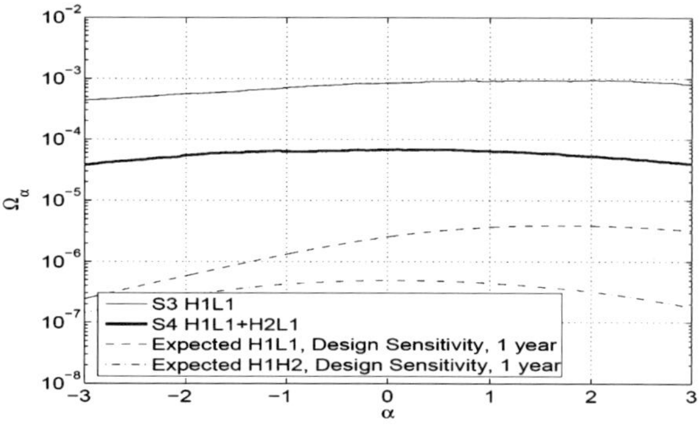

Fig. 68 Upper bounds on stochastic background

13 The Future

In Table 2 the totality of future GW experiments is presented, grouped by status of approval level with a brief explanation of forecast upgradings.

Running Interferometers

TAMA intend to operate with new Superattenuators (see Fig. 69); timing is outlined in Table 2.

GEO600

GEO group intend to push high frequency sensitivity (new name GEO HF) to much higher level as shown in Fig. 70.

Purpose of GEO HF is also the study of advanced techniques to be applied in future large interferometers with particular reference to signal recycling and squeezing. Timing is outlined in Table 2.

Virgo [24]

Virgo Plus

The expected sensitivity of the upgrading to Virgo Plus is presented in Fig. 71; the main items are the power increase to 50 w, implementation of fused silica suspensions (very much depending on how Virgo removes low frequency noise) and

Table 2 The Future

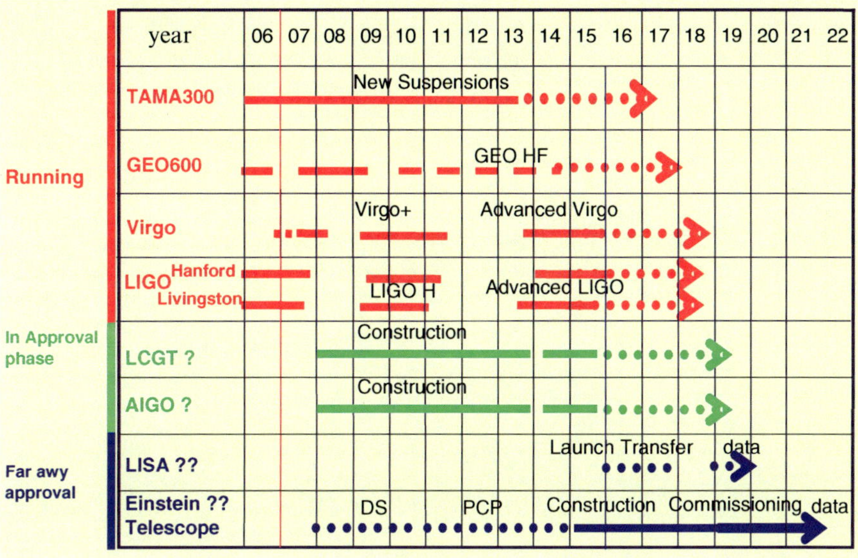

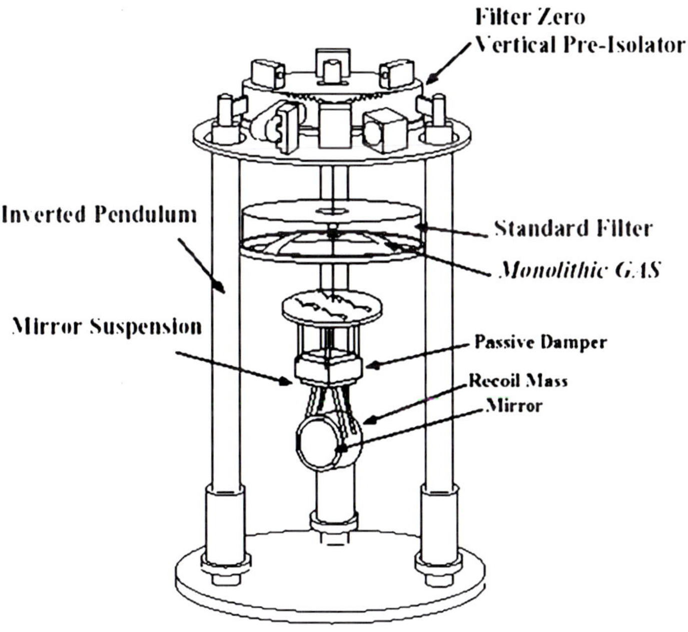

Fig. 69 TAMA new Superattenuators mounted for better control and seismic isolation

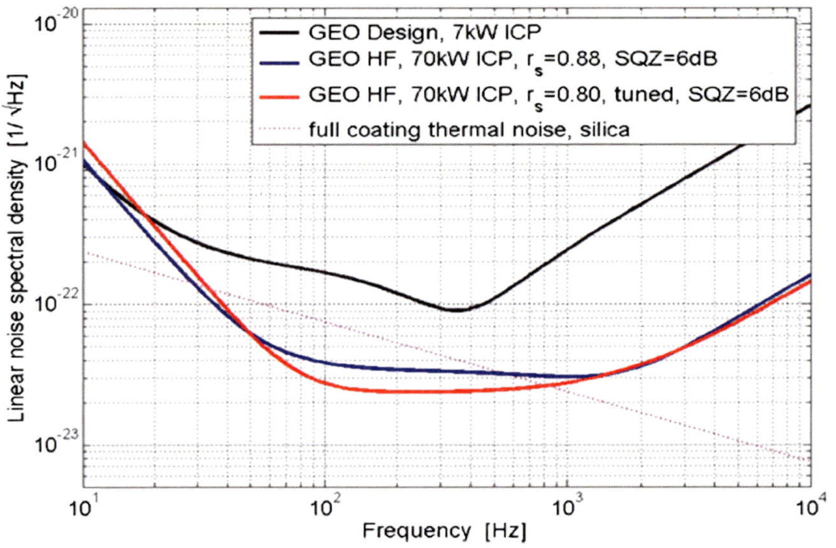

Fig. 70 GEO HF high frequency sensitivity is strongly improved by increasing Intra Cavity Power and by optimizing squeezing

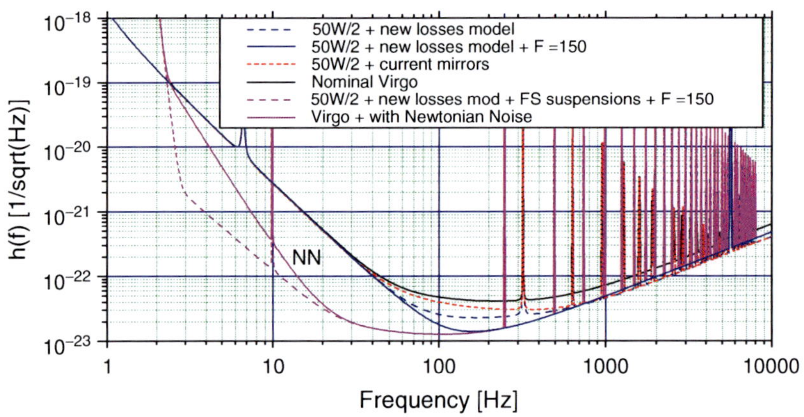

Fig. 71 The expected sensitivity increase for Virgo Plus

increase of Fabry–Perot cavities finesse to about 150 in such a way to produce firm sensitivity increase in the 100–400 Hz region.

The sensitivity increase in this frequency interval will improve a lot coalescing binary system detection sensitivity range up to 28 Mpc. Expected to start runing in June 2009.

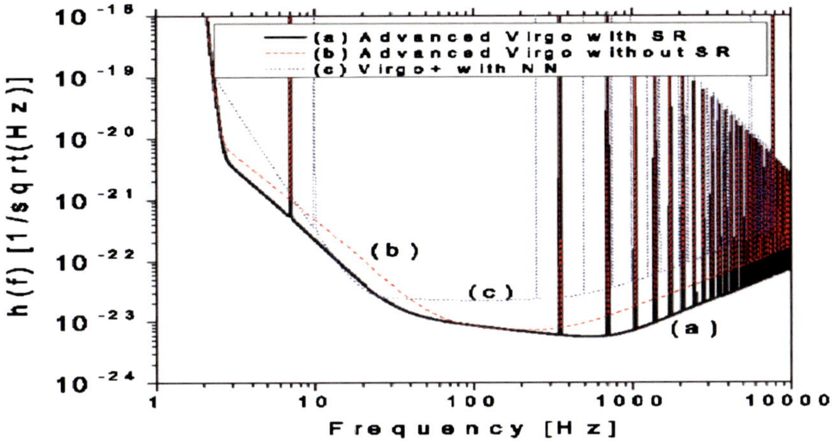

Fig. 72 Virgo Advanced expected sensitivity

Virgo Advanced

Optical configuration will be changed in such a way to have the optical waist displaced to 1,500 m, in this way thermal noise of the cavity entrance mirrors will be reduced. Advanced Virgo sensitivity is shown in Fig. 72.

A further reduction of thermal noise will, hopefully, come from coatings losses reduction by using other less lossy metals than Titanium in the coating recepies. Laser power will also be increased and optical read out changed from heterodyne to DC.

Virgo Advanced will also be equipped with signal recycling mirror for implementing radiation pressure fluctuations noise reduction.

LIGO [25]

LIGO Henanced

The first set of improvements will be set in the LIGO Henanced configuration for starting scientific runs together with Virgo Plus in June 2009.They are Higher laser power, DC readout and insertion of an Output modecleaner; LIGO H will also serve as test bench of Advanced LIGO hardware and techniques.

Target is a factor of 2 improvement in sensitivity (factor of 8 in event rate).

LIGO Advanced will implement active anti-seismic system operating to lower frequencies, lower thermal noise suspensions and optics, higher laser power and more sensitive and more flexible optical configuration; expected sensitivity is shown in Fig. 73 and a pictorial representation of sensitivity increase is shown in Fig. 74.

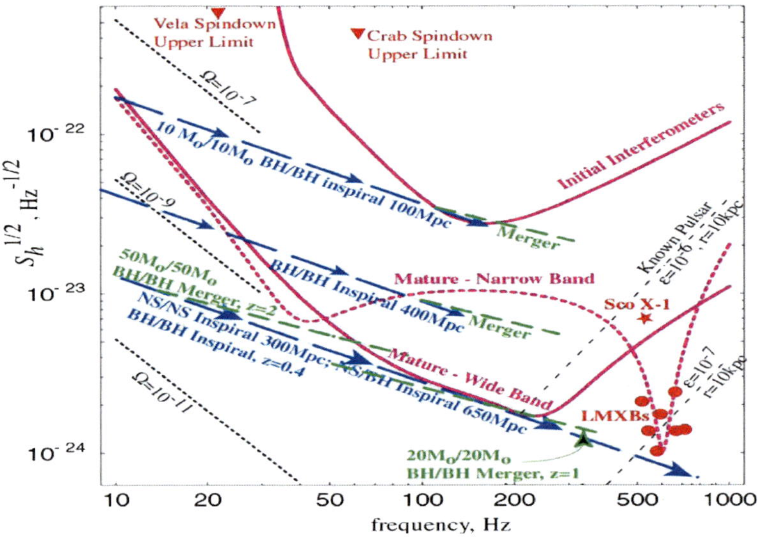

Fig. 73 LIGO Advanced sensitivity

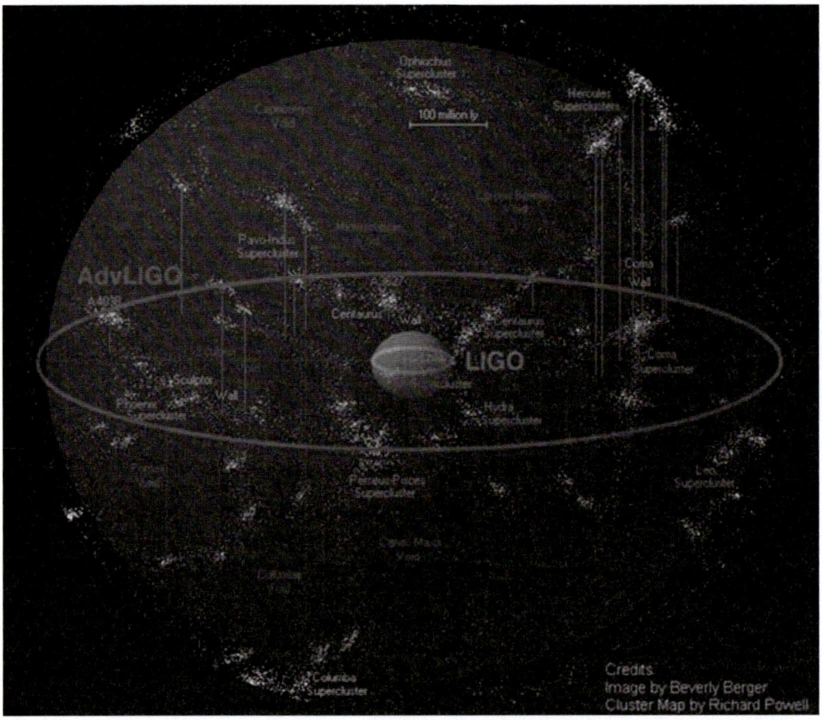

Fig. 74 With LIGO and Virgo Advanced it is expected to observe ten times the present maximal observable distance, i.e. a factor 1,000 in event rate

Table 3 Technical specifications of LIGO Advanced

Parameter	*LIGO*	Advanced *LIGO*
Input Laser Power	10 W	180 W
Mirror Mass	10 kg	40 kg
Interferometer Topology	Power-recycled-Fabry–Perot arm cavity Michelson	Dual-recycled Fabry–Perot arm cavity Michelson
GW Readout Method	RF heterodyne	DC homodyne
Optimal Strain Sensitivity	3×10^{-23}/rHz	Tunable, better than 5×10^{-24}/rHz in broadband
Seismic Isolation Performance	$f_{low} \sim 50$ Hz	$f_{low} \sim 10$ Hz
Mirror Suspensions	Single pendulum	Quadruple pendulum

In Table 3 main technical improvements with respect to initial LIGO, are presented:

Interferometers in Approval Phase

Large-scale Cryogenic Gravitational wave Telescope (LCGT) [26].

This detector has a very relevant impact in the technology of next generation interferometers for GW detection, in fact it is the first one to be designed with Fabry Perot cavity mirrors brought to 20 K. Cold mirror technology seems to be the only one able to win, at lest in part, thermal noise limits.

As it was shown by Numata (*) coating losses seems to be quite insensitive to temperature, in fact a loss angle of about 4 10^{-4} was measured to remain constant at various temperature and down to 4 K. This mirror coating anomalous and unfortunate behavior is creating a problem for next generation machines where thermal noise limits will be met. As far as we know, at present, the only known possibilities for reducing mirror thermal noise are coating loss angle reduction by using new coating materials and cooling mirrors. The latter benefits from the fact that thermal stochastic force is decreasing with $T^{1/2}$ (T is the temperature), hence bringing mirrors to 4 K there is a factor about 10 expected reduction of mirror thermal noise.

LCGT, expected to work at 20 K, is a 3 km interferometer with cryogenic mirror. Constructed underground in the depth of the Kamioka mine, has two beams (see Fig. 75) in the same pipe and the upper beam is used for a suspension point interferometer; the optical diagram is shown in Fig. 76 and the expected sensitivity in Fig. 77.

AIGO [27]

AIGO is a project for the construction of a 5 + 5 km interferometer in Gin, 80 km east of Perth (see Fig. 78); it a very long time since Australian Consortium for

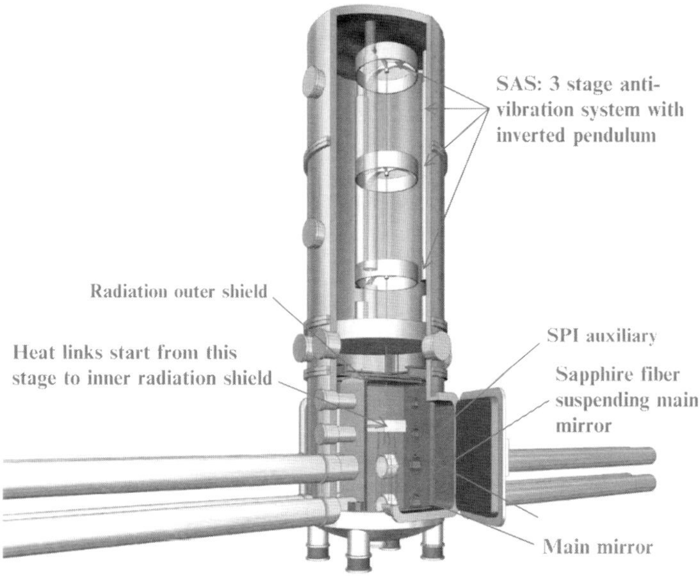

Fig. 75 Layout of LCGT cryogenic mirror suspension

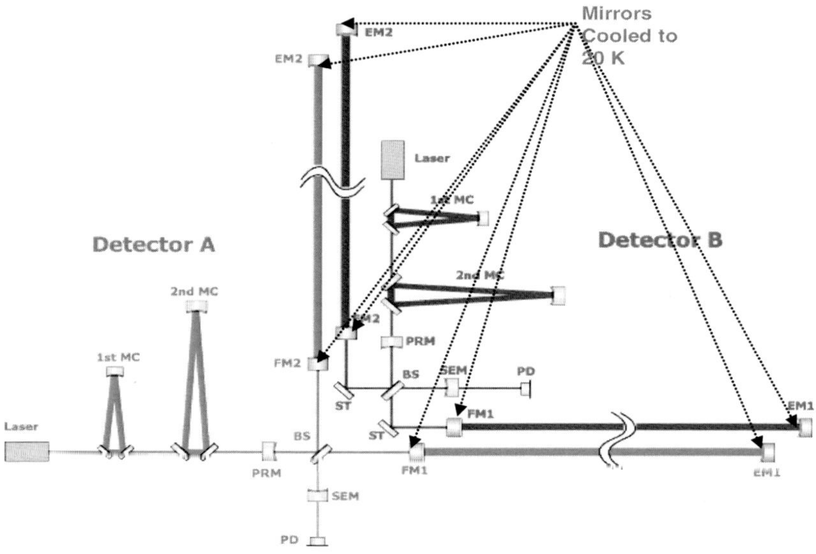

Fig. 76 LCGT Optical diagram

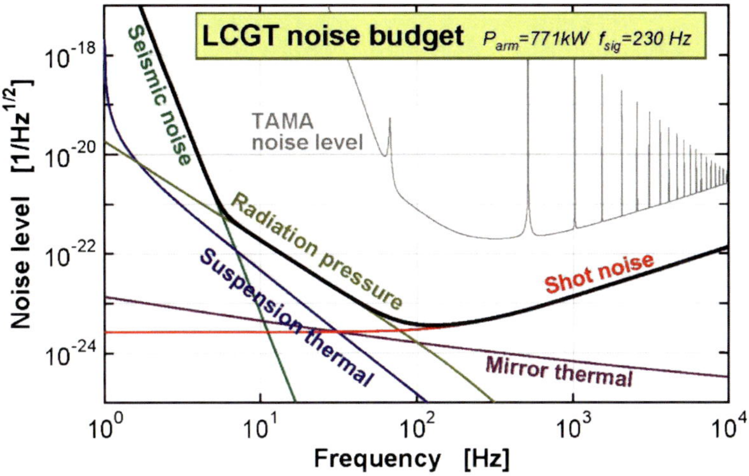

Fig. 77 LCGT sensitivity

Fig. 78 Gin Gin (Western Australia) is the location for the 5 + 5 km australian detector AIGO

Interferometric Gravitational Astronomy (ACIGA) is trying to get approval from Minister for the construction of AIGO.

The importance of AIGO, beyond the very relevant fact that a large GW detector is adding to the existing ones, is its geographical position on Earth; in fact a very sensitive GW detector in the southern hemisphere will allow to measure GW sources

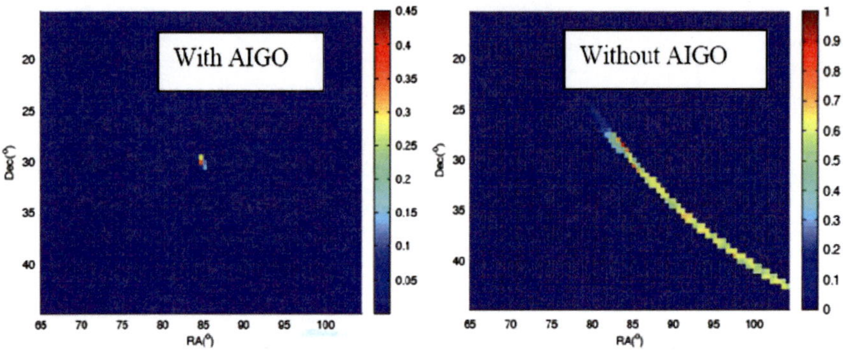

Fig. 79 AIGO allow to measure GW sources position angle with outstanding precision

angular positions with greater accuracy than the one obtained by Earth northern hemisphere detectors, as shown in Fig. 79. The detector, designed for being sensitive to very low frequency GW sources, is equipped with superattenuators, in some way similar to the Virgo ones.

GW detector community is really looking forward for AIGO approval.

GW detectors whose running start date is still far.

LISA [28]

LISA consists of three spacecraft mutually separated by 510^6 km (see Fig. 80). It is aimed to detect GW in the range $10^{-4} - 10^{-3}$ Hz, as shown in Figs. 81 and 82. One of the of LISA key points is the possibility to measure GW emitted by galactic "Standard Candles", such as cataclysmic variables, per ex. AmCvn at 2 mHz and by other binary systems whose GW amplitude can be evaluated knowing the binary system physical parameters. If GW will not be detected before LISA being in air, then LISA can give an inconfutable check of General Relativity validity. Let us hope this will not be the case.

Einstein Telescope [29] (ET)

ET will be the future challenge for GW detection since it should attain a factor 100 more in sensitivity with respect to Virgo and LIGO (see Fig. 83).

It is at least a 10+10 km machine underground mainly for reducing seismic noise and Newtonian noise, but also for avoiding strong surface environmental impact.

It is possible that the lower frequency should be less than 10 Hz; this is a real challenge since noises at very low frequency are very rugged and hard to eliminate.

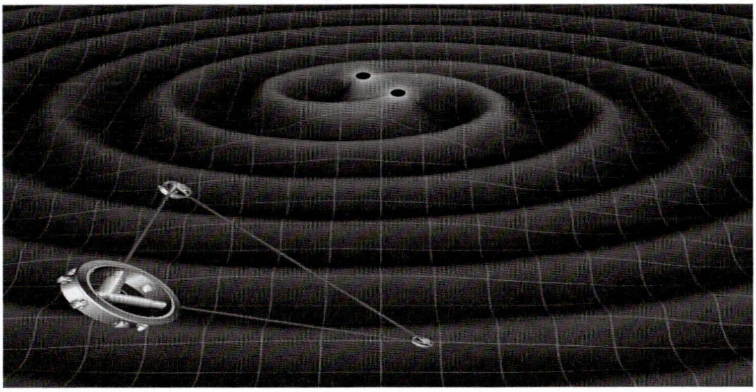

Fig. 80 Artist view of LISA together with a strong GW emitter

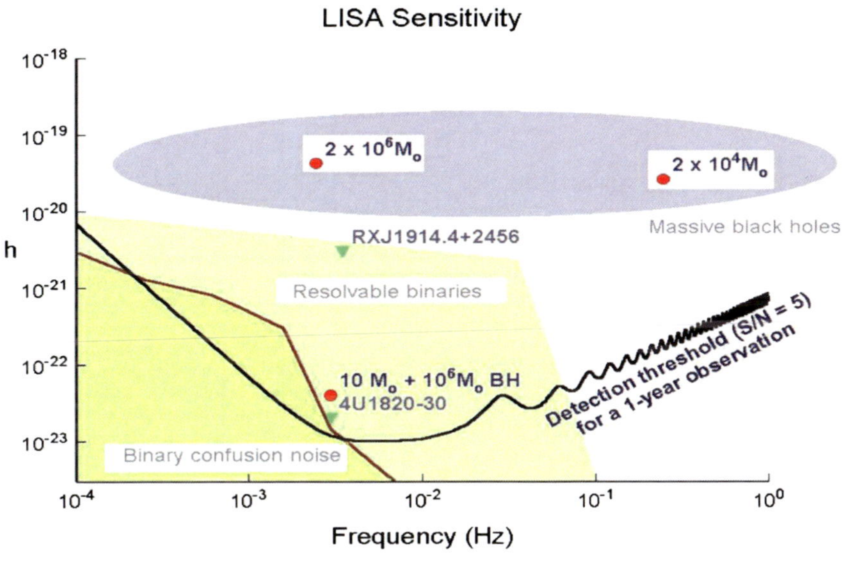

Fig. 81 LISA sensitivity plot

Optical noises will be reduced either by using signal recycling or by injecting squeezed vacuum in beam splitter black port; still to understand how to get rid of vacuums introduced by photodiode inefficiencies.

ET will be a cryogenic detector with mirrors kept at 4 K; it is not obvious if it is necessary to cool also power recycling, signal recycling and beam splitter mirrors.

ET will be formed by at least 4 detectors separated by some thousand km for solving the inverse problem i.e. determination of source 2 polarizations and 2 angles. It is likely that the best configuration should contain at least one detector in the southern hemisphere, as it is suggested by Fig. 79.

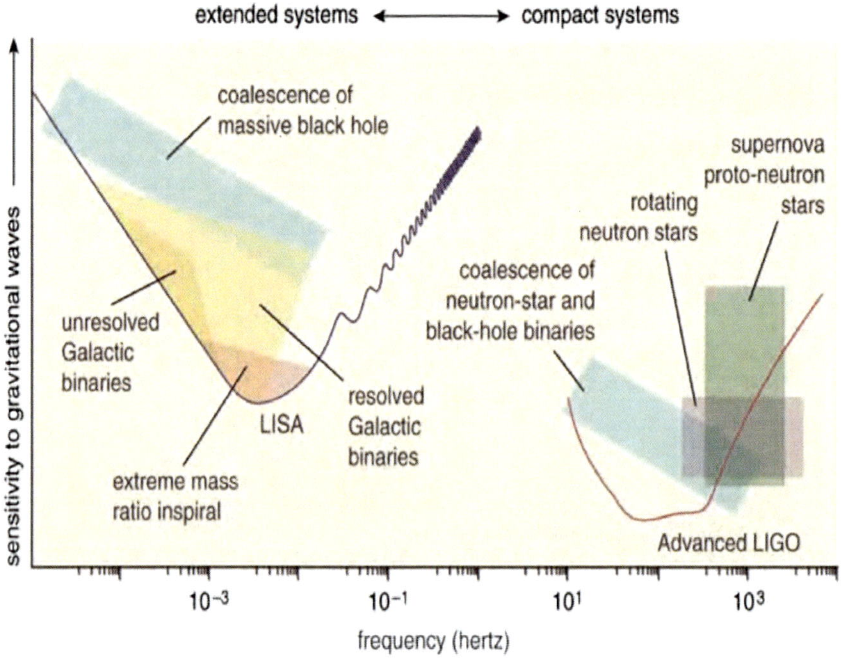

Fig. 82 LISA sensitivity and bandwidth compared with advanced LIGO

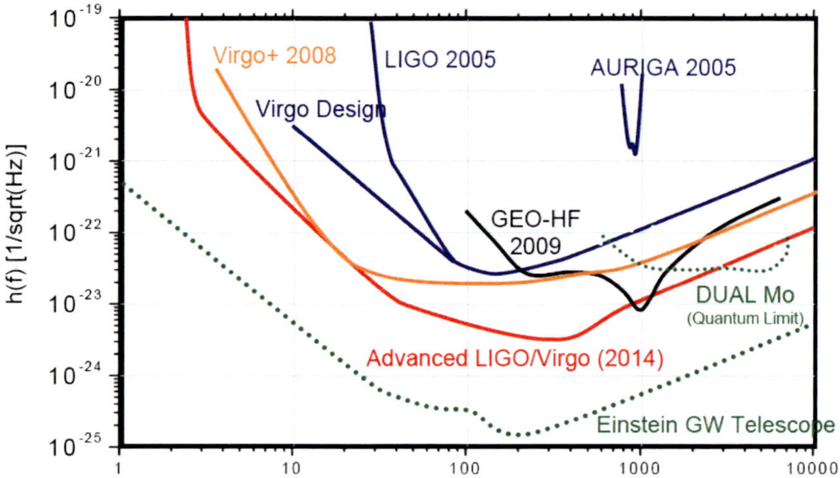

Fig. 83 Einstein Telescope GW detector should have a tremendous sensitivity, 100 times the present one of Virgo and LIGO. One day of data taking of ET is equivalent to 10^6 days of data taking of Virgo and LIGO machines

After 2020 ET will be the only operative project; it is likely that Virgo and LIGO will be obsolete.

GW detector noises are almost non-modelizable and in general it is difficult to understand them by means of an R&D:

(a) Diffused light is one of major sources of noise and can not be easily predicted. (b) Line and power supply e.m. fields are not predictable. (c) Ground Loops are not predictable ....

Since in ET noise should be 100 times smaller than the one in the existing machines, the only way to reach design sensitivity in 2020–2024 is to stars as soon as possible with a one-to-one scale prototype machine which will become the first detector of ET array.

Final Considerations

Cryogenic Bar Detectors have reached high sensitivity ($h > 10^{-21}$) and a duty cycle so high that the first network of GW has been created, i.e. IGEC.

A big step forward in the last 10 years has been in the Interferometer technology. Design sensitivity, above 100 Hz, has been reached ($h > 5\ 10^{-23}$) and stability is so good that a very efficient detector network has been set up. The big challenge will be in the operation of Advanced Interferometers for which sensitivity should reach $h > 4\ 10^{-24}$.

I believe, after what we have learned from Virgo and LIGO construction and operation, that class Einstein detectors (third generation) having sensitivity $h > 10^{-25}$, seems feasible with very high probability of success. This seems to be the right way for starting GW astrophysics.

Acknowledgments I am deeply grateful to Stan Withcomb whose presentation at Amaldi 7 has been very much of help for the preparation of this paper; I am also deeply grateful to Giancarlo Cella for carefully reading the manuscript.

References

1. R.A. Hulse and J.H. Taylor, **Astrophys. J. Lett**. 195, 151 (1975)-H.V. Shapiro, S.A. Teukolsky, 1984, "*The physics of compact objects*" (J.Wiley and Sons, Eds.), New York.
2. A. Einstein, Sitz. Ber. Kon.Preus. **Ak. Wiss**, 688(1916); id. 154(1918).
3. C.W. Misner, K.S. Thorne, J. Wheeler, 1973, "*Gravitation*" (W.H. Freeman Eds.), S. Francisco. - K.S. Thorne, 1987, "Gravitational radiation" in "*300 years of gravitation*" (Edited by S.W. Hawking and W. Israel, Cambridge University Press), New York (9) 330–458.
4. Pirani, F.A.E., 1956, Acta Physica Polonica 15, 389.
5. Coalescing binary neutron stars-Frederic A Rasio et al 1999 Class. Quantum Grav. **16** R1-R29 doi:10.1088/0264–9381/16/6/201.
   - Matched filters for coalescing binaries detection on massively parallel computers - E. Calzavarini, L. Sartori, F. Schifano, R. Tripiccione and A. Viceré - Computer Physics Communications Volume 152, Issue 3, 15 May 2003, Pages 295–306.

   - Coalescing binaries—Probe of the universe - A. Krolak and Bernard F. Schutz-General Relativity and Gravitation - vol. 19, n. 12, 1987
   - Gravitational waves from coalescing binaries: detection strategies and Monte Carlo estimation of parameters-R. Balasubramanian, B.S. Sathyaprakash, S.V. Dhurandhar-arXiv.org > gr-qc > arXiv:gr-qc/9508011.
6. Supernova theory and the nuclear equation of state-E Muller 1990 J. Phys. G: Nucl. Part. Phys. 16 1571–1598 doi:10.1088/0954–3899/16/11/009.
   - Type Ia Supernovae: Theory and Cosmology - Niemeyer, J.C., Truran, James W., Franco, Jose-Cambridge University Press-01 May 2000.
7. Gravitational waves from hot young rapidly rotating neutron stars-Benjamin J. Owen, Lee Lindblom, Curt Cutler, Bernard F. Schutz, Alberto Vecchio, and Nils Andersson-Phys. Rev. D 58, 084020 (1998).
   - A.Giazotto, S. Bonazzola, E. Gourgoulhon-On Gravitational waves emitted by an ensemble of rotating neutron stars - Phys. Rev. D 55, 2014 (15 Feb. 1997).
8. Michele Maggiore "GRAVITATIONAL WAVE EXPERIMENTS AND EARLY UNIVERSE COSMOLOGY" **Phys. Rept.** 331:283–367, 2000.
   - Danilo Babusci (Frascati), Massimo Giovannini (Lausanne U.) "SENSITIVITY OF A VIRGO PAIR TO RELIC GW BACKGROUNDS". Aug 2000. 25 pp. **Class. Quant. Grav**. 17:2621–2633, 2000.
   - David Coward and Tania Regimbau-Detection regimes of the cosmological gravitational wave background from astrophysical sources- New Astronomy Reviews-Volume 50, Issue 6, September 2006.
9. J. Weber, **Phys. Rev**. 117, 306 (1960).
   - J. Weber, **Phys. Rev**. Lett. 22, 1302(1969).
10. Eugenio Coccia - STATUS OF THE BAR DETECTORS EXPLORER AND NAUTILUS - Edoardo Amaldi Conference on Gravitational Waves-8 – 14 July 2007, Sydney.
   - Luca Taffarello - STATUS REPORT OF AURIGA DETECTOR - Edoardo Amaldi Conference on Gravitational Waves-8 – 14 July 2007, Sydney.
   - Status of ALLEGRO - W. Johnson - Edoardo Amaldi Conference on Gravitational Waves - June 20–24, 2005, Okinawa.
   - D. Blair et al. "Niobe: improved noise temperature and Background Noise Suppression" in *Gravitational Waves*, third Amaldi Conference, Pasadena Ca. 12–16 july 1999, AIP Conference Proceedings # 523, Melville, New York).
   - J.J. Johnson et al. in *Gravitational Wave Experiments*, Proceedings of the First Edoardo Amaldi Conference, Frascati 1994, (World Scientific, Singapore, 1995)
11. Oleksandr Usenko-MINIGRAIL PROGRESS; TEST OF COMPLETE TRANSDUCER CHAINS DOWN TO mK TEMPERATURE - Edoardo Amaldi Conference on Gravitational Waves-8 – 14 July 2007, Sydney.
   - Odylio D. Aguiar-THE SCHENBERG SPHERICAL GRAVITATIONAL WAVE DETECTOR:THE FIRST COMMISSIONING RUNS - Edoardo Amaldi Conference on Gravitational Waves-8 – 14 July 2007, Sydney.
   - L. Gottardi, A. de Waard, O. Usenko, G. Frossati, M. Podt, J. Flokstra, M. Bassan, V. Fafone, Y. Minenkov, and A. Rocchi, "Sensitivity of the spherical gravitational wave detector MiniGRAIL operating at 5K"-Phys. Rev. D 76, 102005 (2007).
12. A. Giazotto "Interferometric Detection of of Gravitational Waves", **Physics Rept**, 182, 6, 1989.
   - J.Y. Vinet, B. Meers, C.N. Man, A. Brillet, **Phys. Rev.** *D* 38 (1988) 433.
   - A. Brillet, *Ann. de Phys.* 10 (1985) 219.
   - H.B. Callen, R.F. Greene, **Phys. Rev.** 86 (1952) 702.
13. Two books give a lot of important informations in this research field; they are:
    1. Marlon O. Scully and M. Suhail, Zubary-Quantum Optics Cambridge University Press (Sept. 4, 1997).
    2. S.M. Barnett and P.M. Radmore, Methods in theoretical quantum optics Oxford University Press (Nov. 1, 1997).

The following papers are instrumental:

- C.M. Caves, 1980, Phys. Rev. Lett. 45, 75.
- C.M. Caves, B.L. Schumaker, New formalism for two-photon quantum optics.I. Quadrature phases and squeezed states. - Phys. Rev. A 31,5, may 1985, 3086.
- C.M. Caves, B.L. Schumaker, New formalism for two-photon quantum optics.II. Mathematical foundations and compact notations - Phys. Rev. A 31, 5, may 1985, 3093
- J.M. Courty, A. Heidmann and M. Pinard, Back action cancellation in interferometers by quantum locking.-gr-cq/0301068 v1, jan 2003
- H.J. Kimble et al., Conversion of conventional gravitational-wave interferometers into QND interferometers by modifying their input and/or output optics.-gr-qc/0008026 v2, 3 sept. 2001

14. Daniel Sigg, STATUS OF THE LIGO DETECTORS -Edoardo Amaldi Conference on Gravitational Waves-8 – 14 July 2007, Sydney
15. Frederique Marion, STATUS OF VIRGO - Edoardo Amaldi Conference on Gravitational Waves-8 – 14 July 2007, Sydney.
16. The seismic noise challenge to the detection of low frequency gravitational waves: A. Giazotto "7th congresso di relatività generale e fisica della gravitazione" - Rapallo, Italy, Sept. 1986 - **General relativity and gravitational physics** - World Scientific Pub. (449) 1987.
17. R. Del Fabbro et al., 1988, Phys. Lett. A132, 237.
    - S. Braccini et al., 1997, "*Design of the superattenuators for the VIRGO construction*" VIRGO Internal Report (VIR-TRE-PIS-4600–134).
18. Hartmut Grote, THE STATUS OF GEO 600 - Edoardo Amaldi Conference on Gravitational Waves-8 – 14 July 2007, Sydney.
19. Ryutaro Takahashi, OPERATIONAL STATUS OF TAMA300 WITH THE SEISMIC ATTENUATION SYSTEM (SAS) - Edoardo Amaldi Conference on Gravitational Waves-8 – 14 July 2007, Sydney.
20. Francesco Salemi, RESULTS OF THE SEARCH FOR GRAVITATIONAL WAVE BURSTS BY THE NETWORK OF RESONANT BAR DETECTORS. - Edoardo Amaldi Conference on Gravitational Waves-8 – 14 July 2007, Sydney.
21. LIGO Scientific Collaboration: B. Abbott, et al-All-sky search for periodic gravitational waves in LIGO S4 data-arXiv:0708.3818v1 [gr-qc], submitted 28 Aug 2007.
22. LIGO Scientific Collaboration: B. Abbott, et al -Search for gravitational waves from binary inspirals in S3 and S4 LIGO data-s- arXiv:0704.3368v2 [gr-qc], submitted 25 Apr 2007.
23. LIGO Scientific Collaboration-Search for gravitational-wave bursts in LIGO data from the fourth science run-Classical and Quantum Gravity 24, 5343–5369 (2007).
24. Michele Punturo, VIRGO UPGRADE PLANS - Edoardo Amaldi Conference on Gravitational Waves-8 – 14 July 2007, Sydney.
25. Carol Wilkinson, STATUS AND PLANS FOR ADVANCED LIGO INSTRUMENTS - Edoardo Amaldi Conference on Gravitational Waves-8 – 14 July 2007, Sydney.
26. Kazauki Kuroda, CURRENT STATUS OF LCGT PROJECT- Edoardo Amaldi Conference on Gravitational Waves-8 – 14 July 2007, Sydney.
27. David Blair, THE AIGO PROJECT - Edoardo Amaldi Conference on Gravitational Waves-8 – 14 July 2007, Sydney.
28. Scott Hughes, ASTROPHYSICS AND RELATIVITY USING LISA-Edoardo Amaldi Conference on Gravitational Waves-8–14 July 2007, Sydney.
29. Harald Lueck PLANS FOR E.T. (Einstein Telescope), A EUROPEAN 3rd GENERATION GRAVITATIONAL WAVE OBSERVATORY-Edoardo Amaldi Conference on Gravitational Waves-8 – 14 July 2007, Sydney.

Search for Gravitational Waves with Resonant Detectors

Guido Pizzella

Abstract The search for Gravitational Waves began more than 40 years ago by initiative of Joe Weber. Since then two types of instrumentation have been developed: the large interferometers and the resonant detectors. In this review paper we deal with the resonant detectors: ALLEGRO, AURIGA, EXPLORER, NAUTILUS and NIOBE. These detectors established upper limits for gravitational wave bursts. Since 1998 up to 2004 coincident events were searched for between EXPLORER and NAUTILUS. Excess coincidences are found when the detectors are favorably oriented with respect to the Galactic Disk.

1 Introduction

The existence of gravitational waves (GW) was hypothesized for the first time by O. Heaviside in his book *Electromagnetic Theory* (1893) and, independently, by H.A. Lorentz (1900) and by H. Poincaré (1905). A mathematical derivation of the possible existence of GW was given by Albert Einstein in 1916. Einstein not only foresee the existence of GW, but also presents a complete mathematical survey of the properties of what we call today *weak GW*, deriving results which are the best at our disposal even today.

Einstein shows in his General Relativity Theory (GRT) that the equations of the gravitational field can easily be solved in the weak field approximation, i.e. when the deviation of the geometry of space–time from the four dimensional (pseudoeuclidian) flat-space are so small that their squares and products can be neglected. In this approximation the equations of GR are reduced to second order linear differential equations. In vacuum, by adopting an appropriate gauge, these equations become D'Alembert equations with velocity of propagation equal to the velocity of light c. The problem of absorption and emission of GW and their main properties were

G. Pizzella (✉)
Dipartimento di Fisica, Università di Roma "Tor Vergata" INFN Laboratori Nazionali di Frascati
e-mail: guido.pizzella@roma2.infn.it

I. Ciufolini and R.A. Matzner (eds.), *General Relativity and John Archibald Wheeler*, Astrophysics and Space Science Library 367, DOI 10.1007/978-90-481-3735-0_12,

re-examined more thoroughly by Einstein in a paper of 1918. The absorbed fraction of the incident power is found to be extremely small and so is the emission of GW by any system of bodies; he points out that the irradiated power *in all possible thinkable cases must have a practically vanishing value.*

General Relativity was, at the beginning, mainly a field of mathematical developments, with the only exception of the classical experimental tests which were always affected by large experimental errors. Only in the second half of the last century a number of experimental technologies became sufficiently advanced to allow accurate experimental tests of the GRT. An overall picture of the achievements reached in this field and of their more recent extensions can be found in the enlightening book by C.M. Will [1].

In the GRT Einstein describes the gravitational field by means of a deformation of the space–time geometry, using the metric tensor g_{ik}:

$$g_{ik} = g^{o}_{ik} + h_{ik} \tag{1}$$

where g^{o}_{ik} is the minkowskian metric tensor and h_{ik} is the GW perturbation tensor whose components are much smaller than unity [2–4]. It can be shown that for a plane wave propagating in the x direction, the only non zero components are

$$h_{yz} = h_{zy} = h_{\times}, \qquad h_{yy} = -h_{zz} = h_{+} \tag{2}$$

These quantities are functions of the argument $(t - x/c)$, thus in GRT the GW propagate with the speed of light c, are transverse and have two polarization states. It can be also shown that, because of the conservation laws, the GW can be associated to a particle with spin 2 (the mass is obviously zero): the graviton.

As soon as Einstein published these results, not all scientists believed that GW were a physical entity. Some scientists believed that they were just a mathematical fiction, ripples in the space–time of mathematics. Among them was Levi–Civita,[1] who exchanged many letters with Einstein on this subject. The argument remained vague until in 1950, when H. Bondi demonstrated that in invariant, coordinate-free terms they are physically observable and finally in 1962 that they carry energy and momentum away from the systems which generate them.

GW are generated by the acceleration of gravitational masses. The irradiated power has a null dipole term, because of the momentum conservation law. The first non-zero term is the quadrupole one. Partially for this reason but essentially because of the small value of the gravitational constant G, the irradiated power of GW, given by the third time derivative of the mass quadrupole tensor, is so feeble that, at present, it is not possible to detect GW that could be generated in an earth laboratory. It is necessary that the source be constituted by enormous masses moving with extreme accelerations. Astrophysical sources [2, 5, 7, 8] are the best candidates for

[1] Still today a few scientists believe that GW are just ripples in the space–time of mathematics. Einstein himself, in collaboration with N. Rosen, for a while doubted about the physical existence of GW (see Ref. [6]).

Table 1 Astrophysical sources of gravitational waves

Source	Distance	Rate per year	h
Pulsar	1 kpc	Several sources	Continuous $\sim 10^{-26}$
Binary coalescence	100 Mpc	1	$\sim 10^{-21}$
Galactic Supernovae	10 kpc	0.1	$\sim 10^{-18}$

GW, for instance: supernovae, coalescence of binary systems, pulsars. Some estimation of the strength of the GW signals (in terms of the perturbation h of the metric tensor) and their occurrence rate, as reported in the review by Thorne (1987) [2], are given in Table 1.

Example of continuous sources are the pulsars and the binary systems. In the first case it is necessary that the rotating collapsed star be non-axisymmetric, in order to have a time varying mass quadrupole. Possible values for the metric tensor perturbation could be of the order of $h \sim 10^{-26} - 10^{-27}$, time-varying with twice the pulsar frequency.

Detailed calculations on possible sources can be found in [9]. In the case of binary systems we have two neutron stars or black-holes, spiraling one with respect to the other, until they collide. The final stage, lasting a few seconds, generates a powerful signal of GW with frequency sweeping rapidly (CHIRP). The interesting feature of this source is that it allows, from the detected amplitude and its variation with frequency, to determine the distance of the source and the mass of the two orbiting collapsed objects. Knowing the distance one could also determine the Hubble constant with great accuracy, in this way setting an accurate scale for measuring the cosmic distances.

For the impulsive sources one of the most interesting is the supernova. It is very difficult to guess the degree of non-sphericity during the collapse. If we assume that a fraction of all available energy $M_{star}c^2$ is entirely converted into GW, it is found, for a GW burst with a duration of 1 ms, due to the conversion of 10^{-3} solar masses, $h \sim 10^{-18}$ for a source located at the center of our Galaxy and $h \sim 10^{-21}$ for a source located in the Virgo Cluster.

Among possible sources of GW we have the superposition of many continuous waves generated by the pulsars, the overlapping bursts due to gravitational collapses and to coalescence of binary systems, the stochastic waves due to very massive objects formed in the early Universe or to phase transition in the early Universe. All these sources provide a stochastic cosmological background. Nucleosynthesis considerations put an upper limit on the ratio Ω of the GW energy density to the critical density needed for a closed universe. The upper limit is $\Omega \leq 10^{-5}$.

Recently, sources based on the string theory have been investigated [10]. The model considers a pre-big-bang phase and GW are generated during the transition from this phase. The time scale is close to the Planck time, about 10^{-43} s. Therefore the observation of these stochastic GW would give important information on the extreme early time of the Universe.

2 Beginning of the Experimental Activity

The problem of detecting GW is fascinating for many reasons. Although all metric theories of gravity that embody Lorentz invariance predict gravitational radiation, the detailed properties of such radiation depend on the particular theory. They may predict [1]: a difference between the speed of GW and the speed of light, different polarization states for generic GW, and different multipolarity (monopole, dipole, quadrupole, etc) of GW emitted by a given source.

Another very important reason why the detection of GW is so fascinating is due to the same reason why it is so difficult to detect them. That is, the extreme weakness of the gravitational interaction, which makes it possible for the GW to travel through enormous amount of matter without suffering strong attenuation. So they potentially are a very effective tool for investigating the Universe, reaching places that are completely inaccessible with the electromagnetic radiation.

But, the chance to be able to detect GW appeared so small that the experimental activity had to wait until the early sixties, when the very ingenious Joseph Weber [11, 12] at the University of Maryland invented instruments for their search, and announced evidence [13, 14] for the discovery of GW. Although, during the successive years, his initial results were not confirmed by other experiments with essentially the same technique (resonant bars at room temperature), the search for GW is now pursued by many experimental groups all over the world. The detection technique has greatly improved over the last 30 years and further remarkable steps can be foreseen in the near and far future. Some years ago the existence of GW, foreseen in all the metric theories of gravity, has received an experimental confirmation, although indirectly, by the observation of the astrophysical binary system PSR 1913 + 16 made by Hulse and Taylor [15].

One of the most important final goals for all groups which have engaged in this research is to reach such a sensitivity to be able to detect gravitational collapses occurring in the Virgo Cluster, or the final merge of coalescence binaries up to a distance of at least 100 Mpc. Although much progress has been made, this goal appears to be still far in future because of the extreme weakness of the signal due to the gravitational force as compared with the noise, which is due to the electromagnetic force.

The milestones that mark the experimental progress in the search for GW with resonant detectors are, after the initial work of Weber:

- In 1982 the sensitivity reached by the Stanford group with their cryogenic antenna [16].
- In 1986 and finally in 1990 the steady operation reached by the Rome group with the cryogenic antenna EXPLORER installed at CERN [17] (the gravitational wave experiment began in Rome in the year 1970 on the initiative of E. Amaldi and G. Pizzella).
- In May 1991 two large cryogenic antennas entered in simultaneous and continuous operation: the EXPLORER antenna of the Rome group and the ALLEGRO antenna of the Louisiana State University. They were joined in 1993 by the

antenna NIOBE of the University of Western Australia group, in 1995 by NAUTILUS of the Rome group and in 1997 by AURIGA of the Padua group.
- Coincidences between EXPLORER and NAUTILUS in the years 1998, 2001, 2003 and 2004 show evidence of a coincidences excess fron the Galaxy.

3 Interaction of GW with Free Masses

As well known, in the GRT theory it is always possible to choose a reference system that nullifies the gravitational field at a given point. This makes it impossible to measure the properties of a gravitational field using only one test mass.

Thus we need two test masses. The experimental problem consists in measuring the change of their mutual distance under the action of a gravitational field. The distance between them can be determined from the two geodesic lines. It can be shown that for any vector u^i the following formula holds

$$u^i_{;kl} - u^i_{;lk} = -R^i_{mkl} u^m \tag{3}$$

where R^i_{mkl} is the curvature tensor and the symbol ; indicates the covariant derivative. From this formula, identifying the vector u^i with the four-velocity of one of the test masses we obtain the *geodesic deviation equation* written for the case of velocity much smaller than light and in absence of other forces but gravity

$$\frac{\partial^2 \eta^i}{\partial s^2} + \eta^k R^i_{0k0} = 0 \tag{4}$$

where η^i is the distance between the two test masses and s is the coordinate $ds = cd\tau$, with τ the proper time along the particle trajectory.

Equation (4) is at the basis of all detectors for GW. From this equation we obtain

$$\frac{\partial^2 \eta^i}{\partial t^2} = \eta^k \frac{\partial^2 h^i_k}{\partial t^2} \tag{5}$$

Applying this to the case of two test masses along the y-axis at positions $\pm\frac{l}{2}$ with the GW travelling along the x-axis, indicating with ξ^i the change in the distance η^i, we get for given initial conditions

$$\xi^o = 0, \qquad \xi^1 = 0, \qquad \xi^2 = \frac{l}{2} h_+, \qquad \xi^3 = \frac{l}{2} h_\times \tag{6}$$

More in general for a ring of test masses in the (y,z) plane with radius r, indicating with θ the angle of a diameter with the y-axis we have the following deformation of the ring

$$\xi^2 = r(h_+ \cos\theta + h_\times \sin\theta) \tag{7}$$

$$\xi^3 = r(h_\times \cos\theta - h_+ \sin\theta) \tag{8}$$

These equations clarify the relative effect of the two different polarizations. We notice that the effect produced by the wave h_+ is identical to that produced by the wave $h_\times$ with a rotation by an angle of 45°.

From the measurement of the geodesic deviation ξ^i, using Eqs. (7) and (8), one obtain h_+ and $h_\times$.

4 The Cross-Section for a Resonant Bar

The cross-section for the interaction of gravitational waves (GW) with a bar resonant detector has been calculated on very general bases (see Refs. [18, 19]), obtaining for GW bursts, averaged over all possible polarizations and directions:

$$\Sigma = \frac{32}{15\pi}\frac{G}{c}\frac{v^2}{c^2}\frac{M}{n^2} \quad [m^2\ Hz] \tag{9}$$

that becomes, for optimum polarization and incident angle (perpendicular to the bar axis)

$$\Sigma_o = \frac{8}{\pi}\frac{G}{c}\frac{v^2}{c^2}\frac{M}{n^2} \quad [m^2\ Hz] \tag{10}$$

where M is the mass of the detector, v le velocity of sound and n, only odd numbers, is the order of the longitudinal oscillation. For the lowest mode $n = 1$ and for a bar length of 3 m the oscillation has frequency of the order of 900 Hz. In the following we shall refer to as the *classical cross-section.*

The meaning of this cross section is the following: given a GW flux with spectral density Φ (energy per unit frequency and unit area), the energy absorbed by the resonant detector is

$$\epsilon = \Phi\,\Sigma \tag{11}$$

It is important to recall briefly the reasoning for obtaining Eq. (9). It starts from the Breit–Wigner formula, which we rewrite for the case of GW (Ref. [18])

$$\sigma = \pi\lambda^2\xi\frac{A_{grav}A}{(\omega-\omega_o)^2+\left(\frac{A}{2}\right)^2} \tag{12}$$

where σ is the cross-section near resonance, xi a statistical factor, A_{grav} the decay rate due to emission by the detector of GW, A is the decay rate due to damping ($A \gg A_{grav}$).

The central point here is that, *in thermal equilibrium*, the detector will loose energy by radiation as rapidly as it gains energy by absorption at resonance. One finally obtains [18] the *average* cross-section:

$$\Sigma = \frac{\pi}{2}\lambda^2 A_{grav} \tag{13}$$

which leads to Eqs. (9) and (10).

Other derivations of the cross-section based on detailed considerations of the effect of GW on the various parts of a metallic cylinder considered at rest, divided an infinite number of slices [20], yield exactly the same result.

5 Algorithms for the Search of Short Bursts

We limit ourselves here to the search of short bursts of GW, bursts which can be modeled with a delta-function.

5.1 The ZOP Filter

We consider the simple case of a non-resonant transducer mounted on a bar. The actual case of a resonant transducer can be [21] reconducted to the simpler case, the only difference being a better adaption of the transducer to the bar (like a transformer), that is a larger coupling factor. The signal from the low noise amplifier is send to a lock-in amplifier which extracts the in phase and in quadrature components, $x(t)$ and $y(t)$, of the Fourier transform at the bar resonance frequency f_o. The lock-in amplifier has integration time t_o and both components are sampled with a sampling time t_o, in order to minimize the amount of data to be recorded. The simplest algorithm for extracting a signal due to a GW short bursts from the noise has been suggested by Gibbons and Hawking[2] [22] and called by the Rome group [24] the ZOP algorithm (zero-order prediction), consists in taking the difference between two successive samplings

$$z(t)^2 = [x(t) - x(t - \Delta t)]^2 + [y(t) - y(t - \Delta t)]^2 \tag{14}$$

The key idea is that a short burst will produce a jump in the data, like a hammer hit, while the fluctuations due to the noise have a relative long time constant. Let us now estimate the SNR for this algorithm. The noise is basically due to the sum of the narrow-band Brownian noise (in units of $volt^2$) in the bar

$$V_{nb}^2 = \alpha^2 S_f(f_o)\beta_1 \tag{15}$$

(increased by the back action from the amplifier) and of the white noise S_o from the amplifier. The auto correlation for this noise process at the lock-in output is

$$R(\tau) = V_{nb}^2 \left[\frac{e^{-\tau/\tau_v} - \frac{t_o}{\tau_v} e^{-\tau/t_o}}{1 - t_o^2/\tau_v^2} \right] + \frac{S_o}{t_o} e^{-\tau/t_v} \tag{16}$$

[2] Although this algorithm was already known [23].

for each components, $x(t)$ and $y(t)$, of the signal, where $\tau_v = \frac{1}{\beta_1}$ is related to the merit factor $Q = \tau_v \pi f_o$. From the above autocorrelation we derive the variance for the variable $z(t)$:

$$\sigma^2 = 4[R(0) - R(\Delta t)] = \frac{4V_{nb}^2}{e}\frac{\Delta t}{\tau_v} + 4\frac{e-1}{e}\frac{S_o}{\Delta t} \tag{17}$$

where we notice that there are two noises in competition, one increasing linearly with Δt and the other one with the inverse of Δt . The optimum Δt is given by

$$\Delta t_{opt} = \tau_v \sqrt{(e-1)\Gamma} \tag{18}$$

and

$$\sigma_{opt}^2 = 8V_{nb}^2 \frac{\sqrt{(e-1)\Gamma}}{e} \tag{19}$$

where we have made use of the quantity

$$\Gamma = \frac{S_o \tau_v}{V_{nb}^2} \tag{20}$$

We have now to calculate the signal for this algorithm. An incoming GW short burst will produce at the low noise amplifier output a jump in the signal from the noise level to a value

$$V_s = \alpha \frac{2L}{\pi^2} \omega_o h(\omega_o) \tag{21}$$

slowly decaying with the time constant τ_v. At the lock-in amplifier output (after the lock-in integration with time constant t_o) we take the difference between two next samples. If we assume that the signal arrives exactly at the time of a sampling the difference with the next sampling will give $V_s(1 - 1/e)$. We introduce the signal energy

$$E_s = \frac{1}{2} m\omega_o^2 \left(\frac{V_s}{\alpha}\right)^2 \tag{22}$$

Recalling that $V_{nb}^2 = \frac{\alpha^2 k T_e}{m\omega_o^2}$, we get

$$SNR = \frac{V_s^2(1-1/e)^2}{\sigma_{opt}^2} = \frac{E_s}{4kT_e\sqrt{\Gamma}}\frac{1}{1.21} \tag{23}$$

It must be remarked that this results is valid for signals arriving exactly at the sampling times. If one considers signals arriving at random times he gets a SNR that, on the average, is smaller by several per cent.

5.2 The Wiener Filter

The ZOP filter can be extended by including not just two samplings but many more. This is done by using the Wiener filter [23] that is based on the fact that the data samplings are processed after they have been recorded, in such a way one can make use of past as well *future* data. The measured variables $x(t)$ and $y(t)$ include signal and noise. The best estimation $u(t)$ of the signal for the variable $x(t)$ is

$$\tilde{u}(t) = \int x(t-\tau)W(t)d\tau \tag{24}$$

where $W(t)$ is the filter function which will be estimated with the linear mean square method. It can be demonstrated [23] that the Fourier transform $W(f)$ of $W(t)$ which minimize the average difference $< u(t) - \tilde{u}(t) >^2$ is

$$W(f) = \frac{S_{ux}(f)}{S_{xx}(f)} \tag{25}$$

where $S_{xx}(f)$ is the power spectrum of $x(t)$, and $S_{ux}(f)$ is the cross spectrum of $u(t)$ and $x(t)$.

The power spectrum of the brownian noise reported at the antenna input (in units of $\frac{V^2}{Hz}$) is white,

$$S_{uu} = \frac{2V_{nb}^2}{\beta_1} \tag{26}$$

and so is white the noise

$$S_{ee} = 2S_o \tag{27}$$

at the lock-in output, due to the electronic amplifier that enters in the electronic chain before the lock-in. We notice that

$$\Gamma = \frac{S_{ee}}{S_{uu}} \tag{28}$$

We can then consider the overall electronic chain as made by two pieces. The first piece with transfer function $W_a(f)$ is the bar which, together with the selective part of the lock-in, acts as a low-pass filter with time constant τ_v, the second piece is the integrating part of the lock-in which is again a low-pass filter with transfer function W_e and time constant t_o. We get

$$S_{xx}(f) = S_{uu}W_a^2W_e^2 + S_{ee}W_e^2 \tag{29}$$

$$S_{ux}(f) = S_{uu}W_a^*W_e^* \tag{30}$$

from which we obtain

$$W(f) = \frac{1}{W_a W_e} \frac{1}{1 + \dfrac{\Gamma}{W_a^2}} \tag{31}$$

which solves the problem. If one wants to apply the Wiener filter in the time-domain, he can calculate the coefficients $a_k = \int W(t)dt$ over several sampling intervals, $(k - 1/2)\Delta t$ to $(k + 1/2)\Delta t$, and the estimate

$$\tilde{u}_i \simeq \sum x_{i-k} a_k \tag{32}$$

Applying this filter to the data we get the power spectrum of the noise after the Wiener filter

$$S_{\tilde{u}\tilde{u}}(f) = S_{xx} W(f)^2 = \frac{S_{uu}}{1 + \dfrac{\Gamma}{W_a}^2} \tag{33}$$

From this we get the autocorrelation of the noise

$$R_{\tilde{u}\tilde{u}} = \frac{S_{uu}}{2\tau_v \sqrt{\Gamma(\Gamma + 1}} e^{-\beta_3 t} \tag{34}$$

with

$$\beta_3 = \frac{1}{\tau_v \sqrt{(\Gamma + 1)\Gamma}} \simeq \frac{1}{\tau_v \sqrt{\Gamma}} \tag{35}$$

Later we shall show that β_3 is the antenna bandwidth.

The signal reported at the antenna input has Fourier transform V_s (white spectrum because we consider a GW short burst). For simplicity, we consider the signal in phase with the lock-in reference frequency. The application of the Wiener filter gives

$$u(f) = V_s W_a(f) W_e(f) W(f) = \frac{V_s}{1 + \dfrac{\Gamma}{W_a}^2} \tag{36}$$

formally identical to Eq. (14). We notice that in absence of electronics noise ($\Gamma = 0$) the estimation is perfect, in the sense that its Fourier transform is equal to the Fourier transform of the GW signal. From the above equations we get the maximum SNR at time $t = 0$, when the GW burst arrives, considering that there is an equal contribution to the noise both from the in-phase and from the in-quadrature responses of the lock-in,

$$SNR = \frac{u^2(t=0)}{2 \cdot R_{\tilde{u}\tilde{u}}(t=0)} = \frac{V_s^2}{2\tau_v S_{uu} \sqrt{\Gamma(\Gamma + 1)}} = \frac{E_s}{4kT_e \sqrt{\Gamma}} \tag{37}$$

This shows that the improvement over the optimum ZOP filter seems to be just a factor of 1.21. Actually, the advantage is that, for the Wiener filter, one can sample

Table 2 Data analysis for a 20-kg Aluminium bar, loaded with a large piezoelectric ceramic. We note the excellent agreement between computed and measured noise parameters

Run	T Kelvin	Q	T_e Kelvin	σ_{th}^2 volt^2	σ_{ex}^2 volt^2	T_{eff} Kelvin
1	4.5	64,100	6.4	23	22	0.83
2	4.5	64,700	5.9	21	24	0.89
3	1.5	89,300	4.3	15	15	0.59

faster than the optimum sampling needed to optimize the ZOP filter. In this way there is no loss in SNR due to the random arrival time of the GW bursts.

From Eq. (37), putting SNR = 1, we introduce the *effective noise temperature* T_{eff}

$$T_{eff} = 4T_e\sqrt{\Gamma} \tag{38}$$

This result has been obtained in this form for the first time in [25].

Measurements were made with a small [26] and an intermediate [27] bar to test our algorithms for the data analysis. The results for the small bar are shown in Table 2 and show excellent agreement between computed and measured noise parameters.

This filter [28, 29] has been used for some time in the Rome group [17].

6 The Matched Filter

Let us consider a signal $s(t)$ in presence of noise $n(t)$ [23]. The available information is the sum

$$x(t) = s(t) + n(t) \tag{39}$$

$x(t)$ is the measurement at the output of the low noise amplifier and $n(t)$ is a random process with known properties. Let us start by applying to $x(t)$ a linear filter which must be such to maximize the signal to noise ratio SNR at a given time t_o (we emphasize the fact that we search the signal at a given time t_o).

Indicating with $w(t)$ the impulse response of the filter (to be determined) and with $y_s(t) = s(t) * w(t)$ and $y_n(t) = n(t) * w(t)$ respectively the convolutions of the signal and of the noise we have

$$SNR = \frac{|y_s(t_o)|^2}{E[|y_n(t_o)|^2]} \tag{40}$$

The expectation of the noise after the filter, indicating with $N(\omega)$ the power spectrum of the noise $n(t)$ is

$$E[|y_n(t_o)|^2] = \frac{1}{2\pi}\int_{-\infty}^{\infty} N(\omega)|W(\omega)|^2 d\omega \tag{41}$$

where $W(\omega)$ is the Fourier transform of the unknown $w(t)$.

At $t = t_o$ the output due to $s(t)$ with Fourier transform $S(\omega)$ is given by

$$y_s(t_o) = \frac{1}{2\pi}\int_{-\infty}^{\infty} S(\omega)W(\omega)e^{j\omega t_o}d\omega \tag{42}$$

We apply now [23] the Schwartz' inequality to the integral 42 and using the identity

$$S(\omega)W(\omega) = \frac{S(\omega)}{\sqrt{N(\omega)}}W(\omega)\sqrt{N(\omega)} \tag{43}$$

we obtain

$$SNR \leq \frac{1}{2\pi}\int_{-\infty}^{\infty}\frac{|S(\omega)|^2}{N(\omega)}d\omega \tag{44}$$

It can be shown [23] that the equal sign holds if and only when

$$W(\omega) = constant\frac{S(\omega)^*}{N(\omega)}e^{-j\omega t_o} \tag{45}$$

Applying this optimum filter to the data we obtain the maximum SNR

$$SNR = \frac{1}{2\pi}\int_{-\infty}^{\infty}\frac{|S(\omega)|^2}{N(\omega)}d\omega \tag{46}$$

where $S(\omega)$ and $N(\omega)$ as already specified are, respectively, the Fourier transform of the signal and the power spectrum of the noise at the end of the electronic chain where the measurement $x(t)$ is taken.

Let us apply [30] the above result to the case of measurements $x(t)$ done at the end of a chain of two filters with transfer functions W_a (representing the bar) and W_e (representing the electronics).

Be S_{uu} the white spectrum of the brownian noise entering the bar and S_{ee} the white spectrum of the electronics noise. The total noise power spectrum is

$$N(\omega) = S_{uu}|W_a|^2|W_e|^2 + S_{ee}|W_e|^2 \tag{47}$$

The Fourier transform of the signal is

$$S(\omega) = S_g(\omega)W_aW_e. \tag{48}$$

where S_g is the Fourier transform of the GW signal at the bar entrance.

The optimum filter will have, applying Eq. (45), the transfer function

$$W(\omega) = \frac{S_g^*e^{-j\omega t_o}}{S_{uu}}\frac{1}{W_aW_e}\frac{1}{1+\frac{\Gamma}{|W_a|^2}} \tag{49}$$

where

$$\Gamma = \frac{S_{ee}}{S_{uu}} \tag{50}$$

Using Eqs. (47) and (48), from Eq. (46) we obtain

$$SNR = \frac{1}{2\pi S_{uu}} \int_{-\infty}^{\infty} \frac{|S_g(\omega)|^2 d\omega}{1 + \frac{\Gamma}{|W_a|^2}} \tag{51}$$

We apply now the above result to a delta GW with Fourier transform S_g independent on ω. The integral of Eq. (51) can be easily solved if we make use of a lock-in device which translates the frequency, bringing the resonant frequency ω_o to zero. Then the total noise becomes [23]

$$N(\omega) => N(\omega - \omega_o) + N(\omega + \omega_o) \tag{52}$$

with the bar transfer function given by [29]

$$W_a = \frac{\beta_1}{\beta_1 + i\omega} \tag{53}$$

We estimate now signal and noise just after the first transfer function, before the electronic wide-band noise. For the signal due to a delta excitation we have

$$V(t) = \frac{1}{2\pi} \int_{-\infty}^{\infty} \frac{\beta_1}{\beta_1 + i\omega} S_g e^{i\omega t} d\omega = S_g \beta_1 e^{-\beta_1 t} = V_s e^{-\beta_1 t} \tag{54}$$

where V_s is the maximum signal. For the thermal noise we have

$$V_{nb}^2 = \frac{1}{2\pi} \int_{-\infty}^{\infty} \frac{\beta_1^2}{\beta_1^2 + \omega^2} S_{uu} d\omega = \frac{S_{uu}\beta_1}{2} \tag{55}$$

where $V_{nb}^2 = \frac{\alpha^2 k T_e}{m\omega_o^2}$ is the mean square narrow-band noise (see Eqs. (61) and (62)).

Introducing the signal energy $E_s = \frac{1}{2} m\omega_o^2 (\frac{V_s}{\alpha})^2$ we calculate from Eq. (51) (with the variable $y = \frac{\omega}{\beta_1}$)

$$SNR = \frac{S_g^2 \beta_1}{2\pi S_{uu}} \int_{-\infty}^{\infty} \frac{dy}{1 + \Gamma(1 + y^2)} = \frac{V_s^2}{4V_{nb}^2 \sqrt{\Gamma}} = \frac{E_s}{2kT_e\sqrt{\Gamma}} \tag{56}$$

The effective noise temperature is

$$T_{eff} = 2T_e\sqrt{\Gamma} \tag{57}$$

a factor of two better than the previous Wiener filter because here we have taken into consideration the phase as well as the amplitude of the signal. This filter is at present used in the GW detectors of the Rome group [31, 32].

7 Sensitivity and Bandwidth

We go now to discuss the sensitivity of a resonant GW detector in terms of its *bandwidth*.

The resonant detectors of gravitational waves (GW) now operating use resonant transducers (and therefore there are two resonance modes coupled to the gravitational field) in order to obtain high coupling and high Q. But for the discussion on the detectors sensitivity and frequency bandwidth it is sufficient to consider the simplest resonant antenna, a cylinder of high Q material, strongly coupled to a non resonant transducer followed by a very low noise electronic amplifier. The equation for the end bar displacement ξ is

$$\ddot{\xi} + 2\beta_1\dot{\xi} + \omega_o^2\xi = \frac{f}{m} \tag{58}$$

where f is the applied force, m the oscillator reduced mass (for a cylinder $m = \frac{M}{2}$) and $\beta_1 = \frac{\omega_o}{2Q}$ is the inverse of the decay time of an oscillation due to a delta excitation.

We consider here only the noise which can be easily modeled, the sum of two terms: the thermal (Brownian) noise and the electronic noise. The power spectrum due to the thermal noise is

$$S_f = \frac{2\omega_o}{Q} mkT_e \tag{59}$$

where T_e is the equivalent temperature which includes the effect of the back-action from the electronic amplifier.

By referring the noise to the displacement of the bar ends, we obtain the power spectrum of the displacement due to the Brownian noise:

$$S_\xi^B = \frac{S_f}{m^2} \frac{1}{(\omega^2 - \omega_o^2)^2 + \frac{\omega^2\omega_o^2}{Q^2}} \tag{60}$$

From this we can calculate the mean square displacement

$$\bar{\xi^2} = \frac{kT_e}{m\omega_o^2} \tag{61}$$

that can be also obtained, as well known, from the equipartition of the energy.

To this noise we must add the wide-band noise due to the electronic amplifier (the contribution to the narrow-band noise due to the amplifier has been already included in T_e).

For sake of simplicity we consider an electromechanical transducer that converts the vibration of the detector in a voltage signal

$$V = \alpha\xi \tag{62}$$

with transducer constant α (typically of the order of 10^7 V/m). Thus the electronic wide-band power spectrum, S_o , is expressed in units of $\frac{V^2}{Hz}$ and the overall noise power spectrum referred to the bar end is given by

$$S_\xi^n = \frac{2kT_e\omega_o}{mQ}\frac{1}{(\omega^2-\omega_o^2)^2+\frac{\omega^2\omega_o^2}{Q^2}}+\frac{S_o}{\alpha^2} \tag{63}$$

We calculate now the signal due to a gravitational wave with amplitude h and optimum polarization impinging perpendicularly to the bar axis. The bar displacement corresponds [20] to the action of a force

$$f = \frac{2}{\pi^2}mL\ddot{h} \tag{64}$$

The bar end spectral displacement due to a flat spectrum of GW (as for a delta-excitation) is similar to that due to the action of the Brownian force. Therefore, if only the Brownian noise were present, we would have a nearly infinite bandwidth,[3] in terms of signal to noise ratio (SNR). For a GW excitation with power spectrum $S_h(\omega)$, the spectrum of the corresponding bar end displacement is

$$S_\xi = \frac{4L^2\omega^4S_h}{\pi^4}\frac{1}{(\omega^2-\omega_o^2)^2+\frac{\omega^2\omega_o^2}{Q^2}} \tag{65}$$

We can then write the SNR

$$SNR = \frac{S_\xi}{S_\xi^n} = \frac{4L^2\omega^4S_h}{\pi^4\frac{S_f}{m^2}}\frac{1}{1+\Gamma(Q^2(1-\frac{\omega^2}{\omega_o^2})^2+\frac{\omega^2}{\omega_o^2})} \tag{66}$$

where for the quantity Γ we make use of Eq. (20)

$$\Gamma = \frac{S_o\beta_1}{\alpha^2\bar{\xi}^2} \sim \frac{T_n}{\beta QT_e} \tag{67}$$

[3] Already several years ago Joe Weber pointed out this feature of an oscillator with a very low noise amplifier.

T_n is the noise temperature of the electronic amplifier and β indicates the fraction of energy which is transferred from the bar to the transducer. It can be readily seen that $\Gamma \ll 1$.

The GW spectrum that can be detected with SNR = 1 is:

$$S_h(\omega) = \pi^2 \frac{kT_e}{MQv^2} \frac{\omega_o^3}{\omega^3} \frac{1}{\omega} (1 + \Gamma \left(Q^2 \left(1 - \frac{\omega^2}{\omega_o^2}\right)^2 + \frac{\omega^2}{\omega_o^2} \right) \tag{68}$$

where v is the sound velocity in the bar material ($v = 5{,}400$ m/s in aluminum). For $\omega = \omega_o$ we obtain the highest sensitivity

$$S_h(\omega_o) = \pi^2 \frac{kT_e}{MQv^2} \frac{1}{\omega_0} \left[\frac{1}{Hz}\right] \tag{69}$$

having considered $\Gamma \ll 1$.

We remark that the best spectral sensitivity, obtained at the resonance frequency of the detector, only depends, according to Eq. (69), on the temperature $T \sim T_e$, on the mass M and on the quality factor Q of the detector. Note that this condition is rather different from that required for optimum pulse sensitivity. The bandwidth of the detector is found by imposing that $S_h(\omega)$ be equal to twice the value $S_h(\omega_o)$. We obtain , in terms of the frequency $f = \frac{\omega}{2\pi}$

$$\Delta f = \frac{f_o}{Q} \frac{1}{\sqrt{\Gamma}} = \frac{f_o}{Q} \frac{4T_e}{T_{eff}} \tag{70}$$

Another useful quantity often used is the *spectral amplitude*

$$\tilde{h} = \sqrt{S_h} \tag{71}$$

From Eqs. (69) and (70) one can derive the antenna sensitivity for various types of GW. For delta-like bursts with duration τ_g ($\sim$1 ms) the sensitivity is given by

$$h \simeq \frac{1}{\tau_g} \sqrt{\frac{2S_h}{\pi \Delta f}} \tag{72}$$

From Eqs. (69) and (70) we get also:

$$h \simeq \frac{L}{\tau_g v^2} \sqrt{\frac{kT_{eff}}{M}}. \tag{73}$$

A resonant detector can measure the gravitational stochastic background. The relationship between the density ratio Ω and the power spectrum S_h at frequency f is [33]

$$\Omega = \frac{4\pi}{3} \frac{f^3}{H^2} S_h \tag{74}$$

where H is the Hubble constant.

8 Initial Experiments and the IGEC Collaboration

After the initial controversial results by Weber [14, 34] and an early null search for triple coincidences among Stanford, Louisiana and Rome [35] other results have been obtained with the resonant detectors in operation.

Coincidences were searched among the three detectors EXPLORER, NAUTILUS and NIOBE during the years 1995 and 1996. No double coincidence excess was found between EXPLORER and NIOBE and NAUTILUS and NIOBE, but some intriguing result was found when searching for coincidences between EXPLORER and NAUTILUS [36]. During a period of 29.2 days a number $n_c = 19$ of coincidences was found while expecting on the average $\bar{n} = 11$. This result alone, because of the very weak statistics, does not justify a possible claim of a new phenomenon, but it is worth to mention for consideration with other experimental results.

Later a complete analysis of the data recorded by EXPLORER and ALLEGRO in 1991 was done [37]. The result was of $n_c = 70$ coincidences with a coincidence window $w = \pm 1\ s$ and an average background of $\bar{n} = 59$ coincidences. The coincidence excess was not such as to justify any claim.

On July 1997 the IGEC (International Gravitational Event Collaboration) was established among the resonant detector groups: ALLEGRO, AURIGA, EXPLORER, NAUTILUS and NIOBE (see Refs. [17, 38–41]). Each group has put on a common WEB site the list of the events extracted independently from the data of its own detector. The five detectors were in operation only part of the time. The coincidence search during this period gave no excess over the accidental coincidences [42], and established a new upper limit, about $0.04 \frac{GW\ events}{day}$ with $h \geq 6\ 10^{-18}$.

9 The EXPLORER and NAUTILUS Experiment

9.1 Calibration of the Rome Detectors

A very important problem is the detector calibration. A real calibration of a GW detector is not possible, since no GW source is available. But we can do alternative calibrations to test the proper operation of the apparatus.

The Rome group has done it in two ways. At CERN with EXPLORER, the continuos near gravitational field generated by a rotor in the vicinity of the detector has been measured [43]. The measurement was in very good agreement with the calculation made using the Newton law of gravitation. New measurements were made with the same rotor a few years later [44] for testing the inverse square law at distances order of meters. These measurements gave an upper limit to a possible deviation of the Newton's law.

As far as short bursts, we have calibrated the apparatuses by means of known small forces applied to a piezoelectric ceramic. We tested the calibration by making

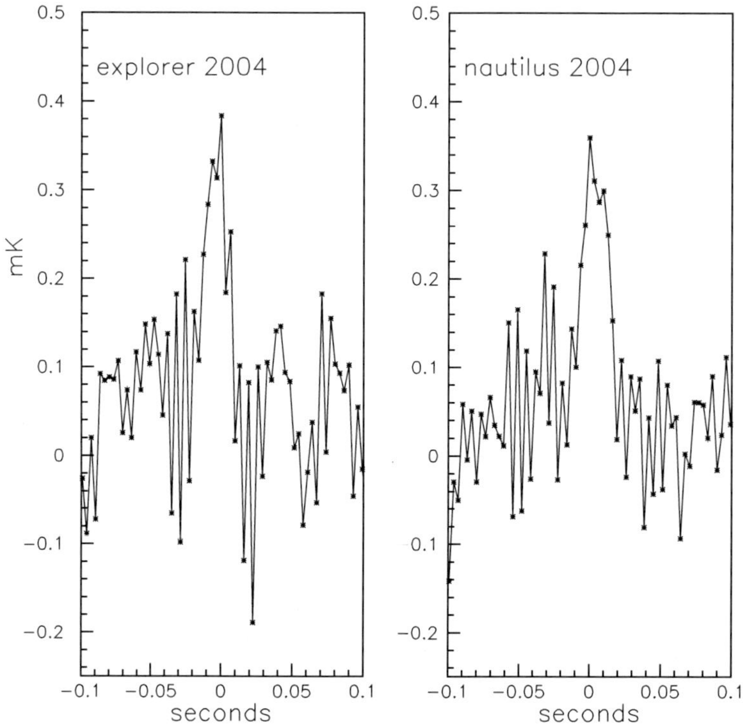

Fig. 1 2,508 small signals for EXPLORER and 1,189 small signals for NAUTILUS. Cumulative response to CRS for the multiplicity interval $400 < \Lambda < 1,500\frac{secondaries}{m^2}$. The background has been subtracted (this accounts for small negative values)

use of the cosmic ray showers (CRS), as follows [45]. We measure the response of each GW detector at the time of the corresponding CRS arrival: 2,508 showers for EXPLORER and 1,189 showers for NAUTILUS in the year 2004, with multiplicity (number of secondaries measured at the bottom of the bar) below $1{,}500\frac{secondaries}{m^2}$. We eliminate a few large signals (see later) and we take the averages of the responses of the two apparatuses referred to the time of each CRS. The result is shown in Fig. 1. We note that both EXPLORER and NAUTILUS respond in the same way to the CRS verifying our calibration, so for each coincidence we can compare the event energy of EXPLORER with that of NAUTILUS.

It is important to remark that the signals shown in Fig. 1, after the elimination of a few large signals, have amplitude in agreement with the thermo-acoustic model [46, 47]. This model has been verified experimentally with an electron beam produced by the $Da\Phi ne$ facility in Frascati impinging on a small resonant bar [48]. Our results shows that both EXPLORER and NAUTILUS (the only GW detectors in the world equipped with cosmic rays apparatuses) are indeed properly working and, in particular, the algorithms used for extracting small signals from the noise are also very effective.

In a small percentage of cases the CRS generate signals much larger then those predicted by the thermo-acoustic model. In one case with NAUTILUS a very large signal escaped to an initial analysis because it was too large (!) and the GW detector was saturated. Searching for coincidences in 2001 between the NAUTILUS events and the signals from the CRS detectors installed around NAUTILUS, we found for showers with particle density greater than 300 particles/m^2 a number of 18 coincidences, while expecting 2.1, and for particle density greater than 600 particles/m^2 a number of 12 coincidences was found while expecting 0.8. Among the coincident events the largest one corresponded to an energy release in the bar of 86 TeV. A Monte Carlo simulation, using the known cross-sections, excludes that these large events be due to the electromagnetic and to the hadronic components of the CRS. Work is in progress for studying this new phenomenon related the CRS [49–51].

9.2 Experimental Apparatus

EXPLORER and NAUTILUS each one consists of an aluminium 2,300 kg bar cooled with liquid helium. The mechanical vibrations are converted by means of an electromechanical resonant transducer into an electrical signal which is amplified by a dcSQUID. The bar and the resonant transducer form a coupled oscillator system with two resonant modes. Their main characteristics are given in Table 3.

During the year 1998 the EXPLORER bandwidth was the same as NAUTILUS, while in 2001 the EXPLORER bandwidth increased, because a new transducer [52] was mounted (see Table 3).

The data, sampled at intervals between 3.2 and 12 ms, are filtered with an adaptive filter matched to delta-like signals for the detection of short bursts [31]. This search for bursts is suitable for any transient GW which shows a nearly flat Fourier spectrum at the two resonant frequencies of each detector. The metric perturbation $h(t)$ can either be a millisecond pulse, a signal made by a few millisecond cycles, or a signal sweeping in frequency through the detector resonances. The search is therefore sensitive to different kinds of GW sources, such as a stellar gravitational collapse, the last stable orbits of an inspiraling neutron star or black hole binary, its merging and its final ringdown.

Table 3 Main characteristics of EXPLORER and NAUTILUS. The axes of the two detectors are aligned to within a few degrees of one other, the chances of coincidence detection thus being maximized. The bar temperature of NAUTILUS was 0.14 K in 1998 and 1.5 K later

Detector	Lat.	Long.	Azimuth	Mass kg	Temp K
EXPLORER	46.45 N	6.20 E	39° E	2270	2.6
NAUTILUS	41.82 N	12.67 E	44° E	2270	(a) 0.14
					(b) 1.5

Let $x(t)$ be the filtered output of the detector. This quantity is normalized, using the detector calibration, such that its square gives the energy innovation E of the oscillation for each sample, expressed in kelvin units.

For well behaved noise due only to the thermal motion of the oscillators and to the electronic noise of the amplifier, the distribution of $x(t)$ is normal with zero mean. Its variance (average value of the square of $x(t)$) is called *effective temperature* and is indicated with T_{eff}. The distribution of $x(t)$ is

$$f(x) = \frac{1}{\sqrt{2\pi T_{eff}}} e^{-\frac{x^2}{2T_{eff}}} \tag{75}$$

In order to extract from the filtered data sequence *events* to be analyzed we set a threshold for x^2. The threshold is set at $E_t = 19.5\ T_{eff}$ in order to obtain, in the presence of thermal and electronic noise alone, a reasonable low number of events per day (see Ref. [37]). When x^2 goes above the threshold, its time behaviour is considered until it falls back below the threshold for longer than one second. The maximum amplitude E_s and its occurrence time define the *event*.

Computation of the GW amplitude h from the energy signal E_s requires a model for the signal shape. A conventionally chosen shape is a short pulse lasting a time of τ_g, resulting (for optimal orientation, see later) in the relationship

$$h = \frac{L}{v_s^2} \frac{1}{\tau_g} \sqrt{\frac{kE_s}{M}} \tag{76}$$

where v_s is the sound velocity in the bar, L and M the length and the mass of the bar and τ_g is conventionally assumed equal to 1 ms (for instance, for $E_s = 1\ mK$ we have $h = 2.5\ 10^{-19}$, for both EXPLORER and NAUTILUS).

For the search of coincidences we have determined the accidentals by time shifting one of the two event list with respect to the other one by time steps of 2 seconds from -100 to $+100$ s. The number of coincidences n_c at zero delay is then compared with the average number $\bar{n}$ of the accidentals obtained with the one hundred time-shifts. For a Poissonian distribution we expect:

$$\bar{n}_{th} = \frac{N_{expl} N_{naut} 2w}{totaltime} \tag{77}$$

Before searching for coincidences we must make two important choices: (a) the threshold SNR_t used for the event definition, (b) the coincidence window. We must also consider whether to apply the energy filter that we adopted in [53, 54], which eliminates the coincidences between events whose energies are not compatible, taking into account the uncertainty due to the noise contribution to the measured event energy.

9.3 The Coincidence Window

We could estimate the time window for finding the coincidence, starting from the frequency bandwidth of the apparatus. For example, for EXPLORER 2004, the bandwidth is $\Delta f = 8.7\,\text{Hz}$ and we can make use of the optimal integration time (optimal for minimizing the sum of the thermal and electrical noise)

$$\Delta t_{ott} \sim \frac{1}{\pi \Delta \nu} \sim 30\ \text{ms} \tag{78}$$

But we can do much better by making use of the cosmic ray apparatus. For EXPLORER 2004, putting a threshold of $1{,}000\frac{secondaries}{m^2}$ on the lower array of the cosmic ray apparatus and selecting signals with signal-to-noise ratio $SNR \geq 20$, we obtain in 1 year 40 signals detected by the GW detector in the time window ± 100 ms with three average accidentals. The distribution of the time differences between the cosmic ray apparatus and the GW apparatus is shown in Fig. 2.

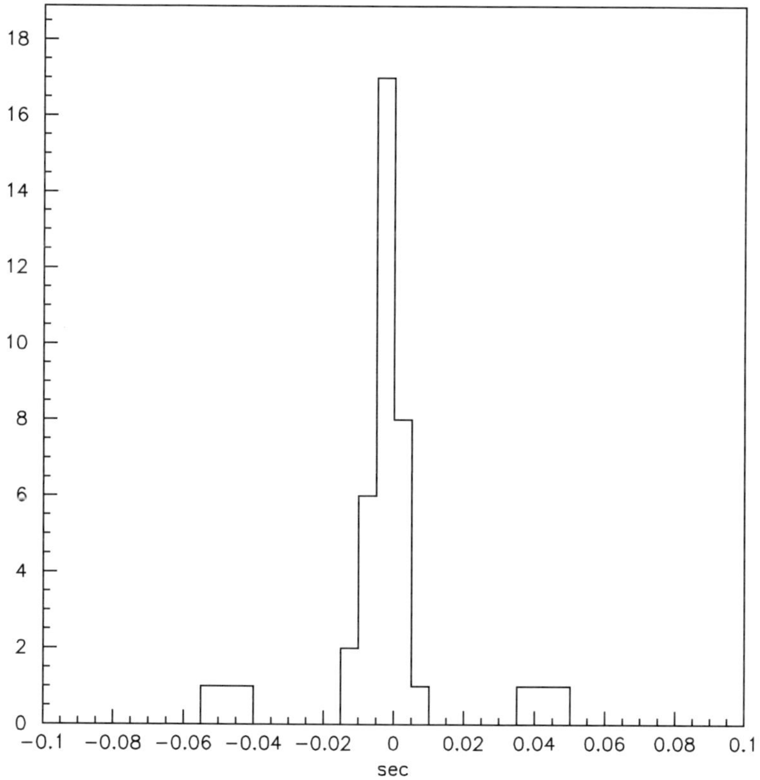

Fig. 2 EXPLORER 2004. The distribution of the time differences between the cosmic ray apparatus and the GW apparatus. Forty coincidences and three average accidentals

From this figure we gather that for one detector a time window of the order of ± 20 ms would be reasonable. Thus for the coincidences between the two detectors EXPLORER and NAUTILUS we have decided to use, from 2004 on, the coincidence window $w = \pm 30$ ms.

9.4 Experimental Results

In 2001 and 2002, the ROG collaboration presented the results of searches for GW bursts with the EXPLORER and NAUTILUS cryogenic bar detectors operating for six months in 1998 [53] and nine months in the year 2001 [54]. In those papers a sidereal time analysis was performed in order to look for specific galactic signatures. A small excess of events with respect to the expected background was found,[4] concentrated around sidereal hour four. At this sidereal hour the two bars, which are oriented parallel to each other, are perpendicular to the galactic plane, and therefore their sensitivity for galactic sources of GW is maximal.

This observation has been debated in the literature and international workshops [55,56], and in spite of different evaluations of its statistical significance, the general opinion was that further work was necessary to put the reported indication on a firmer ground. After an upgrade of the detectors, other data of EXPLORER and NAUTILUS from the 2003 and 2004 runs were analyzed.

We report here these new analyses and compare the new results with the 1998 and 2001 results, being aware that, in the process of combining experimental data obtained in different situations, as in our case because of the continuos upgrades of the apparatuses, we are faced with the danger to make, perhaps unwilling, choices which would affect the final statistical significance. Thus, we have been careful to apply to the coincidence searches the same procedure whenever possible, in order to verify the initial result obtained in 1998 (see footnote).

We recall that during 2001 we first [54] used a variable coincidence window as suggested by simulations [32], later the use of the cosmic ray apparatuses has shown that a fixed coincidence window was to be preferred and we used $w = \pm 0.5$ s (see Ref. [57]).

For the 2003 data we found [45] that, by using the cosmic rays, the EXPLORER event energy needs to be multiplied by the factor 3.3 (because of a mis-calibration of the SQUID apparatus during 2003) and so we apply this factor in the present paper, using a coincidence window $w = \pm 30$ ms.

For the 2004 data we apply the energy filter without any correction, see Fig. 1, and we use the coincidence window $w = \pm 30$ ms.

We show, in Table 4, the characteristics of the detectors during the years 1998, 2001, 2003 and 2004 and the results of the coincidence search in Fig. 3. The only

[4] In the conclusions of the 1998 paper that was presented to the IGEC collaboration in 1999 and 2000:. . . *we find an excess of coincidences at zero time delay in the direction of the galactic centre.*

Table 4 Main characteristics of the detectors for the coincidence search in the 4 years. The hours refer to the common time of operation. The coincidence window has been determined with the cosmic ray apparatuses when available

Year	Detector	Hours	Frequencies	Bandwidth	Window
1998	EXPLORER	94.5 days	904.7, 921.3 Hz	~0.4 Hz	±1 s
	NAUTILUS		907.0, 922.5 Hz	~0.4 Hz	
2001	EXPLORER	90 days	904.7, 921.3 Hz	~9 Hz	$3\sigma \sim 0.5$ s
	NAUTILUS		907.0, 922.5 Hz	~0.4 Hz	
2003	EXPLORER	148.7 days	904.7, 921.3 Hz	8.7 Hz	±30 ms
	NAUTILUS		926.3, 941.5 Hz	9.6 Hz	
2004	EXPLORER	216.5 days	904.7, 921.3 Hz	8.7 Hz	±30 ms
	NAUTILUS		926.3, 941.5 Hz	9.6 Hz	

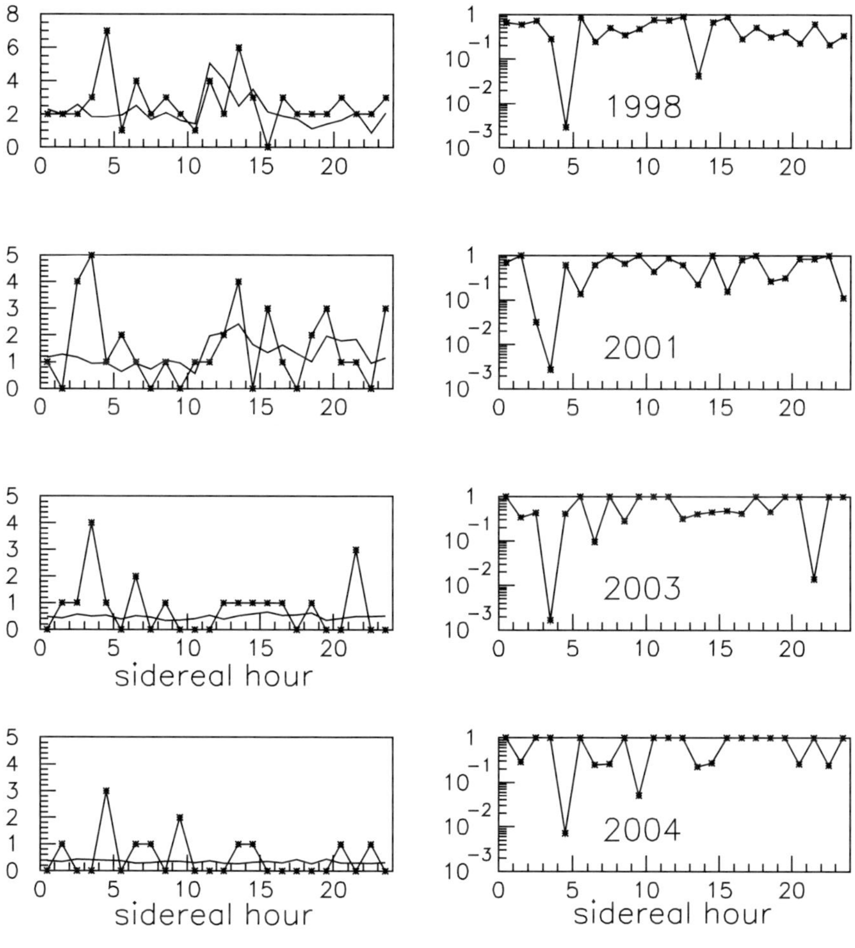

Fig. 3 Coincidences and Poisson probabilities for 1998, 2001, 2003 and 2004. The *left graphs* show the hourly number n_c of coincidences (*continuos line*) and the average number $\bar{n}$ of accidentals (*dashed line*) versus the sidereal and solar hour. The *right graphs* show the corresponding Poisson probability to obtain a number of coincidences n_c greater than or equal to n_c

Table 5 In the second column we give the threshold corresponding to $SNR_t = 19.5$. In the third, fourth and fifth columns the total number n_c of coincidences, average accidentals $\bar{n}$ and the Poisson probability. In the remaining columns the number of coincidences in the sidereal hour range 2–5 (see Ref. [58]) and relative Poisson probability. In total we have 132 coincidences and 103.7 average accidental for a Poisson probability of $1.2\ 10^{-2}$. In the 2–5 sidereal hour range we have 29 coincidences and 12.4 average accidental for a Poisson probability of $1.9\ 10^{-4}$

Year	Threshold	n_c	$\bar{n}$	Poisson	n_c	$\bar{n}$	Poisson
1998	$4.3\ 10^{-18}$	64	52.1	$6.1\ 10^{-2}$	12	6.26	$2.7\ 10^{-2}$
2001	$1.6\ 10^{-18}$	37	31.4	$18\ 10^{-2}$	8	3.06	$1.3\ 10^{-2}$
2003	$1.9\ 10^{-18}$	19	12.1	$4.1\ 10^{-2}$	6	1.60	$6.0\ 10^{-3}$
2004	$1.2\ 10^{-18}$	12	8.1	$12\ 10^{-2}$	3	1.23	$13\ 10^{-2}$

change applied to the 1998 and 2001 data analysis has been to present the results in terms of the sidereal time at the Greenwich longitude.

In Table 5 we give the number of coincidences and average number of accidentals for the 4 years. We also give the same information for the sidereal hour range when we expect signals due to sources in the galactic disk. This range, for a two-detector coincidence search has been calculated in Ref. [58] and is $\sim 3.5 \pm 1.5$ *sidereal hours.*

We now must attempt to combine all data in a single result. We do this by applying the following formula [59]

$$P = p_1 p_2 p_3 p_4 \sum_{j=0}^{3} \frac{1}{j!} |log(p_1 p_2 p_3 p_4)|^j \tag{79}$$

where p_i, $i = 1, 2, 3, 4$ are the Poisson probabilities obtained for the 4 years 1998, 2001, 2003 and 2004.

We get the result shown in Fig. 4.

10 Conclusion

We must conclude that in each year a small coincidence excess, a small excess during each year, is present at sidereal hours compatible with gravitational wave sources in the galactic disk. All together they give a reasonable evidence of signals arriving from the Galaxy.

Furthermore we have evidence that the coincidence excesses occur in limited periods of time, as clearly seen, in particular, with the 1998 data [60].

The physical interpretation appears difficult with our present knowledge, also in consideration of the fact that the sensitivity of our apparatuses has changed during the years. Gravitational waves would require a cross-section larger by at least two orders of magnitude for producing the signals. But, one should not rule out, also, the possibility that *dark matter* be the cause of the observed coincidence excess.

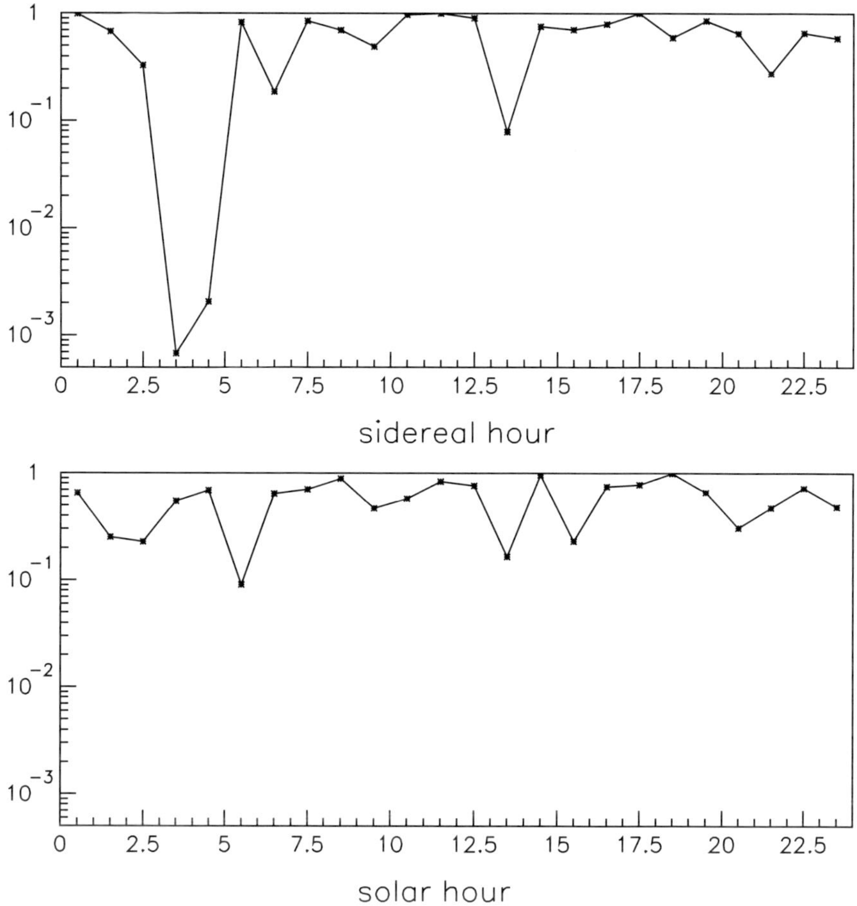

Fig. 4 Combining the probabilities for the 4 years 1998, 2001, 2003 and 2004, according to Eq. (79)

Acknowledgment I am indebted with the collaborators of the ROG group, who, in many years of hard work, have brought a significant contribution to the production of the experimental data and to their interpretation. I am also indebted with William Hamilton and David Blair for useful discussions.

References

1. Will C M 1981 *Theory and Experiment in Gravitational Physics* (Cambridge University Press, Cambridge, United Kingdom)
2. Thorne K S 1987 "Gravitational Radiation", in Hawking, S W, and Israel, W., eds., *300 Years of Gravitation*, 330–458, (Cambridge University Press, Cambridge, United Kingdom)
3. Misner C W, Thorne K S Wheeler J A, *Gravitation* W.H. Freemn and Company (1970)

4. Ciufolini I. and Wheeler J.A. *Gravitation and Inertia*, Princeton Series in Physics (1995)
5. Amaldi E, Pizzella G in De Finis, F., ed. 1979 *Relativity, Quanta, and Cosmology in the development of the scientific thought of Albert Einstein*, 1–230, (Johnson Reprint Corp., Academic Press)
6. Kennefick D., *Einstein versus Physical Review*, Physics Today September 2005.
7. Schutz B F 1995 *Sources of Gravitational Radiation for Detectors of the 21st Century*, Coccia, E., Pizzella, G., Ronga, F., eds., *Proceedings of the First Edoardo Amaldi Conference on Gravitational Wave Experiments*, 3–17, (World Scientific, Singapore)
8. Blair D G ed. 1991 *The Detection of Gravitational Waves*, (Cambridge University Press, Cambridge, United Kingdom)
9. Ferrari V, Palomba C 1998 Int. J. Mod. Phys. **D7** 825–848
   Ferrari V *et al* 1999 Mon. Not. Roy. Astron. Soc. 303:247
10. Brustein R *et al* 1995 *Phys. Lett. B*, **361**, 45–51
11. Weber J 1961 *"General Relativity and Gravitational Waves"*, (Interscience, New York)
12. Weber J 1960 *Phys. Rev.*, **117** (1), 306–313
13. Weber J 1968 *Phys. Rev. Lett.*, **20** (23), 1307–1308
14. Weber J 1969 *Phys. Rev. Lett.*, **22** (24), 1320–1324
15. Hulse R A, Taylor J H 1975 *Astrophys. J.*, **195**, L51–L53
    Taylor J H, Weisberg J M 1982 *Astrophys. J.*, **253**, 908–920
16. Boughn S P *et al* 1982 *Astrophys. J.*, **261**, L19–L22
17. Astone P *et al* 1993 *Phys. Rev. D*, **47** (2), 362–375
18. Ruffini R, Wheeler J A 1969 "Relativistic Cosmology and Space Plattforms", Moore, A.F., Hardy, V., eds., *Proceedings of an ESRO Colloquium*, 45–174, (European Space Research Organisation)
19. Weinberg S 1972 *Gravitation and Cosmology*, (John Wiley & Sons)
20. Pizzella G 1975 *Rivista del Nuovo Cimento*, **5** (3), 369–397
21. Pallottino G V, Pizzella G Nuovo Cim. C4:237–283, 1981
22. Gibbons G W, Hawking S W Phys. Rev. D4, 2191 (1971)
23. Papoulis A "Probability, Random variables and stochastic process" McGrow-Hill Book Co., New York (1965)
24. Bonifazi P el al., Il Nuovo Cimento 1c, 465 (1978)
25. Pizzella G Il Nuovo Cimento 2C, 209 (1979).
26. Amaldi E et al. Nuovo Cim. C1:341–359,1978
27. Amaldi et al., Nuovo Cim. C1:497–509,1978
28. Astone P Bonifazi P Pallottino G V Pizzella G Il Nuovo Cimento 17, 713 (1994)
29. Pallottino G V, Pizzella G in "Data Analysis in Astronomy III" pag. 361, Ed. Di Gesu V, Scarsi L, Crane P, Friedman J H, Levialdi S, Maccarone M C Plenum Press 1998.
30. Pizzella G in "Gravitational Waves" Ed. Ciufolini, Gorini, Moschella, Fre' Institute of Physics, pag. 91 (2001)
31. Astone P *et al, Il Nuovo Cimento* **20,** 9 (1997)
32. Astone P, D'Antonio S Pizzella G 2000 *Phys. Rev. D* **62** 042001
33. Astone P, Pallottino G V, Pizzella G 1997 *Class. Quantum Grav.*, **14**, 2019–2030
34. Ferrari V, Weber J *et al* 1982 *Phys. Rev. D*, **25**, 2471–2486
35. Amaldi E *et al* 1989 *Astron. Astrophys.*, **216**, 325–332
36. Astone P *et al* 1999 *Astroparticle Physics* **10** 83–92
37. Astone P *et al* 1999 *Phys. Rev. D*, **59**, 122001
38. Mauceli E *et al* 1996 *Phys. Rev. D* **54** 1264
39. Blair D G *et al* 1995 *Phys. Rev. Lett.* **74** 1908
40. Astone P *et al* 1997 *Astroparticle Physics* **7** 231–243
41. Cerdonio M *et al* 1997 *Class. Quant. Grav.* **14** 1491–1494
42. Allen Z A *et al* 2000 *Phys. Rev. Lett.* **85** 5046
43. Astone P *et al* 1991 *Z. Phys. C*, **50**, 21–29
44. Astone P et al., Eur. Phys. J. C5:651–664,1998
45. Modestino G, Pizzella G and Ronga F LNF - 05 / 27(IR)
46. Allega A M and Cabibbo N Lett. Nuovo Cimento **38**, 263 (1983)

47. Bernard V, De Rujula A and Lautrup B Nucl. Phys. B **242**, 93(1984)
48. Buonomo B *et al.*, Astropart. Phys. **24** 65–74 (2005)
49. Astone P *et al.*, Phys. Rev. Lett. **84** 14–17 (2000)
50. Astone P *et al.* , Phys. Lett. **B499** 16–22 (2001)
51. Astone P *et al.*, Phys. Lett. **B540** 179–184 (2002)
52. Astone P et al., Phys. Rev. Lett. 91:111101, 2003.
53. Astone P *et al, Class. Quant. Grav.* **18,** 243 (2001)
54. Astone P *et al, Class. Quant. Grav.* **19,** 5449 (2002)
55. Finn, L S Class. Quant. Grav. **20**, L37 (2003)
56. Astone P *et al.* Class. Quant. Grav **20**, S785 (2003)
57. Pizzella, G. : 2003 *Tenth Marcel Grossmann Meeting on General Relativity*, (M. Novello, S. Perez-Bergliaffa, R. Ruffini, Eds.)
58. Babusci D et al., Astron. Astrophys. 421:811–813, 2004.
59. Roe B P 2001 Springer, *Probability and Statistics in Experimental Physics*, pag. 164
60. Modena I and Pizzella G Int. J. Mod. Phys. D15:485–491, 2006.

Gravitational Fields with 2-Dimensional Killing Leaves and the Gravitational Interaction of Light

Gaetano Vilasi

Abstract Gravitational fields invariant for a non Abelian Lie algebra generating a 2-dimensional distribution, are explicitly described. When the orthogonal distribution is integrable and the metric is not degenerate along the orbits, these solutions are parameterized either by solutions of a transcendental equation (the tortoise equation), or by solutions of Darboux equation. Metrics, corresponding to solutions of the tortoise equation, are characterized as those that admit a 3-dimensional Lie algebra of Killing fields with 2-dimensional leaves. It is shown that the remaining metrics represent nonlinear gravitational waves obeying to two nonlinearsuperposition laws. The energy and the polarization of this family of waves are explicitly evaluated; it is shown that they have spin-1 and their possible sources are also described. Old results by Tolman, Ehrenfest, Podolsky and Wheeler on the gravitational interaction of photons are naturally reinterpreted.

1 Introduction

Some decades ago, Professor J.A. Wheeler [53] nicely suggested me to apply our extension of Liouville theorem on complete integrability for searching new solutions of Einstein equations. Thus, for many years it has been a Kantian imperative for me to try to find at least a new solution.

In the some period, by using a suitable generalization of the *Inverse Scattering Transform*, solutions of Einstein equations were found [8] which are invariant for an Abelian 2-dimensional Lie algebra of Killing vector fields such that the distribution $\mathcal{D}^{\perp}$ orthogonal to the one, say $\mathcal{D}$, generated by the Killing fields is transversal to $\mathcal{D}$ and Frobenius-integrable.

Integrable systems, beyond a *Lax pair*, generally exhibit a *recursion operator* which is responsible for the construction of a sequence of conserved functionals and which can be naturally interpreted as a mixed tensor field on the carrier phase space.

G. Vilasi (✉)
Dipartimento di Fisica, Università di Salerno, Istituto Nazionale di Fisica Nucleare, Sezione di Napoli, Via S. Allende, I-84081 Baronissi (Salerno), Italy
e-mail: vilasi@sa.infn.it

I. Ciufolini and R.A. Matzner (eds.), *General Relativity and John Archibald Wheeler*, Astrophysics and Space Science Library 367, DOI 10.1007/978-90-481-3735-0_13,

Such a tensor field has a vanishing *Nijenhuis torsion* and special spectral properties which allow to generalize [18–20, 29], to infinite dimensional manifolds, the classical Liouville theorem on complete integrability. How many conserved functionals do exist in this integrable model and what is their geometric significance?

Thus, as a first step, it has been natural to consider [44] the problem of characterizing all gravitational fields g admitting a Lie algebra $\mathcal{G}$ of Killing fields such that:

I The distribution $\mathcal{D}$, generated by vector fields of $\mathcal{G}$, is 2-dimensional.
II The distribution $\mathcal{D}^{\perp}$, orthogonal to $\mathcal{D}$ is integrable and transversal to $\mathcal{D}$.

The condition of transversality can be relaxed [13, 14]. This case, when the metric g restricted to any integral (2-dimensional) submanifold (Killing leaf) of the distribution $\mathcal{D}$ is degenerate, splits naturally into two sub-cases according to whether the rank of g restricted to Killing leaves is 1 or 0. Sometimes, in order to distinguish various cases occurring in the sequel, the notation $(\mathcal{G}, r)$ will be used: metrics satisfying the conditions I and II will be called *of* $(\mathcal{G}, 2)$ – *type*; metrics satisfying conditions I and II, except the transversality condition, will be called *of* $(\mathcal{G}, 0)$ – *type* or *of* $(\mathcal{G}, 1)$ – *type* according to the rank of their restriction to Killing leaves.

According to whether the dimension of $\mathcal{G}$ is 3 or 2, two qualitatively different cases can occur. Both of them, however, have in common the important feature that all manifolds satisfying the assumptions I and II are in a sense fibered over ζ-complex curves [46].

When dim $\mathcal{G} = 3$, assumption II follows from I and the local structure of this class of Einstein metrics can be explicitly described. Some well known exact solutions [41, 47, 48], e.g. Schwarzschild, belong to this class.

A 2-dimensional $\mathcal{G}$, is either Abelian ($\mathcal{A}_2$) or non-Abelian ($\mathcal{G}_2$) and a metric g satisfying I and II, with $\mathcal{G} = \mathcal{A}_2$ or $\mathcal{G}_2$, will be called $\mathcal{G}$-*integrable*.

The study of $\mathcal{A}_2$-integrable Einstein metrics goes back to Lewis [31], Einstein and Rosen [21], Kompaneyets [28], Ernst [22], Geroch [24], Belinsky, Khalatnikov, Zakharov [6, 8], Verdaguer [50]. Recent results can be found in [14].

The greater rigidity of $\mathcal{G}_2$-integrable metrics, for which some partial results can be found in [2, 16, 26], allows an exhaustive analysis. It will be shown that the ones of $(\mathcal{G}, 2)$-type are parameterized by solutions of a linear second order differential equation on the plane which, in its turn, depends linearly on the choice of a ζ-*harmonic* function (see later). Thus, this class of solutions has a *bilinear structure* and, as such, admits two *superposition laws.*

All the possible situations, corresponding to a 2-dimensional Lie algebra of isometries, are described by the following table

	$\mathcal{D}^{\perp}$, $r = 0$	$\mathcal{D}^{\perp}$, $r = 1$	$\mathcal{D}^{\perp}$, $r = 2$
$\mathcal{G}_2$	***integrable***	***integrable***	***integrable***
$\mathcal{G}_2$	***semi – integrable***	***semi – integrable***	*semi – integrable*
$\mathcal{G}_2$	***non – integrable***	***non – integrable***	*non – integrable*
$\mathcal{A}_2$	***integrable***	***integrable***	*integrable*
$\mathcal{A}_2$	*semi – integrable*	*semi – integrable*	***semi – integrable***
$\mathcal{A}_2$	*non – integrable*	*non – integrable*	*non – integrable*

,

where a non integrable 2-dimensional distribution which is part of a 3-dimensional integrable distribution has been called *semi-integrable* and in which the cases indicated with bold letters have been essentially solved [4, 13, 14, 44–46].

In Section 1, 4-dimensional metrics of $(\mathcal{G}_2, 2)$-*type* invariant for a non Abelian 2-dimensional Lie algebra are characterized from a geometric point of view. The solutions of corresponding Einstein field equations are explicitly written. The construction of global solutions is described in Section 2 and some examples are given in Section 3. Section 4 is devoted to some examples. In Section 5 the case in which the commutator of generators of the Lie algebra is of *light-type* is analyzed from a physical point of view. Harmonic coordinates are also introduced. Moreover, the wave-like character of the solutions is checked through the Zel'manov and the Pirani criterion. The canonical, the Landau-Lifchitz and the Bel energy-momentum pseudo-tensors are introduced and a comparison with the linearised theory is performed. Realistic sources for such gravitational waves are also described. Eventually, the analysis of the polarization leads to the conclusion that these fields are spin-1 gravitational waves.

2 Geometric Aspects

In the following, we will consider 4-dimensional manifolds and Greek letters take values from 1 to 4; the first Latin letters take values from 3 to 4, while i, j from 1 to 2. Moreover, $\mathcal{K}il\,(g)$ will denote the Lie algebra of all Killing fields of a metric g while *Killing algebra* will denote a sub-algebra of $\mathcal{K}il\,(g)$. Moreover, an integral (2-dimensional) submanifold of $\mathcal{D}$ will be called a *Killing leaf*, and an integral (2-dimensional) submanifold of $\mathcal{D}^{\perp}$ *orthogonal leaf*.

- *Semiadapted coordinates*
  Let g be a metric on the space–time $\mathcal{M}$ (a connected smooth manifold) and $\mathcal{G}_2$ one of its Killing algebras whose generators X, Y satisfy $[X, Y] = sY, \quad s = 0, 1$ The Frobenius distribution $\mathcal{D}$ generated by $\mathcal{G}_2$ is 2-dimensional and in the neighborhood of a non singular point a chart $(x^1,\ x^2,\ x^3,\ x^4)$ exists such that

$$X = \frac{\partial}{\partial x^3}, \quad Y = \exp\left(s x^3\right) \frac{\partial}{\partial x^4}$$

  From now on such a chart will be called *semiadapted* (to the Killing fields).
- *Invariant metrics*
  It can be easily verified [44, 45] that in a semiadapted chart g has the form

$$\begin{aligned} g = {} & g_{ij} dx^i dx^j + 2\left(l_i + s m_i x^4\right) dx^i dx^3 - 2 m_i dx^i dx^4 \\ & + \left(s^2 \lambda \left(x^4\right)^2 - 2 s \mu x^4 + \nu\right) dx^3 dx^3 \\ & + 2\left(\mu - s\lambda x^4\right) dx^3 dx^4 + \lambda dx^4 dx^4, \quad i = 1, 2; j = 1, 2 \end{aligned}$$

  with g_{ij}, m_i, l_i, λ, μ, ν arbitrary functions of $\left(x^1, x^2\right)$.

- *Killing leaves*
  Condition II allows to construct semi-adapted charts, with new coordinates (x, y, x^3, x^4), such that the fields $e_1 = \partial/\partial x$, $e_2 = \partial/\partial y$, belong to $\mathcal{D}^{\perp}$. In such a chart, called from now on *adapted*, the components l_i's and m_i's vanish. As it has already said, we will call *Killing leaf* an integral (2-dimensional) submanifold of $\mathcal{D}$ and *orthogonal leaf* an integral (2-dimensional) submanifold of $\mathcal{D}^{\perp}$. Since $\mathcal{D}^{\perp}$ is transversal to $\mathcal{D}$, the restriction of g to any Killing leaf, S, is non-degenerate. Thus, $(S, g|_S)$ is a homogeneous 2-dimensional Riemannian manifold. Then, the Gauss curvature $K(S)$ of the Killing leaves is constant (depending on the leave). In the chart $(p = x^3|_S, q = x^4|_S)$ one has

$$g|_S = \left(s^2\widetilde{\lambda}q^2 - 2s\widetilde{\mu}q + \widetilde{\nu}\right) dp^2 + 2\left(\widetilde{\mu} - s\widetilde{\lambda}q\right) dpdq + \widetilde{\lambda}dq^2,$$

where $\widetilde{\lambda}, \widetilde{\mu}, \widetilde{\nu}$, being the restrictions to S of λ, μ, ν, are constants, and

$$K(S) = \widetilde{\lambda}s^2\left(\widetilde{\mu}^2 - \widetilde{\lambda}\widetilde{\nu}\right)^{-1}.$$

2.1 Einstein Metrics When $g(Y, Y) \neq 0$

In the considered class of metrics, vacuum Einstein equations, $R_{\mu\nu} = 0$, can be completely solved [44]. If the Killing field Y is not of *light type*, i.e. $g(Y, Y) \neq 0$, then in the adapted coordinates (x, y, p, q) the general solution is

$$g = f(dx^2 \pm dy^2) + \beta^2[(s^2k^2q^2 - 2slq + m)dp^2 + 2(l - skq)dpdq + kdq^2] \quad (1)$$

where $f = -\Delta_{\pm}\beta^2/2s^2k$, and $\beta(x, y)$ is a solution of the *tortoise equation*

$$\beta + A\ln|\beta - A| = u(x, y),$$

where A is a constant and the function u is a solution either of Laplace or d'Alembert equation, $\Delta_{\pm}u = 0$, $\Delta_{\pm} = \partial^2_{xx} \pm \partial^2_{yy}$, such that $(\partial_x u)^2 \pm (\partial_y u)^2 \neq 0$. The constants k, l, m are constrained by $km - l^2 = \mp 1, k \neq 0$ for Lorentzian metrics or by $km - l^2 = \pm 1, k \neq 0$ for Kleinian metrics.

Ricci flat manifolds of Kleinian signature possess a number of interesting geometrical properties and undoubtedly deserve attention in their own right. Some topological aspects of these manifolds were studied for the first time in [32, 33] and then in [30]. In recent years the geometry of these manifolds has seen a revival of interest. In part, this is due to the emergence of some new applications in physics.

2.1.1 Canonical Form of Metrics When $g(Y, Y) \neq 0$

The gauge freedom of the above solution, allowed by the function u, can be locally eliminated by introducing the coordinates (u, v, p, q), the function $v(x, y)$ being conjugate to $u(x, y)$, i.e. $\Delta_{\pm} v = 0$ and $u_x = v_y, u_y = \mp v_x$. In these coordinates the metric g takes the form

$$g = \frac{\exp\frac{u-\beta}{A}}{2s^2 k\beta}\left(du^2 \pm dv^2\right) + \beta^2\left[\left(s^2k^2q^2 - 2slq + m\right)dp^2 \right. \\ \left. +2(l - skq)dpdq + kdq^2\right]$$

with $\beta\,(u)$ a solution of $\beta + A\ln|\beta - A| = u$.

2.1.2 Normal Form of Metrics When $g(Y, Y) \neq 0$

In *geographic coordinates* (ϑ, φ) along Killing leaves one has

$$g|_S = \beta^2\left[d\vartheta^2 + \mathcal{F}(\vartheta)\,d\varphi^2\right],$$

where $\mathcal{F}(\vartheta)$ is equal either to $\sinh^2\vartheta$ or $-\cosh^2\vartheta$, depending on the signature of the metric. Thus, in the *normal coordinates*, $\left(r = 2s^2k\beta, \tau = v, \vartheta, \varphi\right)$, the metric takes the form (local "Birkhoff's theorem")

$$g = \varepsilon_1\left(\left[1 - \frac{A}{r}\right]d\tau^2 \pm \left[1 - \frac{A}{r}\right]^{-1} dr^2\right) + \varepsilon_2 r^2\left[d\vartheta^2 + \mathcal{F}(\vartheta)\,d\varphi^2\right] \quad (2)$$

where $\varepsilon_1 = \pm 1, \varepsilon_2 = \pm 1$.

The geometric reason for this form is that, when $g(Y, Y) \neq 0$, a third Killing field exists which together with X and Y constitute a basis of $so(2, 1)$. The larger symmetry implies that the geodesic equations describe a *non-commutatively integrable system* [43], and the corresponding geodesic flow projects on the geodesic flow of the metric restricted to the Killing leaves.

The above local form does not allow, however, to treat properly the singularities appearing inevitably in global solutions. The metrics (1), although they all are locally diffeomorphic to (2), play a relevant role in the construction of new global solutions as described in [45,46].

2.2 *Einstein Metrics When* $g(Y, Y) = 0$

If the Killing field Y is of *light type*, then the general solution of vacuum Einstein equations, in the adapted coordinates (x, y, p, q), is given by

$$g = 2f(dx^2 \pm dy^2) + \mu\left[(w\,(x, y) - 2sq)dp^2 + 2dpdq\right], \quad (3)$$

where $\mu = A\Phi + B$ with $A, B \in \mathbb{R}$, Φ is a non-constant harmonic function of x and y, $f = (\nabla\Phi)^2\sqrt{|\mu|}/\mu$, and $w\,(x, y)$ is solution of the μ-*deformed Laplace equation*:, also known as *Darboux equation*:

$$\Delta_{\pm} w + (\partial_x \ln |\mu|)\, \partial_x w \pm \left(\partial_y \ln |\mu|\right) \partial_y w = 0,$$

where Δ_+ (Δ_-) is the Laplace (d'Alembert) operator in the (x, y) −plane. Metrics (3) are Lorentzian if the orthogonal leaves are conformally Euclidean, i.e. the positive sign is chosen, and Kleinian if not. Only the Lorentzian case will be analyzed and these metrics will be called of $(\mathcal{G}_2, 2)$ −*isotropic type*.

In the particular case $s = 1$, $f = 1/2$ and $\mu = 1$, the above (Lorentzian) metrics are locally diffeomorphic to a subclass of the vacuum Peres solutions [39], that for later purpose we rewrite in the form

$$g = dx^2 \pm dy^2 + 2dudv + 2(\varphi_{,x} dx + \varphi_{,y} dy)du. \tag{4}$$

The correspondence between (3) and (4) depends on the special choice of the function $\varphi(x, y, u)$ (which, in general, is harmonic in x and y arbitrarily dependent on u); in our case

$$x \to x, \quad y \to y, \quad u \to u, \quad v \to v + \varphi\,(x, y, u)$$

with $h = \varphi_{,u}$.

In the case $\mu = const$, the μ-deformed Laplace equation reduces to the Laplace equation; for $\mu = 1$, in the harmonic coordinates system (x, y, z, t) defined [11], for $|z - t| \neq 0$, by

$$\begin{cases} x = x \\ y = y \\ z = \dfrac{1}{2}\left[(2q - w\,(x, y)) \exp\,(-p) + \exp\,(p)\right] \\ t = \dfrac{1}{2}\left[(2q - w\,(x, y)) \exp\,(-p) - \exp\,(p)\right]., \end{cases}$$

the Einstein metrics (3) take the particularly simple form

$$g = 2f(dx^2 \pm dy^2) + dz^2 - dt^2 + d\,(w)\, d\,(\ln |z - t|)\,. \tag{5}$$

This shows that, when w is constant, the Einstein metrics given by Eq. (5) are static and, under the further assumption $\Phi = x\sqrt{2}$, they reduce to the Minkowski one. Moreover, when w is not constant, gravitational fields (5) look like a *disturbance* propagating at light velocity along the z direction on the Killing leaves (integral two-dimensional submanifolds of $\mathcal{D}$).

3 Global Solutions

Here, we will give a coordinate-free description of previous local Ricci-flat metrics, so that it becomes clear what variety of different geometries, in fact, is obtained. We will see that with any of the found solutions a pair, consisting of a *ζ-complex curve*

$\mathcal{W}$ and a ζ-*harmonic* function u on it, is associated. If two solutions are equivalent, then the corresponding pairs, say $(\mathcal{W}, u)$ and $(\mathcal{W}', u')$, are related by an invertible ζ-holomorphic map $\Phi : (\mathcal{W}, u) \longrightarrow (\mathcal{W}', u')$ such that $\Phi^* (u') = u$. Roughly speaking, the *moduli space* of the obtained geometries is surjectively mapped on the *moduli space* of the pairs $(\mathcal{W}, u)$.

Further parameters, distinguishing the metrics we are analyzing, are given below; before that, however, it is worth to underline the following common peculiarities of these metrics:

- They have, in the adapted coordinates, a block diagonal form whose upper block does not depend on the last two coordinates so that orthogonal leaves are totally geodesic.
- They possess a non trivial Killing field. Geodesic flows, corresponding to metrics, admitting 3-dimensional Killing algebras, are non-commutatively integrable.The existence of a non trivial Killing field is obvious from the description of model solution given in next section. For what concerns geodesic flows, they are integrated explicitly for model solution in next section, and the general result follows from the fact that any solution is a pullback of a model one.

Solutions of the Einstein equations previously described manifest an interesting common feature. Namely, each of them is determined completely by a choice of

1. A solution of the wave, or the Laplace equation
   and either by

2′ A choice of the constant A and one of the branches, for β as function of u, of the tortoise equation

$$\beta + A \ln |\beta - A| = u, \tag{6}$$

if $g(Y, Y) \neq 0$, or by

2″) A choice of a solution of one of the two equations

$$\left[\mu\left(\partial_y^2 - \partial_x^2\right) + \mu_y \partial_y - \mu_x \partial_x\right] w = 0 \qquad \Box\mu = 0, \tag{7}$$

$$\left[\mu\left(\partial_y^2 + \partial_x^2\right) + \mu_y \partial_y + \mu_x \partial_x\right] w = 0 \qquad \Delta\mu = 0, \tag{8}$$

in the case $g(Y, Y) = 0$.

They have a natural *fibered structure* with the Killing leaves as fibers. The wave and Laplace equations, mentioned above in 1, are in fact defined on the 2-dimensional manifold $\mathcal{W}$ which parameterizes the Killing leaves. These leaves themselves are 2-dimensional Riemannian manifolds and, as such, are geodesically complete.

For this reason the problem of the extension of described local solutions, is reduced to that of the extension of the base manifold $\mathcal{W}$. Such an extension should carry a geometrical structure that gives an intrinsic sense to the notion of the wave or the Laplace equation and to Eqs. (7) and (8) on it. A brief description of how this can be done is the following.

3.1 ζ-Complex Structures

It is known there exist three different isomorphism classes of 2-dimensional commutative unitary algebras. They are

$$\mathbb{C} = \mathbb{R}[x] / \left(x^2 + 1\right), \qquad \mathbb{R}_{(2)} = \mathbb{R}[x] / \left(x^2\right), \qquad \mathbb{R} \oplus \mathbb{R} = \mathbb{R}[x] / \left(x^2 - 1\right)$$

Elements of this algebra can be represented in the form $a + \zeta b$, $a, b \in \mathbb{R}$, with $\zeta^2 = -1, 0$, or 1, respectively. For a terminological convenience we will call them *ζ-complex numbers*. Of course, ζ-complex numbers for $\zeta^2 = -1$ are just ordinary complex numbers. Furthermore, we will use the unifying notation $\mathbb{R}^2_\zeta$ for the algebra of ζ-complex numbers. For instance $\mathbb{C} = \mathbb{R}^2_\zeta$ for $\zeta^2 = -1$.

In full parallel with ordinary complex numbers, it is possible to develop a *ζ-complex analysis* by defining *ζ-holomorphic* functions as $\mathbb{R}^2_\zeta$-valued differentiable functions of the variable $z = x + \zeta y$. Just as in the case of ordinary complex numbers, the function $f(z) = u(x, y) + \zeta v(x, y)$ is ζ-holomorphic *iff* the ζ-Cauchy–Riemann conditions hold:

$$u_x = v_y, \qquad u_y = \zeta^2 v_x, \tag{9}$$

The compatibility conditions of the above system requires that both u and v satisfy the *ζ-Laplace equation* , that is

$$-\zeta^2 u_{xx} + u_{yy} = 0, \qquad -\zeta^2 v_{xx} + v_{yy} = 0.$$

Of course, the ζ-Laplace equation reduces for $\zeta^2 = -1$ to the ordinary Laplace equation, while for $\zeta^2 = 1$ to the wave equation. The operator $-\zeta^2 \partial_x^2 + \partial_y^2$ will be called the *ζ-Laplace operator*.

In the following a *ζ-complex structure* on $\mathcal{W}$ will denote an endomorphism $J : D(\mathcal{W}) \to D(\mathcal{W})$ of the $C^\infty(\mathcal{W})$ module $D(\mathcal{W})$ of all vector fields on $\mathcal{W}$, with $J^2 = \zeta^2 I$, $J \neq 0, I$,and vanishing Nijenhuis torsion, i.e., $[J, J]^{FN} = 0$, where $[\cdot, \cdot]^{FN}$ denotes for the Frölicher–Nijenhuis bracket. A 2-dimensional manifold $\mathcal{W}$ supplied with a ζ-complex structure is called a *ζ-complex curve*.

Obviously, for $\zeta^2 = -1$ a ζ-complex curve is just an ordinary *1*–dimensional complex manifold (curve).

By using the endomorphism J the ζ-Laplace equation can be written intrinsically as

$$d\left(J^* du\right) = 0,$$

where $J^* : \Lambda^1(\mathcal{W}) \longrightarrow \Lambda^1(\mathcal{W})$ is the *adjoint to J endomorphism* of the $C^\infty(\mathcal{W})$ module of 1-forms on $\mathcal{W}$.

Given a 2-dimensional smooth manifold $\mathcal{W}$, an atlas $\{(U_i, \Phi_i)\}$ on $\mathcal{W}$ is called *ζ-complex iff*

(i) $\Phi_i : U_i \longrightarrow \mathcal{W}$, $\quad U_i$ is open in $\mathbb{R}^2_\zeta$.
(ii) The transition functions $\Phi_j^{-1} \circ \Phi_i$ are ζ-holomorphic.

Two ζ-complex atlases on $\mathcal{W}$ are said to be *equivalent* if their union is again a ζ-complex atlas.

A class of ζ-complex atlases on $\mathcal{W}$ supplies, obviously, $\mathcal{W}$ with a ζ-complex structure. Conversely, given a ζ-complex structure on $\mathcal{W}$ there exists a ζ-complex atlas on $\mathcal{W}$ inducing this structure. Charts of such an atlas will be called *ζ-complex coordinates* on the corresponding ζ-complex curve. In ζ-complex coordinates the endomorphism J and its adjoint J^* are described by the relations

$$J(\partial_x) = \partial_y, \quad J(\partial_y) = \zeta^2\partial_x$$
$$J^*(dx) = \zeta^2 dy, \quad J^*(dy) = dx.$$

If $\zeta^2 \neq 0$, the functions u and v in the Eq. (9) are said to be *conjugate*.

Alternatively, a ζ-complex curve can be regarded as a 2-dimensional smooth manifold supplied with a specific atlas whose transition functions

$$(x, y) \longmapsto (\xi(x, y), \eta(x, y))$$

satisfy to ζ-Cauchy–Riemann relations (9).

As it is easy to see, the ζ-Cauchy–Riemann relations (9) imply that

$$\partial_\eta^2 - \zeta^2\partial_\xi^2 = \frac{1}{\xi_x^2 - \zeta^2\xi_y^2}\left(\partial_y^2 - \zeta^2\partial_x^2\right),$$

and also

$$\mu\left(\partial_\eta^2 - \zeta^2\partial_\xi^2\right) + \mu_\eta\partial_\eta - \zeta^2\mu_\xi\partial_\xi = \frac{1}{\xi_x^2 - \zeta^2\xi_y^2}\Big[\mu\left(\partial_y^2 - \zeta^2\partial_x^2\right) + \mu_y\partial_y - \zeta^2\mu_x\partial_x\Big].$$

This shows that Eq. (7) (respectively, (8)) is well-defined on a ζ-complex curve with $\zeta^2 = 1$ (respectively, $\zeta^2 = -1$). The manifestly intrinsic expression for these equations is

$$d\left(\mu J^* dw\right) = 0.$$

We will refer to it as the *μ-deformed ζ-Laplace equation.*

A solution of the ζ-Laplace equation on $\mathcal{W}$ will be called *ζ-harmonic*. We can see that in the case $\zeta^2 \neq 0$ the notion of *conjugate ζ-harmonic function* is well defined on a ζ-complex curve. In addition, notice that the metric field $d\xi^2 - \zeta^2 d\eta^2$, η being ζ-conjugate with ξ, is canonically associated with a ζ-harmonic function ξ on $\mathcal{W}$.

A map $\Phi : \mathcal{W}_1 \longrightarrow \mathcal{W}_2$ connecting two ζ-complex curves will be called *ζ-holomorphic* if $\varphi \circ \Phi$ is locally ζ-holomorphic for any local ζ-holomorphic function φ on $\mathcal{W}_2$. Obviously, if Φ is ζ-holomorphic and u is a ζ-harmonic function on $\mathcal{W}_2$, then $\Phi^*(u)$ is ζ-harmonic on $\mathcal{W}_1$.

It is worth noting that the *standard ζ-complex curve* is $\mathbb{R}^2_\zeta = \{(x + \zeta y)\}$, and the *standard ζ-harmonic function* on it is given by x, whose conjugated is y. The pair $\left(\mathbb{R}^2_\zeta, x\right)$ is *universal* in the sense that for a given ζ-harmonic function u on a ζ-complex curve $\mathcal{W}$ there exists a ζ-holomorphic map $\Phi : \mathcal{W} \longrightarrow \mathbb{R}^2_\zeta$ defined uniquely by the relations $\Phi^*(x) = u$ and $\Phi^*(y) = v$, v being conjugated with u.

3.2 Global Properties of Solutions

The above discussion shows that any global solution, that can be obtained by matching together local solutions described in Section 1, is a solution whose base manifold is a ζ-complex curve $\mathcal{W}$ and which corresponds to a ζ-harmonic function u on $\mathcal{W}$.

A solution of Einstein equations corresponding to $\mathcal{W} \subseteq \mathbb{R}^2_\zeta$, $u \equiv x$ will be called a *model*. Notice that there exist various model solutions due to various options in the choice of parameters appearing in 2′) and 2″) at the beginning of this section. An important role played by the model solutions is revealed by the property [46] that

Any solution of the Einstein equation which can be constructed by matching together local solutions described in section 1 is the pullback of a model solution via a ζ-holomorphic map from a ζ-complex curve to $\mathbb{R}_{\zeta^2}$.

We distinguish between the two following qualitatively different cases:

I Metrics admitting a normal *3*-dimensional Killing algebra with 2-dimensional leaves

II Metrics admitting a normal 2-dimensional Killing algebra that does not *extend* to a larger algebra having the same leaves and whose distribution orthogonal to the leaves is integrable

It is worth mentioning that the distribution orthogonal to the Killing leaves is automatically integrable in *Case I* [45]. In *Case II* the 2-dimensionality of the Killing leaves is guaranteed by proposition *2* of [45].

Any Ricci-flat manifold (M, g), we are analyzing, is fibered over a ζ-complex curve $\mathcal{W}$

$$\pi : M \longrightarrow \mathcal{W},$$

whose fibers are the Killing leaves and as such are 2-dimensional Riemann manifolds of constant Gauss curvature.

Below, we shall call π the *Killing fibering* and assume that its fibers are *connected and geodesically complete*. Therefore, maximal (i.e., *non-extendible*) Ricci-flat manifolds, of the class we are analyzing in the paper, are those corresponding to maximal (i.e., *non-extendible*) pairs $(\mathcal{W}, u)$, where $\mathcal{W}$ is a ζ-complex curve and u is ζ-harmonic function on $\mathcal{W}$.

4 Examples

In this section, we illustrate the previous general results with a few examples using the fact that any solution can be constructed as the pullback of a model solution *via* a ζ-holomorphic map Φ of a ζ-complex curve $\mathcal{W}$ to $\mathbb{R}^2_\zeta$. Recall that in the pair $(\mathcal{W}, u)$, describing the so obtained solution, $u = \mathcal{R}e\Phi$.

4.1 Algebraic Solutions

Let $\mathcal{W}$ be an algebraic curve over $\mathbb{C}$, understood as a ζ-complex curve with $\zeta^2 = -1$.

With a given meromorphic function Φ on $\mathcal{W}$ a pair $(\mathcal{W}_\Phi, u)$ is associated, where $\mathcal{W}_\Phi$ is $\mathcal{W}$ deprived of the poles of Φ and u the real part of Φ.

A solution (metric) constructed over such a pair will be called *algebraic*. Algebraic metrics are generally singular. For instance, such a metric is degenerate along the fiber $\pi_1^{-1}(a)$ (see Section 3) if $a \in \mathcal{W}$ is such that $d_a u = 0$.

4.2 Info-Holes

Space–times corresponding to algebraic metrics and, generally, to metrics with signature equal to 2 constructed over complex curves ($\zeta^2 = -1$), exhibit the following interesting property: *for a given observer there exists another observer which can be never contacted.*

By defining an *info-hole* (information-hole) of a given point a to be the set of points of the space–time whose future does not intersect the future of a, the above property can be paraphrased by saying that *the info-hole of a given point of such a space–time is not empty.*

In fact, consider a metric of the form $g = \pi_1^*(g_{\{u\}}) + \pi_1^*\left(\beta^2\right)\pi_2^*(g_\Sigma)$ constructed [16] over a complex curve whose standard fiber Σ is a 2-dimensional manifold supplied with an indefinite metric of constant Gauss curvature equal to 1. For our purpose, it is convenient to take for Σ the hyperboloid $x_1^2+x_2^2-x_3^2 = 1$, supplied with the induced metric $g|_\Sigma = dx_1^2 + dx_2^2 - dx_3^2|_\Sigma$. The *light-cone* of $g|_\Sigma$ at a given point b is formed by the pair of rectilinear generators of Σ passing through b. Since the geodesics of g project *via* π_2 into geodesics of $g|_\Sigma$, it is sufficient to prove the existence of info-holes for the 2-dimensional space–time $(\Sigma, g|_\Sigma)$. To this purpose, consider the standard projection π of $\mathbb{R}^3 = \{(x_1, x_2, x_3)\}$ onto $\mathbb{R}^2 = \{(x_1, x_2)\}$. Then, $\pi|_\Sigma$ projects Σ onto the region $x_1^2 + x_2^2 \geq 1$ in $\mathbb{R}^2$ and the rectilinear generators of Σ are projected onto tangents to the circle $x_1^2 + x_2^2 = 1$. Suppose that the time arrow on Σ is oriented according to increasing value of x_3. Then the future region $F(b) \subset \Sigma$ of the point $b \in \Sigma$, $b\,(1, 1, \beta)$ $\beta > 0$, projects onto the domain defined by $D = \{x \in \mathbb{R}^2 : x_1 > 1, x_2 > 1\}$, and the future region of any point $b' \in \Sigma$, such that $\pi(b') \in D'$ and $x_3(b') > 0$, does not intersect $F(b)$. By obvious symmetry arguments the result is valid for any point $b \in \Sigma$.

4.3 A Star "Outside" the Universe

The Schwarzschild solution shows a "star" generating a space "around" itself. It is an $so(3)$-invariant solution of the vacuum Einstein equations. On the contrary, its $so(2,1)$-analogue shows *a "star" generating the space only on "one side of itself". More exactly, the fact that the space in the Schwarzschild universe is formed by a 1-parametric family of "concentric" spheres allows one to give a sense to the adverb "around". In the* $so(2,1)$-case the space is formed by a 1-parameter family of "concentric" hyperboloids. The adjective "concentric" means that the curves orthogonal to hyperboloids are geodesics and metrically converge to a singular point. This explains in what sense this singular point generates the space only on "one side of itself".

5 Physical Properties

From a physical point of view, only Lorentzian metrics will be analyzed in the following, even if Ricci flat manifolds of Kleinian signature appear in the 'no boundary' proposal of Hartle and Hawking [27] in which the idea is suggested that the signature of the space–time metric may have changed in the early universe. So, assuming that particles are free to move between Lorentzian and Kleinian regions some surprising physical phenomena, like time travelling, would be observable (see [1] and [42]). Some other examples of Kleinian geometry in physics occur in the theory of heterotic $N = 2$ string (see [37] and [5]) for which the target space is 4-dimensional. The analysis will be devoted to metrics of $(\mathcal{G}_2, 2) - type$, when the vector field Y, i.e. the commutator $[X, Y]$, is of light-type: $g(Y, Y) = 0$. The wave character of gravitational fields (3) has been checked by using covariant criteria. In the following we will shortly review the most important properties of these waves which will turn out to have spin-1.

In the first part of the section the standard theory of linearized gravitational waves will be shortly described. In the second part, the theoretical reality of spin-1 gravitational waves will be discussed.

In the particular case $s = 1$, $f = 1/2$ and $\mu = 1$, the above (Lorentzian) metrics are locally diffeomorphic [9] to a subclass of the vacuum Peres solutions [39,40,48], that for later purpose we rewrite in the form

$$g = dx^2 + dy^2 + 2dudv + 2(\varphi_{,x}dx + \varphi_{,y}dy)du. \tag{10}$$

The correspondence between (3) and (4) depends on the special choice of the function $\varphi(x, y, u)$ (which, in general, is harmonic in x and y); in our case

$$x \to x, \quad y \to y, \quad u \to u, \quad v \to v + \varphi(x, y, u), \quad h = \varphi_{,u}.$$

In the case $\mu = const$, the Euler–Darboux equation reduces to the Laplace equation; for $\mu = 1$, in the harmonic coordinates system (x, y, z, t) defined in [11], the above Einstein metrics take the particularly simple form

$$g = 2f(dx^2 + dy^2) + dz^2 - dt^2 + d\,(w)\,d\,(\ln|z - t|]).$$

This shows that, when w is constant, the Einstein metrics given by Eq. (5) are static and, under the further assumption $\Phi = x\sqrt{2}$, they reduce to the Minkowski one. Moreover, when w is not constant, gravitational fields (5) look like a *disturbance* propagating at light velocity along the z direction on the Killing leaves. In the following we will only consider the case $\Phi = x\sqrt{2}$.

5.1 Spin-1 Gravitational Waves

In the following, physical properties of metrics (3) will be analyzed only in the case of Lorentzian signature.[1] The wavelike nature of gravitational fields (3) has been checked [11] by using covariant criteria. We will shortly review the most important properties of these waves which will turn out to have spin-1.

Let us remark that the definition and the meaning of spin or polarization for a theory, such as general relativity, which is non-linear and possesses a much bigger invariance than just the Poincaré one, deserve a careful analysis. It is well known that the concept of particle together with its degrees of freedom like the spin may be only introduced for linear theories (for example for the Yang–Mills theories, which are non linear, it is necessary to perform a perturbative expansion around the linearized theory). In these theories, when Poincaré invariant, the particles are classified in terms of the eigenvalues of two Casimir operators of the Poincaré group, P^2 and W^2 where P_μ are the translation generators and $W_\mu = \frac{1}{2}\epsilon_{\mu\nu\rho\sigma}P^\nu M^{\rho\sigma}$ is the *Pauli-Ljubanski polarization vector* with $M^{\mu\nu}$ Lorentz generators. Then, the total angular momentum $J = L + S$ is defined in terms of the generators $M_{\mu\nu}$ as $J^i = \frac{1}{2}\epsilon^{0ijk}M_{jk}$. The generators P_μ and $M_{\mu\nu}$ span the Poincaré algebra, $\mathcal{ISO}(3,1)$. When $P^2 = 0$, $W^2 = 0$, W and P are linearly dependent of each other

[1] Ricci flat manifolds of Kleinian signature appear in the 'no boundary' proposal of Hartle and Hawking [27] in which the idea is suggested that the signature of the space–time metric may have changed in the early universe. Some other examples of Kleinian geometry in physics occur in the theory of heterotic $N = 2$ string (see [5, 37]) for which the target space is 4-dimensional.

$W_\mu = \lambda P_\mu$; the constant of proportionality λ is given by $\lambda = \overrightarrow{P} \cdot \overrightarrow{J} / P_0$ and defines the *helicity* for massless particles like photons.

Let us turn now to the gravitational fields represented by Eq. (5). As it has been remarked, they represent gravitational waves moving at the velocity of light, that is, in the would be quantised theory, particles with zero rest mass. Thus, if a classification in terms of Poincaré group invariants could be performed, these waves would belong to the class of unitary (infinite-dimensional) representations of the Poincaré group characterized by $P^2 = 0$, $W^2 = 0$. Recall that, in order for such a classification to be meaningful P^2 and W^2 have to be invariants of the theory. This is not the case for general relativity, unless we restrict to a subset of transformations selected for example by some physical criterion or by experimental constraints. For the solutions of the linearized vacuum Einstein equations the choice of the harmonic gauge does the job [51]. There, the residual gauge freedom corresponds to the sole Lorentz transformations.

5.2 *The Standard Linearized Theory*

The standard analysis of linearized theory and the issue of the polarization will be analyzed. In particular, the usual transverse-traceless gauge in the linearized vacuum Einstein equations and the (usually implicit) assumptions needed to reduce to this gauge play an important role: the generality of the usual claim "the graviton has spin-2" (that, of course, is strictly related to the possibility of achieving this special gauge in any "reasonable" physical situation) is strictly related to these assumptions.

The gravitational field is said to be *weak* (in M') if there exists a (harmonic) coordinates system and a region $M' \subset M$ of space–time in which the following conditions hold:

$$g_{\mu\nu} = \eta_{\mu\nu} + h_{\mu\nu}, \quad \left|h_{\mu\nu}\right| \ll 1, \left|h_{\mu\nu,\alpha}\right| \ll 1. \tag{11}$$

As it is known, in the weak field approximations in a harmonic coordinates system the Einstein equations read

$$\eta_{\alpha\beta}\partial^\alpha \partial^\beta h_{\mu\nu} = 0. \tag{12}$$

The choice of the harmonic gauge plays a key role in derive equation (12); no other special assumption either on the form or on the analytic properties of the perturbation h has been done. It is commonly believed that, with a suitable gauge transformation preserving the harmonicity of the coordinate system and the "weak character" of the field, one can always kill the "spin-1" components of the gravitational waves. However, even if not explicitly declared, the standard textbook analysis of the polarization is performed for globally *square integrable* solutions of the wave-equation (12) (that is, solutions which are square integrable everywhere on M) but, as we will see in the following, some very interesting solutions do not belong to this class. To make this point clear, now we will briefly describe the standard analysis to kill the spin-1 components stressing the role played by the global *square integrability* assumption.

Indeed, *globally square integrable* solutions of equations (12) can be always Fourier expanded in terms of plane-wave functions with a *real* light-like vector wave k_μ. The standard plane wave solutions of Eq. (12) are $h_{\mu\nu} = e_{\mu\nu}e^{ir} + e^*_{\mu\nu}e^{-ir}$ with $r = k_\mu x^\mu$, k_μ being the propagation direction vector fulfilling $k_\mu k^\mu = 0$, the harmonicity condition reduces to

$$\frac{1}{2}k_\lambda \eta^{\mu\nu} e_{\mu\nu} = \eta^{\mu\nu} k_\nu e_{\mu\lambda}\ , \tag{13}$$

while the *gauge transformations of the linearized theory* are in this case:

$$e_{\mu\nu} \to e'_{\mu\nu} = e_{\mu\nu} + k_\mu l_\nu + k_\nu l_\mu, \quad k^\mu l_\mu = 0 \tag{14}$$

l_μ being a real vector too. A *real* propagation direction vector k_μ can always be chosen in the following form $k_\mu = (1, 0, 0, -1)$, so that the gravitational perturbation is propagating along the z-axis. Then, it is trivial to show (see, for example, [51]) that the spin-1 components of $e_{\mu\nu}$ are e_{it} (where $i, j = x, y$) while the spin-2 are e_{ij} (that is, the spin-2 components are the ones with both index in the plane orthogonal to the propagation direction). It can be easily shown (using standard linear algebra arguments) that with a suitable transformation (14) the e_{it} components can be killed. Formally, this means that the linear system of equations for the Fourier components of l_μ

$$e_{it} + k_i l_t + k_t l_i = 0$$

has always a solution.

However, for complex k_μ such that $k_\mu k^\mu = 0$ (for example $(1, a, \pm ia, 1)$ with $a \in R$), this is not true anymore, that is, the above system of equations has in general no real solution for l_μ: as it can be easily checked, the determinant of the non homogeneous linear system for l_μ (15) vanishes in this case. It is worth to note that, in order to represent an infinitesimal diffeomorphism, the Fourier components of l_μ have to be real otherwise the transformation generated by l_μ would not be infinitesimal.[2]

What is lacking in this case is, obviously, the global *square integrability* due to the presence of the harmonic function solution of the 2-dimensional Laplace equation.[3] Therefore, non-globally square integrable spin-1 perturbations are not pure gauge because they cannot be killed by infinitesimal diffeomorphisms. Even if

[2] Be square integrable and, consequently, it would be unbounded somewhere. In this case, l_μ could not represent an *infinitesimal diffeomorphism*.

[3] Harmonic functions are not globally square integrable, usually there are two possibilities: either they do not decrease at infinity or they have singularities. In the second case however, they can be square integrable if one excludes the singularities from the spacetime.

global *square integrability* is lacking, there exist solutions of this form that far away the singularities[4] are perfectly well-behaved. In other words, spin-1 perturbations which are square integrable on a submanifold $M' \subset M$ of the whole spacetime can be found: thus, in order to exist, spin-1 perturbations necessarily need some singularities and/or some region with non trivial topology.

A transparent method to determine the spin of a gravitational wave is to look at its physical degrees of freedom, i.e. the components which contribute to the energy [17]. One should use the Landau-Lifshitz (pseudo)-tensor t_ν^μ which, in the asymptotically flat case, agrees with the Bondi flux at infinity [12].

It is worth to remark that the canonical and the Landau Lifchitz energy-momentum pseudo-tensors are tensors for Lorentz transformations. Thus, any Lorentz transformation will preserve the form of these tensor; this allows to perform the analysis according to the Dirac-procedure.[5] A globally square integrable solution $h_{\mu\nu}$ of the wave equation is a function of $r = k_\mu x^\mu$ with $k_\mu k^\mu = 0$. With the choice $k_\mu = (1, 0, 0, -1)$, we get for the energy density t_0^0 and the energy momentum t_0^3 the following result:

$$16\pi t_0^0 = \frac{1}{4}(u_{11} - u_{22})^2 + u_{12}^2 \qquad t_0^0 = t_0^3,$$

where $u_{\mu\nu} \equiv dh_{\mu\nu}/dr$. Thus, the physical components which contribute to the energy density are $h_{11} - h_{22}$ and h_{12}. Following the analysis of [17], we see that they are eigenvectors of the infinitesimal rotation generator $\mathcal{R}$, in the plane $x-y$, belonging to the eigenvalues $\pm 2i$. The components of $h_{\mu\nu}$ which contribute to the energy thus correspond to spin-2.

In the case of the prototype of spin-1 gravitational waves (5), for $f = 1/2$, we have:

$$\tau_0^0 \sim c_1 (h_{0x,x})^2 + c_2 (h_{0y,x})^2, \quad t_0^0 = t_0^3$$

where c_1 e c_2 constants, so that the physical components of the metric are h_{0x} and h_{0y}. Following the previous analysis one can see that these two components are eigenvectors of $i\mathcal{R}$ belonging to the eigenvalues ± 1. In other words, metrics (5), which are not pure gauge since the Riemann tensor is not vanishing, represent spin-1 gravitational waves propagating along the z-axis at light velocity.

This is related to the harmonic function of the transverse coordinates: in order to have an asymptotically flat wave, singularities or some sort of non triviality in the spacetime topology are necessary. The question is, can be these solutions declared unphysical? Can reasonable sources be found to smooth out the singularities? The answer is positive as we will see in more details in the next sections. Now we will show a simple and interesting example of such solutions.

[4] As it will be now shown, the singularities are needed in order to have spatially asymptotically flat solutions.

[5] Of course the usual alternative procedure described in almost all textbooks could be followed.

A simple example – Let us consider perturbations, as in Eq. (5), of the form $h = dw(x, y) \cdot df(z - t)$ which are not globally square integrable. The metric

$$g = \eta + dw(x, y) \cdot df(u), \quad u = z - t, \quad \left(\partial_x^2 + \partial_y^2\right) w = 0 \tag{15}$$

being spatially asymptotically flat for a wide choice of harmonic functions w. Indeed, it represents a physically interesting gravitational field: gravitational waves propagating along the z-axis at light velocity. Besides to be a solution of the linearized Einstein equations on flat background, it is an exact solution of Einstein equations too.

It is trivial to verify that metric (15) is written in harmonic coordinates and has an *off-diagonal* form, that is, the perturbation h has only one index in the plane $x-y$ orthogonal to the propagation direction z: for this reason the above gravitational wave has spin equal to 1 and is not a pure gauge [11]. With a suitable transformation it is possible to bring the above gravitational wave in the standard *transverse-traceless* form, however one can check that the new coordinates are not harmonic anymore.

Summarizing: *globally square integrable spin−1 gravitational waves propagating on a flat background are always pure gauge. Spin−1 gravitational waves which are not globally square integrable are not pure gauge.*

5.3 Asymptotic Flatness and Matter Sources

In the vacuum case, the coordinates (x, y, z, t) of the metrics (15) are harmonic. Being z the propagation direction, the physical effects manifest themselves in the $x - y$ planes orthogonal to the propagation direction. In order these metrics be *asymptotically Minkowski* for $x^2 + y^2 \to \infty$, the function w is required to satisfy the condition

$$\lim_{x^2+y^2\to\infty} (w - c_1 x + c_2 y - c_3) = 0,$$

where c_1, c_2 and c_3 are arbitrary constants and the behaviour of w can be easily recognized by looking at the Riemann tensor of the metrics (15):

$$R_{uiuj} = f_{,u} w_{,ij} \tag{16}$$

which depends on the second derivatives of the 2-dimensional harmonic function w.

Therefore, to have an asymptotically Minkowski metric, the function w must be asymptotically close to a linear functions. But, due to standard results in the theory of linear Partial Differential Equations, this is impossible unless w is a linear function everywhere and this would imply the flatness of the metrics (15). However, if we admit δ-like singularities in the $x - y$ planes, non trivial spatially asymptotically Minkowski vacuum solutions with $w \neq const$ can exist [9]. Of course, it is not necessary to consider δ-like singularities: it is enough to take into account matter sources. For example, in the presence of an electromagnetic wave propagating along the z axis, with energy density equal to ρ which vanishes outside a compact region of the $x-y$ planes, the exact non vacuum Einstein equations for metrics (15) read (see, for example, [9])

$$f_{,u}\left(\partial_x^2+\partial_y^2\right)w=\kappa\rho$$

where κ is the gravitational coupling constant.[6] Thus, one can have non singular spin-1 gravitational waves by considering suitable matter sources which smooth out the singularities;

From the phenomenological point of view, it is worth to note that these kind of wave-like gravitational fields, unlike standard spin-2 gravitational waves which can be singularities free even in the vacuum case, have to be coupled to matter sources in order to represent reasonable gravitational fields. The observational consequence of this fact is that spin-1 gravitational waves are naturally weaker than spin-2 gravitational waves [34]: typically, if the characteristic velocity of the matter source is v, the spin-1 wave is suppressed by factors $(v/c)^n$ with respect to a spin-2 wave. It is worth to note that a gravitational field may also have a repulsive character; for instance, a Kerr black hole is "more repulsive" than a Schwarzschild black hole with the same mass. This is obviously related to the angular momentum. Roughly speaking, this effect may be attributed to the "gravitomagnetic" part of the Kerr metric which, in our terminology, is the "spin-1" part. On the other hand, the Kaluza–Klein mechanism allows to construct in pure 5-dimensional gravity, solutions with spin-1 excitations (which in 4-dimensions may be interpreted as electromagnetic-and therefore *repulsive*-degrees of freedom. Of course, the Kaluza–Klein mechanism also works when reducing from 4- to 3-dimensions. Solutions we are calling *spin*-1 *gravitational waves* when reduced to 3-dimensions (considering as extra dimension the propagation direction of the wave) give rise to purely electromagnetic fields.

5.4 More on the Wave Character of the Field

Up to now, it has been assumed that metrics (15) indeed represent wave-like gravitational fields. Even if from a "linearized" perspective this is obvious, being the above metrics solutions of the exact Einstein equations too, one should try to use covariant criteria in order to establish their wave character. Here gravitational fields (4) will be considered, where:

$$\partial_u\left(\partial_x^2+\partial_y^2\right)\varphi=0,\quad\left(\partial_x^2+\partial_y^2\right)\varphi=0$$

are, respectively, the vacuum Einstein equations and the harmonicity conditions.

The wave character and the polarization of these gravitational fields can be analyzed in many ways. For example, we could use the Zel'manov criterion [55] to show that these are gravitational waves and the Landau–Lifshitz pseudo-tensor to

[6] Aichelburg-Sexl solution [3] belongs to this class of solutions.

find their propagation direction [11, 12]. However, the algebraic Pirani's criterion is easier to handle since it determines the wave character of the solutions and the propagation direction both at once. Moreover, it has been shown that, in the vacuum case, the two methods agree [12]. To use this criterion the Weyl scalars must be evaluated according to the Petrov–Penrose classification [38, 41].

To perform the Petrov–Penrose classification, one has to choose a *tetrad* basis with two real null vector fields and two real spacelike (or two complex null) vector fields. Then, according to the Pirani's criterion, if the metric belongs to type **N** of the Petrov classification, it is a gravitational wave propagating along one of the two real null vector fields. Since ∂_u and ∂_v are null real vector fields and ∂_x and ∂_y are spacelike real vector fields, the above set of coordinates is the right one to apply for the Pirani's criterion.

Since the only nonvanishing components of the Riemann tensor, corresponding to the metric (4), are

$$R_{iuju} = -\partial^2_{ij}\partial_u\varphi, \quad i, j = x, y,$$

this gravitational fields belong to Petrov type **N** [15, 55]. Then, according to the Pirani's criterion, the metric (4) does indeed represent a gravitational wave propagating along the null vector field ∂_u.

6 Final Remarks

It is well known that linearized gravitational waves can be characterized entirely in terms of the linearized and gauge invariant Weyl scalars. The non vanishing Weyl scalar of a typical spin-2 gravitational wave is Ψ_4. Metrics (4) also have as non vanishing Weyl scalar Ψ_4. What truly distinguishes spin-1 from spin-2 gravitational waves is the fact that in the spin-1 case the Weyl scalar has a non trivial dependence on the transverse coordinates (x, y) due to the presence of the harmonic function. This could led to observable effects on length scales larger than the *characteristic length scale* where the harmonic function changes significantly. Indeed, the Weyl scalar enters in the geodesic deviation equation implying a non standard deformation of a ring of test particles breaking the invariance under rotation of π around the propagation direction. Eventually, one can say that there should be distinguishable effects of spin-1 waves on suitably large length scales.

It is also worth to stress that the results of [25] suggest that the sources of asymptotically flat *pp*-waves (which have been interpreted as spin-1 gravitational waves [11, 12]) repel each other. Thus, in a field theoretical perspective, "*pp*-gravitons" must have spin-1 [10].

This is also supported from the following remarks

A long time ago Tolman, Ehrenfest and Podolsky [49]analysed the gravitational field of light beams and the corresponding geodesics in the linear approximation of Einstein equations. They discovered that null rays behave differently according whether they propagate parallely or antiparallely to a steady, long, straight beam of light, but they din't provide a physical explanation of this fact. The result has been explained by Faraoni and Dumse [23] using an approach based on a generalization to null rays of the gravitoelectromagnetic Lorentz force of linearized gravity. They have also extended the analysis to the realm of exact *pp*-wave solutions of the Einstein equations, and gave a physical explanation of the superposition property of parallel beams of light in the strong gravity regime.

Theory [7, 49] and physical observation [54] have shown that photons are attracted by mass by twice the amount expected if they were instead massive particles. Parallel photons interacting gravitationally in antiparallel orientation attract one another by overall factor four [23]. It seems that photons in parallel orientation do not attract [49, 52]. Wheeler, starting the construction of his electromagnetic *geon model*, generalized the Tolman, Ehrenfest and Podolski analysis to the case of two light beams, stating that *"two nearly parallel pencils of light attract gravitationally with twice the strenght one might have thought when their propagation vectors are oppositely directed, and when similarly directed attract not at all"*. Wheeler's proposition is important for the confinement of electromagnetic radiation and was proven by [23] in linearized gravity. However, the lacking of attraction could be ascribed to the linear approximation since, according to our results, photons generate spin-1 gravitational fields and, as a consequence, two photons with same helicity must repel one another.

7 Appendix: The Petrov Classification

In the study of the electrodynamics, the algebraic properties of the Maxwell and Faraday tensors play a key role. In fact, by looking at the eigenvalues and eigenvectors of the Maxwell stress-energy tensor, it is possible to understand if a given electromagnetic field is of wave type, Coulomb type, magnetostatic type, etc. Thus, since many properties of electrodynamics are shared by general relativity, it is a useful job to try to characterize the different algebraic types of the Riemann tensor. In the fifties, Petrov was able to classify the different algebraic types of the Riemann tensor of Lorentzian manifold. This classification is a good tool to understand the physical nature of a given gravitational field. Of course, the situation in general relativity is by far more complicated then in electrodynamics because in general relativity there is not a natural stress–energy tensor for the gravitational field. Nevertheless, the Petrov classification allows us to find reasonable intrinsic definitions of wave-like gravitational field that do not refer to the linearized theory. Moreover, the blackhole solutions also fit very well in this picture. In the following, we will adopt the Debever–Sachs formulation of the Petrov classification because, in the physical problems we will be interested in, it is easier to handle.

Let us consider a Lorentzian manifold (M, g_{ab}) whose Weyl tensor is C_{abcd}:

$$C_{abcd} = R_{abcd} - \frac{1}{2}\left(g_{ca}R_{bd} - g_{ad}R_{bc} + g_{bd}R_{ac} - g_{bc}R_{ad}\right) + \frac{1}{6}R\left(g_{ca}g_{bd} - g_{ad}g_{bc}\right),$$

Then, when the vacuum Einstein equations are fulfilled, $C_{abcd} = R_{abcd}$. It is easy to see that the Weyl tensor has the same symmetries of the Riemann tensor and, moreover the following identity holds:

$$g^{bd} C_{abcd} = 0.$$

Let l^a be a null vector field: $g_{ca} l^c l^a = 0$. Let us consider the following equation for l^a:

$$l_{[h} C_{a]bc[d} l_{f]} l^b l^c = 0, \qquad (17)$$

where square brakets mean antisymmetrization. It can be shown that, for general (M, g_{ab}), there are at least one and at most four different solutions of the Eq. (17): $l^a_{(i)}, i = 1, \ldots, 4$.

The type **I** of the Petrov classification is characterized by the fact that all the four solutions $l^a_{(i)}$ of Eq. (17) are different;

For type **D** they are identical in pairs (so there are two independent vectors)
For type **II** there are three independent vectors (two out of four are identical)
For type **III** there are two independent vectors (three out of four are identical)
Lastly, type **N** is characterized by the fact that all four vector are identical.

In the following, these vector fields will be called Debever vectors. The mutual orientation of the Debever vectors is determined by Eq. (17). The general form of this equation characterizes the most general orientation of such vectors, that is, type **I**. For the other types Eq. (17) transforms into more stringent relations:

$$\mathbf{N} \Longrightarrow C_{abcd} l^a = 0, \qquad (18)$$

$$\mathbf{III} \Longrightarrow C_{abc[d} l^c l_{g]} = 0, \qquad (19)$$

$$\mathbf{II}, \mathbf{D} \Longrightarrow C_{abc[d} l_{g]} l^b l^c = 0, \qquad (20)$$

$$\mathbf{I} \Longrightarrow l_{[h} C_{a]bc[d} l_{f]} l^b l^c = 0. \qquad (21)$$

It is easy to show that if a vector satisfies any one of the Eqs. (18), (19), (20), (21) then automatically it will satisfy all the following as well. Thus, a gravitational field belongs to a given Petrov type if the Debever vector(s) satisfies the related equation and none of the preceding ones.

7.1 Newman–Penrose Formalism

Now we will give a brief description of the Newman–Penrose formalism and its relations with the Petrov classification.

The physical idea of this formalism is to shed light in a direct way on the casual structure of the spacetime encoded in the null-cones. The Newman–Penrose formalism is a tetrad formalism with a suitable choice of the basis vectors. This basis is made of four null vectors $e^{(a)}_{\mu}$ (where $a = 1, \ldots, 4$ is a tetrad index, i.e. it counts the basis vectors): l, n, m and $\widetilde{m}$ of which l and n are real and m and $\widetilde{m}$ are complex conjugates. The orthogonality conditions are:

$$l \cdot m = l \cdot \widetilde{m} = n \cdot m = n \cdot \widetilde{m} = 0.$$

Usually, one imposes on this null basis the following normalization conditions too:

$$\begin{aligned} l \cdot m &= 1 \\ m \cdot \widetilde{m} &= -1. \end{aligned}$$

Thus the tetrad metric $\eta_{(a)(b)} = e^{\mu}_{(a)} e_{\mu(b)}$ has the following structure:

$$\eta_{(a)(b)} = \begin{pmatrix} 0 & 1 & 0 & 0 \\ 1 & 0 & 0 & 0 \\ 0 & 0 & 0 & -1 \\ 0 & 0 & -1 & 0 \end{pmatrix},$$

with the correspondence

$$e^{(1)} = l, e^{(2)} = n, e^{(3)} = m, e^{(4)} = \widetilde{m}. \tag{22}$$

The Weyl Scalars In the Newman–Penrose formalism the ten (in four spacetime dimension) independent components of the Weyl tensor are represented by the five complex *Weyl scalars*:

$$\Psi_0 = -C_{\mu\nu\rho\sigma} l^{\mu} m^{\nu} l^{\rho} m^{\sigma} \tag{23}$$
$$\Psi_1 = -C_{\mu\nu\rho\sigma} l^{\mu} n^{\nu} l^{\rho} m^{\sigma} \tag{24}$$
$$\Psi_2 = -C_{\mu\nu\rho\sigma} l^{\mu} m^{\nu} \widetilde{m}^{\rho} n^{\sigma} \tag{25}$$
$$\Psi_3 = -C_{\mu\nu\rho\sigma} l^{\mu} n^{\nu} \widetilde{m}^{\rho} n^{\sigma} \tag{26}$$
$$\Psi_4 = -C_{\mu\nu\rho\sigma} n^{\mu} \widetilde{m}^{\nu} n^{\rho} \widetilde{m}^{\sigma}. \tag{27}$$

For a recent discussion see [35].

7.2 The Petrov Classification and the Newman–Penrose Formalism

Petrov worked out his classification before the formulation of the Newman–Penrose formalism. However, a very clear formulation of the Petrov classification is given by using the Newman–Penrose formalism and the Goldberg–Sachs theorem [36]. Now we will simply state the final results describing the various Petrov types, referring for details to [15] and [55].

Type **I**: this type is characterized by the vanishing of Ψ_0 and Ψ_4. More exactly, Ψ_0 and Ψ_4 can be made to vanish with a rotation of the null tetrad basis that does not change the other scalars.

Type **II**: this type is characterized by the vanishing of Ψ_0, Ψ_1 and Ψ_4. More exactly, Ψ_0, Ψ_1 and Ψ_4 can be made to vanish with a rotation of the null tetrad basis that does not change the other scalars. Thus only Ψ_2 and Ψ_3 will be left nonvanishing.

Type **D**: this type is characterized by the vanishing of Ψ_0, Ψ_1, Ψ_3 and Ψ_4. Thus Ψ_2 is the only nonvanishing Weyl scalar for the type **D**.

Type **III**: this type is characterized by the vanishing of Ψ_0, Ψ_1, Ψ_2 and Ψ_4. Thus Ψ_3 is the only nonvanishing Weyl scalar for the type **III**.

Type **N**: this type is characterized by the vanishing of Ψ_0, Ψ_1, Ψ_2 and Ψ_3. Thus Ψ_4 is the only non-vanishing Weyl scalar for the type **N**.

A very interesting fact (namely a corollary of the Goldberg-Sachs theorem) is that all the black-hole solutions of general relativity are of type **D** (see, for example, [15]). Thus, since the black-hole solutions represent gravitational fields of isolated body, we could say that the type **D** represents Coulomb-like gravitational field. However we will be mainly interested in the wave-like solutions of the Einstein equations.

Acknowledgment The results here exposed have been obtained in collaboration with F. Canfora, G. Sparano, A. Vinogradov and P. Vitale.

References

1. Alty L. J., Kleinian signature change. *Class. Quantum. Grav.* **11**(1994) 2523.
2. Aliev B.N. and Leznov A.N., Exact solutions of the vacuum Einstein's equations allowing for two noncommutative killing vectors Type G_2II of Petrov classification, *J . Math. Phys.* **33** n. 7, (1992) 2567.
3. Aichelburg P. C. and Sexl R. U., Gen. Rel. Grav. 2(1971) 303
4. Bächtold M., Ricci Flat Metrics with bidimensional null Orbits and non-integrable orthogonal Distribution, *Preprint DIPS–3/2005*.
5. Barret J., Gibbons G.W., Perry M. J., Pope C. N. and Ruback P., Kleinian geometry and N = 2 superstring. *Int. J. Mod. Phys. A* **9**, (1994) 1457.
6. Belinsky V. A. and Khalatnikov I. M., General solution of the gravitational equations with a physical singularities, *Sov. Phys. JETP* **30**, (1970) 6.
7. Bonnor, W. B. *Commun. Math. Phys.* **13**, (1969) 163
8. Belinsky V. A. and Zakharov V. E., Integration of the Einstein equations by means of the inverse scattering problem technique and construction of the exact soliton solutions. *Sov. Phys. JETP* **48**, (1978) 6; Stationary gravitational solitons with axial symmetry. *Phys. JETP* **50**, (1979) 1.
9. Canfora F. and Vilasi G., Spin-1 gravitational waves and their natural sources. *Phys. Lett. B* **585**, (2004) 193.
10. Canfora F and Vilasi G, Class. Quantum Grav. **22**, (2005) 1193.
11. Canfora F., Vilasi G. and Vitale P., Nonlinear gravitational waves and their polarization, *Phys. Lett. B* **545,** (2002) 373.
12. Canfora F., Vilasi G. and Vitale P., Spin-1 gravitational waves, *Int. J. Mod. Phys. B*, **18**, (2004) 527.

13. Catalano Ferraioli D. and Vinogradov A.M., Ricci-flat 4-metrics with bidimensional null orbits, Part I. General aspects and nonabelian case. *Acta Appl. Math.* 92 (3)(2006) 209–225.
14. Catalano Ferraioli D. and Vinogradov A.M., Ricci-flat 4-metrics with bidimensional null orbits, Part II. Abelian case. *Acta Appl. Math.* 92 (3)(2006) 226–239.
15. Chandrasekar S., *The mathematical theory of black holes* (Clarendon Press, Oxford, 1983).
16. Chinea F.J., New first integral for twisting type-N vacuum gravitational fields with two non-commuting Killing vectors. *Class. Quantum Grav.* **15**, (1998) 367.
17. Dirac P. A. M., *General Theory of Relativity*, (Wiley, N.Y. 1975).
18. De Filippo S., Marmo G. and Salerno M., Vilasi G., *On the Phase Manifold Geometry of Integrable Nonlinear Field Theories.* IFUSA (1982) n. 2.
19. De Filippo S., Marmo G., Salerno M., Vilasi G., A New Characterization of Completely Integrable Systems. *Il Nuovo Cimento B* **83**, 2 (1984) 97.
20. De Filippo S., Salerno M. and Vilasi G., A Geometrical Approach to Integrability of Soliton Equations. *Letters in Math. Phys.* **9**, (1985) 85.
21. Einstein A. and Rosen N., On gravitational waves, *J.Franklin Inst.* **223**, (1937) 43.
22. F. Ernst, New formulation of the axially symmetric gravitational field problem, *Phys. Rev.*, **167**, (1968) pp. 1175–8; New formulation of the axially symmetric gravitational field problem. II, *Phys. Rev.*, **168**, (1968) pp. 1415–17.
23. Faraoni V. and Dumse R. M., The strong Intercation of Light: From Weak to Strong Fields, *Gen. Rel. Grav.* **31, No 1** (1999) 91
24. Geroch R., A Method for Generating New Solutions of Einstein's Equation. II. *J. Math. Phys.* **13**, (1972) 394.
25. Gibbons G. W. Ruback P. J., *Phys. Rev. Lett.* **57**, (1986) 1492.
26. Hallisoy M., Studies in space-times admitting two spacelike Killing vectors. *J. Math. Phys.* **29**, (1988) 320.
27. Hartle J. B. and Hawking S. W., Wave function of the universe. *Phys. Rev. D* **28**, (1983) 2960.
28. Kompaneyets A. S., Strong Gravitational Waves in free space. *Sov. Phys. JETP* **7**, (1958) 659.
29. Landi G., Marmo G. and Vilasi G., Recursion Operators: Meaning and Existence. *J. Math. Phys.* **35**, (1994) 2.
30. Law P. R., Neutral Einstein metrics in four dimensions. *J. Math. Phys.* **32**, (1991) 3039.
31. T. Lewis, Some special solutions of the equations of axially symmetric gravitational fields, *Proc. Roy. Soc. London A* **136**, (1932) 176–192.
32. Matsushita Y., On Euler characteristics of compact Einstein 4-manifolds of metric signature $(++--)$. *J. Math. Phys.* **22**, (1981) 979.
33. Matsushita Y., Thorpe-Hitchin inequality for compact Einstein 4-manifolds of metric signature $(++--)$ and the generalized Hirzebruch index formula. *J. Math. Phys.* **24**, (1983) 36.
34. Neto ECD, Phys. Rev. D **68**, (12)(2003) 124013.
35. Nerozzi A., Beetle C., Bruni M., Burko L. M. and Pollney D., Towards a Wave-Extraction Method for Numerical Relativity: II. The quasi-Kinnersley Frame. *Phys. Rev. D* **72**, (2005) 024014.
36. Newman E. and Penrose R. J., An Approach to Gravitational Radiation by a Method of Spin Coefficients. *J. Math. Phys.* **3**, (1962) 566.
37. Ooguri H. and Vafa C., $N = 2$ heterotic strings. *Nucl. Phys. B* **367**, (1991) 83.
38. Penrose R. J., A spinor approach to general relativity. *Ann. of Phys.* **10**, (1960) 171.
39. Peres A., Some Gravitational Waves. *Phys. Rev. Lett.* **3**, (1959) 571.
40. Peres A., *Theory Phys. Rev.* **118**, (1960) 1105.
41. Petrov A.Z., *Einstein spaces,* (Pergamon Press, New York, 1969).
42. Sakharov A. D., Cosmological transitions with changes in the signature of the metric. *Sov. Phys. JETP* **60**, (1984) 214.
43. Sparano G. and Vilasi G., Noncommutative integrability and recursion operators. *J. Geom. Phys.* **36**, (2000) 270.
44. Sparano G., Vilasi G. and Vinogradov A. M., Gravitational fields with a non-Abelian, bidimensional Lie algebra of symmetries. *Phys. Lett. B* **513**, (2001) 142.
45. Sparano G., Vilasi G. and Vinogradov A. M., Vacuum Einstein metrics with bidimensional Killing leaves. I. Local aspects. *Diff. Geom. Appl.* **16**, (2002) 95.

46. Sparano G., Vilasi G. and Vinogradov A. M., Vacuum Einstein metrics with bidimensional Killing leaves. II. Global aspects. *Diff. Geom. Appl.* **17**, (2002) 1.
47. Kinnersley W., Recent progress in exact solutions, G. Shaviv (ed.) and J. Rosen (ed.), General Relativity and Gravitation , J. Wiley and Sons Ltd (New York 1975).
48. Stephani H., Kramer D., MacCallum M., Honselaers C. and Herlt E., *Exact solutions of Einstein field equations*, Cambridge University Press, (Cambridge, 2003).
49. Tolman R. C., Ehrenfest P. and Podolsky B., *Phys. Rev.* **37,** (1931) 602.
50. Verdaguer E., Soliton solutions in spacetimes with two spacelike killing fields. *Phys. Rep.* **229**, (1993) 1.
51. Weinberg S., *Gravitation and Cosmology* (J. Wiley & Sons, N. Y., 1972).
52. Wheeler J. A., Phys. Rev. **97**, (1955) 511
53. Wheeler J. A., "*...The Lord praise you if by this beautiful method you can get all of the solutions of Einstein empty space field equations...*" Private letter (1984).
54. Will C. M., *Theory and Experiment in Gravitational Physics* (revised ed. Cambridge University Press, Cambridge 1993).
55. Zakharov V. D., *Gravitational waves in Einstein's theory,* (Halsted Press, N.Y. 1973).

Part IV
Frame Dragging and Gravitomagnetism

Rotation and Spin in Physics*

Robert F. O'Connell

Abstract We delineate the role of rotation and spin in physics, discussing in order Newtonian classical physics, special relativity, quantum mechanics, quantum electrodynamics and general relativity. In the latter case, we discuss the generalization of the Kepler formula to post-Newtonian order (c^{-2}) including spin effects and two-body effects. Experiments which verify the theoretical results for general relativistic spin–orbit effects are discussed as well as efforts being made to verify the spin–spin effects.

1 Introduction

This article is essentially a companion to an article which I recently prepared for the 2007 Varenna summer school [1]. The latter deals in detail with spin effects in general relativity at the post-Newtonian level (order c^{-2}) with particular emphasis on the theory, and related experiments, of the two-body Kepler problem extended to include spin. Here, we use "spin" in the generic sense of meaning "internal spin" in the case of an elementary particle and "rotation" in the case of a macroscopic body.

My work with Barker on the general relativistic theory [2–4] brought to the fore some interesting conceptual matters dealing with the relation between velocity and momentum (especially for particles with spin), the non-uniqueness of spin supplementary conditions and the choice of coordinates even at the classical special relativistic level, the fact that a spinning particle necessarily has a minimum radius, the corresponding concepts in quantum theory (relating to such topics as the Foldy–Wouthuysen transformation and position operators) and the fact that the spin effects in quantum electrodynamics (obtained from one-photon exchange) have

*This paper is dedicated to the memory of Professor J.A. Wheeler.

R.F. O'Connell (✉)
Department of Physics and Astronomy, Louisiana State University,
Baton Rouge, LA 70803-4001, USA
e-mail: oconnell@phys.lsu.edu

I. Ciufolini and R.A. Matzner (eds.), *General Relativity and John Archibald Wheeler*, Astrophysics and Space Science Library 367, DOI 10.1007/978-90-481-3735-0_14,

their analogy in general relativity (obtained from one-graviton exchange or purely classical calculations) to the extent that except for (important) numerical factors, the latter results may be obtained from the former by simply letting $e^2 \rightarrow Gm_1m_2$.

Thus, we are motivated to present in a systematic way the role of rotation and spin in physics generally. So, we discuss, in order, Newtonian classical physics, special relativity, quantum mechanics, quantum electrodynamics and general relativity, in Sections 2 to 6, respectively. Our conclusions are presented in Section 7. The closest discussion of this nature in the literature is the book by Corben [5]. However, whereas his list of references proved a useful resource, his emphasis is on elementary particles and he does not consider general relativity, which is our emphasis.

2 Newtonian Classical Physics

Newton's second law is generally written in the form

$$\vec{F} = m\frac{d\vec{v}}{dt}, \tag{1}$$

where $\vec{v}$ is the velocity. Next, the momentum $\vec{p}$ of a particle is essentially defined as $m\vec{v}$ so that we have

$$\vec{F} = \frac{d\vec{p}}{dt}. \tag{2}$$

As French has emphasized [6], Newton in the *Principia* did not write (1) but in essence, wrote (2). Also, only in certain frames of reference, so-called inertial frames, are Newton's laws valid. All frames, moving uniformly in a straight line relative to a particular inertial frame, constitute an infinity of inertial frames where "–the properties of space and time are the same, and the laws of mechanics are the same–" [7], and such "–frame(s) of reference can always be chosen in which space is homogenous and isotropic and time is homogeneous" [7]. This leads us to a new concept which assigns a broader independent meaning to momentum. It follows from the introduction of the Lagrangian $L(q, \dot{q}, t)$ of a system, where q refers to position and $\dot{q} = (dq/dt)$ is the velocity. As a result, the associated Euler–Lagrange equations for a closed system lead to conservation laws of energy (resulting from the homogeneity of time), momentum (from the homogeneity of space) and angular momentum (from the isotropy of space) where, in particular, the momentum is referred to as the canonical momentum $\vec{P}$, as distinct from the mechanical momentum $\vec{p} = m\vec{v}$, where

$$\vec{P} = \frac{\partial L}{\partial \vec{v}}. \tag{3}$$

In general, $\vec{P}$ and $\vec{p}$ are not equal, an example being the case of a particle with charge q in an electromagnetic field, where the force is velocity dependent, in which case

$$\vec{P} = \vec{p} + \frac{q\vec{A}}{c}. \tag{4}$$

Thus, whereas $\vec{P}$ is conserved, $\vec{p}$ is not conserved in general. Moreover, in the passage to quantum mechanics, it is the canonical variables which are of importance in forming operators. In addition, it is $\vec{P}$ that enters into the conserved angular momentum $\vec{L} = \vec{r} \times \vec{P}$. Of course, in this case, the reason why $\vec{P}$ and $\vec{v}$ are not proportional is due to the presence of an external field. However, in special relativity, we shall see that, even for a free particle, even $\vec{p}$ and $m\vec{v}$ are, in general, not equal to each other.

Turning to the rotation of macroscopic bodies, we define $\vec{S}$ as the spin or intrinsic angular momentum. In general,

$$S_i = I_{ij}\omega_j \quad (i, j = 1, 2, 3), \tag{5}$$

where $\vec{\omega}$ is the angular velocity and I_{ij} are the components of the inertia tensor.

Based on observations, the earth is an inertial frame if one neglects its rotation. However, taking the "fixed stars" as an inertial system with "space axes", the earth rotates with an angular velocity $\vec{\omega}$ relative to these axis, giving rise to Coriolis and centrifugal forces. Thus, following Goldstein [8] and writing

$$\vec{v}_s = \vec{v}_v + \vec{\omega} \times \vec{r}, \tag{6}$$

where $\vec{v}_s$ and $\vec{v}_r$ are the velocities of a particle relative to the space and the earth's rotating set of axis, the equation

$$\vec{F}_s = ma_s, \tag{7}$$

in the space system translates to

$$\vec{F}_r = ma_r = ma_s - 2m(\vec{\omega} \times \vec{v}_r) - m\vec{\omega} \times (\vec{\omega} \times \vec{r}), \tag{8}$$

in the rotating system. The second and third terms are the familiar Coriolis and centrifugal forces, respectively. Of particular interest is a result, not found in [8] but derived in [7], that the momentum of a particle is the same in both frames of reference so that, in the rotating system,

$$\vec{p} = m\vec{v} + m(\vec{\omega} \times \vec{r}). \tag{9}$$

Thus, here we have a case where the momentum is different from $m\vec{v}$ purely due to the non-inertial nature of the rotating system. We also remark that the angular momenta are also equal in both systems but the energy in the rotating system $E_r = E_s - \vec{L} \cdot \vec{\omega}$ [7].

Finally, we remark that the earth is not a perfect sphere but is in fact an "asymmetric top", so that $\vec{\omega}$, the angular velocity of the earth, as distinct from $\vec{S}$ does not remain fixed in space but, instead, it executes what is known as polhode motion [7, 8]. This force-free precession of the earth's axis is a key to the explanation of Chandler's precession of the earth's axis and is also proving to be a major bugbear in the analysis of the GP-B gyroscope experiment [9, 10].

3 Special Relativity

Just as the angular momentum $\vec{L}$ generalizes to an anti-symmetric second rank tensor so does the 3-vector $\vec{S}$ generalizes to $S_{\alpha\beta} = -S_{\beta\alpha}$. Another possibility is to define an axial 4-vector S^{μ} which reduces to the 3-vector $\vec{S}$ in the rest-frame of the particle. In fact, this can be achieved by defining

$$S_\alpha \equiv \frac{1}{2}\epsilon_{\alpha\beta\sigma\tau} \quad S^{\beta\sigma}U^{\tau}, \tag{10}$$

where $\epsilon_{\alpha\beta\sigma\tau}$ is the completely antisymmetric Levy–Civita tensor, $U^{\tau} = (\gamma, \vec{v}/c)$ is the familiar 4-velocity, and $S^{\alpha} = (0, \vec{S})$ in the rest frame where $\vec{v} = 0$. Hence, using the fact that $\epsilon_{\alpha\beta\sigma\tau}$ is antisymmetric in α and τ, we obtain

$$U^{\alpha}S_{\alpha} = 0, \tag{11}$$

so that the 4-vectors U^{α} and S_{α} are not only orthogonal in the rest frame as constructed but are also orthogonal in all frames. In addition, using the properties of the Levi–Citiva symbol (Ref. [5], p. 79) and $U^{\alpha}U_{\alpha} = -1$, (10) may be inverted to give

$$S^{\alpha\beta} = \epsilon^{\alpha\beta\sigma\tau}S_{\sigma}U^{\tau}. \tag{12}$$

Equation (11) is a spin supplementary condition (SSC) which ensures that even when $\vec{v}$ is non-zero, S^{μ} has only three independent components and similarly for $S^{\alpha\beta}$. For a free particle (no external forces or torques)

$$\frac{dS^{\mu}}{dt} = 0, \tag{13}$$

but, nevertheless, as detailed in [5],the particle moves in a circle in a plane normal to $\vec{S}$ with a radius

$$\vec{r} = -\frac{\vec{v} \times \vec{S}}{mc^2}. \tag{14}$$

Such a motion is reminiscent of Zitterbewegung in relativistic quantum mechanics [11], as we shall discuss below. We will also see how such a quantity as $\vec{r}$

appears in the discussion of spin effects on orbital motion in post-Newtonian general relativistic theory. For now, we note that such a problem can be circumvented by choosing a different SSC based on the fact that there are essentially two basic rest systems for the particle, corresponding to either $\vec{v} = 0$ or $\vec{p} = 0$. This is also connected to the fact that, as Moller has shown [12, 13], a spinning body has a minimum radius equal to $|\vec{S}|/mc$. As a result, the definition of a rest frame is related to a choice of SSC which, in turn, is related to the choice of a center-of-mass for the spinning particle [14]. Thus, switching from one SSC (or rest-frame) to another is exactly the same as shifting the center of mass by a Lorentz transformation.

Turning to the case where a particle is acted on by a force f^{μ} without experiencing any torque, Weinberg [15] has shown that

$$\frac{dS^{\alpha}}{d\tau} = \left(S_{\beta} \frac{f^{\beta}}{m} \right) U^{\alpha}, \tag{15}$$

which leads to the famous Thomas precession [16, 17]. The usual discussion of Thomas precession is concerned with precession of a 3-vector, leading to a reduction by $\frac{1}{2}$ in the spin–orbit energy of an atomic electron (taking the gyromagnetic ratio g to be 2). This analysis was generalized by Bargmann et al. [18] to obtain $dS^{\alpha}/d\tau$ in its most simple form by assuming from the outset that the momentum and velocity of a spinning particle are proportional [5]. Thus, they were led to the so-called BMT equation for the classical relativistic equation of motion for spin in uniform or slowly varying external fields and its various applications. Details of this work are given in Jackson's well-known book [17], along with the derivation of Thomas's equation of motion of the spin vector from which Thomas precession emerges. Important applications of the BMT equation are listed in [5].

4 Quantum Mechanics

As detailed in all the textbooks, spin was introduced into non-relativistic quantum mechanics by Goudsmit and Uhlenbeck [19]. It should also be mentioned that the Wigner distribution, which represents an alternative formulation of quantum mechanics [20, 21] has also been extended to incorporate spin [22].

Relativistic quantum mechanics was initiated by Dirac's equation for the electron. However, the Dirac equation for a free particle, leads to the conclusion that the momentum and velocity are not simply related [11]. Thus, whereas the plane-wave solutions are eigenfunctions of the momentum operator, the velocity operator is not a constant of the motion with the implication that the free electron not only displays uniform rectilinear motion but also executes very rapid oscillations, called Zitterbewegung by Schrodinger who, as early as 1930, first discussed such motion. However, since we noted earlier that all classical spinning bodies have a minimum radius $|\vec{S}|/mc$, it is reasonable to assume that the electron (for which $|\vec{S}| = \hbar/2$)

cannot be localized to a size smaller than $\lambda\!\!\!^{-}_c/2$, where $\lambda\!\!\!^{-} = \hbar/mc$ is the Compton wavelength. Hence, analogous to the Lorentz transformations which shift the center of mass of a classical particle with spin, for an electron we can change the coordinate operator by means of a unitary Foldy–Wouthuysen transformation [23, 24]. In this representation the velocity operator is proportional to the momentum operator but, as emphasized by Sakurai (p. 177 of [11]), "nonlocality – is the price we must pay." By contrast, "–a well-localized state contains, in general, plane-wave components of negative energy" [11]. In earlier work and in a similar vein, Newton and Wigner [25] created the most localized state possible for a free particle without spin, and also the corresponding position operator, by using only plane-wave components of positive energies (solutions of the Klein–Gordon equation) and concluded that it is not a delta function as in the nonrelativistic case; instead, the wave function goes as $r^{-5/2}$ for small r and falls off exponentially for large r and, in general its characteristic extension in space is $\approx \lambda\!\!\!^{-}_c$. This work was then extended to particles with spin for which position operators and quantum states were also obtained. The results obtained are unique subject to the restriction that the orbital angular momentum of the localized state has the value $\ell = 0$. However, for spin 1/2 particles, it was shown by the present author and Wigner [26] that another choice of position operator is possible, given by

$$\vec{Q} = \vec{q} + (\vec{p} \times \vec{\sigma})/p^2, \tag{16}$$

where $\vec{q}$ is the original position operator, $\vec{Q}$ is the new position operator, and where the components of $\vec{\sigma}$ denote the Pauli spin matrices. In addition, it was shown [27] that

$$\langle \vec{q}(t) \rangle = \langle \vec{q}(0) \rangle + t \langle \vec{v} \rangle, \tag{17}$$

where here the velocity operator is $\vec{v} = (c^2 \vec{p}/E)$, where E is the total energy, including the rest mass. Thus, the movement of the mean position of a free particle obeys the classical Newtonian equation except for $m \to (E/c^2)$. Moreover, a similar result was obtained [26] for $\left\langle \vec{Q}(t) \right\rangle$. Even for the case of a spin zero particle, (17) is a non-trivial result because, as emphasized in [27], "–the situation in quantum mechanics is not immediately obvious because although $\vec{p}$ has a natural definition – it is the generator of spatial translations of the state vector – this is not so for $\vec{v}$ – the "velocity" does depend on the definition of the position–". Finally, we note that the choice of state vectors given by (16) and (17) leads to motion free from Zitterbewegung, just as we saw for the Foldy–Wouthuysen position operator.

5 Quantum Electrodynamics (QED)

The initial (semi-classical) treatment of particles and atoms with the electromagnetic field considered the latter to be classical. This was dramatically changed in 1927 when Dirac quantized the electromagnetic field [28]. Thus, for fields in general, the emphasis then switched from trying to construct relativistic potentials to the consideration of suitable Lagrangians and the use of Feynman diagrams, S-matrix theory [29] and other techniques. These are particularly useful in carrying out perturbation calculations. In contrast to nonrelativistic quantum theory (where potentials are used) in relativistic field theory, interactions are considered to arise from the exchange of quanta. Thus, in QED, the quanta are photons of spin 1 (whereas, as we shall discuss at length in the next section, gravitational interactions can be considered to be mediated by spin 2 gravitons). In essence, each diagram gives rise to a covariant matrix element whose Fourier transform "–enables us to construct an effective three-dimensional potential (to be used in connection with the Schrodinger equation)–" (Ref. [11], p. 259). A detailed description of this technique is given in [30]; in particular, they consider the electromagnetic interaction of two particles with masses m, and m_2 (such as an electron and a muon) for which they first calculate the particle interaction operator in the momentum representation. Next, by taking Fourier transforms, they obtain the corresponding operator in the coordinate representation (Ref. [30], p. 283, Eq. (83.15)); the latter contains terms of purely orbital origin as well as spin–orbit and spin–spin terms, and Darwin-like terms. As we shall see in the next section, an analogous method for the gravitational interaction of two spinning bodies leads to similar type terms.

6 General Relativity

For a free particle with spin S^{α}, in special relativity we had the result given by (13). The covariant generalization of this formula is [15]

$$\frac{DS^{\mu}}{D\tau} = 0, \tag{18}$$

where D denotes covariant differentiation. Also, in the presence of a gravitational field, (15) generalizes to [15, 31]

$$\frac{DS^{\mu}}{D\tau} = S_{\nu}\frac{DU^{\nu}}{D\tau}U^{\mu} \equiv S_{\nu}a^{\nu}U^{\mu}, \tag{19}$$

where $a^{u} = \frac{f^{\mu}}{m}$ is the 4-acceleration. This is referred to as Fermi-Walker transport of the spin vector in its motion along a curve x^{α} with tangent vector U^{α}. In the special case where $f^{\mu} = 0$ (free fall), the right-side of (19) is zero and the Fermi-Walker transport reduces to parallel transport along the geodesic, given by (18), which may be written explicitly as

$$\frac{dS_{\mu}}{d\tau} = \Gamma^{\lambda}{}_{\mu\nu}S_{\lambda}\frac{dx^{\nu}}{d\tau}, \tag{20}$$

where $\Gamma^{\lambda}_{\mu\nu}$ is the Christoffel symbol. Next, Papapetrou [32, 33] considered the motion of a small mass with spin in a gravitational field in order to obtain the deviation from geodesic motion. This deviation is expressed in both orbital and spin equations of motion, given by

$$\frac{D}{Ds}\left(mu^{\alpha}+u_{\beta}\frac{DS^{\alpha\beta}}{Ds}\right)+\frac{1}{2}S^{\mu\nu}u^{\sigma}R^{\alpha}_{\nu\sigma\mu}=0, \tag{21}$$

and

$$\frac{DS^{\alpha\beta}}{Ds}+u^{\alpha}u_{\rho}\frac{DS^{\beta\rho}}{Ds}-u^{\beta}u_{\rho}\frac{DS^{\alpha\rho}}{Ds}=0, \tag{22}$$

respectively, where $R^{\alpha}_{\nu\sigma\mu}$ is the Riemann curvature tensor and $S^{\alpha\beta}$ is given by (12). This was the starting-point used by Schiff [34] in his derivation of the theory underlying the GP-B experiment [10], which involved calculations to post-Newtonian order. However, Barker and the present author [2] applied a completely different approach to the problem than the method used by Papapetrou and Schiff, obtaining results which agreed with those of Schiff for the spin precession but apparently different for the orbital equations of motion.

The method used in [2] was based on a quantum approach to gravitation developed by Gupta [35] and his collaborators [36]. In particular, the gravitational interaction between two spin 1/2 Dirac particles was calculated by means of the exchange of a spin 2 graviton, analogous to what we discussed in Section 5 in the case of the electromagnetic interaction of two charged spin 1/2 Dirac particles. In fact, comparing the results in the latter case for the interaction of, say, an electron and muon (Ref. [30], Eq. (83.15)) with these obtained in the former case [2], we were able to conclude that all such effects in QED have their analogy in general relativity and, except for (important) numerical factors, the latter results may be obtained from the former by simply letting $e^2 \to Gm_1m_2$. Based on the universality of gravitational interactions, the transition from the microscopic to the macroscopic is achieved by the simple replacement $\frac{1}{2}\,\hbar\,\vec{\sigma} \to \vec{S}$ [2]. In a later publication [37], we traced the apparent difference between our results and those of Schiff for the orbital equations of motion to the use of two different coordinate vectors, reflecting the choice of two different SSC's which in turn imply different choices for the center-of-mass of the gyroscope and also imply different relationships between the momentum and velocity vectors. Thus, as outlined in [37], Corinaldesi–Papapetrou and Schiff used $S^{i0} = 0$ as their SSC; Pirani used $S^{\alpha\beta}U_{\beta} = 0$ whereas Moller, Tulczyjew and Dixon used $S^{\alpha\beta}P_{\beta} = 0$. In particular, the difference in the coordinate vectors used by all investigators is always of the order $(\vec{v}\times\vec{s})/mc^2$, the same quantity which arose in the discussion of anomalous motion in special relativity [see (14)] and not unrelated to the choice of position operators in relativistic quantum theory [see (16)]. However, regardless of the choice of SSC, in all cases identical results are obtained for orbital precessions. We also remark that, in contrast to the orbital case, the form of the spin equations of motion is the same for all choices of SSC's; the reason is that spin contributions are always order c^{-2} compared to the lowest-order Newtonian term appearing in the equation of motion.

Turning to experimental tests, Schiff's 1960 paper stimulated the birth of the GP-B experiment. However, because results are only now emerging, the initial euphoria which this experiment generated has waned [38] due to the fact that other investigators have verified the existence of spin–orbit effects in gravitation but, at least, the GP-B initial report [10] claims increased accuracy for the results obtained for the same quantity. The other investigations (which are discussed in more detail in [1], where the emphasis is on the Kepler problem and its post-Newtonian generalization, including spin) include (a) lunar-laser-ranging measurements of the lunar perigee [39, 40] based on the fact that the earth–moon system is essentially a gyroscope in the field of the sun [41]; (b) the Ciufolini Lageos experiment [42], based on the determination of earth–satellite distances with a precision of a few mm; (c) the binary pulsar PSRB1534 + 12 observations [43] based on a determination of time evolutions involving the spin direction.

The advantage of these experiments is that they are on-going (in contrast to the GP-B experiment) and thus improved accuracy is likely to be achieved. In particular, Ciufolini et al. whose *Nature* paper [42] is the first very credible demonstration of gravitomagnetism, are making use of the continual improvement in accuracy of earth gravity models [44].

The observations in [43] while limited in precession also provide an initial test of a 2-body result for the spin–orbit precession [3, 4]. What we found is that, for the spin precession of body 1, the replacement in the one-body formula is $m_2 \to m_2 + (\mu/3)$ and, in the case of the spin precession of body 2, the replacement is $m_1 \to m_1 + (\mu/3)$, where μ is the reduced mass. This result has now been verified by at least 7 other authors, the simplest being a calculation which makes use of the one-body result, followed by a transformation to the center-of-mass of the two-body system [45]. A potentially more promising candidate is the double binary system [46] for which spin precession angles of 4.8^o/yr (pulsar A) and 5.1^o/yr (pulsar B) have been predicted, based on our theoretical results [3, 4].

Finally, we remark on what we consider to be the most theoretically appealing alternative to Einstein's theory in which the spin of matter, as well as its mass, plays a dynamical role. This idea was initiated by Cartan [47] and extended by Sciama [48] and Kibble [49] using gauge theory. Utiyama initially investigated Einstein's theory and the Cartan theory within the framework of the homogeneous and inhomogeneous (Poincaré) groups, respectively [50]. For a review, we refer to [51]. We investigated possible additional spin interactions based on this so-called torsion theory and we found that spin–spin contact interactions, additional to these associated with Einstein's theory, occur [52,53]. However, as it turns out, they do not contribute to the macroscopic experiments presently being carried out.

7 Conclusions

In surveying the role of rotation and spin over the broad range of physics, several important concepts were considered. In non-relativistic classical physics, we recalled that Newton's second law only holds in an inertial system (in which case the mechanical momentum $\vec{p} = m\vec{v}$) but does not hold in a rotating system which is non-inertial. In the latter case, $\vec{p}$ is not proportional to $\vec{v}$. Also, the introduction of a Lagrangian L for a system brings into consideration a new type of momentum, the canonical momentum $\vec{P}$, which is the derivative of L with respect to $\vec{v}$ and which, in the case of external fields, is different than $\vec{p}$. The intrinsic angular momentum (spin) $\vec{S}$ also finds a natural definition within the Newtonian framework. However, in special relativity, one must treat spin as either a 4-vector S^{μ} or else as a second rank antisymmetric tensor and define a relationship between them which requires the introduction of a spin supplementary condition (SSC). This led to the realization that there are essentially two basic rest systems, corresponding to either $\vec{v} = 0$ or $\vec{p} = 0$, with the possibility of considering a variety of choices for a SSC instead of the choice given in (11). As a result, different results for $\vec{p}$ in terms of $\vec{v}$ emerge and, concomitantly, this also implies that the position vector of even a microscopic spinning particle is not an observable in special relativity.

Next, turning to quantum mechanics, we encountered similar phenomena. In particular, Zitterbewegung was found to be a feature of a particular choice of coordinate operator associated with Dirac's formulation of relativistic electron theory. It could be removed by carrying out a unitary transformation to a different coordinate for which Zitterbewegung disappears and simple proportionality between $\langle\vec{p}\rangle$ and $\langle\vec{v}\rangle$ holds. However, even for a free particle with or without spin, it was found necessary to choose a very specific position operator if one uses only plane–wave components of positive energy, in which case the particle is non-local and has a characteristic extension $\approx \lambda\!\!\!^{-}_c$. Extending consideration to the QED domain, we saw that the modus operandi is to consider the interactions as arising from the exchange of spin1 photons, leading to spin–orbit, spin–spin and other terms. A further extension of these ideas to the realm of general relativity called for the exchange of spin 2 gravitons. This led to a systematic derivation of spin–orbit and spin–spin terms, to order c^{-2}, for a two-body binary system. We discussed experiments which have already verified the spin–orbit effects. In fact, given the latter, the results for the spin–spin effects must conform to the requirement that the total angular momentum (orbital + spin of body 1 + spin of body 2) is conserved.

Our result for spin-orbit precession in a 2-body system [3] has been verified, to an accuracy of 13%, by Breton et al. [54] for the precession of pulsar B in the double binary system. In addition, in [55], we presented, in Table 2, a comparison of numerical results for a variety of one-body and two-body systems.

References

1. R.F. O'Connell, "Gravito-Magnetism in one-body and two-body systems: Theory and Experiment", in *Atom Optics and Space Physics*, Proc. of Course CLXVIII of the International School of Physics "Enrico Fermi", Varenna, Italy, ed. E. Arimondo, W. Ertmer, W. Schleich, and E.M. Rasel (IOS, Amsterdam, 2009).
2. B.M. Barker and R.F. O'Connell, "Derivation of the Equations of Motion of a Gyroscope from the Quantum Theory of Gravitation", *Phys. Rev. D* **2**, 1428 (1970).
3. B.M. Barker and R.F. O'Connell, "The Gravitational Two Body Problem with Arbitrary Masses, Spins, and Quadrupole Moments", *Phys. Rev. D* **12**, 329 (1975).
4. B.M. Barker and R.F. O'Connell, "General Relativistic Effects in Binary Systems", in *Physics and Astrophysics of Neutron Stars and Black Holes: Proceedings of Course 65 of the International School of Physics "Enrico Fermi"*, Varenna, Italy, 1975, ed. R. Giacconi and R. Ruffini (North-Holland, 1976), p. 437.
5. H.C. Corben, "Classical and Quantum Theories of Spinning Particles", (Holden-Day, Inc., San Francisco, 1968).
6. A. P. French, "Newtonian mechanics", (W.W. Norton, New York, 1971).
7. L.D. Landau and E.M. Lifshitz, "Mechanics", 3rd ed. (Pergamon Press, Oxford, 1976).
8. H. Goldstein, "Classical Mechanics", 2nd ed. (Addison-Wesley, Reading, MA, 1992).
9. G. Brumfiel, "Gravity probe falters", *Nature* **444**, 978 (2006).
10. Stanford, "GP-B Mission Status Update", Dec. 22, 2006.
11. J.J. Sakurai, "Advanced Quantum Mechanics", (Addison-Wesley, Reading, MA, 1967).
12. C. Moller, Commun. Dublin Inst. Adv. Stud., **A5**, 1 (1949).
13. C. Moller, "The Theory of Relativity", 2nd ed. (Oxford University Press, 1972), p. 176.
14. B.M. Barker and R.F. O'Connell, "Nongeodesic Motion in General Relativity", *Gen. Relativ. Gravit.* **5**, 539 (1974).
15. S. Weinberg, "Gravitation and Cosmology" (Wiley and Sons, New York, 1972), p. 124.
16. L.H. Thomas, *Nature* **117**, 514 (1926).
17. J.D. Jackson, "Classical Electrodynamics", 3rd ed. (Wiley, New York, 1998).
18. V. Bargmann, L. Michel and V.L. Telegdi, "Precession of the Polarization of Particles Moving in a Homogeneous Electromagnetic Field", *Phys. Rev. Lett.* **2**, 435 (1959).
19. G.E. Uhlenbeck and S. Goudsmit, *Nature* **117**, 264 (1926).
20. M. Hillery, R.F. O'Connell, M. Scully, and E. Wigner, "Distribution Functions in Physics: Fundamentals", *Phys. Rep.* **106**, 121–167 (1984).
21. R.F. O'Connell, "The Wigner Distribution" in *Compendium of Quantum Physics*, ed. by D. Greenberger, B. Falkenburg, K. Hentschel, and F. Weinert (Springer, New York, 2009).
22. R.F. O'Connell and E.P. Wigner, "Manifestations of Bose and Fermi Statistics on the Quantum Distribution Function for Systems of Spin Zero and Spin One-Half Particles", *Phys. Rev. A* **30**, 2613 (1984).
23. L. Foldy and S.A. Wouthuysen, "On the Dirac Theory of Spin 1/2 Particles and Its Non-Relativistic Limit", *Phys. Rev.* **78**, 29 (1950).
24. M.E. Rose, "Relativistic Electron Theory" (Wiley, New York, 1961).
25. T.D. Newton and E.P. Wigner, "Localized States for Elementary Systems", *Rev. Mod. Phys.* **21**, 400 (1949).
26. R.F. O'Connell and E.P. Wigner, "Position Operators for Systems Exhibiting the Special Relativistic Relation Between Momentum and Velocity", *Phys. Lett.* **67A**, 319 (1978).
27. R.F. O'Connell and E.P. Wigner, "On the Relation Between Momentum and Velocity for Elementary Systems", *Phys. Lett.* **61A**, 353 (1977).
28. P.A.M. Dirac, *Proc. Roy. Soc. A* **114**, 243 (1927).
29. J.A. Wheeler, "On the Mathematical Description of Light Nuclei by the Method of Resonating Group Structure", *Phys. Rev.* **52**, 1107 (1937).
30. V.B. Berestetskii, E.M. Lifshitz, and L.P. Pitaevskii, *Relativistic Quantum Theory* (New York, N.Y., 1971), p. 280.
31. I. Ciufolini and J.A. Wheeler, "Gravitation and Inertia" (Princeton University Press, 1995).

32. A. Papapetrou, "Spinning Test-Particles in General Relativity I", *Proc. R. Soc. London* **A209**, 248 (1951).
33. E. Corinaldesi and A. Papapetrou, "Spinning Test-Particles in General Relativity II", *Proc. R. Soc. London* **A209**, 259 (1951).
34. L.I. Schiff, *Proc. Nat. Acad. Sci.* **46**, 871 (1960); *Proceedings of the Theory of Gravitation*, Jablonna, Poland, 1962, edited by L. Infeld (Paris, 1964), p. 71.
35. S.N. Gupta, "Einstein's and Other Theories of Gravitation", *Rev. Mod. Phys.* **29**, 334 (1957).
36. B.M. Barker, S.N. Gupta, and R.D. Haracz, "One-Graviton Exchange Interaction of Elementary Particles", *Phys. Rev.* **149**, 1027 (1966).
37. B.M. Barker and R.F. O'Connell, "Nongeodesic Motion in General Relativity", *Gen. Rel. Grav.* **5**, 539 (1974).
38. A. Lawler, "NASA Orders Make-or-Break Tests for Gravity Probe", *Science* **300**, 880 (2003).
39. J.G. Williams, X.X. Newhall, and J.O. Dickey, "Relativity parameters determined from lunar laser ranging", *Phys. Rev. D* **53**, 6730 (1996).
40. J.G. Williams, S.G. Turyshev and D.H. Boggs, "Progress in Lunar Laser Ranging Tests of Relativistic Gravity", *Phys. Rev. Lett.* **93**, 261101 (2004).
41. R.F. O'Connell, "A Note of Frame Dragging", *Class. Quantum Grav.* **22**, 3815 (2005).
42. I. Ciufolini and E.C. Pavlis, "A Confirmation of the General Relativistic Prediction of the Lense-Thirring Effect", *Nature* **431**, 958 (2004).
43. I.H. Stairs, S.E. Thoresett and Z. Arzoumanian, "Measurement of Gravitational Spin-Orbit Coupling in a Binary-Pulsar System", *Phys. Rev. Lett.* **93**, 141101 (2004).
44. I. Ciufolini, E.C. Pavlis and R. Peron, "Determination of frame-dragging using Earth gravity models from CHAMP and GRACE", *New Astronomy* **11**, 527 (2006).
45. L. Chan and R.F. O'Connell, "Two-Body Problems–A Unified, Classical, and Simple Treatment of Spin-Orbit Effects", *Phys. Rev. D* **15**, 3058 (1977).
46. M. Kramer et al., "Texts of General Relativity from Timing the Double Pulsar", *Science* **314**, 97 (2006).
47. E. Cartan, *C. R. Acad. Sci* **174**, 593 (1972), and *Ann. Ed. Norm. Sup.* **40**, 325 (1923), and **41**, 1 (1924), and **42**, 17 (1925).
48. D.W. Sciama, "The Physical Structure of General Relativity", *Rev. Mod. Phys.* **36**, 463, 1103 (1964).
49. T.W.B. Kibble, "Lorentz Invariance and the Gravitational Field", *J. Math. Phys. (N.Y.)* **2**, 212 (1961).
50. R. Utiyama, "Invariant Theoretical Interpretation of Interaction", *Phys. Rev.* **101**, 1597 (1956).
51. F.W. Hehl et al., "General Relativity with Spin and Torsion: Foundations and Prospects", *Rev. Mod. Phys.* **48**, 393 (1976).
52. R.F. O'Connell, "Contact Interactions in the Einstein and Einstein-Cartan-Sciama-Kibble (ECSK) Theories of Gravitation", *Phys. Rev. Lett.* **37**, 1653 (1976) and **38**, 298(E) (1977).
53. R.F. O'Connell, "Attractive Spin-Spin Contact Interactions in the ECSK Torsion Theory of Gravitation", *Phys. Rev. D* **16**, 1247 (1977).
54. R.P. Breton, V.M. Kaspi, M. Kramer, M.A. McLaughlin, M. Lyutikov, S.M. Ransom, I.H. Stairs, R.D. Ferdman, F. Camilo, and A. Possenti, "Relativistic Spin Precession in the Double Pulsar," *Science* **321**, 104 (2008).
55. C. Feiler, M. Buser, E. Kajari, W.P. Schleich, E.M. Rasel, and R.F. O'Connell, "New Frontiers at the Interface of General Relativity and Quantum Optics," *Space Sci. Rev.* **148**, 123–147 (2009).

The Gravitomagnetic Influence on Earth-Orbiting Spacecrafts and on the Lunar Orbit

Sergei M. Kopeikin

Abstract Gravitomagnetic field is covariantly split in the *intrinsic* and *extrinsic* parts, which are generated by rotational and translational currents of matter respectively. The *intrinsic* component has been recently discovered in the LAGEOS spacecraft experiment. We discuss the method of detection of the *extrinsic* tidal component with the lunar laser ranging (LLR) technique. Analysis of the gauge residual freedom in the relativistic theory of three-body problem demonstrates that LLR is currently not capable to detect the *extrinsic* gravitomagnetic effects which are at the ranging level of few millimeters. Its detection requires further advances in the LLR technique that are coming in the next 5–10 years.

Detection of gravitomagnetic components as predicted by Einstein's theory of general relativity is one of the primary goals of experimental gravitational physics. The paper [1] states that the gravitomagnetic interaction plays a part in shaping the lunar orbit readily observable by LLR. The authors picked up a "gravitomagnetic" term from the parameterized post-Newtonian (PPN) equation of motion of massive bodies [2] and proved that it correctly reproduces the Lense–Thirring precession of the GP-B gyroscope. The paper [1] argues that the very same term in the equations of motion of the Moon, derived in the solar-system barycentric (SSB) frame, perturbs the lunar orbit with a radial amplitude $\simeq$6 m that was observed with the lunar laser ranging (LLR). We have explained [3] and confirm in more detail in the present paper that the gauge freedom of the relativistic three-body problem suppresses the gravitomagnetic effects in the lunar motion, that depend on the Earth's velocity V around the Sun, to the level $\leq$1 cm. It makes LLR currently insensitive to the gravitomagnetic interaction.

There are two types of mass currents in gravity [4,5]. The first type is produced by the intrinsic rotation of matter around body's center of mass. It generates an *intrinsic* gravitomagnetic field tightly associated with body's angular momentum (spin) and most research in gravitomagnetism has been focused on the discussion of its various properties. Textbook [6] gives a comprehensive review of various aspects of the

S.M. Kopeikin (✉)
Department of Physics and Astronomy, University of Missouri, Columbia, MO 65211
e-mail: kopeikins@missouri.edu

I. Ciufolini and R.A. Matzner (eds.), *General Relativity and John Archibald Wheeler*, Astrophysics and Space Science Library 367, DOI 10.1007/978-90-481-3735-0_15,

intrinsic gravitomagnetism. It is interesting to note that the *intrinsic* gravitomagnetic field can be associated with the holonomy invariance group [7]. Some authors [8, 9] have proposed to measure the *intrinsic* gravitomagnetic field by observing quantum effects of coupling of fermion's spin with the angular momentum of the Earth. It might be worthwhile to explore association of the *intrinsic* gravitomagnetism with the classic Hannay precession phase [10–12].

The first classic experiment to test the *intrinsic* gravitomagnetic effect of the rotating Earth has been carried out by observing LAGEOS in combination with other geodetic satellites [13–16] which verified its existence with a remarkable precision as predicted by Einstein's general relativity. Independent experimental measurement of the *intrinsic* gravitomagnetic field of the rotating Earth is currently under way by the Gravity Probe B mission [17].

The second type of the mass current is caused by translational motion of matter. It generates an *extrinsic* gravitomagnetic field that depends on the frame of reference of observer and can be either completely eliminated in the rest frame of the matter or significantly suppressed by the transformation to the local inertial reference frame of observer. This property of the *extrinsic* gravitomagnetic field is a direct consequence of the gauge invariance of Einstein's gravity field equations for an isolated astronomical system [18] embedded to the asymptotically flat space–time. Experimental testing of the *extrinsic* gravitomagnetic field is as important as that of the *intrinsic* gravitomagnetic field. The point is that both the *intrinsic* and the *extrinsic* gravitomagnetic fields obey the same equations and, therefore, their measurement would essentially complement each other [5]. Furthermore, detection of the *extrinsic* gravitomagnetic field probes the time-dependent behavior of the gravitational field which is determined by the structure of the gravity null cone (the domain of causal influence) on which the gravity force propagates. Experimental verification of the gravitomagnetic properties of gravity is important for the theory of braneworlds [19] and for setting other, more stringent limitations on vector–tensor theories of gravity [20].

Ciufolini [4] proposed to distinguish the rotationally induced gravitomagnetic field from the translationally induced gravitomagnetic effects by making use of two scalar invariants of the curvature tensor

$$I_1 = R_{\alpha\beta\mu\nu} R^{\alpha\beta\mu\nu}, \tag{1}$$

$$I_2 = R_{\alpha\beta}{}^{\mu\nu} R^{\alpha\beta\rho\sigma} E_{\mu\nu\rho\sigma}, \tag{2}$$

where $R_{\alpha\beta\mu\nu}$ is the curvature tensor, $E_{\mu\nu\rho\sigma}$ is the fully anti-symmetric Levi–Civita tensor with $E_{0123} = +\sqrt{-g}$, and $g = \det(g_{\mu\nu}) < 0$ is the determinant of the metric tensor. Ciufolini [4] notices that a weak gravitational field of an isolated astronomical system yields $I_2 = 0$ if the *intrinsic* gravitomagnetic field is absent. However, one should not confuse the invariant I_2 with the gravitomagnetic field itself. The gravitomagnetic field is generated by *any* current of matter. Hence, $I_2 = 0$ does not mean that any gravitomagnetic field is absent as has been erroneously interpreted in [21]. Equality $I_2 = 0$ only implies that the gravitomagnetic field is of the *extrinsic* origin ($I_1 \neq 0$), that is generated by a translational motion of matter. The translational gravitomagnetic field can be measured, for instance, by observing the gravitational deflection of light by a moving massive body like Jupiter [22,23]. This gravitomagnetic frame-dragging effect on the light ray was indeed observed in a dedicated radio-interferometric experiment [24].

Paper [1] makes an attempt to demonstrate that the *extrinsic* gravitomagnetic field can be measured by making use of the LLR observations of the lunar orbit. This must not be confused with the measurement of the *intrinsic* gravitomagnetic field by means of the satellite laser ranging technique applied to LAGEOS [13–16]. The LAGEOS experiment measures the Lense–Thirring precession of the satellite's orbit caused by the Earth's angular momentum entering g_{0i} component of the metric tensor. The authors of [1] have been trying to measure the gravitomagnetic precession of the lunar orbit caused by the orbital motion of the Earth–Moon system around the Sun. They used the barycentric coordinates of the solar system (BCRS) to derive the equation of motion of the Moon relative to the Earth. The equation is effectively obtained as a difference between the Einstein–Infeld–Hoffmann (EIH) equations of motion for the Earth and for the Moon with respect to the barycenter of the solar system, and repeats the original derivation by Brumberg [25,26], which later was independently derived by Baierlein [27].

The barycentric equation of motion of the Moon formally includes the gravitomagnetic perturbation in the following form [1]

$$\boldsymbol{a}_{\rm GM} = \boldsymbol{a} + \frac{\gamma - 1}{2}\boldsymbol{a}, \tag{3}$$

where γ parameterizes a deviation from general relativity, the bold letters denote spatial vectors, and the dot between two vectors means their Euclidean dot product. The general-relativistic post-Newtonian acceleration

$$\boldsymbol{a} = \frac{4Gm}{c^2 r^2}\left[\hat{\boldsymbol{r}}\left(\boldsymbol{V}\cdot\boldsymbol{u}\right) - \boldsymbol{V}\left(\boldsymbol{u}\cdot\hat{\boldsymbol{r}}\right)\right], \tag{4}$$

where m is mass of the Earth, r is radius of the lunar orbit, $\hat{\boldsymbol{r}}$ is the unit vector from the Earth to the Moon, $\boldsymbol{V}$ is the Earth's velocity around the Sun, and $\boldsymbol{u}$ is the Moon's velocity around the Earth.

We notice [3] that the barycentric coordinate frame referred to the geocenter by a simple, Newtonian-like spatial translation (the time coordinate is unchanged)

$$\boldsymbol{r} = \boldsymbol{x} - \boldsymbol{x}_E(t), \tag{5}$$

as it is obtained in [1,28], is *not* in a free fall about the Sun, and does *not* make a local inertial frame. Thus, perturbations in Eqs. (3)–(4) cannot be interpreted as physically observable and, in fact, represent a spurious gauge-dependent effect that is canceled by transformation to the local-inertial frame of the geocenter. This transformation is a generalized Lorentz boost with taking into account a number of additional terms due to the presence of the external gravitational field of the Sun [29,30].

The gauge freedom of the lunar equations of motion must be analyzed to eliminate all gauge-dependent, non-observable terms. Only the terms in the equations of motion, which cannot be eliminated by the gauge transformation to the local inertial frame can be physically interpreted. The analysis of the gauge freedom in the three body-problem had been done in [30–32]. It proves that all non-tidal and $\boldsymbol{V}$-dependent terms, including the first term in the right side of Eq. (3), are pure coordinate effects that disappear from the lunar equations of motion after transformation to the geocentric, locally inertial frame. This is because the Lorentz invariance and the principle of equivalence reduce the relativistic equation of motion of the Moon to the covariant equation of the geodesic deviation between the Moon's and the Earth's world lines [31, 32], where gravitomagnetic effects appear only as tidal relativistic forces with amplitude smaller than 1 cm. The covariant nature of gravity tells us [2] that if some effect is not present in the local frame of observer, it cannot be observed in any other coordinate system. This means that besides physically observable terms, the barycentric LLR model [1, 28] also operates with terms having the gauge-dependent origin, which mathematically nullify each other in the data-processing computer code irrespectively of the frame of reference. The mutually annihilating terms enter different parts of the barycentric LLR model with opposite signs [3, 31] but, if taken separately, can be erroneously interpreted as really observable. This is what exactly happened with the misleading analysis given in [1].

General relativity indicates that the barycentric EIH lunar equations of motion may admit the observable gravitomagnetic acceleration only in the form of the second term in the right side of Eq. (3) that is proportional to $\gamma - 1$. Radio experiments set a limit on $\gamma - 1 \leq 10^{-3}$ [2] that yields $|\boldsymbol{a}_{\rm GM}| \leq 1$ millimeter. The current half-centimeter accuracy of LLR is insufficient to measure such negligible effect. Moreover, our analysis of the gauge invariance of the scalar-tensor theory of gravity [30] points out that the term being proportional to γ in Eq. (3) is also eliminated in the locally inertial, geocentric reference frame. We conclude that LLR is currently insensitive to the gravitomagnetism and, yet, cannot compete with the LAGEOS and/or GP-B experiments.

Recent paper by Soffel et al. [33] is another attempt to prove that the *extrinsic* gravitomagnetic acceleration (4) can be measured with the LLR technique in a locally inertial reference frame. The authors of [33] accept our criticism [3] but continue to believe that one can measure the *extrinsic* gravitomagnetic field by making use of the preferred frame parametrization of the gravimagnetic terms. To this end, Soffel et al. [33] introduce the preferred-frame generalization of Eq. (3) by replacing

$$\gamma - 1 \to \gamma - 1 + \eta_G/4, \tag{6}$$

where η_G is a parameter labeling the gravitomagnetic effects in the barycentric equations of motion of the Moon. Parameter $\eta_G = -\alpha_1/2$ in the framework of the PPN formalism [2] but unless this is not stated explicitly one can regard η_G as an independent parameter and fit it to LLR data irrespectively of α_1. This was done by Soffel et al. [33] who had obtained $\eta_G = (0.9 \pm 0.7) \times 10^{-3}$.

We argue that this measurement says nothing about the *extrinsic* gravitomagnetic field. This is because η_G has no any fundamental significance. Its value is not invariant and crucially depends on the choice of the preferred reference frame, which one uses for processing LLR data. In general theory of relativity $\eta_G \equiv 0$ in any frame, that means a vanishing (non-observable) effect. Indeed, analytic calculations given in appendix of the paper [33] and in [34, 35] confirm that in the PPN framework with $\alpha_1 = 0$ all gravitomagnetic effects in the motion of the solar system are nullified by the corresponding gravitoelectric effects from other well-established post-Newtonian gravitational potentials. Those calculations reveal that the gravitomagnetic effect described by Eq. (3) is nothing but a symbolic property of the particular coordinate system used for calculations. LLR data fit and the coordinate-dependent limits on η_G obtained in paper [33] confirm that the gravitational model used in the data processing, is self-consistent. However, it neither means that the *extrinsic* gravitomagnetic field was measured nor that general theory of relativity was tested. This is because the gauge-invariance is the main property of a large class of the metric-based gravitational theories and testing the self-consistency of the LLR equations does not single out a specific gravitational theory.

In order to measure the *extrinsic* gravitomagnetic field, one has to find the real gravitomagnetic effect in the motion of the Moon, which does not vanish in the framework of general theory of relativity in the locally inertial frame of observer on the Earth. For this reason, the primary goal of the relativistic theory of the lunar motion is to construct an inertial reference frame along the world-line of the geocenter and to identify the gravitomagnetic effects in this frame. This task was solved in our paper [31] in the post-Newtonian approximation in the case of a three-body problem (Sun, Earth, Moon) under assumption that the Moon is considered as a test particle. In the quadrupole approximation the metric tensor in the locally inertial geocentric reference frame $X^\alpha = (cT, \boldsymbol{X})$ has the following form

$$G_{00}(T, \boldsymbol{X}) = -1 + \frac{2}{c^2}\left[U(T, \boldsymbol{X}) + Q_p X^p + \frac{3}{2} Q_{pq} X^p X^q\right] + O\left(\frac{1}{c^4}\right), \quad (7)$$

$$G_{0i}(T, \boldsymbol{X}) = -\frac{4}{c^3}\left[U^i(T, \boldsymbol{X}) + \epsilon_{ipk} C_{pq} X^k X^q\right] + O\left(\frac{1}{c^5}\right), \quad (8)$$

$$G_{ij}(T, \boldsymbol{X}) = \delta_{ij} + \frac{2}{c^2}\left[U(T, \boldsymbol{X}) + Q_p X^p + \frac{3}{2} Q_{pq} X^p X^q\right]\delta_{ij} + O\left(\frac{1}{c^4}\right), \quad (9)$$

where $U(T, \boldsymbol{X})$ is the Newtonian potential of the Earth, $U^i(T, \boldsymbol{X})$ is the post-Newtonian vector-potential of the Earth, Q_p is the acceleration of the geocenter with respect to the geodesic world line, Q_{pq} and C_{pq} are tidal gravitational-force gradients from the Sun, and ϵ_{ipk} is the fully anti-symmetric symbol.

We notice that $U^i(T, \boldsymbol{X})$ represents the *intrinsic* gravitomagnetic field of the Earth, which has been measured in the LAGEOS experiment [14]. It can be shown [36] that gravitomagnetic vector potential $U^i(T, \boldsymbol{X})$ produces a negligibly small acceleration on the lunar orbit, and can be discarded. The tensor potential C_{pq} has the *extrinsic* gravitomagnetic origin as it is generated in the geocentric frame by the motion of an external mass (the Sun) with respect to the Earth. This potential is expressed in terms of the orbital velocity of the Earth, V^i, and the Newtonian tidal matrix Q_{pq} [31]

$$C_{pq} = \epsilon_{ikp}\left(V_i Q_{kq} - V_k Q_{iq} + \frac{1}{2}\delta_{kq} V^j Q_{ij} - \frac{1}{2}\delta_{iq} V^j Q_{kj}\right), \tag{10}$$

where

$$Q_{pq} = \frac{GM}{R^5}\left(3X_p X_q - \delta_{pq} X^2\right), \tag{11}$$

M is mass of the Sun, and R is the distance between the Sun and Earth.

The gravitomagnetic tidal force causes a non-vanishing gravitomagnetic acceleration of the Moon, which reads in the locally inertial geocentric frame as follows (see Eq. 7.12 from [31])

$$A^i_{GM} = 12\epsilon_{ijk} C_{jq} u^q r^k, \tag{12}$$

where u^i is the geocentric velocity of the Moon, and r^i is the Earth–Moon radius-vector. This gravitomagnetic acceleration causes a radial oscillations of the lunar orbit that can be estimated by making use of Eqs. (8) and (9). Noticing that the term $Q_{ij} r^j$ is proportional to the Newtonian tidal force from the Sun, which produces the *variation* inequality of the lunar orbit [37], one obtains

$$|\boldsymbol{A}_{GM}| \simeq 10 \times (\text{variation}) \times \frac{Vu}{c^2}. \tag{13}$$

The variation amounts to $\simeq$3,000 km in radial oscillation [37]. Hence, the amplitude of the oscillation of the lunar orbit caused by the gravitomagnetic tidal force has an amplitude of about 1 cm. This is unmeasurable with the current LLR data but can be measured in the next 5–10 years with the advent of a millimeter-range LLR technology [38].

References

1. T. W. Murphy, Jr., K. Nordtvedt, and S.G. Turyshev, (2007) *Phys. Rev. Lett.*, **98**, 071102
2. C. M. Will, (2006) "The Confrontation between General Relativity and Experiment", Living Rev. Relativity **9**, 3
3. S. M. Kopeikin, (2007), *Phys. Rev. Lett.*, **98**, 229001
4. I. Ciufolini, (2001), *Gravitomagnetism, Lense-Thirring Effect and De Sitter Precession*, In: Reference Frames and Gravitomagnetism, Proc. XXIII Spanish Relativity Meeting. Eds. J. F. Pascual-Sánchez, L. Floriá, A. San Miguel and F. Vicente. (Singapore: World Scientific), pp. 25–34
5. S. M. Kopeikin, (2006), *Int. J. Mod. Phys. D*, **15**, 305
6. I. Ciufolini & J. A. Wheeler, (1995), *Gravitation and Inertia* (Princeton: Princeton University Press, 1995)

7. R. Maartens, B. Mashhoon & D. R. Matravers, (2002), *Class. Quant. Grav.*, **19**, 195
8. A. Camacho, (2002), *Quantum Zeno effect and the detection of gravitomagnetism*, In: Recent developments in general relativity. 14th SIGRAV Conference on Gen. Rel. and Grav. Phys., Eds. R. Cianci, R. Collina, M. Francaviglia, P. Fré. (Milano: Springer), pp. 347–351
9. A. Camacho, (2002), *Gen. Rel. Grav.*, **34**, 1963
10. J. H. Hannay, (1985), *J. Phys. A: Math. Gen.*, **18**, 221
11. A. Spallicci, (2004), *Nuov. Cim. B.*, **119**, 1215
12. A. Spallicci, A. Morbidelli & G. Metris, (2005), *Nonlinearity*, **18**, 45
13. I. Ciufolini, (1986), *Phys. Rev. Lett.*, **56**, 278
14. I. Ciufolini & E. C. Pavlis, (2004), *Nature*, **31**, 958
15. I. Ciufolini & E. C. Pavlis, (2005), *New Astron.*, **10**, 636
16. I. Ciufolini, (2007), *Nature*, **449**, 41
17. http://einstein.stanford.edu/
18. V. Fock, (1964), *Theory of Space, Time and Gravitation* (Clifton: Reader's Digest Young Families)
19. X, Bekaert, N. Boulanger & J. F. Vázquez-Poritz, (2002), *J. High Energy Phys.*, **10**, 53
20. W.-T. Ni, (2005), *Int. J. Mod. Phys. D*, **14**, 901
21. J.-F. Pascual-Sánchez, (2004), *Int. J. Mod. Phys.*, **13**, 2345
22. S.M. Kopeikin, (2003), *Phys. Lett. A.*, **312**, 147
23. S. M. Kopeikin & V. V. Makarov, (2007), *Phys. Rev. D*, **75**, 062002
24. Fomalont, E. B. & Kopeikin, S., (2008), "Radio interferometric tests of general relativity", In: Proc. of IAU Symposium, **248**, 383–386
25. V.A. Brumberg, (1958), Bull. Inst. Theor. Astron. of Acad. of Sci. USSR, **6**, 733 (in Russian)
26. V.A. Brumberg, (1972), *Relativistic Celestial Mechanics*, (Moscow: Nauka) (in Russian)
27. R. Baierlein, (1967), *Phys. Rev. D*, **162**, 1275
28. J.G. Williams, S.G. Turyshev, and D.H. Boggs, arXiv:gr-qc/0507083
29. S. M. Kopejkin, (1988), *Cel. Mech.*, **44**, 87
30. S. Kopeikin & I. Vlasov, (2004) *Phys. Rep.*, **400**, 209
31. V.A. Brumberg, and S.M. Kopeikin, (1989) *Nuovo Cim. B*, **103**, 63
32. T. Damour, M. Soffel, and C. Xu, (1994) *Phys. Rev. D*, **49**, 618
33. M. Soffel, S. Klioner, J. Müller & L. Biskupek, (2008), *Phys. Rev. D*, 78, 024033
34. K. Nordtvedt, (1988), *Phys. Rev. Lett.*, **61**, 2647
35. K. Nordtvedt, (1988), *Int. J. Theor. Phys.*, **27**, 1395
36. I. Ciufolini, (2008), *Lunar Laser Ranging, Gravitomagnetism and Frame-Dragging*, e-print http://xxx.lanl.gov/abs/0809.3219
37. J. Kovalevsky, (1967), *Introduction to Celestial Mechanics*, (Dordrecht: Reidel)
38. T. Murphy, (2008), APS Meeting Abstracts, 7002

Quasi-inertial Coordinates

Neil Ashby

Abstract We discuss the close conceptual connections between Geodetic Precession of a gyroscope, and Lense–Thirring precession (usually called Frame-Dragging). The origin of a quasi-inertial coordinate system may be allowed to fall freely or to be accelerated with non-gravitational forces, but the directions of the coordinate axes are maintained fixed with respect to the background metric, which may consist of contributions from several moving or rotating masses. Transformation of the space–time metric into quasi-inertial coordinates provides an instructive way of viewing the precession of a gyroscope placed at the origin. We shall show that in quasi-inertial coordinates Geodetic precession and Frame-Dragging can be viewed equivalently as arising from either orbital or spin angular momentum of the mass sources. The Geodetic and Frame-Dragging precessions, and Thomas precession as well, are gravitomagnetic effects that are related to spatial curvature only when adopting a particular point of view. We also illustrate the application of quasi-inertial coordinates by considering gyroscopic precession in the Gödel universe.

1 Introduction

The purpose of this lecture is to examine the question, to what extent are Frame Dragging and Geodetic Precession physically equivalent? Many examples of such equivalence can be found in relativity. The Shapiro time delay, and the deflection of light rays passing near a massive body, are equivalent in a similar sense: although experimental procedures for measurement of the two effects differ, one can derive light deflection by an appropriate differentiation of the theoretical prediction for time delay with respect to the impact parameter. One may further observe that in leading order, both of these phenomena are proportional to the quantity $(\gamma + 1)$,

N. Ashby (✉)
Department of Physics, University of Colorado Boulder, UCB 390, CO 80309-0390
e-mail: ashby@boulder.nist.gov

I. Ciufolini and R.A. Matzner (eds.), *General Relativity and John Archibald Wheeler*, Astrophysics and Space Science Library 367, DOI 10.1007/978-90-481-3735-0_16,

where γ is the PPN parameter that appears as a coefficient in the leading relativistic contribution to the spatial part of the metric. The two phenomena are so intertwined that if the experiments should disagree with each other, metric theories of gravity would have to be entirely reworked.

Another example can be found in special relativity: consider an unstable particle passing at high speed down a tube. If the particle had the same lifetime as it has when it is at rest, the tube could be made sufficiently long that the particle could be expected to decay while inside the tube. However the moving particle suffers slowing of its proper time (time dilation), so that at very high speed the decay could occur after completing its passage down the tube. For an observer at rest in the particle's rest frame, there is no change in the particle's lifetime, but the tube suffers Lorentz contraction. Either way the decay occurs after the particle passes out of the tube, but different observers attribute the results to different physical effects. The phenomena are related in such a fundamental way in special relativity, that if the experimental results did not agree, the edifice of special relativity would have to be reconstructed on a different basis.

In the following sections we shall develop the connection between the several different kinds of gyroscopic precessions by introducing quasi-inertial coordinates. These will be defined in Section 4. All the calculations given here are approximate, but are of sufficiently high order that gyroscopic precession rates of $O(1/c^2)$ are calculated. We use upper-case Latin letters such as (X^0, X^k) to denote quantities measured in the underlying reference frame in which a solution to the field equations is assumed. Repeated Greek indices run from 0 to 3; repeated Latin indices run from 1 to 3. The background metric tensor for the solar system will be assumed to be of the form:

$$\begin{aligned} G_{00} &= -\left(1 + \frac{2\Phi}{c^2}\right) \\ G_{0i} &= -2(\gamma + 1)\sum_A \frac{GM_A \mathbf{V}^i_A}{c^3|\mathbf{X} - \mathbf{X}_A|} \\ G_{ij} &= \left(1 - \frac{2\gamma\Phi}{c^2}\right), \end{aligned} \tag{1}$$

where Φ is the Newtonian gravitational potential arising from the mass sources M_A at positions $\mathbf{X}_A$:

$$\Phi(X_0, \mathbf{X}) = -\sum_A \frac{GM_A}{|\mathbf{X} - \mathbf{X}_A|}. \tag{2}$$

Here, the index A runs over all the mass sources. Note that the space–time components of the metric are of order c^{-3}, while the other components are only given to order c^{-2}. This is consistent to order c^{-3} since the contributions to G_{00} and G_{ij} are of even order in c^{-1}. Contributions of order c^{-3} to G_{0i} are needed to calculate spin precession rates to order c^{-2}, as will be seen. We include the factors of c^{-1} explicitly since that is helpful in performing various expansions.

In the following section we obtain solutions of the Fermi–Walker transport equations for a tetrad that is accelerated along an arbitrary path in the fields given by Eq. (1).

2 Fermi–Walker Transport of a Tetrad

For description of changes in direction of the spin axis of an accelerated gyroscope, we use Fermi–Walker transport of the spin. Let us construct a tetrad of four 4-vectors, denoted by

$$\Lambda^{\mu}_{(k)} \tag{3}$$

in such a way that the spatial members of the tetrad are in the directions of three mutually orthogonal accelerating spins that are very close together. Here the subscripts (k) label the four 4-vectors. For convenience we let the labels range from 0 to 3 but these are not tensor indices. If the spins were freely falling, then they could be transported parallel to themselves along a geodesic, but here we are interested in the more general case in which a non-gravitational acceleration may be applied. (Although spins do not fall exactly along geodesics, the difference is negligible [12]). The equation of Fermi–Walker transport is

$$\frac{D\Lambda^{\mu}_{(i)}}{Ds} = \left(G_{\lambda\nu}\Lambda^{\lambda}_{(i)}\frac{DU^{\nu}}{Ds}\right)U^{\mu} - \left(G_{\lambda\nu}\Lambda^{\lambda}_{(i)}U^{\nu}\right)\frac{DU^{\mu}}{Ds}, \tag{4}$$

where U^{μ} is the four-velocity along the accelerated path and the operator D/Ds represents the covariant derivative:

$$\frac{DV^{\mu}}{Ds} = \frac{dV^{\mu}}{ds} + \Gamma^{\mu}_{\alpha\beta}V^{\alpha}U^{\beta}. \tag{5}$$

The Christoffel Symbol of the second kind is:

$$\Gamma^{\mu}_{\alpha\beta} = \frac{1}{2}G^{\alpha\nu}\left(G_{\nu\alpha,\beta} + G_{\nu\beta,\alpha} - G_{\alpha\beta,\nu}\right), \tag{6}$$

and a comma represents partial differentiation with respect to the indicated variable.

Let the tangent to the path along which the spins are carried be described by the unitless four-vector

$$U^{\mu} = \frac{dX^{\mu}}{ds}. \tag{7}$$

We let the "time" member of the tetrad be tangent to the path by choosing:

$$\Lambda^{\mu}_{(0)} = U^{\mu}. \tag{8}$$

We also impose orthonormality conditions on the tetrad:

$$G_{\mu\nu}\Lambda^{\mu}_{(\alpha)}\Lambda^{\nu}_{(\beta)} = \eta_{\alpha\beta} \tag{9}$$

where $\eta_{\alpha\beta}$ is the Minkowski metric,

$$\eta_{\alpha\beta} = \text{diag}(-1,1,1,1). \tag{10}$$

We assume that the four-velocity U^{μ} and the path are given, and we seek solutions of the Fermi–Walker Transport equations subject to the orthogonality conditions (9). The time-components of the spatial members of the tetrad can be found from the orthogonality between the four-velocity and the other three tetrad members. The result is:

$$\Lambda^{0}_{(i)} = G_{0i}|_0 + \frac{V^k}{c}\left(1 - \frac{2(\gamma+1)\Phi|_0}{c^2}\right)\Lambda^{l}_{(i)}\delta_{kl}. \tag{11}$$

All quantities on the right side of this expression are evaluated on the trajectory of the gyroscope, indicated by the notation "$|_0$". The calculation is carried to order c^{-3}, and the ds in the definition of the four-velocity has been replaced by its expression in terms of the metric:

$$ds = \sqrt{-G_{\mu\nu}dX^{\mu}dX^{\nu}}. \tag{12}$$

We seek a solution for the spatial members of the tetrad. To do this we propose a solution and then check it by substituting into the transport equation. The proposed solution is of the form [1]

$$\Lambda^{k}_{(i)} = \delta^{k}_{i}\left(1 + \frac{\gamma\Phi|_0}{c^2}\right) + \frac{V^iV^k}{2c^2} + \Omega^{ik}. \tag{13}$$

The usual correspondence between positions of the Latin indices on both sides of this equation has been violated – but on the left i is not a tensor index. Here the term multiplying δ^{k}_{i} represents a change in scale arising from the external potential. The velocity terms represent Lorentz contraction, and the 3×3 antisymmetric matrix Ω^{ik} represents a rotation of the spatial members of the tetrad, which is the precession we are ultimately interested in. The solution is thus separated into a symmetric part and an antisymmetric part describing rotation.

We find the equation of motion of the rotation matrix by substitution into the transport equation (4). The second term on the right in (4) vanishes because of the orthogonality condition. Further, U^k is of order c^{-1}, DU^k/Ds is of order c^{-2}, and DU^0/Ds is of order c^{-3}, so the transport equation reduces to

$$\frac{D\Lambda^{k}_{(i)}}{Ds} = \frac{DU^i}{Ds}U^k. \tag{14}$$

Computation of all the contributions to (14) is tedious but straightforward. The calculation is taken to third order. The Christoffel symbols are needed using the given background metric, and evaluated along the path. We shall not give all the details here. The path derivative of Φ evaluated along the path is needed. This is

$$\frac{d\Phi|_0}{ds} = \Phi|_{0,0} + \Phi|_{0,k}u^k; \tag{15}$$

There is considerable simplification and the Fermi–Walker transport equation becomes:

$$\begin{aligned}\frac{d\Omega^{ki}}{ds} &= \frac{1}{2}\left(G_{0i,k} - G_{0k,i}\right) + \frac{\gamma+1}{c^3}\left(V^k\Phi|_{0,i} - V^i\Phi|_{0,k}\right)\\ &\quad + \frac{1}{2c^3}\left(V^k\frac{dV^k}{dT} - V^i\frac{dV^k}{dT}\right),\end{aligned} \tag{16}$$

where all quantities on the right side are evaluated along the path and where dV^i/dT is the total acceleration, gravitational as well as non-gravitational. With this understanding, we shall drop notations such as "$|_0$."

A necessary and sufficient condition that the tetrad be transported along the path by Fermi–Walker transport is that the rotation matrix satisfy this equation. The rate of change of the antisymmetric rotation matrix divides neatly into three contributions. The first of these is usually attributed to frame-dragging. The second contribution is usually connected with geodetic precession, and the third with Thomas precession. The frame-dragging contributions arise from gravitomagnetic components of the metric tensor, G_{0i} and have a coefficient $(\gamma + 1)$. The geodetic precession terms carry the same coefficient, $(\gamma + 1)$, but in this case the γ comes from G_{ij} and the 1 comes from G_{00}. Sometimes this leads to an assertion that part of the geodetic precession is due to spatial curvature. We shall argue that this is because of the adoption of a particular point of view, that of observers in the reference frame in which the metric (1) is specified.

The Thomas precession terms, involving dV^i/dT, can be combined with the geodetic precession contributions if the non-gravitational part of the acceleration vanishes. Then since these terms are already of order c^{-3}, we can substitute from the leading contributions to the geodesic equations of free fall:

$$\frac{dV^i}{dT} = -\Phi_{,i}. \tag{17}$$

The Thomas precession contribution then has a contribution of the same form as the geodetic contribution, except the coefficient is $-1/2$. The combined geodetic and Thomas contributions are then

$$\frac{\gamma + \frac{1}{2}}{c^3}\left(V^k\Phi_{,i} - V^i\Phi_{,k}\right). \tag{18}$$

This form for the sum of Thomas and geodetic precessions is confusing, which is perhaps why in the literature one can find the assertion that a freely falling gyroscope does not suffer Thomas Precession [13]. Of course it would be incorrect to add Thomas Precession to the "geodetic" precession when it already carries the coefficient $(\gamma + \frac{1}{2})$.

3 Fixed and Orbiting Gyroscopes

To illustrate the connection between frame-dragging and geodetic precession, consider first a gyroscope held fixed, having neither velocity nor acceleration in the given reference frame. Only the frame-dragging term will contribute and we will have

$$\frac{d\Omega^{ki}}{ds} = \frac{1}{2}(G_{0i,k} - G_{0k,i}). \tag{19}$$

The rotation rate of the matrix Ω^{ki} can easily be converted into an angular precession rate for any one of the spinning gyroscopes used to define the directions of the tetrad. First we note that terms on the right side of (19) are already of order c^{-3}, so ds can be replaced by cdT. Then, introducing the three-index alternating symbol ϵ_{ijk}, which is ± 1 or 0 depending on whether the indices are even/odd permutations of (123), and zero if any two indices are equal, the angular velocity of precession will be

$$\Omega_i = -\frac{c}{2}\epsilon_{ijk}\frac{d\Omega^{jk}}{ds} = \frac{c}{2}\sum_{jk}\epsilon_{ijk}G_{0j,k}. \tag{20}$$

This shows why third-order calculations are needed in order to end up with second-order contributions to the precession.

If we suppose that the source is rotating about its center at $\mathbf{X}_0$ with angular velocity $\boldsymbol{\omega}$, then

$$G_{0i} = -2\frac{(\gamma + 1)}{c^3}\sum_A \frac{GM_A\,[\boldsymbol{\omega} \times (\mathbf{X}_A - \mathbf{X}_0)]_i}{|\mathbf{X} - \mathbf{X}_A|} \tag{21}$$

and then it is straightforward to show that the spin precession rate is

$$\boldsymbol{\Omega} = \frac{\gamma + 1}{c^2}\sum_A \frac{[(\mathbf{X}_A - \mathbf{X}) \times (\boldsymbol{\omega} \times M_A(\mathbf{X}_A - \mathbf{X}_0))]}{|\mathbf{X} - \mathbf{X}_A|^3}. \tag{22}$$

The quantity in square brackets in the numerator of (22) is the angular momentum of the mass element M_A taken relative to the point of observation; this is weighted by the inverse cube of the distance between source and field point.

Rotating Ring It is instructive to evaluate this expression for simple objects as it helps to understand the close connection between frame-dragging and geodetic precession. For example, let the mass source be a rotating ring of mass M and radius a with origin of coordinates at the center. Let the gyroscope be placed at the center of the ring. The resulting precession rate is

$$\boldsymbol{\Omega} = \frac{(\gamma + 1)G\mathbf{J}}{c^2 a^3}. \tag{23}$$

Thus the precession rate is proportional to the angular momentum of the rotating ring and falls off as the cube of the ring radius.

Orbiting Gyroscope Ignoring the Thomas Precession terms, let us consider the geodetic precession terms of (16):

$$\frac{d\Omega^{ki}}{ds} = \frac{\gamma + 1}{c^3}\left(V^k \Phi_{,k} - V^i \Phi_{,i}\right) \tag{24}$$

which gives the spin precession rate

$$\Omega_i = -\frac{c}{2}\epsilon_{ijk}\frac{d\Omega^{jk}}{ds} \tag{25}$$

or

$$\boldsymbol{\Omega} = -\frac{\gamma + 1}{c^2}\mathbf{V} \times \nabla\Phi. \tag{26}$$

If the orbit is circular and of radius X, and the source at the origin of coordinates has mass M, the potential is $-GM/a$ and the precession rate is

$$\boldsymbol{\Omega} = \frac{\gamma + 1}{c^2}\mathbf{V} \times \left(-\frac{GM\mathbf{X}}{X^3}\right). \tag{27}$$

This can be written

$$\boldsymbol{\Omega} = \frac{(\gamma + 1)G}{c^2 X^3}\mathbf{X} \times (M\mathbf{V}). \tag{28}$$

But in terms of the orbital angular momentum $\mathbf{L}$ of the mass source relative to the gyroscope, this is just

$$\boldsymbol{\Omega} = \frac{(\gamma + 1)G\mathbf{L}}{c^2 X^3}. \tag{29}$$

On replacing X by a and comparing this result to (23), we see that whether the gyroscope is surrounded by a spinning ring of angular momentum $\mathbf{J}$ or whether relative to a freely falling gyroscope the mass source has angular momentum $\mathbf{J} = \mathbf{L}$, the spin precession rates would be equal if it were not for the Thomas Precession. This illustrates the close connection between geodetic and Lense–Thirring precession.

Orbiting Gyroscope Outside a Rotating Spherical Shell For a gyroscope fixed outside a rotating spherical shell, evaluation of Eq. (22) yields the well-known result [6]

$$\boldsymbol{\Omega} = \frac{(\gamma + 1)G}{2c^2 X^3}\left[-\mathbf{J} + 3\mathbf{N}(\mathbf{N} \cdot \mathbf{J})\right], \tag{30}$$

where **N** is a unit vector pointing from the shell's center to the gyroscope and X is the distance from the shell's center to the gyroscope. When the observation point is on the rotation axis, or in the equatorial plane, the spin rate is proportional to the shell's angular momentum. Adding up contributions for shells of a rotating sphere, the answer is similar. For a gyroscope in a polar orbit around a rotating sphere, as in GP-B, the time-averaged precession rate is

$$\mathbf{\Omega} = \frac{(\gamma + 1)G}{4c^2 X^3}\mathbf{J}. \tag{31}$$

4 Construction of Quasi-inertial Coordinates

We now introduce coordinates in which the close connection between the types of precession may be further explored. In particular we will show that, in these new coordinates – so-called "quasi-inertial" coordinates, all precessions are gravitomagnetic in origin, arising from the time–space components of the metric.

We construct coordinates whose origin coincides with a gyroscope that may either be freely falling or accelerated, but with coordinate axis directions that maintain their directions in the given background metric. That is, we do not allow the tetrad to have a rotation Ω^{ik}. The construction is otherwise the same as that for normal Fermi coordinates [2]. Referring to Fig. 1, the tetrad is propagated along the path G, which would be a geodesic if the gyroscope were freely falling. The coordinate time variable in the new coordinates is the proper time of a standard clock falling along

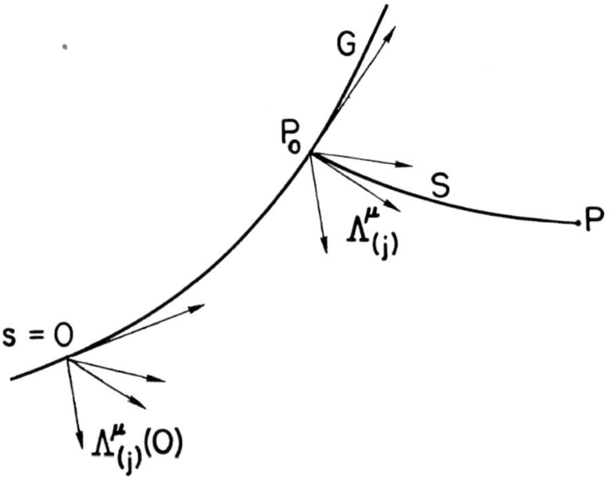

Fig. 1 Construction of quasi-inertial coordinates. G is a timelike base path that may be a geodesic. The spacelike geodesic S from the field point P intersects G orthogonally at P_0. The spatial direction of S at P_0 is characterized by direction cosines l^i relative to the spatial members of the tetrad

with the gyroscope. The coordinates of a point P not too far from G are determined as follows. A space-like geodesic S is dropped from P to a point P_0 on G in such a way that it intersects G orthogonally at P_0. The unit tangent vector to S at P_0 has direction cosines α^i relative to the spatial members of the tetrad

$$\frac{dX^\mu}{d\lambda} = t^\mu = \alpha^i \Lambda^\mu_{(i)} \tag{32}$$

and if λ is the proper distance from P_0 to P, the spatial coordinates of P are taken to be

$$x^i = \lambda \alpha^i, \tag{33}$$

so that

$$\lambda \frac{dX^\mu}{d\lambda} = \lambda \alpha^k \Lambda^\mu_{(k)} = x^k \Lambda^\mu_{(k)}. \tag{34}$$

The time coordinate of points along S is taken to be x^0, the proper time of the standard clock at P_0. Then the coordinate transformation from X^μ to quasi-inertial coordinates may be obtained from a Taylor expansion. We carry this expansion only to quadratic order in local spatial coordinates, which is needed in order to compute the metric tensor to linear order in these coordinates. Thus

$$X^\mu(P) = X^\mu(P_0) + \lambda \frac{dX^\mu}{d\lambda} + \frac{1}{2}\lambda^2 \frac{d^2 X^\mu}{d\lambda^2} + \cdots. \tag{35}$$

The first term in Eq. (35) is obtained in terms of x^k from (34). The second term (and higher terms as well) is expressed in terms of Christoffel symbols of the second kind, from the equation of the spacelike geodesic. This yields the expansion:

$$X^\mu(P) = X^\mu(P_0) + x^k \Lambda^\mu_{(k)} - \frac{1}{2}\Gamma^\mu_{\alpha\beta}\Lambda^\alpha_{(k)}\Lambda^\beta_{(l)} x^k x^l + \cdots, \tag{36}$$

where all coefficients on the right are evaluated at P_0 on G.

In the time transformation, the quantity $X^0(P_0)$ can be expressed in terms of the proper time coordinate x^0 by using the metric:

$$X^0(P_0) = \int^{P_0} \frac{dX^0}{ds} dx^0 = \int^{P_0} K ds, \tag{37}$$

where K is obtained from the metric, evaluated along G:

$$K^{-2} = \frac{ds^2}{(dX^0)^2} = -\left(G_{00} + 2G_{0i}\frac{dX^i}{dX^0} + G_{ij}\frac{dX^i}{dX^0}\frac{dX^j}{dX^0} \right). \tag{38}$$

which may be evaluated along G to the desired order.

The construction of quasi-inertial coordinates proceeds as follows. First we develop appropriate expressions for the tetrad which is propagated along the base path G with acceleration, but *without rotation*. Then we construct the coordinate transformations using (36). Next the usual tensor transformations are applied to compute the metric tensor in quasi-inertial coordinates. This requires computation of transformation coefficients, derivatives of (36). The calculations are algebraically formidable but straightforward; one must be careful at each stage to include all contributions to the appropriate order.

The result will be the metric tensor in a reference frame in which the gyroscope is at rest. The description of the spin precession of the gyroscope is just that of parallel transport of the spin vector in such a frame. Furthermore, since the gyroscope is at rest, only a limited number of terms contribute to the spin precession rate. Once one obtains the metric tensor in such a reference frame, it does not matter where the coordinate transformation came from; experimental implications of a given metric tensor are a matter of correct interpretation within the framework of the General Theory.

For local quasi-inertial coordinates that are not allowed to rotate with respect to reference directions for the background metric, spin precessions must arise from torques contributed by Christoffel symbols of the second kind, evaluated at the position of the gyroscope, that is, at $x^k = 0$. In the case of normal Fermi coordinates, all Christoffel symbols vanish at the origin. They do not vanish at the origin of quasi-inertial coordinates because the basis tetrad is not allowed to rotate. As will be seen, all gyroscopic precessions arise in this picture from gravitomagnetic forces, from space–time components g_{0i} of the quasi-inertial metric tensor.

4.1 The Basis Tetrad

An accelerated tetrad has already been constructed using Fermi–Walker transport; we may take for the tetrad for quasi-inertial coordinates the same expressions but leaving out the rotation term Ω^{ik}:

$$\Lambda^k_{(i)} = \delta^k_i \left(1 + \frac{\gamma\Phi}{c^2}\right) + \frac{V^i V^k}{2c^2}. \tag{39}$$

$$\Lambda^0_{(i)} = G_{0i} + \frac{V^i}{c}\left[1 - \frac{(\gamma+2)\Phi}{c^2} + \frac{V^2}{2c^2}\right]. \tag{40}$$

and finally, for the four-velocity which determines the zero-component of the tetrad,

$$\Lambda^\mu_{(0)} = \frac{dX^\mu}{ds} = \left(K, K\frac{V^k}{c}\right) \tag{41}$$

4.2 Coordinate Transformations

We begin the construction of the coordinate transformations by evaluating some intermediate sums. We have to order c^{-2}

$$\Lambda^i_{(j)}x^j = x^i\left(1+\frac{\gamma\Phi}{c^2}\right)+\frac{V^i\mathbf{V}\cdot\mathbf{r}}{2c^2}, \tag{42}$$

and to order c^{-3}

$$\Lambda^0_{(j)}x^j = G_{0j}x^j + \frac{\mathbf{V}\cdot\mathbf{r}}{c}\left[1-\frac{\Phi(\gamma+2)}{c^2}+\frac{V^2}{2c^2}\right]. \tag{43}$$

Henceforth it will be understood that all such quantities are evaluated on the path G.

Time Transformation The second-order terms that contribute to the time transformation are

$$-\frac{1}{2}\left[\Gamma^0_{00}\Lambda^0_{(k)}\Lambda^0_{(l)} + 2\Gamma^0_{0j}\Lambda^0_{(k)}\Lambda^j_{(l)} + \Gamma^0_{ij}\Lambda^i_{(k)}\Lambda^j_{(l)}\right]x^k x^l. \tag{44}$$

In evaluating these terms to order c^{-3} we find the first term is negligible, the second and third terms give respectively

$$\left[-\frac{1}{c^3}x^j\Phi_{,j}\mathbf{V}\cdot\mathbf{r}\right]+\frac{1}{2}\left[G_{0i,j}x^i x^j + \frac{\gamma r^2\Phi_{,0}}{c^2}\right]. \tag{45}$$

where $r^2 = \delta_{ij}x^i x^j$. The time transformation is thus

$$\begin{aligned} X^0 = \int^{P_0} K dx^0 + \frac{\mathbf{V}\cdot\mathbf{r}}{c}\left[1-\frac{\Phi(\gamma+2)}{c^2}+\frac{V^2}{2c^2}\right] + G_{0j}x^j \\ -\frac{1}{c^3}x^j\Phi_{,j}\mathbf{V}\cdot\mathbf{r} + \frac{1}{2}G_{0i,j}x^i x^j + \frac{\gamma r^2\Phi_{,0}}{2c^2}. \end{aligned} \tag{46}$$

Spatial transformation. We need the terms

$$-\frac{1}{2}\left[\Gamma^k_{00}\Lambda^0_{(k)}\Lambda^0_{(l)} + 2\Gamma^k_{0j}\Lambda^0_{(k)}\Lambda^j_{(l)} + \Gamma^k_{ij}\Lambda^i_{(k)}\Lambda^j_{(l)}\right]x^k x^l. \tag{47}$$

Only the last term contributes the amount

$$-\frac{1}{2}\Gamma^k_{ij}\Lambda^i_{(k)}\Lambda^j_{(l)} = \frac{\gamma x^k x^j\Phi_{,j}}{c^2} - \frac{\gamma}{2}\frac{r^2\Phi_{,k}}{c^2}. \tag{48}$$

The spatial transformation, to quadratic order in x^k and second order in c^{-1}, is therefore

$$X^k = X^k(P_0) + x^k\left(1 + \frac{\gamma\Phi}{c^2} + \frac{\gamma x^j\Phi_{,j}}{c^2}\right) + \frac{V^k\mathbf{V}\cdot\mathbf{r}}{2c^2} - \frac{\gamma r^2\Phi_{,k}}{2c^2}. \tag{49}$$

The coefficients appearing in these coordinate transformations are naturally specified as functions of time X^0 in the frame of the background metric, on the path G where $x^k = 0$. They can be considered functions of x^0 if one uses the time transformation (46) to express X^0 in terms of x^0 at P_0.

Having the coordinate transformations to quadratic order, one may compute transformation coefficients and hence metric tensor components correct to linear order in x^k. This is all that is necessary in order to describe gyroscopic precession in the quasi-inertial frame.

4.3 Metric in the Quasi-inertial Frame

We begin by computing the metric tensor component g_{00} in quasi-inertial coordinates. The calculation is needed to order c^{-2}. The only terms in the tensor transformation that contribute are

$$g_{00} = \frac{\partial X^0}{\partial x^0}\frac{\partial X^0}{\partial x^0}G_{00} + \frac{\partial X^k}{\partial x^0}\frac{\partial X^l}{\partial x^0}G_{kl}, \tag{50}$$

so the needed transformation coefficients are

$$\frac{\partial X^0}{\partial x^0} = 1 - \frac{\Phi}{c^2} + \frac{V^2}{2c^2} + \frac{\mathbf{A}\cdot\mathbf{r}}{c^2}, \tag{51}$$

$$\frac{\partial X^l}{\partial x^0} = \left(1 - \frac{\Phi}{c^2} + \frac{V^2}{2c^2}\right)\frac{V^l}{c} + \frac{\gamma\Phi_{,0}x^l}{c^2} + \frac{A^l\mathbf{V}\cdot\mathbf{r}}{2c^3} + \frac{V^l\mathbf{A}\cdot\mathbf{r}}{2c^3}. \tag{52}$$

Applying the coordinate transformation, the metric tensor components must be evaluated at the observation point P, but only to linear order in the coordinates x^k. This means an expansion for P near P_0 is needed. For example to order c^{-2},

$$G_{00} = -\left[1 + \frac{2(\Phi + \Phi_{,k}(X^k - X^k))}{c^2}\right] = -\left[1 + \frac{2(\Phi + \Phi_{,k}x^k)}{c^2}\right]. \tag{53}$$

Here a time derivative introduces contributions of order one higher, that are negligible. After considerable cancellation of various contributions, the result is:

$$g_{00} = -1 - \sum_k \frac{2(A^k + \Phi_{,k})x^k}{c^2}. \tag{54}$$

This result has a very simple interpretation. If the origin is in free fall, then $\mathbf{A} = -\nabla\Phi$, and there are no terms linear in the local quasi-inertial coordinates. Thus the principle of equivalence in satisfied. If there is an additional non-gravitational acceleration, then there is an induced gravitational potential near the origin, and an effective gravitational force per unit mass $-\mathbf{A}$ acting on the gyroscope there. Such a uniform force cannot cause any gyroscopic precession.

Next we compute the spatial components of the metric tensor. The calculation is to order c^{-2}, and again to linear order in x^k. The only terms that contribute are

$$g_{ij} = \frac{\partial X^0}{\partial x^i}\frac{\partial X^0}{\partial x^j}G_{00} + \frac{\partial X^k}{\partial x^i}\frac{\partial X^l}{\partial x^j}G_{kl}. \tag{55}$$

The new transformation coefficients that are needed are rather complicated. They are

$$\begin{aligned}\frac{\partial X^0}{\partial x^i} &= G_{0i} + \frac{V^i}{c}\left[1 - \frac{\Phi(\gamma+2)}{c^2} + \frac{V^2}{2c^2}\right] - \frac{V^i x^j \Phi_{,j}}{c^3}\\ &\quad - \frac{\mathbf{V}\cdot\mathbf{r}\Phi_{,i}}{c^3} + \frac{1}{2}\left(G_{0i,j} + G_{0j,i}\right)x^j + \frac{\gamma x^i \Phi_{,0}}{c^2},\end{aligned} \tag{56}$$

and

$$\frac{\partial X^k}{\partial x^i} = \delta^k_i\left(1 + \frac{\gamma\Phi}{c^2} + \frac{\gamma\Phi_{,j}x^j}{c^2}\right) + \frac{V^k V^i}{2c^2} + \frac{\gamma x^k \Phi_{,i}}{c^2} - \frac{\gamma x^i \Phi_{,k}}{c^2}. \tag{57}$$

Carrying out the necessary expansions and approximations to obtain results correct to order c^{-2}, we find that *all linear terms cancel*, so that

$$g_{ij} = \delta_{ij} + O(x^i x^j). \tag{58}$$

Thus, there can be no "spatial curvature" contributions to spin precessions in this reference frame; they must come from the only remaining components of the metric tensor that have yet to be evaluated, g_{0i}. This is a surprising feature of quasi-inertial coordinates, and happens because no rotation has been allowed in the construction of coordinates.

In computing g_{0i} we find there are three contributions:

$$g_{0i} = \frac{\partial X^0}{\partial x^0}\frac{\partial X^0}{\partial x^i}G_{00} + \frac{\partial X^0}{\partial x^0}\frac{\partial X^l}{\partial x^i}G_{0l} + \frac{\partial X^k}{\partial x^0}\frac{\partial X^l}{\partial x^i}G_{kl}. \tag{59}$$

We already have all the transformation coefficients that are needed. Here we must find g_{0i} to order c^{-3}. The first term contributes

$$\begin{aligned}&-\frac{V^i}{c}\left[1 - \frac{(\gamma+1)\Phi}{c^2} + \frac{V^2}{c^2} + \frac{\mathbf{A}\cdot\mathbf{r}}{c^2} + \frac{\Phi_{,m}x^m}{c^2}\right]\\ &-G_{0i} - \frac{1}{2}(G_{0i,j} + G_{0j,i})x^j + \frac{\mathbf{V}\cdot\mathbf{r}\Phi_{,i}}{c^3} - \frac{\gamma x^i \Phi_{,0}}{c^2}.\end{aligned} \tag{60}$$

The second term contributes just the series expansion

$$G_{0i} + G_{0i,j}x^j; \tag{61}$$

and the third term contributes

$$+\frac{V^i}{c}\left[1 - \frac{\gamma\Phi}{c^2} - \frac{\gamma\Phi_{,m}x^m}{c^2} - \frac{\Phi}{c^2} + \frac{V^2}{2c^2}\right] + \frac{\gamma\Phi_{,0}x^i}{c^2} + \frac{\mathbf{V}\cdot\mathbf{r}A^i}{2c^3}$$
$$+\frac{\mathbf{A}\cdot\mathbf{r}V^i}{2c^3} + \frac{\gamma\mathbf{V}\cdot\mathbf{r}\Phi_{,i}}{c^3} + \frac{V^2V^i}{2c^2} - \frac{\gamma x^i V^m\Phi_{,m}}{c^2}. \tag{62}$$

There is considerable cancellation and simplification when all contributions are combined. We give the result for the space–time component:

$$g_{0i} = \frac{1}{2}(G_{0i,j} - G_{0j,i})x^j + \frac{\gamma+1}{c^3}(\mathbf{V}\cdot\mathbf{r}\Phi_{,i} - V^i\Phi_{,m}x^m)$$
$$+\frac{1}{2c^3}(\mathbf{V}\cdot\mathbf{r}A^i - \mathbf{A}\cdot\mathbf{r}V^i) - \frac{\gamma x^i V^m\Phi_{,m}}{c^3}. \tag{63}$$

Thus the contributions group naturally into frame-dragging, geodetic and Thomas type contributions along with one additional contribution of as yet unknown significance. In the next section we discuss the implications of this metric. Again we emphasize that both frame-dragging and geodetic precession are gravitomagnetic in this coordinate system – that is, they arise from the time–space components of the metric.

5 Spin Precession in the Quasi-inertial Frame

Consider a gyroscope placed at rest at the origin of the quasi-inertial frame. Since the directions of the axes are non-rotating with respect to the background metric, quasi-inertial coordinates provide a platform for deriving various spin precessions. This is particularly simple since the gyroscope is placed at the origin, so its four-velocity has only the component $u^0 = 1$. Fermi–Walker transport of this spin, which remains at the origin, reduces to parallel transport. Thus,

$$\frac{dS^\mu}{ds} + \Gamma^\mu_{\alpha 0}S^\alpha = 0. \tag{64}$$

Further, for a spin at rest the Pirani condition [10] $S^\mu u_\mu = 0$ implies $S^0 = 0$. Therefore the equation further simplifies to

$$\frac{dS^k}{ds} + \Gamma^k_{i0}S^i. \tag{65}$$

The Christoffel symbol may readily be calculated from the metric tensor which has been obtained. Only terms linear in x^k in g_{0i} will contribute. In this case, because the frame axes are constrained not to rotate, there will exist Christoffel symbols that do not vanish at the origin. We easily find that at $x^k = 0$,

$$\Gamma^k_{i0} = \frac{1}{2}\left(g_{0k,i} - g_{0i,k}\right). \tag{66}$$

It is easy to see that the last term in (63) cannot contribute to any spin precession. Thus we have the three types of spin precessions: Lense–Thirring, Geodetic, and Thomas, all arising from g_{0i}. To complete the calculation we obtain

$$\Gamma^k_{i0} = \frac{1}{2c^3}(G_{0k,i} - G_{0i,k}) + \frac{\gamma + \frac{1}{2}}{c^3}\left(\Phi_{,k}V^i - \Phi_{,i}V^k\right) + \frac{1}{2c^3}\left(A^k_{NG}V^i - A^i_{NG}V^k\right), \tag{67}$$

where $\mathbf{A}_{NG}$ is the non-gravitational part of the total acceleration.

This gives essentially the same precession rate obtained before, in (16), but now described in a reference frame with spatial axes that are locally Minkowskian and point in fixed directions. It is to be expected that the precession angles would be the same, since the quasi-inertial axes are parallel to the axes of the background metric. However in the quasi-inertial frame a number of issues regarding measurement have been accounted for by the transformation. The time variable is the proper time on a standard clock at the origin, and distances and angles are measured using standard lengths in a system of mutually perpendicular coordinate axes. In the next section we shall discuss the measurement of the angle between a spin axis and light from a distant "fixed" star. The quantities that appear (upper-case letters) in (67) are evaluated in the frame of the background metric, but are to be considered functions of time x^0 in the quasi-inertial frame.

For completeness, we list expressions for the angular precession rate of a gyroscope, which follow from (67). The non-gravitational part of the Thomas precession is

$$\mathbf{\Omega}_T = \frac{1}{2c^2}\mathbf{A}_{NG} \times \mathbf{V}; \tag{68}$$

The geodetic precession term, including the gravitational part of the Thomas Precession, can be written

$$\mathbf{\Omega}_G = \frac{\gamma + \frac{1}{2}}{c^2}(\nabla\Phi) \times \mathbf{V}. \tag{69}$$

And the frame-dragging term can be written compactly if one thinks of the three components of G_{0i} as three components of a vector $\mathbf{G}$:

$$\mathbf{\Omega}_{LT} = -\frac{c}{2}\nabla \times \mathbf{G}. \tag{70}$$

Thus the total precession rate is the sum:

$$\mathbf{\Omega} = \mathbf{\Omega}_{LT} + \mathbf{\Omega}_G + \mathbf{\Omega}_T. \tag{71}$$

Finally, for a test particle in orbit about a spherical spinning body with angular momentum **J**, the rate of precession of the line of nodes of the orbit can be shown to be [8]

$$\dot{\Omega} = \frac{(\gamma + 1)GJ}{c^2 a^3 (1 - e^2)^{3/2}}. \tag{72}$$

where a and e are the semi-major axis and eccentricity of the orbit. The altitude of perigee also precesses but we shall not discuss that.

In expressions such as those given above for $\boldsymbol{\Omega}_G$, it is important to note that $\mathbf{V}$ and Φ are the total velocity and total gravitational potential, respectively. This means that when more than one mass source is present, such as in the case of sun and earth together in the model, there will be cross-terms that are possibly significant. We shall discuss this in a later section.

6 Aberration

Quasi-inertial coordinates provide a useful way of separating certain kinematic effects from spin effects. Suppose that one uses light from a distant guide star, or "fixed" star, to measure the precession angles of an orbiting gyroscope, as in Gravity Probe B. The direction cosines of the null tangent vector impinging on a detector at the spacecraft can be computed in the background metric and transformed to quasi-inertial coordinates with the aid of the coordinate transformations developed above. In the background metric we take the direction cosines of the spatial part of this null vector to be L^i and we shall assume these remain constant. (If the null vector passes near a massive source such as the sun it will suffer a deflection that can be calculated.) The tangent vector to the null geodesic at the spacecraft will be

$$N^\mu = (N^0, L^i). \tag{73}$$

We impose the normalization condition $L^i L^j \delta_{ij} = 1$. Then for N^μ to be a null vector, to order c^{-3} the time-component must be

$$N^0 = 1 - \frac{(\gamma + 1)\Phi}{c^2} + G_{0i} L^i \tag{74}$$

We transform this vector to quasi-inertial coordinates by means of:

$$n^\mu = \frac{\partial x^\mu}{\partial X^\nu} N^\nu. \tag{75}$$

Thus, the inverse of the matrix of transformation coefficients is needed; fortunately we only need them at the origin of coordinates where $x^i = 0$, at the position where the null vector impinges on the gyroscope. Calculation of these coefficients is straightforward and the results are:

$$\begin{aligned}
\frac{\partial x^0}{\partial X^0} &= 1 + \frac{\Phi}{c^2} + \frac{V^2}{2c^2} \\
\frac{\partial x^0}{\partial X^i} &= -\frac{V^i}{c}\left[1 - \frac{(2\gamma+1)\Phi}{c^2} + \frac{V^2}{2c^2}\right] - G_{0i} \\
\frac{\partial x^i}{\partial X^0} &= -\frac{V^i}{c}\left(1 - \frac{\gamma\Phi}{c^2} + \frac{V^2}{2c^2}\right) \\
\frac{\partial x^i}{\partial X^j} &= \delta^i_j\left(1 - \frac{\gamma\Phi}{c^2}\right) + \frac{V^i V^j}{2c^2}.
\end{aligned} \tag{76}$$

The tangent vector in quasi-inertial coordinates is thus

$$n^0 = 1 - \frac{\mathbf{L}\cdot\mathbf{V}}{c}\left[1 - \frac{(2\gamma+1)\Phi}{c^2} + \frac{V^2}{2c^2}\right] - \frac{\gamma\Phi}{c^2} + \frac{V^2}{2c^2}; \tag{77}$$

$$n^k = L^k\left(1 - \frac{\gamma\Phi}{c^2}\right) - \frac{V^k}{c}\left[1 - \frac{(2\gamma+1)\Phi}{c^2} + \frac{V^2}{2c^2} - \frac{\mathbf{l}\cdot\mathbf{V}}{2c}\right]. \tag{78}$$

The principal correction is that due to aberration of light, represented by the term $-V^k/c$ in (78).

The scalar product of a spin vector **s** with the null tangent vector can be calculated covariantly. The spin is at rest in this reference frame, so the Pirani condition [10] implies $s^0 = 0$. Then

$$n_\mu s^\mu = n_k s^k = \mathbf{n}\cdot\mathbf{s}, \tag{79}$$

because the spatial part of the metric is δ_{ij} in these coordinates. Suppose that the spin vector has rotated through the vector angle $\boldsymbol{\theta}$. Then

$$\mathbf{s} = \mathbf{s}_0 + \boldsymbol{\theta}\times s_0 \tag{80}$$

so

$$\mathbf{n}\cdot\mathbf{s} = \mathbf{L}\cdot\mathbf{s}_0\left(1 - \frac{\gamma\Phi}{c^2}\right) - \frac{\mathbf{V}\cdot\mathbf{s}_0}{c}\left(1 - \frac{\mathbf{L}\cdot\mathbf{V}}{2c}\right) + \mathbf{L}\cdot\boldsymbol{\theta}\times\mathbf{s}_0 \tag{81}$$

to order c^{-2}, since $\boldsymbol{\theta}$ may be considered to be of order c^{-2}. The first two terms in this equation have nothing to do with spin precession and must be subtracted out in order to obtain the quantities of interest experimentally. The various corrections in this equation represent a number of effects which must be accounted for in order to measure the net rotation. These include time dilation, Lorentz contraction, relativity of clock synchronization, rescaling of lengths due to external gravitational potentials, and aberration.

7 Two Sources – Sun and Earth

In this section we further explore the relationships between geodetic and frame-dragging precessions by considering a simplified model of a gyroscope in free fall about the earth, which in turn orbits the sun. For this model we take the gravitational potential to be

$$\Phi = -\frac{GM_\odot}{X} - \frac{GM_e}{|\mathbf{X} - \mathbf{X}_e|}, \tag{82}$$

where $\mathbf{X}$ is from the (non-moving) sun to the point of observation, and $\mathbf{X}_e$ is from the sun to earth's center. The gravitomagnetic terms in the background metric include contributions from the overall motion of earth, plus rotation of earth:

$$G_{0i} = -\frac{2(\gamma+1)GM_e V_e^i}{c^3|\mathbf{X} - \mathbf{X}_e|} + \frac{(\gamma+1)G}{c^3|\mathbf{X} - \mathbf{X}_e|^3}\left[(\mathbf{X} - \mathbf{X}_e) \times \mathbf{J}\right]_i \tag{83}$$

where $\mathbf{J}$ is earth's angular momentum. For this metric, the general result (16) applies. In particular, after combining the Thomas precession term with the geodetic term, the net geodetic part of the precession is

$$\frac{d\Omega^{ki}}{ds} = \frac{\gamma + \frac{1}{2}}{c^3}\left(V^k\Phi_{,k} - V^i\Phi_{,k}\right) \tag{84}$$

and this corresponds to the angular precession rate given in (69). The total velocity $\mathbf{V}$ is composed of the velocity of the earth relative to the sun, plus the velocity of the satellite relative to the earth, $\mathbf{V}_s$. Since the angular precession rate is already of order c^{-2}, relativistic corrections to the velocity composition law are negligible, and so

$$\mathbf{V} = \mathbf{V}_e + \mathbf{V}_s. \tag{85}$$

Substituting this into (69), there arise four terms:

$$\begin{aligned}\mathbf{\Omega}_G = \frac{\left(\gamma + \frac{1}{2}\right)}{c^2}&\left[(\nabla\Phi_\odot) \times \mathbf{V}_e + (\nabla\Phi_\odot) \times \mathbf{V}_s\right.\\ &\left.+(\nabla\Phi_e) \times \mathbf{V}_e + (\nabla\Phi_e) \times \mathbf{V}_s\right]\end{aligned} \tag{86}$$

The angular velocity of precession for the first term in (86) is normal to the plane of the ecliptic, and is of magnitude 19.2 milliarcseconds (mas) year^{-1}. This precession affects any gyroscope near the earth, including the moon's orbit as the moon revolves about the earth, the LAGEOS satellites and Gravity-Probe-B. This effect has been observed with ever-increasing accuracy by lunar laser ranging [3, 11, 15, 16]; the experimental result now agrees with the general relativity prediction (with $\gamma = 1$) to within a fraction of a percent. Note that this term does not depend on the position or velocity of the gyroscope relative to the earth; it is a precession induced by the motion of earth and its nearby satellites relative to the sun. The last term in (86) is the geodetic precession of an orbiting gyroscope by virtue of its orbital motion about the earth; in the case of GP-B this term is expected to be 6.6 arcsec year^{-1}. In the second term, the velocity $\mathbf{V}_s$ will be rapidly varying and the term should average down to a very small value. Similarly in the third term of (86) the quantity $(\nabla\Phi_e)$ will be rapidly varying and therefore this contribution should average down to a very small value. We shall therefore not consider these cross-terms further.

However, it is clear that every gyroscope in the neighborhood of earth, as well as the orbits in which they revolve around earth, will experience geodetic precession of 19.2 mas year^{-1} (turning in the ecliptic plane), superimposed on other precession effects.

Next consider frame-dragging of an orbiting gyroscope due to earth's angular momentum. The time-averaged precession rate for a gyroscope in a circular polar orbit, Eq. (31) is

$$\boldsymbol{\Omega}_{LT} = \frac{(\gamma + 1)G\mathbf{J}}{4c^2X^3} \tag{87}$$

where X is the orbit radius. For an altitude of 400 km, this is of magnitude 45 mas year^{-1}, while the geodetic precession due to motion about the earth is 6.6 sec year^{-1} at a right angle to earth's angular momentum. Superimposed on these precessions is the geodetic precession of 19.2 mas year^{-1} about an axis normal to the ecliptic plane. This latter contribution was discussed by Will [14] but is not usually mentioned.

Orbits of the LAGEOS and LAGEOS II satellites have been accurately tracked with laser ranging for many years. The orbits of these two satellites are inclined at 110° and 52.64°, respectively, to earth's equatorial plane, and the frame-dragging precessions of their nodal lines should be 31 mas year^{-1} and 31.5 mas year^{-1}, respectively. Earth's quadrupole moment also causes the nodal lines to precess, but because the inclination of LAGEOS I is greater than 90°, the quadrupolar precessions are in opposite directions. Until recently, uncertainty in the knowledge of earth's quadrupole moment coefficient J_2 was enough to completely mask the frame-dragging effect. Use of the nodal precessions of only two satellites was proposed by Van Patten and Everitt [15] and Ciufolini [4] and discussed by Ries and co-workers [12] and Iorio and co-workers [8]. A particular linear combination of nodal precession rates that completely cancels the contribution due to J_2 is:

$$\dot{\Omega}^{\text{LAGEOS}} + 0.546\dot{\Omega}^{\text{LAGEOS II}} = \frac{\gamma+1}{2}48.2 \text{ mas year}^{-1}. \tag{88}$$

This linear combination was used by Ciufolini, Pavlis and Peron [4] to measure frame-dragging due to the earth's angular momentum with a claimed accuracy of 10%.

We note again that using the background metric (1) each of the LAGEOS satellites will experience geodetic precession of 19.2 mas year^{-1} about an axis normal to the ecliptic plane. The projection of this precession rate on the rotation axis of earth is $19.2 \cos e$ year^{-1} = 17.6 mas year^{-1}. Such an effect would contribute 27.2 mas year^{-1} to the linear combination of (88) and this must be removed from the observations in order to measure frame-dragging. A similar comment applies to the gyroscopes of GP-B: they experience geodetic precession due to earth's orbiting the sun, which is of sufficient magnitude that it must be considered in analysing the observations.

One way to clarify the situation with respect to geodetic precession contributions to observation of near-earth gyroscopic or orbital precessions is to introduce a local inertial frame, centered at the earth, as in (13) which includes rotation of gyroscopes arising from geodetic precession as the earth orbits the sun. The construction is similar to the construction of normal Fermi coordinates.

8 Local Inertial Frame of Earth

The 19.2 mas year^{-1} geodetic precession discussed above can be clearly separated from other precession effects by constructing a local *inertial* frame with center at earth's center, falling freely with earth, and hence with a basis tetrad propagated by parallel transport along the orbital path of earth. These axes will rotate due to geodetic precession and hence the geodetic precession – such as that observed with lunar laser ranging – will be unobservable in the reference frame based on this tetrad.

In constructing this local inertial frame, the metric (1) must be "effaced" – that is, terms arising from earth's mass or rotation must be removed first, as they would be singular at the position of earth's center [1]. The coordinate transformation thus constructed is then applied to the *entire* metric. The construction proceeds along the lines discussed in Section 2. The effaced metric includes only the solar potential:

$$G_{00} = -\left(1 + \frac{2\Phi_\odot}{c^2}\right);$$
$$G_{ij} = \delta_{ij}\left(1 + \frac{2\gamma\Phi_\odot}{c^2}\right). \tag{89}$$

We let the time-member of the tetrad be determined by the four-velocity of earth's center:

$$\Lambda^{\mu}_{(0)} = (K_e, K_e V_e^k), \tag{90}$$

where to order $1/c^2$

$$K_e = 1 - \frac{\Phi_\odot}{c^2} + \frac{V_e^2}{2c^2}, \tag{91}$$

evaluated at earth's center. The other components of the tetrad are:

$$\Lambda^{0}_{(i)} = \frac{V_e^i}{c}\left[1 - \frac{(\gamma+2)\Phi_\odot}{c^2} + \frac{V_e^2}{2c^2}\right] + \frac{V_e^k}{c}\delta_{kl}\Omega^{il};$$
$$\Lambda^{k}_{(i)} = \delta_i^k\left(1 + \frac{\gamma\Phi_\odot}{c^2}\right) + \frac{V_e^i V_e^k}{2c^2} + \Omega^{ik}. \tag{92}$$

The equation of parallel transport then gives for the rotation matrix

$$\frac{d\Omega^{ki}}{ds} = \frac{\gamma + \frac{1}{2}}{c^3}\left(V^k \Phi_{\odot,k} - V^i \Phi_{\odot,k}\right) \tag{93}$$

which is basically only the combination of geodetic and Thomas precessions, but in the absence of earth–that is, as though the earth were a test particle.

Coordinate Transformation The time coordinate in the local frame is taken to be that on a standard clock falling along at earth's center, but with the earth absent. The construction of coordinates follows the same general lines as discussed in Section 4, as in (36), except that here the axes are allowed to rotate. The time transformation is

$$X^0 = \int^{P_0} K dx^0 + \frac{\mathbf{V}\cdot\mathbf{r}}{c}\left[1 - \frac{\Phi_\odot(\gamma+2)}{c^2} + \frac{V^2}{2c^2}\right]$$
$$-\frac{1}{c^3} x^j \Phi_{\odot,j}\mathbf{V}\cdot\mathbf{r} + \frac{V_e^k}{c}\delta_{kl}\Omega^{li} x_i. \tag{94}$$

For the spatial part of the transformation, we find

$$X^k = X^k(P_0) + x^k\left(1 + \frac{\gamma\Phi_\odot}{c^2} + \frac{\gamma x^j \Phi_{\odot,j}}{c^2}\right) + \frac{V^k \mathbf{V}\cdot\mathbf{r}}{2c^2} - \frac{\gamma r^2 \Phi_{,k}}{2c^2} + \Omega^{ki} x_i. \tag{95}$$

Transformation Coefficients All transformation coefficients needed to construct the metric tensor to linear order in x^i are summarized here. These are simply obtained by partial differentiation of the coordinate transformations. These are

$$\frac{\partial X^0}{\partial x^0} = 1 - \frac{\Phi_\odot}{c^2} + \frac{V_e^2}{2c^2} + \frac{\mathbf{A}_e\cdot\mathbf{r}}{c^2}; \tag{96}$$

$$\frac{\partial X^l}{\partial x^0} = \left(1 - \frac{\Phi_\odot}{c^2} + \frac{V_e^2}{2c^2}\right)\frac{V_e^l}{c} + \frac{A_e^l \mathbf{V}_e\cdot\mathbf{r}}{2c^3} + \frac{V_e^l \mathbf{A}_e\cdot\mathbf{r}}{2c^3} + \frac{d\Omega^{li}}{ds} x_i; \tag{97}$$

$$\frac{\partial X^0}{\partial x^i} = \frac{V^i}{c}\left[1 - \frac{\Phi_\odot(\gamma+2)}{c^2} + \frac{V_e^2}{2c^2}\right] - \frac{V_e^i x^j \Phi_{\odot,j}}{c^3} - \frac{\mathbf{V}_e\cdot\mathbf{r}\Phi_{\odot,i}}{c^2}; \tag{98}$$

and

$$\frac{\partial X^k}{\partial x^i} = \delta_i^k\left(1 + \frac{\gamma\Phi_\odot}{c^2} + \frac{\gamma\Phi_{\odot,j} x^j}{c^2}\right) + \frac{V_e^k V_e^i}{2c^2} + \frac{\gamma x^k \Phi_{\odot,i}}{c^2} - \frac{\gamma x^i \Phi_{\odot,k}}{c^2}. \tag{99}$$

Metric Tensor The coordinate transformation is now applied to the *entire* metric, including gravitomagnetic contributions from motion and rotation of earth. The full metric is:

$$
\begin{aligned}
G_{00} &= -\left(1 + \frac{2\Phi_\odot}{c^2} + \frac{2\Phi_e}{c^2}\right) \\
G_{0i} &= -2(\gamma + 1)\frac{GM_e V_e^i}{|\mathbf{X} - \mathbf{X}_e|} - 2(\gamma + 1)\sum_A \frac{GM_A V_A^i}{c^3|\mathbf{X} - \mathbf{X}_A|} \\
G_{ij} &= \left(1 - \frac{2\gamma\Phi_\odot}{c^2} - \frac{2\gamma\Phi_e}{c^2}\right),
\end{aligned}
\tag{100}
$$

The second equation in (100) is a sum over all particles in the earth; $\mathbf{V}_A$ is the velocity of particles in the earth relative to earth's center. This last term gives rise to frame-dragging.

The metric tensor in the local frame may now be found by straightforward tensor transformation. Starting with the time–time component, we find after many cancellations

$$g_{00} = -\left[1 + \frac{2\Phi_e}{c^2} + \frac{2(A_{e,m} + \Phi_{\odot,m})x^m}{c^2}\right] \tag{101}$$

and thus because the earth is in free fall in the sun's gravitational field, only the earth's gravitational potential itself survives:

$$g_{00} = -\left(1 + \frac{2\Phi_e}{c^2}\right) + O(x^i x^j) \tag{102}$$

It is remarkable how this construction assembles the pieces from here and there – from various components of the metric, and parts of the transformation coefficients – to cancel out the contribution from the sun's field, and its gradient. This continues with the construction of the spatial part of the local metric:

$$g_{ij} = \delta_{ij}\left(1 + \frac{2\gamma\Phi_e}{c^2}\right). \tag{103}$$

Lastly, in the gravitomagnetic part of the metric tensor, about two dozen terms conspire to completely cancel out the geodetic contributions, leaving only

$$g_{0i} = -2(\gamma + 1)\sum_A \frac{GM_A V_A^i}{c^3|\mathbf{X} - \mathbf{X}_A|} - \frac{\gamma x^i \mathbf{V}\cdot\nabla\Phi_\odot}{c^3}. \tag{104}$$

The first term is the frame-dragging contribution from earth's rotation. The first term in (100) has cancelled out. Thus in this reference frame, one may safely omit consideration of geodetic precession due to the sun. But this frame is steadily rotating with respect to the "fixed" stars.

9 Quasi-inertial Coordinates in the Gödel Universe

Construction of quasi-inertial coordinates within the Gödel universe is an instructive application of these coordinates. In this section we apply this construction to a gyroscope at rest at the origin of coordinates. The metric tensor used is slightly modified with respect to the one introduced by Gödel [5] for the purpose of more easily exhibiting the precession. The metric we use is

$$\begin{aligned} G_{00} &= -1; \quad G_{11} = G_{33} = 1; \\ G_{22} &= -e^{2X^1/a}; \\ G_{02} &= G_{20} = -\sqrt{2}e^{X^1/a}. \end{aligned} \tag{105}$$

Here a is the characteristic length scale of the universe. For uniform density ρ,

$$a = \frac{c}{\sqrt{8\pi G\rho}} \tag{106}$$

and the cosmological constant is

$$\lambda = -\frac{4\pi G\rho}{c^2}. \tag{107}$$

The four-velocity of the matter, which is at rest in this universe, has a single contravariant component,

$$U^0 = c. \tag{108}$$

and its contravariant components, obtained by lowering the index, are

$$U_\mu = \left\{-U^0, 0, -\sqrt{2}U^0 e^{X^1/a}, 0\right\}. \tag{109}$$

The field equations satisfied by this metric are of the form

$$R_{\mu\nu} - \lambda G_{\mu\nu} - \frac{1}{2}G_{\mu\nu}R = \frac{8\pi G\rho}{c^4}U_\mu U_\nu. \tag{110}$$

This form of the equations for the Gödel universe differs from that published in [5] by exhibiting the dependence of the exponentials on the scale factor a (so all the coordinates have units of length), and by rescaling the coordinate X^2 by a factor $\sqrt{2}$ for convenience.

Now we imagine a gyroscope placed at rest at the origin of coordinates, and let it fall freely thereafter. The nonzero Christoffel symbols of the second kind are:

$$\begin{aligned} \Gamma^0_{01} &= \Gamma^0_{10} = \frac{1}{a}; \\ \Gamma^0_{12} &= \Gamma^0_{21} = \Gamma^1_{20} = \Gamma^1_{02} = \frac{e^{X^1/a}}{\sqrt{2}a}; \end{aligned}$$

$$\Gamma^1_{22} = \frac{e^{X^1/a}}{a}; \tag{111}$$

$$\Gamma^2_{01} = \Gamma^2_{10} = -\frac{e^{-X^1/a}}{\sqrt{2}a} \tag{112}$$

For a point gyroscope initially at rest at the origin, it is easy to see from the geodesic equations of motion that the gyroscope will remain at rest. To construct a non-accelerated orthonormal tetrad at this point, that "falls" along the trajectory of the gyroscope, we need the equations of parallel propagation:

$$\frac{d\Lambda^\mu_{(i)}}{ds} + \Gamma^\mu_{\alpha\beta}\Lambda^\alpha_{(i)}\frac{dx^\beta}{ds} = 0. \tag{113}$$

When the point P_0 of interest is at rest, these equations reduce to

$$\frac{d\Lambda^0_{(i)}}{ds} + \frac{1}{a}\Lambda^1_{(i)}\frac{dX^0}{ds} = 0;$$

$$\frac{d\Lambda^1_{(i)}}{ds} + \frac{1}{\sqrt{2}a}\Lambda^2_{(i)}\frac{dX^0}{ds} = 0; \tag{114}$$

$$\frac{d\Lambda^2_{(i)}}{ds} - \frac{1}{\sqrt{2}a}\Lambda^1_{(i)}\frac{dX^0}{ds} = 0; \tag{115}$$

$$\frac{d\Lambda^3_{(i)}}{ds} = 0. \tag{116}$$

The differential equations can thus easily rewritten in terms of X^0 as the independent variable. With the scaling chosen in the metric used here, it is then obvious that the one and two components of each of the members of the tetrad rotate uniformly with angular frequency $\omega = c/(\sqrt{2}a)$. The tetrad can be chosen to satisfy the orthogonality relations, Eq. (9) so that writing the components in columns,

$$\Lambda^\mu_{(i)} = \begin{pmatrix} 1 & -\sqrt{2}\sin(\omega t) & -\sqrt{2}\cos(\omega t) & 0 \\ 0 & \cos(\omega t) & -\sin(\omega t) & 0 \\ 0 & \sin(\omega t) & \cos(\omega t) & 0 \\ 0 & 0 & 0 & 1 \end{pmatrix} \tag{117}$$

Quasi-inertial Coordinates In this case there is no obvious small rotation matrix Ω^{ij}. However the rotation may be removed by simply setting

$$\cos(\omega t) \to 1; \quad \sin(\omega t) \to 0. \tag{118}$$

The tetrad for quasi-inertial coordinates is then:

$$\Lambda^{\mu}_{(i)} = \begin{pmatrix} 1 & 0 & -\sqrt{2} & 0 \\ 0 & 1 & 0 & 0 \\ 0 & 0 & 1 & 0 \\ 0 & 0 & 0 & 1 \end{pmatrix} \tag{119}$$

We then need the coordinate transformations to quadratic order. These are easy to construct using the prescription given in Eq. (19), and are:

$$\begin{aligned} X^0 &= x^0 - \sqrt{2}x^2 + \frac{x^1 x^2}{\sqrt{2}a} \\ X^1 &= x^1 + \frac{(x^2)^2}{2a} \\ X^2 &= x^2 - \frac{x^1 x^2}{a} \qquad (120) \\ X^3 &= x^3. \qquad (121) \end{aligned}$$

Computing the transformation coefficients and then transforming the metric tensor to the local quasi-inertial frame is a straightforward exercise that gives

$$g_{\mu\nu} = \begin{pmatrix} -1 & x^2/(\sqrt{2}a) & -x^1/(\sqrt{2}a) & 0 \\ x^2/(\sqrt{2}a) & 1 & 0 & 0 \\ -x^1/(\sqrt{2}a) & 0 & 1 & 0 \\ 0 & 0 & 0 & 1 \end{pmatrix}. \tag{122}$$

In the local non-rotating frame there are space–time cross terms that are gravitomagnetic in nature; the angular precession frequency of a free gyroscope is then easily obtained from Eq. (19) and it is the same as the rotation frequency of the tetrads implied by the harmonically varying functions in Eq. (117). The result is

$$\omega = \frac{c}{\sqrt{2}a}. \tag{123}$$

These results show that at the origin there exists a locally inertial frame, in which the metric is almost Minkowskian but with small contributions that are linear in local coordinates in the space–time terms. These quasi-inertial coordinates were constructed by removing the rotation of a tetrad that was propagated parallel to itself (Eq. 117). This new metric in the local frame contain terms that cause a gyroscope at rest to precess with angular velocity ω. The physical results are the same independent of the coordinate system in which they are viewed. The angular velocities are identical since at the origin of coordinates where the gyroscope is at rest, the time transformation is $X^0 = x^0 = s$.

Thus in the Gödel universe there exists a cosmological effect causing precession of a gyroscope at rest even though the matter is "at rest" in the universe. It is easy to verify, by rescaling certain coordinates and following the above procedures, that gyroscopes placed at rest anywhere in the universe will precess at the same rate.

Other stationary rotating cosmologies, motivated by cosmic string considerations, have been discussed by Patel and Dadhich [9]. These cosmological metrics do not necessarily cause gyroscopic precession.

10 Summary

In this talk we have shown how to construct a quasi-inertial reference frame, whose origin falls along with some object of interest – such as a gyroscope or earth's center – but whose axes do not rotate with respect to the "fixed" stars. This construction is helpful in separating relativistic effects such as Lorentz contraction, relativity of simultaneity, and length rescaling, from the important gyroscopic precession effects. One can use the same construction with minor modifications to construct a local inertial frame, falling along with earth's center, to separate geodetic precession due to earth's orbital motion, from frame dragging; this separation is relevant for studies of the LAGEOS satellites and for GP-B. The computation of precession rate requires that the metric tensor in quasi-inertial coordinates be calculated to linear order in local coordinates (see Eq. 19). Keeping terms only to this order, it is not possible to compute the Riemann–Christoffel curvature tensor, the Ricci Tensor, or the Einstein tensor because they depend on second derivatives of the metric tensor.

References

1. N. Ashby, B. Bertotti, Phys. Rev. **D34**, 2246 (1986).
2. N. Ashby, B. Shahid-Saless, Phys. Rev. D42, 1118–1122, 1990.
3. B. Bertotti, I. Ciufolini, P. Bender, Phys. Rev. Letts. **58**, 1062 (1987).
4. I. Ciufolini, E. Pavlis, Nature **431**, 958, (2004); I. Ciufolini I, E.C. Pavlis and R. Peron, New Astronomy, 11, 527–550 (2006).
5. K. Gödel, Revs. Mod. Phys. **21**, 447 (1949).
6. J. Hartle, *Gravity: an Introduction to Einstein's Theory*, (Addison Wesley, San Francisco, 2003), p. 307.
7. L. Iorio, A. Morea, Gen. Rel. Grav. **36**, 1321 (2004).
8. J. Lense, H. Thirring, Phys. Z. **19**, 156 (1918).
9. L. K. Patel and Naresh Dadhich, Pramana **47**(5) 387–392 (1996); arXiv:gr-qc/9507030v1, 14 July 1995.
10. F. A. E. Pirani, Acta. Physica Polonica **15**, 389 (1956)
11. I. I. Shapiro, R. Reasenberg, J. F. Chandler, R. W. Babcock, Phys. Rev. Letts **61**, 2643 (1988).
12. H. Stephani, *Introduction to General Relativity*, Second ed., (Cambridge University Press 1990), pp 93–95.
13. S. Vokos, arXiv:hep-ph/9304260 15 April 1993.
14. C. M. Will, Phys. Rev. **D67**, 062003-1, (2003).
15. J. G. Williams, X. X. Newhall, J. O. Dickey, Phys. Rec **D33**, 6730 (1996).
16. J. J. Williams, Boggs, J. O. Dickey, Folkner, Proc. 9th Marcel Grossman Meeting (World Scientific, Singapore, 2002).

Gravitomagnetism and Its Measurement with Laser Ranging to the LAGEOS Satellites and GRACE Earth Gravity Models*

Ignazio Ciufolini, Erricos C. Pavlis, John Ries, Rolf Koenig, Giampiero Sindoni, Antonio Paolozzi, and Hans Newmayer

Abstract Dragging of Inertial Frames and gravitomagnetism are predictions of Einstein's theory of General Relativity. Here, after a brief introduction to these phenomena of Einstein's gravitational theory, we describe the method we have used to measure the Earth's gravitomagnetic field using the satellites LAGEOS (LAser GEOdynamics Satellite), LAGEOS 2 and the Earth's gravity models obtained by the spacecraft GRACE. We then report the results of our analysis with LAGEOS and LAGEOS 2, and with a number of GRACE (Gravity Recovery and Climate Experiment) models, that have confirmed this prediction of Einstein General Relativity and measured the Earth's gravitomagnetic field with an accuracy of approximately 10%. We finally discuss the error sources in our measurement of gravitomagnetism and, in particular, the error induced by the uncertainties in the GRACE Earth gravity models. Here we both analyze the errors due to the static and time-varying Earth

---

*Dedicated to John Archibald Wheeler, one of the masters of physics of the XX century and father of the renaissance of General Relativity.

I. Ciufolini (✉)
Dip. Ingegneria dell'Innovazione, Università del Salento and INFN Sezione di Lecce, via Monteroni, 73100 Lecce, Italy
e-mail: ignazio.ciufolini@unisalento.it

E.C. Pavlis
Joint Center for Earth Systems Technology, University of Maryland, Baltimore County, 1000 Hilltop Circle, Baltimore, Maryland 21250, USA
e-mail: epavlis@umbc.edu

J. Ries
Center for Space Research, University of Texas at Austin, Austin, Texas 78759, USA
e-mail: ries@csr.utexas.edu

R. Koenig and H. Newmayer
GFZ German Research Centre for Geosciences, Potsdam, Germany
e-mail: rolf.koenig@gfz-potsdam.de; neumayer@gfz-potsdam.de

G. Sindoni and A. Paolozzi
Scuola di Ingegneria Aerospaziale, Sapienza Università di Roma, via Salaria 851/881, 00138 Roma, Italy
e-mail: giampiero.sindoni@uniroma1.it; antonio.paolozzi@uniroma1.it

I. Ciufolini and R.A. Matzner (eds.), *General Relativity and John Archibald Wheeler*, Astrophysics and Space Science Library 367, DOI 10.1007/978-90-481-3735-0_17,

gravity field, and in particular we discuss the accuracy of the GRACE-only gravity models used in our measurement. We also provide a detailed analysis of the errors due to atmospheric refraction mis-modelling and to the uncertainties in measuring the orbital inclination. In the appendix, we report the complete error analysis and the total error budget in the measurement of gravitomagnetism with the LAGEOS satellites.

1 Dragging of Inertial Frames and Gravitomagnetism

During the past century the theory of General Relativity [1–3] gave rise to an experimental triumph [3–5]. On the one hand, a number of key predictions of Einstein's gravitational theory have been experimentally confirmed with impressive accuracy [5]. On the other hand, Einstein gravitational theory today has practical applications in space research [3], geodesy, astronomy and navigation in the Solar System, from the global positioning satellite system to the techniques of very long baseline interferometry (VLBI) and satellite laser ranging, and is a fundamental component for the understanding of some astrophysical and cosmological observations.

Despite all the observational tests passed by General Relativity, we have recently discovered supernovae that appear to accelerate away from us [6,7]. This can be explained by dark energy that is accelerating the expansion of the Universe, and may imply a non-zero value to Einstein's cosmological constant, or an exotic new form of energy known as quintessence [8,9], or perhaps a modification of the theory of General Relativity. Together with dark matter, approximately 96% of the mass–energy of the Universe should be of a form unknown to us. A relativistic theory of gravity, combining the theory of General Relativity at some level with quantum mechanics, may help to reveal what dark energy is and whether it might be related to dark matter. Therefore, new effects are being sought on very small length scales [10], on very large scales [11] and even in the Solar System [12] for what they might reveal about theories such as quantum gravity, string and brane-world models of the Universe. Furthermore, even though a breakdown of the general theory of relativity should occur at the quantum level, some viable modifications of Einstein's theory already give different predictions at the classical, non quantum, level [4]. Therefore, every aspect of the General Relativity should be directly tested and the accuracy of the present measurements of Einstein's theory and of the foundations of gravitational theories should be further improved.

Today, the biggest observational efforts in gravitational physics are for direct detection of gravitational waves, although we already know from the rate of change of the orbital period of the binary pulsar PSR B1913116 that this system emits gravitational waves at the predicted level [5], and for the accurate measurement of gravitomagnetism and frame-dragging [14]. In this paper we describe the confirmation of frame-dragging by spin with an accuracy of the order of 10% using the LAGEOS satellites and the GRACE gravity models [13, 15, 129, 135], elsewhere in this book we report on the LARES (LAser RElativity Satellite) space experiment to be launched in 2011 for a 1% test of gravitomagnetism [16] and the Gravity Probe B (GP-B) team reports on GP-B, launched in 2004, that, by further data analysis, might also detect gravitomagnetism. In this section, let us briefly describe the phenomena of frame-dragging and gravitomagnetism [14].

The origin of inertia has intrigued scientists and philosophers for centuries. Inertial frames of reference permeate our daily life. The inertial and centrifugal forces, such as the pull and push that we feel when our vehicle accelerates, brakes and turns, arise because of changes in velocity relative to uniformly moving inertial frames. A classical interpretation ascribed these forces to acceleration relative to some absolute frame independent of the cosmological matter, whereas an opposite view related them to acceleration relative to all the masses and fixed stars in the Universe. An echo and partial realization of the latter idea can be found in Albert Einstein's theory of General Relativity, which predicts that a spinning mass drags inertial frames.

Inertial frames are systems in which any isolated body, not subjected to any interaction, is at rest or moving with constant velocity. However, what determines an inertial frame? And, in general, what is the origin of inertia? In the mechanics of Galileo Galilei and Isaac Newton an inertial frame has an absolute existence, uninfluenced by the matter in the Universe. In Einstein's theory of General Relativity [1–3], the inertial and centrifugal forces are due to our accelerations and rotations with respect to the so-called local inertial frames which, in turn, are determined, influenced and dragged by the distribution and flow of mass energy in the Universe. In particular, they are dragged by the motion and rotation of nearby matter; this general relativistic phenomenon is called frame-dragging ([17], and see page 544 of [3] for example) and represents in Einstein's theory the remnant of the ideas of Ernst Mach on the origin of inertia. Mach thought that centrifugal and inertial forces are due to rotations and accelerations with respect to all the masses in the Universe; this is known as Mach's principle [18–20].

Since 1896 researchers, influenced by these ideas, tried to measure the frame-dragging effects generated by the rotation of Earth on torsion balances [21] and gyroscopes [22]. In 1916, on the basis of General Relativity, de Sitter [23] derived the Mercury perihelion precession due to the Sun angular momentum and, in 1918, Lense and Thirring [24] gave a general weak-field description of the frame-dragging effect on the orbit of a test-particle around a spinning body, today known as Lense–Thirring effect.

Frame-dragging has also an intriguing influence on the flow of time around a spinning body (Fig. 1). Indeed, synchronization of clocks all around a closed path near a spinning body is not possible [25, 26] in any rigid frame not rotating relative to the fixed stars, because light corotating around a spinning body would take less time to return to a starting point (fixed relative to the distant stars) than would light rotating in the opposite direction [25–29].

In General Relativity, a torque-free spinning gyroscope defines an axis non-rotating relative to the local inertial frames, however, the orbital plane of a test particle is also a kind of gyroscope. Since frame-dragging affects clocks, light,

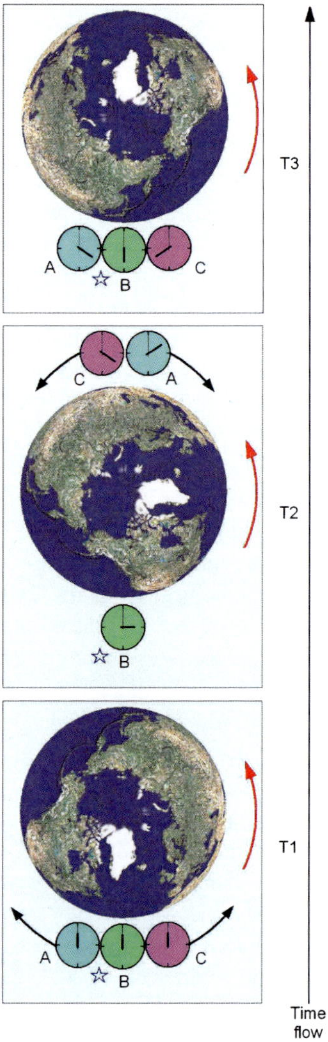

Fig. 1 Frame-dragging effects on clocks by a rotating mass. If two clocks, or twins A and C, fly all around a spinning body, even very slowly, and a third one B awaits them at the starting point, fixed relative to the distant stars (a fixed star is shown in *blue*, and T1, T2 and T3 are three consecutive instants of time), then when they meet again, the twin A that was travelling in the direction opposite to the rotation of the central body, would be younger relative to the twin B awaiting at the starting point. On the other hand twin C, travelling in the same direction of rotation of the body, would be older with respect to the standing twin B and to the twin A rotating in the opposite direction [27–29]. For this time dilation, due to the spin of the central body, to occur, they would not need to move near the speed of light (as in the case of the well-known twin-paradox of special relativity). For example, if two such twins meet again, having flown arbitrarily slowly around the whole Earth in opposite directions on the equatorial plane and exactly at the same altitude, the difference in their ages due to the Earth's spin would be approximately 10^{-16} s (for an altitude of about 6,000 km), which would be in principle detectable if not for the other, much larger, relativistic clock effects. However, frame-dragging does produce relevant effects on light and matter around a rotating black hole [30] (Earth's image by NASA and Google Earth)

gyroscopes [31, 32, 114] (see the GP-B space experiment in this book) and orbiting particles [24] (see the section on the LAGEOS satellites below and the LARES space experiment in this book), it also affects matter orbiting and falling on a spinning body. Indeed, an explanation of the constant orientation of the spectacular jets from active galactic nuclei and quasars, emitted in the same direction during a time that may reach millions of years, is based on frame-dragging of the accretion disk by a super-massive spinning black hole [33–35] acting as a gyroscope (see the cover picture of this book).

Frame-dragging may be usefully described by a formal analogy of General Relativity, in a weak gravitational field and for slow motion, with electrodynamics [3, 34] (Fig. 2).

In addition to frame-dragging due to the rotation of a mass, i.e., the Lense–Thirring effect, there is a frame-dragging effect due to the motion of a test-particle (that is a point-like particle with such a small mass that would not appreciably change the gravitational field) with respect to a static, non-rotating, mass. This effect is known as geodetic precession or de Sitter effect [36–38].

However, since the geodetic precession is a frame-dragging effect, what is the difference between gravitomagnetic effects of the type of the Lense–Thirring precession and those of the type of the geodetic precession? This topic is treated in the next section.

2 An Invariant Characterization of Gravitomagnetism

In Einstein's gravitational theory the local inertial frames have a key role [1–3]. The strong equivalence principle, at the foundations of General Relativity, states that the gravitational field is locally 'unobservable' in the freely falling frames and thus, in these local inertial frames, all the laws of physics are the laws of Special Relativity. However, the local inertial frames are determined, influenced and dragged by the distribution and flow of mass–energy in the Universe, and the axes of these non-rotating, local, inertial frames are determined by free-falling torque-free test-gyroscopes, i.e., sufficiently small and accurate spinning top, these gyroscopes are therefore dragged by the motion and rotation of nearby matter [1–3], i.e., their orientation changes with respect to the distant stars; this is the dragging of inertial frames or frame-dragging.

The precession $\vec{\Omega}_{Spin}$ of the spin axis of a test-gyroscope by the angular momentum $\vec{J}$ of the central body is: $\vec{\Omega}_{Spin} = \frac{3G((\vec{J}\cdot\hat{r})\hat{r}-\vec{J})}{c^2 r^3}$ where $\hat{r}$ is the position unit-vector of the test-gyroscope and r is its radial distance from the central body. Similarly, the orbital plane of a planet, moon or satellite is a huge gyroscope that feels general relativistic effects. The precession of the nodes of a test-particle, i.e., the rate of change of the longitude of the nodal line of the satellite, that is the intersection of its orbital plane with the equatorial plane of the central body, that is, of its orbital angular momentum vector, is known as Lense–Thirring effect and is given

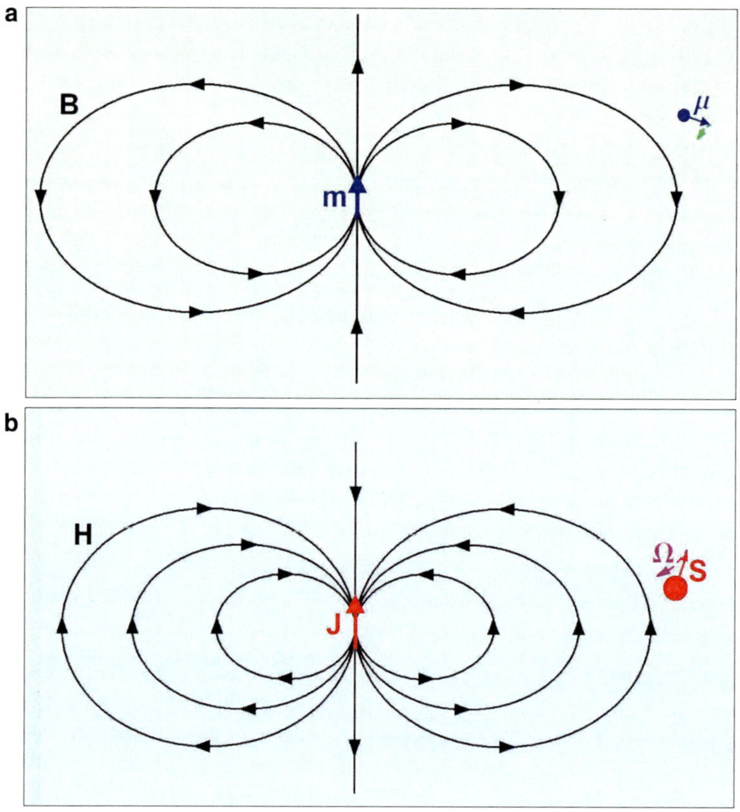

Fig. 2 Frame-dragging and the gravitomagnetic analogy of General Relativity with electrodynamics. In General Relativity, freely falling test-gyroscopes define axes *fixed* relative to the local inertial frames, where the equivalence principle holds, that is, where the gravitational field is locally unobservable; if we would rotate with respect to these gyroscope, we would then feel centrifugal forces, even though we may not rotate at all with respect to the distant stars, contrary to our everyday intuition. Indeed, a gyroscope is dragged by spinning masses, that is, its orientation changes with respect to the distant stars. Frame-dragging phenomena, which are due to mass currents and mass rotation, have been called gravitomagnetism [3, 34] because of a formal analogy of electrodynamics with the general theory of relativity (in the weak field and slow motion approximation). Whereas an electric charge generates an electric field and a current of electric charge generates a magnetic field, in newtonian gravitational theory the mass of a body generates a gravitational field but a current of mass, for example the rotation of a body, would not generate any additional gravitational field. On the other hand, Einstein's gravitational theory predicts that a current of mass would generate a gravitomagnetic field that would exert a force on surrounding bodies and would change the spacetime structure by generating additional curvature [30]. The gravitomagnetic field generates frame-dragging of a gyroscope, in a similar way to the magnetic field producing the change of the orientation of a magnetic needle (magnetic dipole). Indeed, in the general theory of relativity, a current of mass in a loop (that is, a gyroscope) has a behaviour formally similar to that of a magnetic dipole in electrodynamics, which is made of an electric current in a loop. In panel *a*, we show the magnetic field **B** generated by a magnetic dipole **m** and a test magnetic dipole $\boldsymbol{\mu}$, that is, a magnetic needle, which tends to be aligned along **B**. In panel *b* is the gravitomagnetic field [114] **H** generated by the spin **J** of a central body and frame dragging $\boldsymbol{\Omega}$ of a test gyroscope **S**

by: $\vec{\Omega}_{Lense-Thirring} = \frac{2G\vec{J}}{c^2a^3(1-e^2)^{3/2}}$, where $\vec{J}$ is the angular momentum of the central body, a the semi-major axis of the orbiting test-particle, e its orbital eccentricity, G the gravitational constant and c the speed of light. The rate of change $\vec{\omega}_{Lense-Thirring}$ of the longitude of the pericentre of a test-particle, that is, of the so-called Runge-Lenz vector, is: $\vec{\omega}_{Lense-Thirring} = \frac{2G\vec{J}}{c^2a^3(1-e^2)^{3/2}}(\hat{J} + 3 \ \cos I \ \hat{l})$ where I is the orbital inclination, that is, the angle between the orbital plane of the test-particle and the equatorial plane of the central object, $\hat{J}$ is the angular momentum unit-vector of the central body and $\hat{l}$ the orbital angular momentum unit-vector of the test-particle.

However, General Relativity predicts another type of frame-dragging effect and precession of a gyroscope: the geodetic precession or de Sitter effect [3,39]. If a gyroscope is at rest with respect to a non-rotating mass, it does not experience any drag. However, if the gyroscope starts to move with respect to the non-rotating mass it acquires a precession that will again disappear when the gyroscope will stop relative to the non-rotating mass. The geodetic precession, due to the velocity $\vec{v}$ of a test-gyroscope, is: $\vec{\Omega}_{geodetic} = \frac{3}{2}\frac{GM}{c^2r^3}\vec{x}\times\vec{v}$, where M is the mass of the central body and $\vec{x}$ and r are position vector and radial distance of the gyroscope from the central mass.

A fundamental difference between frame-dragging by spin and geodetic precession is that in the case of the former (the Lense–Thirring effect) the frame-dragging effect is due to the additional spacetime curvature produced by the rotation of a mass, whereas in the case of the latter (the de Sitter effect) no spacetime curvature ([39] and Section 6.11 of [3]) is produced by the motion of the test observer and by annulling its velocity with respect to the non-rotating central mass the effect disappears, i.e., this frame-dragging effect is due to the motion of a test-gyroscope on a static background, (see below and Section 6.11 of [3]; for a discussion on frame-dragging and geodetic precession see Refs. [37–39]).

The geodetic precession has been measured on the Moon's orbit by LLR with accuracy of the order of 0.6% [40–42], by Gravity Probe B with approximately 1% accuracy [43] and has been detected on binary pulsars [44,45].

Recently, a number of researchers have debated whether the gravitomagnetic interaction and frame-dragging by spin have also been accurately measured on the Moon orbit by Lunar Laser Ranging [39,46–48]. This is a recent chapter of a long debate on the meaning of frame-dragging and gravitomagnetism [3,37–39,46–51]; a basic issue treated in [46–48] is whether the effect detected by LLR is a frame-dependent effect or not.

In order to answer to this question, we propose here a distinction between gravitomagnetic effects generated by the translational motion of the frame of reference where they are observed, e.g., by the motion of a test-gyroscope with respect to a central mass (not necessarily rotating), and those generated by the rotation of a mass, without any necessary motion of the frame of reference where they are observed. The geodetic precession is a translational effect due to the motion of the 'Earth–Moon gyroscope' in the static field of the Sun. The Lense–Thirring effect measured by the LAGEOS satellites, that might also be detected by further Gravity Probe B data analysis (see, elsewhere in this book) and that will be accurately measured by LARES (see, elsewhere in this book, the paper on LARES), is due to the rotation of a mass, i.e., by the rotation of the Earth mass. In [39] we show that the gravitomagnetic effect discussed in [46] is just a translational effect that is conceptually similar to the Moon's geodetic precession.

Let us then further describe the illuminating formal analogy of General Relativity with electrodynamics. In General Relativity, the gravitomagnetic field, $\vec{H}$, due to the angular momentum $\vec{J}$ of a central body is, in the weak-field and slow-motion approximation:

$$\vec{H} = \vec{\nabla} \times \vec{h} \cong 2\,G\left[\frac{\vec{J} - 3(\vec{J}\cdot\hat{x})\,\hat{x}}{c^3 r^3}\right] \tag{1}$$

where r is the radial distance from the central body, $\hat{x}$ is the position unit-vector and $\vec{h}$ is the so-called 'gravitomagnetic vector potential' (equal to the non-diagonal, space and time, part of the metric), see Fig. 2.

In electromagnetism, in a frame where a test-particle with electric charge is at rest we only observe an electric field E^k but no magnetic field, however, in a frame that is moving relative to the charge we also measure a magnetic field B^k. In General Relativity, in a similar way, in a frame where a non-rotating mass is at rest, the components of the gravitomagnetic vector potential h^k are zero. Nevertheless, if we consider an observer moving relative to the mass, in a local frame moving with the observer, the components of the gravitomagnetic vector potential h^k are non-zero but can of course be annulled by a Lorentz transformation back to the original frame. Indeed, in a frame where a non-rotating mass M is at rest, the only components of the Schwarzschild metric $g_{\alpha\beta}$ different from zero (written in standard Schwarzschild coordinates) are: $g_{00} = -g_{rr}^{-1} = -\left(1 - \frac{2GM}{c^2 r}\right)$, $g_{\theta\theta} = r^2$ and $g_{\phi\phi} = r^2 \sin^2\theta$ and the three non-diagonal components of the metric g_{0k}, i.e., the components of the 'gravitomagnetic vector potential' h^k, are zero. Nevertheless, if we perform a local Lorentz transformation with velocity v^k relative to the mass M, the components of the gravitomagnetic vector potential g_{0k} are non-zero in the new frame. The orbital effects of this gravitomagnetic vector potential, arising from motion of the Earth–Moon system relative to the Sun mass, have been observed by LLR since the first measurements of the geodetic precession of the Moon orbit, i.e., of the Earth–Moon 'gyroscope' moving around the Sun. On the other hand, the angular momentum $\vec{J}$ of a body generates a gravitomagnetic field and produces spacetime curvature that cannot be eliminated by a simple change of frame of reference or by a coordinate transformation. This gravitomagnetic field generates the Lense–Thirring effect on the orbit of the LAGEOS satellites and affected the orientation of the GP-B gyroscopes.

In order to distinguish between 'intrinsic' gravitomagnetic effects (the Lense–Thirring effect) and 'translational' ones (the geodetic precession), we have proposed to use spacetime curvature invariants. In [39] we have shown that the phenomenon discussed in [46, 48] is a translational gravitomagnetic effect. In general, one cannot derive intrinsic gravitomagnetic effects from translational ones unless making additional theoretical hypotheses, such as the linear superposition of the translational gravitomagnetic effects; for example, the magnetic field generated by the intrinsic magnetic moment (Bohr magneton) is an intrinsic phenomenon due to the intrinsic spin of a particle that cannot be explained and derived as a translational effect by any Lorentz and frame transformation.

In electromagnetism, in order to characterize the electromagnetic field, using the electromagnetic field Lorentz-tensor $F_{\alpha\beta}$ we can build the scalar Lorentz-invariant $^*\mathbf{F}\cdot\mathbf{F} \equiv \frac{1}{4}F_{\alpha\beta}\ {}^*F^{\alpha\beta} = \vec{E}\cdot\vec{B}$, where $^*F^{\alpha\beta}$ is the dual of $F^{\alpha\beta}$, defined as: $^*F^{\alpha\beta} = \frac{1}{2}\varepsilon^{\alpha\beta\mu\nu}F_{\mu\nu}$ and $\varepsilon^{\alpha\beta\mu\nu}$ is the Levi–Civita pseudotensor (that is equal to $+\sqrt{-g}$, i.e., plus the square root of minus the determinant, g, of the metric, if the indices are even permutations of (0,1,2,3), $-\sqrt{-g}$ for odd permutations of (0,1,2,3) and 0 if any indices are repeated). $^*\mathbf{F}\cdot\mathbf{F}$ is an invariant for Lorentz transformations (precisely a pseudo-invariant under coordinate reflections), i.e., is either null or not in every inertial frame. For example, in the rest frame of a test-particle with charge q we have an electric field only and no magnetic field, and this invariant is zero, therefore even in any frame moving relative to q, where both $\vec{B} \neq \vec{0}$ and $\vec{E} \neq \vec{0}$, this invariant is zero. However, in a frame where a charge q and a magnetic dipole $\vec{m}$ are at rest, we have in general $^*\mathbf{F}\cdot\mathbf{F} \neq 0$ and therefore this invariant is non-zero in any other inertial frame.

In General Relativity, the gravitomagnetic 'vector' potential h^k can be zero or not depending on the frame where it is calculated. Nevertheless, the curvature of a manifold is a coordinate independent quantity [1–3]. Therefore, in order to test for intrinsic gravitomagnetic effects, i.e., independent of the coordinate system (and not eliminable with a coordinate transformation) we have to use the Riemann curvature tensor $R_{\alpha\beta\mu\nu}$ and the spacetime invariants built with it [3, 51]. Given a metric $g_{\alpha\beta}$ in some coordinate system (with or without the so-called 'magnetic' components g_{0k}), in a way similar to electromagnetism, using the Riemann curvature tensor $R_{\alpha\beta\mu\nu}$ we can build the spacetime curvature invariant $^*\mathbf{R}\cdot\mathbf{R} \equiv {}^*R^{\alpha\beta\mu\nu}\ R_{\alpha\beta\mu\nu}$, where $^*R^{\alpha\beta\mu\nu} \equiv \frac{1}{2}\ \varepsilon^{\alpha\beta\sigma\rho}R_{\sigma\rho}{}^{\mu\nu}$ is the dual of $R_{\alpha\beta\mu\nu}$ [3]. Here below and in [3] the exact explicit expression of the Riemann curvature invariant $^*\mathbf{R}\cdot\mathbf{R}$ is given for some spacetime solutions of the Einstein field equation. For example, in the case of the Kerr metric generated by the angular momentum J and the mass M of a rotating body, this invariant (precisely a pseudo-invariant under coordinate reflections) is equal to [52]:

$$\frac{1}{2}\varepsilon^{\alpha\beta\sigma\rho}R_{\sigma\rho}{}^{\mu\nu}R_{\alpha\beta\mu\nu} = 1536\,J\,M\ \cos\theta\left(r^5\rho^{-6} - r^3\rho^{-5} + \frac{3}{16}r\rho^{-4}\right) \tag{2}$$

where $\rho = \left(r^2 + \left(\frac{J}{M}\right)^2\cos^2\theta\right)$, where θ is the colatitude; this expression of $^*\mathbf{R}\cdot\mathbf{R}$ is then different from zero if $J \neq 0$, e.g., it is zero in the case of the Schwarzschild metric generated by the mass only of a non-rotating body (with

$J = 0$). In the case of Earth with angular momentum $J_\oplus$, the invariant $^*\mathbf{R}\cdot\mathbf{R}$ is at the lowest order: $^*\mathbf{R}\cdot\mathbf{R} \sim \frac{G^2 J_\oplus M_\oplus}{c^5 r^7} + \cdots$, thus the Lense–Thirring effect on the LAGEOS satellites is an intrinsic gravitomagnetic effect [3, 51] that cannot be eliminated by a change of frame of reference.

However, in [39] we show that the effect discussed in [46, 53], accurately measured by Lunar Laser Ranging, is just a 'translational' gravitomagnetic effect which depends on the frame of reference used in the analysis; the invariant $^*\mathbf{R}\cdot\mathbf{R}$ is indeed null on the ecliptic plane (apart from the intrinsic gravitomagnetic terms due to $J_\oplus$ and $J_\odot$) and the gravitomagnetic term discussed in [46, 53], when analyzed in a different frame, is conceptually similar to the geodetic precession. The expression of $^*\mathbf{R}\cdot\mathbf{R}$, as calculated [52] in quasi-cartesian coordinates in a frame with origin at the Sun (x and y, and z are respectively the coordinates on and off the ecliptic plane), is at the lowest order in G/c^2:

$$^*\mathbf{R}\cdot\mathbf{R} \cong 288 \frac{G^2 M_\oplus M_\odot}{c^5 r^4_{Moon} r^4_{Moon\oplus}} z_{Moon} (v^x_\oplus y_\oplus - v^y_\oplus x_\oplus)(\hat{x}_{Moon}\cdot \hat{r}_{Moon\oplus}) \qquad (3)$$

where $\vec{x}_\oplus$ and $\vec{x}_{Moon}$ are the position vectors of Earth and Moon from the Sun and z_{Moon} is the distance of the Moon from the ecliptic plane; this expression is then zero on the ecliptic plane (even by considering that the Moon orbit is slightly inclined of 5° on the ecliptic plane, its z component would only give a contribution to the change of its radial distance from Earth of less than 1% of the total change discussed in [53]).

In conclusion, the frame-dragging effects measured on the Moon orbit by LLR should be interpreted as a translational gravitomagnetic effect due to the motion of the Moon in the solar system, different from the intrinsic gravitomagnetic effects due to the Earth's angular momentum.

3 Measurement of Gravitomagnetism with the LAGEOS Satellites and the GRACE Earth Gravity Models

Since 1995–2001, the Lense–Thirring effect was observed using the LAGEOS and LAGEOS 2 satellites [54, 55, 58–61].

The basic components used in this older approach to detect the Lense–Thirring effect were: (1) the very accurately tracked orbits of the laser-ranged satellites LAGEOS and LAGEOS 2, (2) the accurate Earth gravity models JGM-3 and EGM96 and (3) three observables: the two nodes of LAGEOS and LAGEOS 2 and the perigee of LAGEOS 2, in order to remove the error due to the uncertainties in Earth's quadrupole moment, J_2 and in the next largest even zonal harmonic, J_4, and thus to measure the Lense–Thirring effect, i.e., in order to use three observables for the three unknowns: Lense–Thirring effect, δJ_2 and δJ_4. We recall that the equation describing the nodal rate $\dot{\Omega}^{Class}$ of a satellite, due to the even zonal harmonics, is:

$$\dot{\Omega}^{Class} = -\frac{3}{2} n \left(\frac{R_\oplus}{a}\right)^2 \frac{\cos I}{(1-e^2)^2} \left\{ J_2 + J_4 \left[\frac{5}{8} \left(\frac{R_\oplus}{a}\right)^2 \times (7 \sin^2 I - 4) \frac{\left(1+\frac{3}{2}e^2\right)}{(1-e^2)^2} \right] + \Sigma N_{2n} \times J_{2n} \right\}, \quad (4)$$

where the J_{2n} are the non-normalized even zonal harmonic coefficients, N_{2n} their coefficients (in the equation for the nodal rate) functions of the orbital parameters a, I and e, $n = 2\pi / P$ is the orbital mean motion, P the orbital period and $R_\oplus$ the Earth equatorial radius.

Unfortunately: (a) The Earth gravity models JGM-3 and EGM96, even though representing the state of the art in 1995–1996, were not accurate enough for a precise measurement of the Lense–Thirring effect and (b) the third observable, i.e., the perigee of a satellite, is not so stable under non-gravitational perturbations as it is its node. Indeed, in classical mechanics the node (orbital angular momentum) is conserved under any central force, however, the perigee (Runge–Lenz vector) is conserved only under a central force of the type $\sim 1/r^2$.

Therefore, since 1996, our effort was to find a new observable to replace the perigee of LAGEOS 2 (see the LARES satellite paper, elsewhere in this book), in order to use the nodes only, and to use a more accurate Earth gravity field model (see the discussion on the GRACE Earth gravity models below).

3.1 Method of the 2004 Analysis with LAGEOS, LAGEOS 2 and the GRACE Models

The accurate measurement of the Lense–Thirring effect, obtained in 2004 and described in this paper, has been obtained by using the laser-ranging data of the satellites LAGEOS (LAser GEOdynamics Satellite) and LAGEOS 2 [62, 63] and the Earth gravity field models EIGEN-GRACE02S [64, 65]. This analysis has been extended to a number of GRACE gravity models: EIGEN-GRACE03S, JEM03G, GGM02S, EIGEN-CG03C, GIF22a, JEM04G, EIGEN-GL04C, JEM01-RL03B, GGM03S, ITG-GRACE03S, EIGEN-GL05C [66–73]; most of these gravity models can be found at: http://icgem.gfz-potsdam.de/ICGEM/ICGEM.html. The period of analysis covered an observational period between about 11 years and 14 years, i.e. more than 2.5 times longer than in any previous analyses.

The LAGEOS satellites are heavy brass and aluminum satellites, of about 406 kg weight, completely passive and covered with retro-reflectors, orbiting at an altitude of about 6,000 km above the surface of Earth. LAGEOS, launched in 1976 by NASA (National Aeronautics and Space Administration), and LAGEOS 2, launched by NASA and ASI (Italian Space Agency) in 1992, have an essentially identical structure but they have different orbits. The semimajor axis of LAGEOS is $a \cong 12{,}270$ km, the period $P \cong 3.758$ h, the eccentricity $e \cong 0.004$ and the inclination $I \cong 109.9°$. The semimajor axis of LAGEOS 2 is $a_{II} \cong 12{,}163$ km, the eccentricity $e_{II} \cong 0.014$ and the inclination $I_{II} \cong 52.65°$. We have analyzed the laser-ranging data using the principles described in [74] and adopted the underlying IERS conventions in our modeling, except that, in the 2004 analysis and following ones, we used the GRACE Earth's static gravity models listed above. Our analysis was performed using 15-day arcs. For each 15-day arc, initial state vector (position and velocity), coefficient of reflectivity (C_R) and polar motion were adjusted. Solar radiation pressure, Earth albedo, and anisotropic thermal effects were modeled according to [75–78]. In modeling the thermal effects, the orientation of the satellite spin axis was obtained from [79]. We have applied a $\dot{J}_4 = -1.41 \cdot 10^{-11}$ correction [65, 110]; see Section 9.1.3. Lunar, solar, and planetary perturbations were also included in the equations of motion, formulated according to Einstein's general theory of relativity with the exception of the Lense–Thirring effect, which was purposely set to zero. Polar motion was adjusted and Earth's rotation was modeled from the very long baseline interferometry-based series SPACE [80] which are extended annually. We analyzed the laser-ranging data and the orbits of the LAGEOS satellites using the orbital analysis and data reduction softwares GEODYN II (NASA) [81], EPOS-OC (GFZ, GeoForschungsZentrum) and UTOPIA (CSR-UT). The models used in the GEODYN II and EPOS-OC analysis are listed in Table 1.

In Section 2, we have seen how the node and the perigee of a test-particle are dragged by the angular momentum of a central body. From the Lense–Thirring formula (Section 2), we get for the nodes of the satellites LAGEOS and LAGEOS 2:

Table 1 Models used in the orbital analysis with GGM01S, EIGEN-GRACE02S, EIGEN-GRACE03S, JEM03G

Geopotential (static part)	EIGEN-GRACE02S, EIGEN-GRACE03S, JEM03G
Geopotential (tides)	Ray GOT99.2 and FES2002, FES2004
Lunisolar and planetary perturbations	JPL ephemerides DE-403
General relativistic corrections	PPN except L–T
Lense–Thirring effect	Set to zero
Direct solar radiation pressure	Cannonball model
Albedo radiation pressure	Knocke–Rubincam model
Yarkovsky–Rubincam effect	GEODYN model
Spin axis evolution of LAGEOS satellites	Farinella–Vokhroulicky–Barlier model
Station positions (ITRF)	ITRF2000
Ocean loading	Scherneck model with GOT99.2 and FES2002 tides
Polar motion	Estimated
Earth rotation	VLBI + GPS

$\dot{\Omega}_I^{Lense-Thirring} \cong 31$ mas/year and $\dot{\Omega}_{II}^{Lense-Thirring} \cong 31.5$ mas/year, where mas is a millisecond of arc. The argument of pericenter (perigee in our analysis), ω, of the LAGEOS satellites has also a Lense–Thirring drag [61]. *However, whereas in a previous observation of the Lense–Thirring effect [61] we used both the nodes of LAGEOS and LAGEOS 2 and the perigee of LAGEOS 2, in the present analyses we only use the nodes of LAGEOS and LAGEOS 2.* Indeed, the perigee of an Earth satellite such as LAGEOS 2 is affected by a number of perturbations whose impact in the final error budget is not easily assessed and this was one of the two main points of concern of Ries et al. [82]. The other point of concern was some favorable correlation of the errors of the Earth's spherical harmonics for the EGM96 model that might lead to some underestimated error budget. However, these points of concern do not exist in the present analyses, see below Sections 4 and 5. Indeed, using the previous models JGM-3 and EGM96, we were forced to use three observables and thus we also needed to use the perigee of LAGEOS 2. However, for a measurement of the Lense–Thirring effect with accuracy of the order of 10%, using a number of 2004–2008 GRACE models, thanks to the dramatic improvement in the determination of the Earth's gravity field due to GRACE, it is just enough to eliminate the uncertainty in the Earth's quadrupole moment and thus to just use two observables, i.e. the two nodes of the LAGEOS satellites. However, for a measurement of the Lense–Thirring effect with accuracy of the order of 1%, it is necessary to use one additional observable that will be provided by the node of the LARES satellite (see the LARES paper elsewhere in this book).

To precisely quantify and measure the gravitomagnetic effects we have introduced the parameter μ that is by definition 1 in general relativity [3] and zero in Newtonian theory (thus, our approach is not based on the metric gravitational theories described by the PPN, Post-Newtonian-Parametrized, approximation).

The main error in this measurement is due to the uncertainties in the Earth's even zonal harmonics and their time variations. The un-modeled orbital effects due to the harmonics of lower degree are of order of magnitude comparable to the Lense–Thirring effect (see Section 8). However, by analyzing the GRACE models and their uncertainties in the even zonal harmonics, and by calculating the secular effects of these uncertainties on the orbital elements of LAGEOS and LAGEOS 2, we find that by far the main source of error in the determination of the Lense–Thirring effect is just due to the first even zonal harmonic, J_2 (see Section 8).

We can therefore use the two observable quantities $\dot{\Omega}_I$ and $\dot{\Omega}_{II}$ to determine μ [54, 55, 58, 83], thereby avoiding the largest source of error arising from the uncertainty in J_2. We do this by solving the system of the two equations for $\delta\dot{\Omega}_I$ and $\delta\dot{\Omega}_{II}$ in the two unknowns μ and J_2, obtaining for μ:

$$\begin{aligned}&\delta\,\dot{\Omega}^{Exp}_{LAGEOS\,I} + c\,\delta\,\dot{\Omega}^{Exp}_{LAGEOS\,II}\\ &\quad = \mu\,(31 + c\,31.5)\,mas/year + other\ errors \cong \mu\,(48.2\,mas/year),\end{aligned} \tag{5}$$

where $c = 0.545$.

The use of the nodes of two laser ranged satellites of LAGEOS type to measure the Lense–Thirring effect, by eliminating in this way the Earth spherical harmonics uncertainties, was first proposed and published in [54, 84] and further studied in [56, 57]. The use of a combination of the nodes only of a number of laser ranged satellites was first explained in [55], p. 3102, see also [3], p. 336, where is written "...A solution would be to orbit several high-altitude, laser-ranged satellites, similar to LAGEOS, to measure J_2, J_4, J_6, etc, and one satellite to measure $\dot{\Omega}_{Lense-Thirring}$ ....". The *use of the nodes of LAGEOS and LAGEOS 2*, together with the explicit expression of the LAGEOS satellites nodal equations, was first proposed in [58]; the explicit expression of this combination, that is however a trivial step on the basis of the explicit equations given in [58], was presented by one of us during the 2002 I-SIGRAV school and published in its proceedings [83]. A detailed study of the various possibilities to measure the Lense–Thirring effect using LAGEOS and other laser-ranged satellites was presented in [85]. The use of the GRACE-derived gravitational models, when available, to measure the Lense–Thirring effect with accuracy of a few percent was first published by Ries et al. [82, 86] and Pavlis [87]. The calculation of the standard relativistic perigee precession of LAGEOS was carried out in [88] and the proposal to use laser ranging to artificial satellites to detect relativistic effects, among which the Lense–Thirring effect, was published in [89], in this paper the LAGEOS Lense–Thirring precession was calculated to be 4 arcsec/century, i.e. 40 milliarcsec/year, instead of the correct 31 milliarcsec/year figure calculated in [54] furthermore the problem of the Earth's even zonal harmonics errors was not treated in [89]. One solution to the problem of the Earth's spherical harmonics using polar satellites was proposed in [90] and then in [91].

Equation (5) for μ does not depend on J_2 nor on its uncertainty, thus, the value of μ that we obtain is unaffected by the largest error, due to δJ_2, and it is sensitive only to the smaller uncertainties due to δJ_{2n}, with $2n \geq 4$.

The next largest error source is due to uncertainty in J_4 that may be as large as 10% of the Lense–Thirring effect. To eliminate this error source we need a new observable that will indeed be provided by the node of LARES, allowing a measurement of the Lense–Thirring effect at the 1% level.

Similarly, regarding tidal, secular and seasonal changes in the geopotential coefficients, the main effects on the nodes of LAGEOS and LAGEOS 2, caused by tidal and other time variations in Earth's gravitational field [55, 92], are due to changes in the quadrupole coefficient J_2, e.g., by the 18.6 year Lunar tide with the period of the Moon node and by the post-glacial rebound, and due to anomalous variations of J_2 (see Section 8.1). However, the tidal errors in J_2 and the errors resulting from other un-modeled, medium and long period, time variations in J_2, including its secular and seasonal variations, are removed by the combination (5) of the node residuals (8.1). In particular, most of the errors resulting from the 18.6 and 9.3 years tides, associated with the lunar node, are removed in our measurement. The various error sources that can affect the measurement of the Lense–Thirring effect using the nodes of the LAGEOS satellites have been extensively treated in a large number of papers by several authors, see, e.g., [54,55,58,60,75–79,92–94,104,135]; the main error sources are treated here in the appendix (8).

4 GRACE and Its Gravity Field Models

For many decades geodesy facilitated the testing of some basic laws of physics. An area that space geodesy contributes the most, is the precise determination of the terrestrial gravity field and its temporal variations. This provides the precise, stable and free-of-gravitational-noise environment where very delicate experiments in gravitational physics are conducted. Until the mid-1990s, the geodesists attention was centered on the static field, mainly because of the need to compute precise orbits for satellites carrying various instruments [95]. Since the emergence of climate change as one of the most important research topics at the international level [96], the temporal variations of the gravity field have become a prime topic and the focus of several space missions with clearly international nature. These proposed missions generated great expectations within the fundamental physics community [97]. Some of the proposed missions are already in orbit now: CHAMP, launched July 15, 2000; GRACE, launched March 17, 2002 and some are still under development: GOCE, scheduled for launch in 2009.

The first mission, CHAMP, has delivered a number of global models of the static gravitational field: EIGEN-1S, EIGEN-2S, EIGEN-3p [64, 98, 100]. The innovative aspect of this mission is its precise tracking from the Global Positioning System (GPS) constellation of satellites and the simultaneous measurement of non-gravitational forces from an on-board precise accelerometer [99]. This is the first time that such an accelerometer is used for an extended mission. Due to the high altitude of the GPS spacecrafts (12 h orbits) and the low altitude of CHAMP (400 km), the configuration is called high–low (H–L) satellite-to-satellite tracking (SST). The benefit of this configuration is that the entirety of the unaccounted perturbations in the observed range-rate between the two spacecraft can be now attributed to un-modeled gravitation sensed by the low orbiter (non-gravitational forces are accounted for using the observed accelerometer measurements). The model enhancement from CHAMP is primarily the accuracy improvement of the very low degree terms (long wavelengths) up to about 40°, to roughly two orders of magnitude better than what was available from EGM96.

The second mission, GRACE, is a pair of CHAMP-like spacecraft flying in tandem formation (nominally 250 km apart), enhanced with higher quality accelerometers and a one-way radio frequency range-difference measuring system on both of them. This configuration of satellite-to-satellite tracking between two low orbiters is called low–low (L–L SST). The L–L configuration is characterized from its ability to resolve features proportional to the distance between the two spacecraft. GRACE therefore, since it also carries GPS receivers on both spacecraft, contributes to the observation of the long, as well as the medium wavelengths (up to 120) of the gravitational spectrum. Moreover, the mission is nearly polar ($i = 89.5°$) and it samples the field well enough to produce monthly snapshots. GRACE has acquired enough such monthly snapshots to be able to determine the temporal variations in the long wavelengths down to at least seasonal frequencies. This achievement alone is a unique contribution compared to todays status quo. To date, only the secular variations of the first few zonal terms up to 6° have been studied on the basis of perturbation analysis from ground-based tracking data.

Annual and semi-annual signals have been determined only for J_2 and J_3, and these are not as definitive and robust results as we would like to have. The models based on GRACE data alone [100, 101] show significant improvements in the accuracy of the static part of the field and especially so for the long wavelength terms, reaching about two orders of magnitude improvements compared to EGM96 [100].

The third mission, GOCE, launched in 2009, is really the true "geopotential mapping" mission that geodesists always hoped for since the early 1970s [102]. Instead of relying on perturbations from which to infer the gravitational signal (as in H–L and L–L), GOCE will carry a gradiometer, a device that will measure directly the gravitational tensor components in space. Along with precise knowledge of the time and location of these measurements (GOCE will also be tracked with GPS), we can then construct a gravitational model solving a geodetic boundary problem by means of one of the standard geodetic techniques. Although this task is not as simple as it sounds, gradiometry is definitely the cleanest and most direct measurement of gravitation that we could possibly make in space. With a two 6-month observing periods scenario, GOCE promises to deliver a model that will extend our precise knowledge to much higher resolution, somewhere in the neighborhood of maximum degree 220–250. It is expected that by that time, the release of additional observations on land and over the oceans, from national data bases and terrestrial campaigns (airborne and ship-borne), will allow the development of global models to maximum degree >200. While these models will not have the same uniform quality for the very high degree portions (above 250) on a global scale, their overall improvement will remove all of the ambiguities associated with the static long and medium wavelengths. The temporal signals will also be precisely observed thanks to GRACE. This multitude of gravitational modeling missions is the reason for calling the present decade the International Decade for Geopotential Research.

The analysis of the SLR data set reported here selected the models based on GRACE-only [64, 65, 98, 100, 141] as the underlying gravitational models, primarily because of their pedigree, i.e., the independence from any tracking data other than that from GRACE, and the high quality low degree performance of the models. It is worth to briefly describe some of these models, in order to appreciate their contribution and significance to the success of this experiment.

The GGM01S model is the first GRACE-only static gravity model from the UT/CSR group [116], based on a significant number of GRACE observations (111 days). The model is complete to degree and order 120, however, it is recommended that for precise work one should limit it to degree and order 95. The associated error model predicts a geoid accuracy of 2 cm at degree 70. This model is now superseded by the release of GGM02S (November 2004), a model based on 363 days of GRACE data, complete to degree and order 160, although it is again recommended to limit its use to the 120 × 120 part and by the release of GGM03S. The release of the second GGM02S model allowed for a validation of the error estimates quoted for the first GGM01S model. This gives us some confidence that our error propagation in estimating our uncertainty in the Lense–Thirring estimate is also correct. The release of the second GGM02S model indicates that the unprecedented gravity modeling improvement that we saw in the release of the first GGM01S model, in comparison to the previously used EGM96 model, was the definitive improvement from GRACE. Subsequent releases of models based on increasingly more extensive GRACE data sets will only marginally improve upon the quality of the previous models. Nevertheless, the addition of more data, spread over increasingly longer time intervals, will definitely drive the errors associated with all harmonics describing the field to even lower levels, thereby increasing their accuracy and consequently the accuracy of other results that are derived on that basis (e.g. precise orbits, geophysical parameters and the Lense–Thirring effect).

The EIGEN-GRACE02S model is the second GRACE-only static gravity model from the GFZ group [65], derived from 110 days of GRACE only satellite-to-satellite observations. The model is complete to degree and order 150 and claims an accuracy of 1 mm in geoid heights at half-wavelength of 800 km, reaching 1 cm at degree 75. The errors associated with this model have been calibrated using surface gravimetry, oceanographic data and precise satellite tracking data. The improvement over previous solutions (e.g. GGM01S and EIGEN-GRACE01S) is attributed to the better data reduction strategy in the case of EIGEN-GRACE02S. The EIGEN-GRACE02S model is discussed and used in [13, 129]. Here, we have also used the gravity models EIGEN-GRACE03S, EIGEN-GRACE03C, EIGEN-GL04C and EIGEN-GL05C based on longer GRACE observations.

5 Results: Measurement of the Lense–Thirring Effect and Its Uncertainty

5.1 Error Budget

In the measurement of the Lense–Thirring effect, that is a secular change of the nodal longitude of a satellite, the largest error sources are due to the other secular and long-period orbital perturbations affecting the node. Among the secular perturbations of the node of the LAGEOS satellites there are: the shift of the nodal line due to the even zonal harmonics of the Earth's gravitational field [103], the geodetic precession and the gravitomagnetic or Lense–Thirring effect. The geodetic precession has been measured with an accuracy of about 6×10^{-3} and since its effect on the node of the LAGEOS satellites is only 17.6 mas [55], its uncertainty is negligible in the Lense–Thirring measurement. However, the uncertainty in the even zonal harmonics is a critical factor for the determination of frame-dragging, indeed an error in one of the lower even zonal harmonics may be large enough to be indistinguishable from the Lense–Thirring nodal drag; this error source is treated below. However, the periodic perturbations of the node of the LAGEOS satellites may also be a crucial factor in the determination of frame-dragging, especially critical is the uncertainty in the perturbations with long period compared to the period of observation. The periodic effects with period much shorter than the observational period are averaged out. In the present determination of the Lense–Thirring effect, we have three basic factors that make negligible the error due to periodic effects in the measurement of frame-dragging and also make this error easy to be assessed in the final error budget. These basic factors are: (i) the period of the present analyses is about 11 years, or more, and thus all the periodic perturbations of the nodes are basically averaged out apart from the 18.6 years tide associated with the lunar node, however, as explained below, the main effect of this tide is a change of the even zonal harmonic of degree 2, i.e., of the J_2 coefficient that is however removed out using our combination of observables (5); (ii) since the original proposal of the LAGEOS III experiment [54], numerous researchers [54,55,58,60,75–79,92–94,104], ... have treated, in many papers and reports, the perturbations affecting the LAGEOS node in order to determine the Lense–Thirring effect and have concluded that the only critical perturbations on the nodes of the LAGEOS satellites are the secular ones due to the Earth's even zonal harmonics, however, as shown below in Sections 6 and 8, due to the high accuracy of the recent novel Earth's gravity field models based on GRACE data, the even zonal harmonic uncertainties is at a level of the order of 10% or less; (iii) in regards to the periodic perturbations, in addition to a detailed treatment of the various perturbations affecting the LAGEOS node and their uncertainties given below, a simple but very meaningful test shows that the periodic perturbations cannot introduce an error larger than about 3%, or at most 4%, in our determination of the Lense–Thirring effect. Since the period of the gravitational and non-gravitational orbital perturbations (but not their amplitude) is very well determined, we have also fitted, together with a secular trend, a number of periodic effects. We have done a number of different fits, each with different periodic perturbations, we have then compared the result in the case of the fit with a secular trend only with the various results when, together with a secular trend, we have included a different number of the main periodic perturbations. The result is that the maximum deviation of the secular trend from the benchmark case of its fit with no periodic perturbations does not exceed a few percent of Earth's frame-dragging (at most ~4%), as it is displayed in figure (5) and discussed in the corresponding caption. Of course the RMS of the fit is much smaller when we allow in the fit for a substantial number of periodic perturbations. Therefore, the two points of concern in [82], regarding really not the method or the resulting value of the Lense–Thirring effect that we had obtained in our previous analysis [60, 61] but rather our previous error budgets, are no longer of concern in the present analyses, as explained below and in Section 8. Indeed, (i) the concern regarding the perturbations of the perigee of LAGEOS 2 is clearly absent in the present analysis since here we only use the nodes of the LAGEOS satellites and we do not use the perigee, (ii) the concern regarding the high correlation of some of the even zonal harmonics of the EGM96 model is also absent in the present analyses since the Earth models used here: EIGEN-GRACE02S, EIGEN-GRACE03S, GGM01S, GGM02S, GGM03S, JEM03G, JEM03G, JEM04G, EIGEN-CG03C, GIF22a, EIGEN-GL04C, JEM01-RL03B, ITG-GRACE03S and EIGEN-GL05C, have a low correlation between the even zonal harmonics (see below and related discussion in Section 8), and especially because to evaluate the even zonal harmonic induced error, we did not use any covariance matrix, instead, we simply added the absolute values of the errors on our combined residuals due to the even zonal uncertainties.

In Section 5.2 we report the results of the measurement of the Lense–Thirring effect using the GRACE models EIGEN-GRACE02S, EIGEN-GRACE03S and

JEM03G. The values of the spherical harmonic coefficients describing the Earth's gravity field model may appreciably change with longer periods of observations. However, our analysis is not sensitive to changes in the Earth's quadrupole moment and it is only affected by changes in the higher zonal harmonics. Therefore, in the future, more accurate determinations of the even zonal coefficients and of their uncertainties might lead to slightly different values of μ and consequently to a slightly different and improved total error budget. As a matter of fact using EIGEN-GRACE02S, whose uncertainties in Earth's even zonal harmonics include systematic errors to certain extent, the contribution of the even zonal errors to the total error budget is 4% only, or more realistically 8% by doubling it to take into account some underestimation of the published errors of this model (see discussion in Section 6). Thus, it is crucial to note that when more accurate models using GRACE, CHAMP or GOCE will be available, it will be straightforward to, a posteriori, very accurately re-assess the total error of the present analyses. This will be done by simply taking the differences between the values of the even zonals of the presently used models (EIGEN-GRACE02S, EIGEN-GRACE03S, JEM03G, etc.) with the corresponding values of the future, more accurate, models and by considering the uncertainties of these forthcoming models. Thus, one will be able to easily re-estimate the total error in the present measurement of μ.

In summary, by doubling for each GRACE gravity model the published errors in the static even zonal harmonics (to take into account some underestimation of these published uncertainties, see Sections 6 and 9) and by summing up the absolute value of the error due to each even zonal harmonic, in the analysis with EIGEN-GRACE02S and EIGEN-GRACE03S, we get a total error budget of the order of respectively 9% and 13% of the Lense–Thirring effect, however, in the case of JEM03G, only the formal uncertainties of the J_{2n} were available to us but *not* their calibrated uncertainties, i.e., the uncertainties including the systematic errors, and we have then tentatively used the uncertainties of the GRACE model GGM02S; then, using the published, calibrated, uncertainties of GGM02S we obtain a total error budget of the order of 10%, or by doubling the static even zonal harmonics uncertainties we get a total error of the order of 18% (see more details in Sections 6 to 9 below). The main perturbations in our determination of the Lense–Thirring effect are described and examined in detail in the appendix, Section 8.

John Ries et al. (2008) [15] have extended the measurement of the Lense–Thirring effect to the models GGM02S, EIGEN-CG03C, GIF22a, JEM04G, EIGEN-GL04C, JEM01-RL03B, GGM03S, ITG-GRACE03S and EIGEN-GL05C using the orbital estimator UTOPIA, the results are presented in Section 5.2, Fig. 7. Using a number of GRACE models and the LAGEOS and LAGEOS 2 orbital data, Ries et al. have confirmed the general relativistic prediction for the Lense–Thirring effect, they have determined its mean value to be 99% of the prediction of General Relativity and have concluded that the total error budget in the measurement of the Lense–Thirring effect using the latest GRACE Earth gravity models is of the order of 10% (see Section 8.4 and Table 9).

5.2 *Results of the Measurement of the Lense–Thirring Effect*

In this section, we present the results in the measurement of the Lense–Thirring effect on the basis of a number of different GRACE Earth gravitational models, through the analysis of the nodal rates of the LAGEOS and LAGEOS 2 satellites. We first report the results we obtained using EIGEN-GRACE02S, EIGEN-GRACE03S and JEM03G (Figs. 4, 5, and 6) over a period of about 11 years. We also separately plot the integrated residuals of the node of LAGEOS, of the node of LAGEOS 2 and of their J_2-free combination (Fig. 3). In Fig. 4d, we report the result of the measurement of the Lense–Thirring effect with EIGEN-GRACE02S using the orbital estimator EPOS-OC by the German GRACE group of GFZ, this program is independent of GEODYN; thus this independent result confirms our previous analyses obtained with GEODYN. Then, in Fig. 6, we present the results with these three models, including an error bar representing the total estimated error in our measurement including systematic errors. Finally, in Fig. 7 we report the result of Ries et al. [15] for the measurement of the Lense–Thirring effect using a number of GRACE models, including EIGEN-GRACE02S, GGM02S, EIGEN-CG03C, GIF22a, JEM04G, EIGEN-GL04C, JEM01-RL03B, GGM03S, ITG-GRACE03S and EIGEN-GL05C.

Figure 3 shows (in blue) the integrated residuals of the node of LAGEOS from January 1993 to October 2003 using the model EIGEN-GRACE02S, the integrated

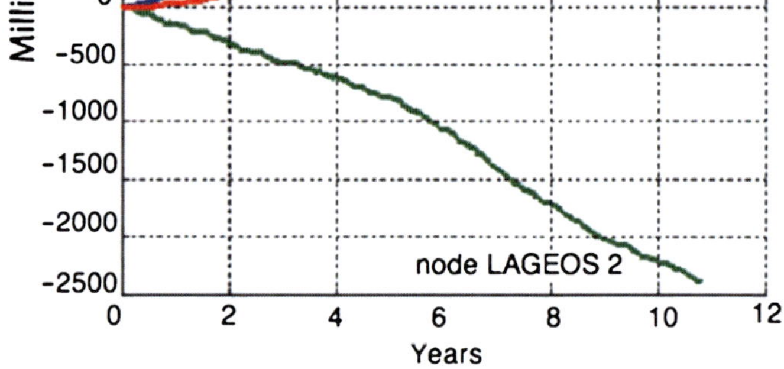

Fig. 3 Residuals of the node of LAGEOS (in *blue*), node of LAGEOS 2 (in *green*) and combination of the nodes of LAGEOS and LAGEOS 2 using formula (5) (in *red*), JEM03G is the GRACE Earth gravity model used here

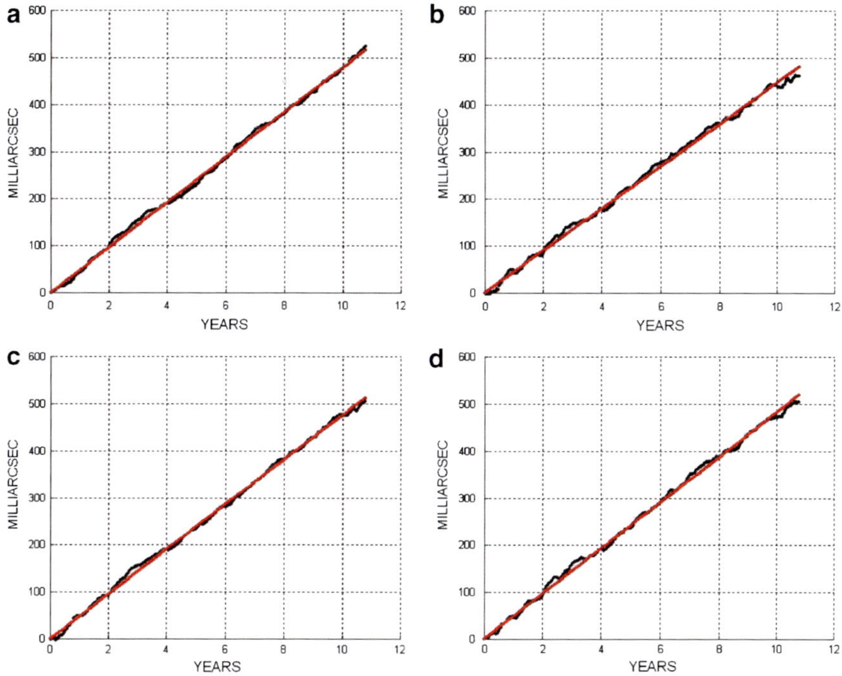

Fig. 4 Fit of the residuals of the nodes of LAGEOS and LAGEOS 2, using the combination (5) with: (**a**) the model EIGEN-GRACE02S, (**b**) EIGEN-GRACE03S , (**c**) JEM03G. In (**d**) is shown the fit of the nodes of LAGEOS and LAGEOS 2, using *the orbital estimator EPOS-OC* of the German GRACE team of GFZ and the model EIGEN-GRACE02S. The fits in (a) to (d) are with a secular trend plus *six* periodic terms. The slope in (**a**) is $\mu \simeq 0.99$, in (**b**) $\mu \simeq 0.93$, in (**c**) $\mu \simeq 0.99$ and in (**d**) $\mu \simeq 1.0$. The scale of the axes is different with respect to Fig. 3

node residuals of LAGEOS 2 (in green) using EIGEN-GRACE02S and the integrated combination of the nodes residuals of LAGEOS and LAGEOS 2 (in red), according to formula (5), using EIGEN-GRACE02S; this figure displays that the large residuals of the node of each satellite (in blue and green) due to the error in J_2 are eliminated in the combined residuals (in red).

Figure 4 shows the result of the measurement of the Lense–Thirring effect using the three GRACE Earth models EIGEN-GRACE02S (Fig. 4a), EIGEN-GRACE03S (Fig. 4b) and JEM03G (Fig. 4c) by fitting the orbital residuals with a secular trend together with *six* periodic effects. Figure 4d represents the result of the measurement of the Lense–Thirring effect using *the orbital estimator EPOS-OC* of the German GRACE team of GFZ, with EIGEN-GRACE02S and by fitting the orbital residuals with a secular trend together with *six* periodic effects.

Figure 5 shows our measurement of the Lense–Thirring effect using JEM03G by fitting the orbital residuals with a secular trend only (Fig. 5a) and with a secular

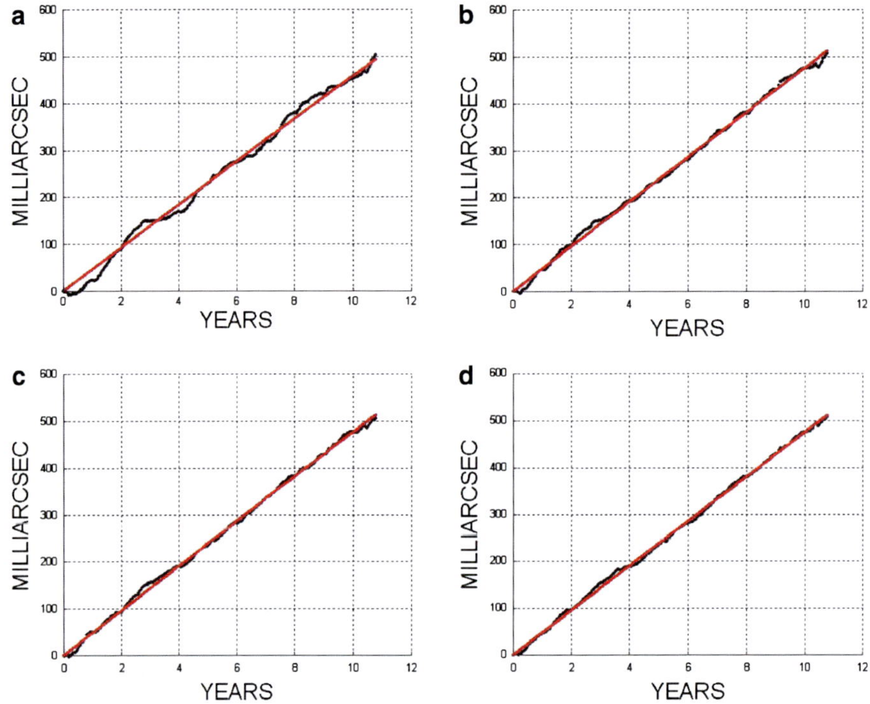

Fig. 5 Fit of the residuals of the nodes of LAGEOS and LAGEOS 2, using the combination (5) and the Earth gravity model JEM03G, with: (**a**) a *secular trend only*, (**b**) a secular trend plus *two* periodic terms, (**c**) a secular trend plus *six* periodic terms, (**d**) with a secular trend plus *ten* periodic terms. The slope in (**a**) is $\mu \simeq 0.95$ with RMS of the post-fit residuals equal to 12.5 mas, in (**b**) $\mu \simeq 0.986$ with RMS $\simeq 6.9$ mas, in (**c**) $\mu \simeq 0.987$ with RMS $\simeq 5.9$, and in (**c**) $\mu \simeq 0.984$ with RMS $\simeq 5.2$. The scale of the axes is different with respect to Fig. 3

trend together with two (Fig. 5b), six (Fig. 5c) and ten (Fig. 5d) periodic effects. The slopes of Fig. 5a–d are respectively: 0.95 μ, 0.986 μ, 0.987 μ, and 0.984 μ.

In conclusion, by fitting our combined residuals with a secular trend plus six periodic signals, using EIGEN-GRACE02S, we found over an observational period of 11 years:

$$\mu_{EIGEN-GRACE02S} = 0.994 \pm 0.10$$

with RMS of the post-fit residuals of 5.98 mas using GEODYN II, however using EPOS-OC and the corresponding GFZ set-up describing the satellites orbital perturbations we found: $\mu_{EIGEN-GRACE02S} = 1.0 \pm 0.10$ (with RMS of the post-fit residuals of 6.92 mas); this small 0.6% difference for μ using the two orbital estimators is due to a different modeling of the orbital perturbations in the two cases.

By fitting our combined residuals with a secular trend plus six periodic signals, using EIGEN-GRACE03S and JEM03G, we found respectively:

$$\mu_{EIGEN-GRACE03S} \simeq 0.93 \pm 0.13$$

and:

$$\mu_{JEM03G} \simeq 0.99 \pm 0.18,$$

where these uncertainties include all systematic errors. The static gravitational errors for the models EIGEN-GRACE02S, EIGEN-GRACE03S and JEM03G have been calculated by simply adding the absolute values of the errors in our combined residuals due to the even zonal harmonics uncertainties and by then multiplying this total error for a factor 2 to take into account for possible underestimations of the published uncertainties of each model (see next Section 6). However, in the case of JEM03G, only the formal uncertainties of the J_{2n} were available to us but *not* their calibrated uncertainties, i.e., the uncertainties including the systematic errors, and we have then tentatively used the uncertainties of the GRACE model GGM02S; then, using the published, calibrated, uncertainties of GGM02S, we obtain a total error budget of the order of 10%, or by doubling the static even zonal harmonics uncertainties, we get a total error of the order of 18%.

In the case of EIGEN-GRACE02S, by fitting our combined residuals with 2, 6, or 10 periodic terms we practically got the same value for the Lense–Thirring effect and by analyzing the data with the NASA orbital estimator GEODYN II and with the GFZ orbital estimator EPOS-OC with their corresponding different set-up for the orbital perturbations, we substantially obtained the same result. Furthermore, these different measurements of the Earth frame-dragging effect obtained with EIGEN-GRACE02S, EIGEN-GRACE03S and JEM03G are in agreement with each other within their uncertainties. Therefore, our measured value of the Lense–Thirring effect with the Earth gravity models EIGEN-GRACE02S, EIGEN-GRACE03S and JEM03G agree with the general relativistic prediction, and the corresponding uncertainty of our measurement using the more recent GRACE models is of the order of 10% (see appendix (8)).

John Ries et al. [15] have extended the measurement of the Lense–Thirring effect to a number of more recent models, including EIGEN-GRACE02S, GGM02S, EIGEN-CG03C, GIF22a, JEM04G, EIGEN-GL04C, JEM01-RL03B, GGM03S, ITG-GRACE03S and EIGEN-GL05C, using the orbital estimator UTOPIA and corresponding set-up for the orbital perturbations, the results are presented in Fig. 7. Ries et al. have concluded that the mean value of the Lense–Thirring effect using these models is 0.99 μ with a total error budget in the measurement of the Lense–Thirring effect of the order of 12%.

In conclusion, the analysis of the University of Salento, Sapienza University of Rome, University of Maryland Baltimore County and GFZ Potsdam/Munich (using the orbital estimators GEODYN and EPOC-OC), and of the Center for Space Research of the University of Texas at Austin [141] have confirmed the general relativistic prediction for Lense–Thirring effect using the LAGEOS and LAGEOS 2 orbital data with an accuracy *of the order of* 10% (see Figs. 6 and 7).

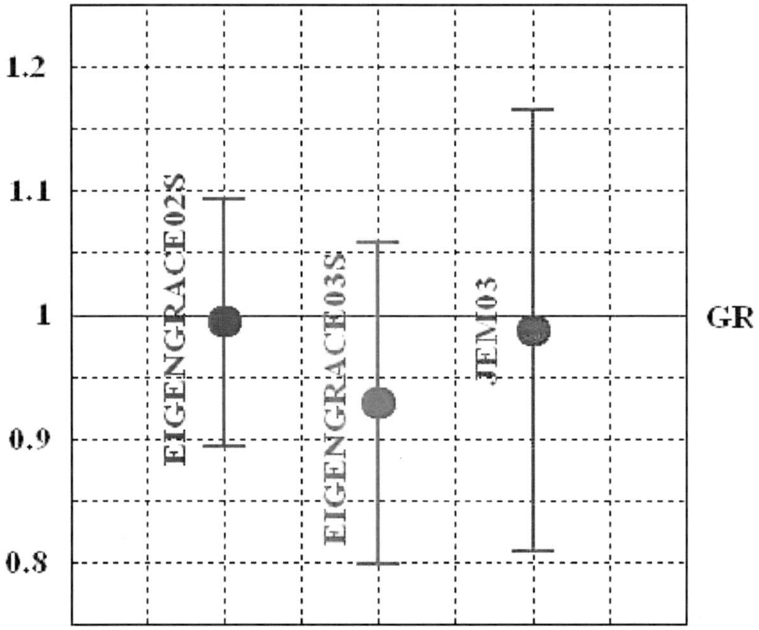

Fig. 6 Measurement of the Lense–Thirring effect with the GRACE models EIGEN-GRACE02S [13], EIGEN-GRACE03S, and JEM03G, obtained using GEODYN and EPOS-OC. The error bar includes systematic errors calculated using the published uncertainties of each model (see Sections 6 and 8)

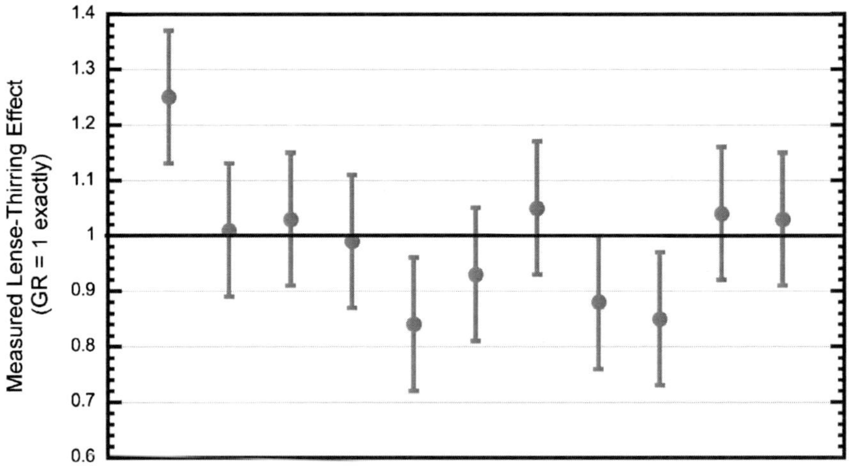

Fig. 7 Measurement of the Lense–Thirring effect with the GRACE models: EIGEN-GRACE02S, GGM02S, EIGEN-CG03C, GIF22a, JEM04G, EIGEN-GL04C, JEM01-RL03B, GGM03S, ITG-GRACE03S and EIGEN-GL05C, obtained using UTOPIA [15]

6 The Error due to Earth's Gravity Field Uncertainties and a Realistic Assessment of the Accuracy of the GRACE Earth Gravity Models

By far, the main source of error in the measurement of the Lense–Thirring effect with the LAGEOS satellites is due to the uncertainties in the Earth's even zonal harmonics; this error can be quantified by simply propagating on the nodes of the LAGEOS satellites the published uncertainties in the Earth's even zonal harmonics of the GRACE models used in our orbital estimation.

In this section, we discuss how the published uncertainties of the Earth's even zonal harmonics of the GRACE models, and therefore the error in the Lense–Thirring measurement, are representing the true errors in our knowledge of these harmonics. The conclusion, based on a comparison carried out by CSR [70] between different GRACE gravity field models, is that the *true errors* in the lower degree even zonal harmonics of the GRACE models are, on average, contained within the published uncertainties of the recent GRACE models multiplied by a factor 2 or at most 3 (see Figs. 9 and 10). Thus, for a realistic assessment of the corresponding error in the Lense–Thirring measurement, it is sufficient to multiply by a factor 2, or at most 3, the error obtained by propagating the published GRACE uncertainties on the nodes of the LAGEOS satellites.

6.1 On the Accuracy of GRACE-Based Models

The GGM01S model is the first GRACE-only static gravity model from the UT/CSR group [116], based on a significant number of GRACE observations (111 days). The model is complete to degree and order 120, however, it is recommended that for precise work, one should limit it to degree order 95. The associated error model predicts a geoid accuracy of 2 cm at degree 70. This model is now superseded by the release of GGM02S (November 2004), a model based on 363 days of GRACE data, complete to degree and order 160, although it is again recommended to limit its use to the 120×120 part. At the 2007 Fall meeting of the American Geophysical Union (AGU) [70] presented results and an accuracy assessment for the new model, GGM03S (Fig. 8) and a preliminary version of a combination field GGM03C. The release of these fields allowed for a validation of the error estimates quoted for the previous models, simply by looking at the successive differences in the coefficients versus the calibrated error estimates for each field (Fig. 9). This gives us confidence that our error propagation in estimating the errors for the Lense–Thirring parameter, derived from LAGEOS SLR data analysis using these gravity models, is also correct.

Subsequent releases of models based on increasingly more extensive GRACE data sets will only marginally improve upon the quality of the current static models. Nevertheless, the addition of more data, spread over increasingly longer time intervals, will definitely drive the errors associated with all harmonics describing the

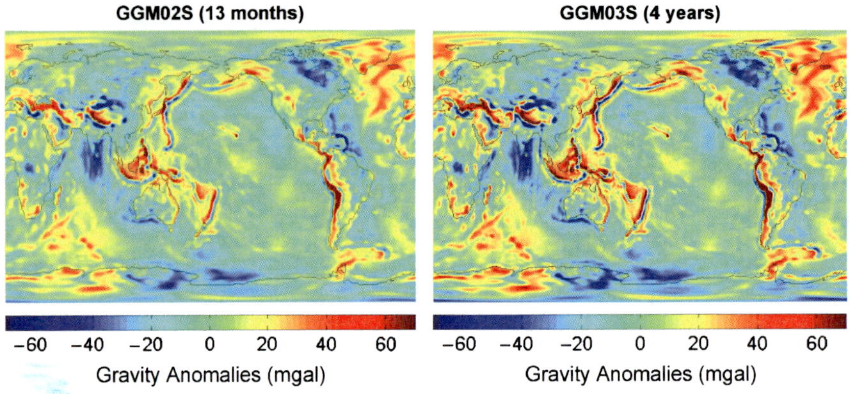

Fig. 8 Global gravity anomaly maps for GGM02S (160 × 160) and GGM03S (180 × 180) illustrating the increased resolution of the latter [70]

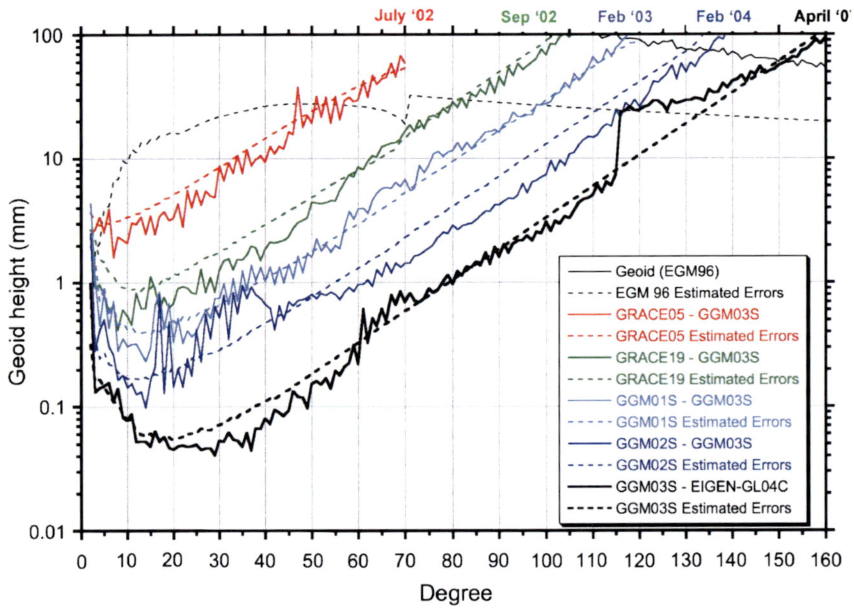

Fig. 9 Successive static gravity model improvement from GRACE data as a function of the data span used in the solution and correlation of the model errors and the assumed calibrated model errors [70]. In this figure different gravity field models are compared by comparing the difference of the corresponding coefficients for each degree (by including the sum over every corresponding order) with their published uncertainties. For example the gravity field model GGM01S, a preliminary GRACE model, less accurate and based on a much short period of observation of GRACE, is compared with GGM03S, one of the best models today available. Clearly the difference between GMM01S and GGM03S, the *solid line in light blue*, is of comparable size with the uncertainty of the less accurate model, i.e., GGM01S, the *dashed line in light blue*, however the same difference is about one order of magnitude larger than the estimated uncertainty of the more accurate model, i.e., GGM03S, the *dashed black line*

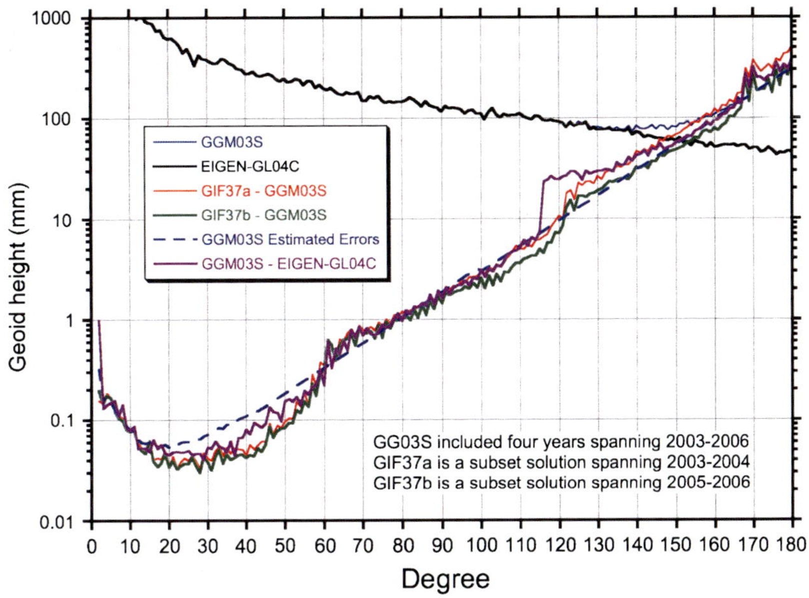

Fig. 10 Comparison of the calibrated error estimates of GGM03S to the coefficient differences from two subset (non-overlapping) solutions, GIF37a and GIF37b, and the independent GFZ/GRGS-produced model EIGEN-GL04C [70]. The runoff above degree 130 is typical of a satellite-only gravity solution that is not constrained or regularized. It provides a useful calibration for the error estimates at the higher degrees

gravity field and their temporal variations, especially the secular ones, to even lower levels, thereby increasing their accuracy, and consequently, the accuracy of other results that are derived on that basis (e.g. precise orbits, geophysical parameters and the Lense–Thirring parameter).

At this point however, with almost seven years of GRACE data and monthly fields, we can look at the level of the errors in the static field models incurred by the variability of the field due to geophysical fluid redistribution in the terrestrial system. GRACE mission is primarily to observe these variations, and the monthly solutions describe primarily this variability due to changes in the hydrological field of the planet, since atmospheric and oceanographic signals are modelled a priori on the basis of Global Circulation Model (GCM) results, the so-called "de-aliasing" product GAC. Using the monthly GRACE-derived fields and adding to them the GAC part, we have a series of monthly fields that (to within their error) describe entirely the average gravity field for each of these months. We then fit a simple mathematical model to these series, including a linear term (secular change), and three harmonics, an annual, a semi-annual and a seasonal frequency term. By doing so we can obtain a set of "mean" harmonics at a selected epoch (2000.0 in our case) and a set of secular and periodic terms that can be used to describe the majority of the GRACE-observed variations in our orbital reductions with GEODYN, EPOS-OC and UTOPIA.

Our tests indicate that even if we ignore entirely the gravitational signals due to the redistribution of mass in the atmosphere and oceans, their effect on the value of the coefficients themselves and their secular rates is at or well below the level of the calibrated accuracy estimates for these quantities.

6.2 The EIGEN-GRACE Series of Models from GFZ

The EIGEN-GRACE02S model was the second GRACE-only static gravity model from the GFZ group [65], derived from 110 days of GRACE satellite-to-satellite observations. The model is complete to degree and order 150, and claims an accuracy of 1 mm in geoid heights at half-wavelength of 800 km, reaching 1 cm at degree 75. The errors associated with this model have been calibrated using surface gravimetry, oceanographic data and precise satellite tracking data. The improvement over previous solutions (e.g. GGM01S and EIGEN-GRACE01S) is attributed to the better data reduction strategy in the case of EIGEN-GRACE02S.

This model was followed by the release, in a collaboration between GFZ (GeoForschungsZentrum) and GRGS (Groupe de Recherche de Geodesie Spatiale), of several other models in the EIGEN series (e.g. EIGEN-CG03, EIGEN-GL04 and recently EIGEN-5C), however, these later models have included at least the LAGEOS and LAGEOS 2 SLR data, and thus they might be unsuitable for our analysis, since we prefer models derived from data independent of SLR to LAGEOS.

6.3 The Independence of Errors and the Quality of Secular Zonal Rates

Long period and secular variations in the zonal harmonics of the spherical harmonic expansion describing Earths gravitational field are generally attributed to post-glacial rebound and ice-sheet mass changes [122, 123, 127] and reflect changes in Earths viscous lower mantle [122]. These changes hold information about the dynamics of Earth and its anelastic response to tidal forcing. Estimation of the secular rates for the zonal harmonics therefore has been a high priority research topic of space geodesy for many years now, with a considerable number of independent results in the literature [105–109, 111–113, 115–128]. The majority of these results were obtained from the analysis of long time intervals spanning several decades of data obtained from several satellites including the two LAGEOS, Starlette, Ajisai, Stella, BE-C and the two ETALONs.

Due to the limited number of satellite orbit configurations, only a handful of the very lowest degree zonal terms rates can be "observed" [130], reaching at best, up to degree 6. The estimation of the zonal rates is in most cases based on perturbation analysis of very long arcs (several years), but not always necessarily so. In particular, the last four studies cited above, involve short-arc techniques, estimating the average "static" value of the zonal harmonics, and deriving the rate from a subsequent analysis of the resulting time series. Furthermore, in most cases, the investigators used their static gravitational model of preference, implying that the adopted epoch values for J_4 were different amongst studies. These differences are well documented in the literature dealing with the static models JGM-3, EGM96, GRIM5-S1, TEG-4, EIGEN-GRACE02S, and GGM01S. It is then remarkable that from a plethora of approaches, analyzed data sets, and diverse analysis software packages used, the resulting estimate for the secular rate of these harmonics, e.g. J_4, which is of interest here, differs by no more than 100% from the commonly accepted value [IERS Conventions 2003], the one that was also used in the development of EIGEN-GRACE02S that was used for our 2004 analysis [13].

As it is discussed in [106], there is considerable correlation between the zonal rate estimates ($\rho_{J2dot,J4dot} = -0.75, \rho_{J2dot,J6dot} = +0.76, \rho_{J4dot,J6dot} = -0.86$), but there is no significant correlation between the static values of the harmonics and their rates. This independence of errors between the static values estimates and the associated rates is further demonstrated in [109], where monthly estimates of the J_2 term obtained from GRACE data and those obtained entirely independently from SLR data analysis are compared. The agreement is remarkable, even though the GRACE data analysis has not reached yet its definitive level. The results from [109] suggest that the current best estimate for $\dot{J}_4$, with all other zonal rates up to $\dot{J}_6$ simultaneously estimated, is -1.99×10^{-11}/year. This value is only 0.58×10^{-11}/year off the currently accepted value and the one adopted in developing EIGEN-GRACE02S, or in terms of a percentage, it is off by only 41%.

A careful collection of all the published values of $\dot{J}_4$ in the literature and subsequent statistical analysis, points to the same conclusion. Since the published estimates can be in most cases considered independent as we explained above, we can perform a simple statistical analysis of these values and their variance in comparison to the quoted error estimates of these values. The population is not large by any statistical standards, but that is what is available. There are two estimates, $+0.2 \times 10^{-11}$ [124] and $+0.3 \times 10^{-11}$ [118], which are clearly outliers and the reasons are explained in [106], so our statistical tests include cases where both of these or either one separately are eliminated from the population of estimates. The results in terms of a mean $\dot{J}_4$ and its associated scatter are shown in Fig. 11.

These tests point to one conclusion very clearly: the uncertainty in our present knowledge of $\dot{J}_4$ is in the neighborhood of $\pm 0.7 \times 10^{-11}$/year and this quantity is only 50% of the $\dot{J}_4$ value that we used in our Lense–Thirring work. It implies that with a 67% probability, the true value of $\dot{J}_4$ could be in the interval $[(-1.697 - 0.7)$ to $(-1.697 + 0.7)]$, that is $[-2.4, -1.0]$, or if we increase the probability to 95% confidence level, then we have the corresponding interval of $[-3.1, -0.3]$, or even at the 99% level, $\dot{J}_4$ will lie in the interval $[-3.8, +0.4]$ (all values are in 10^{-11}/year).

In conclusion, from all the cases studied, there is no reason to expect an error in $\dot{J}_4$ of more than $\pm 0.7\, 10^{-11}$/year (at the 1-σ level of significance, i.e. 67% probability). This possible range of error, with all the statistically acceptable variations in the estimate of $\dot{J}_4$, is well within our Lense–Thirring parameter μ error margin of about (10%) [13, 15, 116, 129], indeed our simulation of an error of 510% in $\dot{J}_4$,

Case Description	Mean value of J_4-dot	Scatter of J_4-dot	References
A. Six estimates from Table 5 (Ref. [1]) and the new estimate from (Ref. [2])	-0.70×10^{-11}/year	$\pm 0.8 \times 10^{-11}$/year	(1): [Cheng et al. 1997] (2): [Cheng and Tapley 2004b]
B. As in A, but without the two outliers of +0.2 and +0.3 $\times 10^{-11}$/year	-1.02×10^{-11}/year	$\pm 0.6 \times 10^{-11}$/year	As in (A)
C. As in A, removing only the estimate of $+0.3 \times 10^{-11}$/year	-0.84×10^{-11}/year	$\pm 0.74 \times 10^{-11}$/year	As in (A)
D. As in A, removing only the estimate of $+0.2 \times 10^{-11}$/year	-0.83×10^{-11}/year	$\pm 0.76 \times 10^{-11}$/year	As in (A)
E. Nine estimates from Table 7 of Ref. [1]	-1.697×10^{-11}/year	$\pm 0.69 \times 10^{-11}$/year	(1): [Cheng et al. 1997]

Fig. 11 Statistical significance of the variations between individual $\dot{J}_4$ estimates

i.e., of a value of $\dot{J}_4 = -8.61\ 10^{-11}$/y, by fitting the residuals with a secular trend *plus a quadratic*, resulted in a corresponding error in the recovered value of μ of only 1% [110] (see Section 8.1.3).

The evaluation of the $\dot{J}_4$ error estimate on the basis of the various results indicates also a perfect agreement with the quoted $\pm 10^{-11}$/year error for the $1.41 10^{-11}$/year rate for J_4, which was adopted in developing EIGEN-GRACE02S that was used in the 2004 analysis [13].

7 Conclusions

We have measured the Earth's gravitomagnetic field and the Lense–Thirring effect predicted by Einstein's theory of General Relativity on a satellite orbit, using a technique that we initially proposed in 1986–1996 described here in Section 3.1. We accomplished that by analyzing the time series of the nodal variations of the LAGEOS and LAGEOS 2 orbits from both 15-day and 7-day arcs fitted to SLR data, using NASA's GEODYN and GFZ's EPOS-OC software programs and the Earth's gravitational field models EIGEN-GRACE02S, EIGEN-GRACE03S and JEM03G and, using CSR-UT software program UTOPIA and the models EIGEN-GRACE02S, GGM02S, EIGEN-CG03C, GIF22a, JEM04G, EIGEN-GL04C, JEM01-RL03B, GGM03S, ITG-GRACE03S and EIGEN-GL05C. Our results [13, 15, 129] for the measured value of the Lense–Thirring effect correspond to ~99% of the prediction of Einstein's gravitational theory, thus agreeing with the General Relativity prediction of gravitomagnetism within our estimate of a total experimental uncertainty of the order of 10%. The measurements of the Lense–Thirring effect obtained with GEODYN and EPOS-OC using EIGEN-GRACE02S, EIGEN-GRACE03S and JEM03G were in agreement with each other within their uncertainties (see Fig. 6). Using the CSR's program UTOPIA, the Lense–Thirring measurement was performed using a number of more recent GRACE models, including EIGEN-GRACE02S, EIGEN-GL05C, ITG-GRACE03S, JEM01-RL03B, EIGEN-GL04C, JEM04G, GIF22a, EIGEN-CG03C and GGM02S, GGM03S; this CSR analysis resulted (see Fig. 7) in a mean value of the Lense–Thirring effect, measured using these ten models, of 99% of the General Relativity prediction with a slightly more conservative estimated error budget of about 12% [15].

In regard to the total error budget, we have treated the error due to the uncertainties in the Earth's gravity field in Section 10, including the uncertainties in the static value and in the secular variations of the even zonal harmonics and a realistic assessment of the uncertainties of the GRACE gravity models, we are also describing atmospheric refraction and orbital inclination errors in Section 10. For a detailed error analysis we refer to the appendix. Here we only stress that when fitting our combined residuals with a secular trend only, or with a trend plus 2, 6, or 10 periodic terms, we have obtained a measured value of the Lense–Thirring effect, with a relative variation of at most $\sim$3%. We finally note that the largest uncertainty in our present error budget is due to the published errors of the GRACE Earth's gravity models that might underestimate the systematic errors in the uncertainties associated with the even zonal harmonics. However, it is possible to evaluate the accuracy of the Earth's gravity models used in our analysis by simply taking the difference between each corresponding spherical harmonic coefficient of two different gravity modes and by comparing these differences with their "calibrated" uncertainties. This comparison, carried out by [70] and described in Section 6, Figs. 9 and 10, has, a posteriori, provided a solid re-assessment of the error in our determination of frame-dragging. The conclusion is that, for a realistic assessment of the error in the Lense–Thirring measurement due to the Earth static gravitational field, it is sufficient to multiply by a factor of about 2, or at most 3, the error obtained by propagating the published uncertainties of the GRACE gravity models on the nodes of the LAGEOS satellites.

As an additional outcome of our analysis and in agreement with other recent findings by other authors [128], we have also clearly observed a variation in the Earth's quadrupole coefficient, J_2, since 1998 [129].

8 Appendix 1: Error Analysis

The perturbations affecting our measurement of the Lense–Thirring effect may be divided in gravitational and non-gravitational perturbations. In particular, among the gravitational perturbations we have:

- Errors due to the uncertainties in the static even zonal harmonic coefficients, J_{2n}, of Earth's gravity field;
- tides (especially long-period tides)
- Un-modeled secular variations in the even zonal harmonics coefficients
- Other periodic and seasonal variations in the coefficients of Earth's gravity field

Among the non-gravitational perturbations, we have:

- Atmospheric drag
- Solar radiation pressure
- Earth's albedo radiation pressure
- Anisotropic thermal radiation (Yarkovsky–Schach effect)
- Anisotropic thermal radiation due to Earth's infrared radiation (Yarkovsky–Rubincam effect)

We then have random and stochastic observational errors and uncertainties in the determination of the orbital parameters and in particular the error due to the uncertainty in the orbital inclination induced by atmospheric refraction mismodelling (treated in the appendix, Section 10). In addition, we have studied and evaluated a conceivable bias in our measurement that we have called the "imprint" of the Lense–Thirring effect in the gravity field model (see below). A complete list of the various perturbations that have been studied in order to measure the Lense–Thirring effect using LAGEOS-type satellites is given in [55, 92].

8.1 Gravitational Perturbations

8.1.1 Static Gravitational Field – Even Zonals

The main source of error in the measurement of the Lense–Thirring effect using the nodes of LAGEOS and LAGEOS 2 is due to the uncertainties in the even zonal harmonics, J_{2n}, of Earth's gravity model. One may calculate the total error using the GRACE models covariance matrices. However, using the covariance matrix we can get some favorable correlation between the harmonics leading to an optimistic error budget [82]. Therefore, in order to get an upper bound to the total error due to the uncertainties in the even zonal harmonics, we can simply add the absolute values of the errors in our combination (5) due the uncertainty in each even zonal harmonic coefficient of, e.g., EIGEN-GRACE02S, EIGEN-GRACE03S, GGM02S, etc; the total error due to the static gravity field of these models is then respectively about 4% for EIGEN-GRACE02S, 6% for EIGEN-GRACE03S, and 8.7% for GGM02S.

We have also estimated the total error with EIGEN-GRACE02S and GGM02S by using their published uncertainties in the even zonal harmonics and by considering a diagonal covariance matrix (i.e. by assuming a negligible correlation among the even zonal coefficients which is certainly justified to a large extent, based on our experience with freely available covariance matrices from these new missions). We then found a tentative total error due to the static gravity field of about 3% of the Lense–Thirring effect for EIGEN-GRACE02S and 6.7% for GGM02S.

Let us further briefly discuss the errors due to the uncertainties of the static value of the J_{2n}. The values of the ten main even zonal harmonic coefficients C_{2n0} of, e.g., both the GRACE models EIGEN-GRACE02S and GGM02S, with their estimated errors δC_{2n0}, are listed in Tables 2 to 6, where the C_{l0} are the normalized

Table 2 Values of the five main even zonal harmonic coefficients of the EIGEN-GRACE02S model, with their estimated errors

l	C_{l0}	$\delta C_{l0}(\times 10^{-10})$
2	$-.484165197888 \times 10^{-3}$	0.5304
4	$0.539992946856 \times 10^{-6}$	0.03921
6	$-.149930382378 \times 10^{-6}$	0.02049
8	$0.494878910262 \times 10^{-7}$	0.01479
10	$0.533212229998 \times 10^{-7}$	0.02101

Table 3 Uncertainties in the nodal rates of LAGEOS and LAGEOS 2 due to the uncertainties in the EIGEN-GRACE02S C_{2n0} coefficients

l	$\delta\dot{\Omega}_I(\delta C_{l0})$	$\delta\dot{\Omega}_{II}(\delta C_{l0})$
2	$1.03\ \dot{\Omega}_I^{L-T}$	$1.89\ \dot{\Omega}_{II}^{L-T}$
4	$0.04\ \dot{\Omega}_I^{L-T}$	$0.01\ \dot{\Omega}_{II}^{L-T}$
6	$0.005\ \dot{\Omega}_I^{L-T}$	$0.008\ \dot{\Omega}_{II}^{L-T}$
8	$0.0003\ \dot{\Omega}_I^{L-T}$	$0.001\ \dot{\Omega}_{II}^{L-T}$
10	$0.0001\ \dot{\Omega}_I^{L-T}$	$0.0004\ \dot{\Omega}_{II}^{L-T}$

Table 4 Values of the five main even zonal harmonic coefficients of the GGM02S model, with their estimated errors

l	C_{l0}	$\delta C_{l0}(\times 10^{-10})$
2	$-0.48416970738820 \times 10^{-3}$	1.1
4	$0.53999162754299 \times 10^{-6}$	0.083
6	$-0.14993995936704 \times 10^{-6}$	0.045
8	$0.49481202573881 \times 10^{-7}$	0.028
10	$0.53320098349330 \times 10^{-7}$	0.02

Table 5 Uncertainties in the nodal rates of LAGEOS and LAGEOS 2 due to the uncertainties in the GGM02S C_{2n0} coefficients

l	$\delta\dot{\Omega}_I(\delta C_{l0})$	$\delta\dot{\Omega}_{II}(\delta C_{l0})$
2	$2.13\ \dot{\Omega}_I^{L-T}$	$3.91\ \dot{\Omega}_{II}^{L-T}$
4	$0.08\ \dot{\Omega}_I^{L-T}$	$0.03\ \dot{\Omega}_{II}^{L-T}$
6	$0.01\ \dot{\Omega}_I^{L-T}$	$0.02\ \dot{\Omega}_{II}^{L-T}$
8	$0.0005\ \dot{\Omega}_I^{L-T}$	$0.003\ \dot{\Omega}_{II}^{L-T}$
10	$0.0003\ \dot{\Omega}_I^{L-T}$	$0.0004\ \dot{\Omega}_{II}^{L-T}$

Table 6 Uncertainties in the combined nodal rates of LAGEOS and LAGEOS 2, according to formula (5), due to the published uncertainties in, respectively, the EIGEN-GRACE02S and GGM02S C_{2n0} coefficients

l	EIGEN-GRACE02S: $\delta\mu$ %	GGM02S: $\delta\mu$ %
2	0 % μ	0 % μ
4	3 % μ	6.4 % μ
6	0.9 % μ	2 % μ
8	0.1 % μ	0.2 % μ
10	0.05 % μ	0.05 % μ

zonal harmonic coefficients, $J_l \equiv -\sqrt{2l+1}C_{l0}$, and the J_l are the non-normalized zonal harmonic coefficients. The uncertainties in the nodal rates of LAGEOS and LAGEOS 2 due to the uncertainties in both the EIGEN-GRACE02S and GGM02S C_{2n0} coefficients are listed in Tables 3 and 5, where $\dot{\Omega}^{L-T}$ is the Lense–Thirring effect on the node. In Table 6 we show the errors in our combination of the nodes of LAGEOS and LAGEOS 2 due to the uncertainties in both the EIGEN-GRACE02S and GGM02S C_{2n0} even zonal coefficients.

From these uncertainties in the nodal rates of LAGEOS and LAGEOS 2, we see that as expected the dominant error source is due to the uncertainty in C_{20}. Smaller errors, compared to the Lense–Thirring effect, are due to the higher even zonal harmonics. Therefore, in order to get a measurement of the Lense–Thirring effect, one needs to at least eliminate the errors arising from C_{20}. This removal, achieved using formula (5), see Table 6, also includes the uncertainties in the temporal and seasonal variations of the even zonal harmonic C_{20} with a period longer than our 15-day arc period.

In summary, the sum of the absolute values of the errors in our combination (5) due to each even zonal's *published* uncertainty is respectively about 4% of the Lense–Thirring effect for EIGEN-GRACE02S, 6% for EIGEN-GRACE03S and 8.7% for GGM02S.

8.1.2 Odd Zonal Harmonics

The errors in the odd zonal harmonics, $J_{(2n+1)}$, have negligible effects on the secular nodal rates of LAGEOS and LAGEOS 2 compared to the Lense–Thirring drag [103].

8.1.3 Tides and Other Variations in the Gravity Field

In regard to what concerns the tidal perturbations and other temporal variations in the Earth gravity field, we used and adapted to the present analysis some detailed studies using the LAGEOS satellite [130] and previous results and error studies of the LAGEOS III/LARES experiment [55, 92]. Of course tidal perturbations have a different effect on the orbit of LAGEOS 2 in comparison to the orbit of the proposed LAGEOS III/LARES. First, we observe that one of the main tidal sources of error is due to the 18.6 year tidal orbital perturbation associated with the Lunar node. A large part of the error due to this 18.6 year tide is due to the uncertainty, δC_{20}, in its $l = 2$, $m = 0$ component. However, using the combination (5) of observables, not only the static part but also the seasonal, tidal, periodic (medium and long period) and secular variations of the largest harmonic coefficient C_{20} are removed in our measurement of μ. Thus, we only need to consider the smaller errors due to the other remaining higher degree coefficients: $C_{40}, C_{60}, \ldots$. Regarding the tidal perturbations, we analyzed the main tidal waves corresponding to: ($l = 2, m = 0$), ($l = 2, m = 1$), ($l = 2, m = 2$), ($l = 3, m = 0$), ($l = 3, m = 1$), ($l = 3, m = 2$), ($l = 3, m = 3$) and ($l = 4, m = 0$) on the node of LAGEOS and node of LAGEOS 2. Of course, as just explained, every effect of all the tides with ($l = 2, m = 0$) cancels out in our measurement of μ. For example, the largest tide on the LAGEOS node is the 18.6 year tide (of type $l = 2$, $m = 0$) associated with the

Table 7 Largest medium-long and long periodic effects on the node of the LAGEOS satellite (other than $l = 2, m = 0$)

P (days)	l	m	p
1,044	2	1	1
905	2	1	1
281	2	2	1

Table 8 Largest medium-long and long periodic effects on the node of the LAGEOS 2 satellite (other than $l = 2, m = 0$)

P (days)	l	m	p
569	2	1	1
111	2	2	1
285	2	2	1

Moon node; the corresponding amplitudes on the node of LAGEOS and node of LAGEOS 2 are [55, 92, 94]

$$\begin{aligned} \delta\,\Omega_I^{18.6y} &\approx -1080\,\text{mas}, \\ \delta\,\Omega_{II}^{18.6y} &\approx 1980\,\text{mas}. \end{aligned} \tag{6}$$

By combining these elements according to our formula (5), we thus have:

$$\delta\,\Omega_I^{18.6y} + 0.545\,\delta\,\Omega_{II}^{18.6y} \approx 0 \tag{7}$$

in agreement with what we expected.

As far as the other tides listed above (other than $l = 2$, $m = 0$), the largest medium-long and long periodic effects on the node of LAGEOS are listed in Table 7, and those on the node of LAGEOS 2 are listed in Table 8 (we simply give here the three largest constituents for each orbital element).

Let us estimate the uncertainty in each of these solid tides and the total uncertainty due to all the tidal perturbations. In order of magnitude, the uncertainty in the solid tides is roughly 3% of the total calculated amplitude. In addition one has to estimate the uncertainties in ocean and atmospheric tides. The response of the solid Earth to the tidal generating potential is roughly 90% of the total effect, the response of the ocean and atmosphere is roughly the remaining 10% of the total effect. However, there is a large uncertainty in the modeling of ocean tides due to the complexity of the phenomenon, then for ocean tides we consider an uncertainty of roughly 30% of the total effect. Thus, the uncertainty due to ocean tides is of the order of 3% of the total tidal effect, that is of the order of magnitude of the uncertainty due to solid tides. Therefore, to estimate the total error in our measurement of μ, we assume an uncertainty in the amplitude of solid and ocean tides of the order of 6% of the total amplitude [56].

Then, we note that all the tidal effects with $m \neq 0$ affecting the LAGEOS and LAGEOS 2 nodes have a period much shorter than our period of observation of about 11 years. Therefore, all the tides with $m \neq 0$ are substantially averaged out over our period of observation. Furthermore, corresponding to the gravity model JEM03G, in Fig. 5a we show the residuals fitted with a secular trend only, in

(Fig. 5b) the residuals fitted with a secular trend contemporarily with two main tidal signals with periods of 1,044 and 569 days, in (Fig.5c) with six larger tidal signals corresponding to periods of 1,044, 905, 281, 569, 111 and 285 days, and in (5d) with ten periodical signals corresponding to 1,044, 905, 281, 221, 522, 569, 111, 285, 621 and 182 days. The best-fit line in (Fig. 5a), with a secular trend only, without any periodic signal, has a trend of $\mu \cong 0.95$, in this case the root mean square of the post-fit residuals is about 12.5 mas. The best-fit line in (Fig. 5b), with two periodic terms, has a slope of $\mu \cong 0.986$ and the root mean square of the post-fit residuals is about 12.5 mas. The best-fit lines of Fig. 5c, d with respectively 6 and 10 periodic terms removed, have respectively a slope of $\mu \cong 0.987$ and $\mu \cong 0.984$, with root mean square of the post-fit residuals of about 5.9 mas and 5.2 mas respectively. Similar small variations, of the order of a few percent, are obtained with the other GRACE models. These small differences in the fit of the secular trend, with and without the periodic terms, reflect the uncertainties in the slope of the best-fit line due to aliasing of tidal effects and to other periodic perturbations, such as the non-gravitational perturbations described below. In the case of JEM03G, the maximum difference in the measured values of the Lense–Thirring effect, fitting different periodic signals, is about 3.7%. In the case of EIGEN-GRACE02S, the difference (in the measured value of the Lense–Thirring effect), between the fit with a secular trend only and the fit with a secular trend plus a number of periodic effects is only of about 1%; the largest difference between the different fits is the one between the fit with a secular trend plus two periodic terms, with slope 0.97, and the fits with a secular trend plus 6 or 10 periodic terms, both with slope 0.99; this difference is only about 2% of the Lense–Thirring effect, however this change also includes the error due to the main non-gravitational perturbations with the same periodicity (see below). Thus, in conclusion, over a long period of observation of about 11 years, there is a substantial reduction of the uncertainty due to tides. We also note that all the tidal effects with $l = 3$ have very short periods. In summary, we have calculated an error of less than 1% of μ from solid and ocean tides uncertainties: $\delta\mu^{Tides} \lesssim 1\%\mu$.

Let us now consider the seasonal variations in the J_{2n} coefficients and also the effects of the inter-annual variability in the seasonal excitation. We observe that: (i) the largest seasonal variations, those corresponding to J_2, are removed using the combination (5); (ii) the periods of observation are multiples of 365.25 days, thus all the perturbations with one year period are averaged out to zero in our measurement of μ; (iii) we fitted our residuals with a signal at 182.6 days and with a signal at 365.25 days that, however, had a negligible effect on the secular trend. We then conclude that the residual effect of seasonal variations in the even zonal harmonics with $2n \geq 4$, i.e., $J_4, J_6, \ldots$ is very small, substantially negligible in our measurement of μ with the nodes of the LAGEOS satellites. Similarly, regarding the effects due to inter-annual variability in the seasonal excitation, we first observe once again that the largest effect, the one associated with J_2, cancels out in our measurement. After removing secular and tidal signals from the LAGEOS node residuals [130], there are remaining signals that have their largest peaks, on the LAGEOS node, at about 248 days and 386 days (calculated in [130] by the FFT of the node residuals over about 19 years of LAGEOS observations). The corresponding amplitudes are about 10 mas/year. Therefore, in our measurement of μ, over a period of observation of about 11 years, the total effect of inter-annual variations is very small, substantially negligible (this is also true for other periodic effects analyzed in [130], such as the 2,317 days period effect) and we can consider this effect already included in our previous 1% tidal error budget.

In regard to un-modeled (or unknown) secular variations in the even zonal harmonics, the time variations of the even zonal harmonics coefficients were studied using the laser-ranged satellites LAGEOS and Starlette. An effective $\dot{J}_2$, or $\dot{J}_2^{eff}$, that is given by the combination of the $\dot{J}_{2n}$ effects on LAGEOS and Starlette, was defined as [130]:

$$\dot{J}_2^{eff}\,(LAGEOS) = \dot{J}_2 + .371\dot{J}_4 + .079\dot{J}_6 + .006\dot{J}_8 - .003\dot{J}_{10} + \cdots,$$

and

$$\dot{J}_2^{eff}\,(Starlette) = \dot{J}_2 + .040\dot{J}_4 - .555\dot{J}_6 - .150\dot{J}_8 + .283\dot{J}_{10} + \cdots.$$

The effective $\dot{J}_2^{eff}$ was then evaluated: $\dot{J}_2^{\,eff}\,(LAGEOS) \cong (-2.6 \pm 0.3) \times 10^{-11}$ year^{-1}, and $\dot{J}_2^{\,eff}\,(Starlette) \cong (-2.9 \pm 0.3) \times 10^{-11}$ year^{-1}.

Let us then estimate the maximum error due to the $\dot{J}_{2n}$ in our measurement of the Lense–Thirring effect. We study this source of error in the case of the analyses that we carried out using the gravity models produced [65] by using the data of the GRACE mission. Therefore, our error analysis relies on the results obtained from the GRACE mission.

In order to estimate the maximum possible error in the measurement of the Lense–Thirring effect due to the $\dot{J}_{2n}$, we first stress that all the effects due to $\dot{J}_2$ are removed using the combination (5) of observables and that the largest contribution to $\dot{J}_2^{eff}$ is due to $\dot{J}_2$ [130], see also [129].

During the development of the EIGEN-GRACE02S model [65], obtained on the basis of data from the GRACE mission only, the gravity field model was measured based on data observed during the period 2002–2003. Corrections due to $\dot{J}_2$ and $\dot{J}_4$ were then applied to this 2002–2003 measurement in order to obtain a gravity field model antecedent to 2002–2003. These values of $\dot{J}_2$ and $\dot{J}_4$, used by the GFZ team, are $\dot{J}_2 = -2.6 \cdot 10^{-11}$ and $\dot{J}_4 = -1.41 \cdot 10^{-11}$ and they were measured on the basis of completely independent 30-year observations before 2002. Any correction to the coefficients J_4, J_6 for the rates $\dot{J}_4$, $\dot{J}_6$ must be such that at the date of the GRACE measurements, e.g., 2003, it has to agree with the GRACE measured coefficients, e.g., at 2003.

In the data reduction of our present measurements of the Lense–Thirring effect [13], obtained using the GRACE Earth gravity models, we have used the value of $\dot{J}_4 = -1.41 \cdot 10^{-11}$, adopted by the GRACE teams [65], and in our analyses we have fitted our combined residuals with a secular trend only plus a number of periodic terms. We can of course introduce $\dot{J}_4$ as a free parameter in our fits (see below). In this case, together with the measurement of the Lense–Thirring effect, we also measure the effect of the secular variations of J_2, J_4 and J_6 on the combination of the nodal longitudes of the LAGEOS satellites; this is described by a $\dot{J}_4^{Effective}$ [130] in our combination, which includes the effect of the secular variation of the higher even zonal harmonics. In our EIGEN-GRACE02S analysis we have indeed observed an effective value of $\dot{J}_4^{Effective} \cong -1.5 \cdot 10^{-11}$ for the combination of the LAGEOS satellites nodes, which is consistent with the EIGEN-GRACE02S model since, on our combination of the nodal longitudes of the LAGEOS satellites, it represents just a 6% variation of its value given with EIGEN-GRACE02S and, however, it includes the effect of any higher $\dot{J}_{2n}$, with $2n \geq 4$; this value is also fully consistent with our measurement of the Lense–Thirring effect with an estimated uncertainty of the order of 10% [13, 15, 129].

Let us describe the result of the orbital analyses using the orbital estimator GEODYN with and without a contribution of $\dot{J}_4 = -1.41 \cdot 10^{-11}$, and the result of a simulation with a $\dot{J}_4$ equal to 611% of the value given in EIGEN-GRACE02S, i.e. $\dot{J}_4 \simeq -8.61 \cdot 10^{-11}$.

First, we stress that in the case of *not* applying the $\dot{J}_4$ correction to the orbital analysis its effect may be identified, at least for large values of $\dot{J}_4$ (and, however, completely fitted for, see below), even by visual inspection, indeed, since $\dot{J}_4$ produces a quadratic term in the integrated residuals, it shows up as a hump in the combined residuals [110, 129], whereas in our fits with EIGEN-GRACE02S (see Fig. 4a and [13]), especially after de-trending the combined residuals, the absence of a large quadratic effect is evident. Indeed, in [110], we display the results of a simulation of the orbital residuals (see Fig. 5 on p. 639 of [110]) obtained using in the data reduction an unrealistic value of $\dot{J}_4$, the presence of a quadratic term in the integrated residuals, induced by this value of $\dot{J}_4$, can be easily seen even by visual inspection and fitted for. Then, since the effect of the time variation $\dot{J}_4$ shows up as a quadratic effect in the cumulative nodal longitude of the LAGEOS satellites, the combined residuals of LAGEOS and LAGEOS 2 may be fitted with a quadratic curve, together with a straight line and with the main periodic terms, by imposing that the fitting parabola has the vertex at the time of the GRACE measurements of J_4, i.e., that the changing values of the J_4 must however *at the time of the GRACE measurements* agree with the GRACE measured value.

In other words, the main relevant secular effect that can mimic a secular Lense–Thirring trend is due to the error in the values of the static J_{2n} coefficients used in our analysis, including however the error from mapping the even zonals to a mean SLR epoch, whereas the $\dot{J}_{2n}$ will show up as quadratic effects. For example, in Fig. 3, in the single, i.e., not combined, residuals of the node of LAGEOS (in blue) and of the node of LAGEOS 2 (in green) is observed a long term variation in the trend that, since it disappears in the nodes combination (in red) of Fig. 3, can be identified as an anomalous increase in the Earth quadrupole moment, corresponding to the effect observed by Cox and Chao [128].

Indeed, by fitting the raw residuals obtained *without* any $\dot{J}_4$ in a *test-analysis* using EIGEN-GRACE02S, see [110], we measured a $\dot{J}_4^{Effective} \cong -1.5 \cdot 10^{-11}$, which includes the effect of $\dot{J}_6$ and of higher even zonal harmonics on the combination of the LAGEOS nodes. On other hand, by fitting the combined residuals obtained with the EIGEN-GRACE02S correction of $\dot{J}_4 = -1.41 \cdot 10^{-11}$, we measured a $\delta \dot{J}_4^{Effective}$ of less than $-0.1 \cdot 10^{-11}$, see [110], and we thus recovered the value of $\dot{J}_4$ determined in the case of the previous test-analysis where we applied no $\dot{J}_4$ correction. In all the cases of inclusion of a different $\dot{J}_4$ in the data analysis, we obtained a measured value of $\mu \cong 0.98$, see [110]. Therefore, the small value of the un-modeled quadratic effects in our nodal combination due to the un-modeled $\dot{J}_{2n}$ effects (with $2n \geq 4$) corresponds to a change in the measured value of the Lense–Thirring effect of about 1% only. As a *consistency test*, using the value $\dot{J}_4^{Effective} = -1.5 \cdot 10^{-11}$, obtained from fitting the combined residuals (which is only about 6% larger than the value $\dot{J}_4 = -1.41 \cdot 10^{-11}$ used in most GRACE models), we have again generated the residuals and fit these with a straight line *only* plus the main periodic terms. It turned out that the change of the measured value of frame-dragging was about 1% only with respect to the case of using $\dot{J}_4 - 1.41 \cdot 10^{-11}$ and fitting the residuals with a straight line only, plus the main periodic terms, i.e., also this test resulted in a measured value of $\mu \cong 0.98$ [110]. The result of this test agrees with the case of fitting the combined orbital residuals with a straight line plus a quadratic curve, indeed in *all cases* we obtained $\mu \cong 0.98$ [110]. As a third case, using in the data reduction a highly unrealistic value of $\dot{J}_4$ equal to the value adopted in EIGEN-GRACE02S *plus* 12 times its published error, i.e., a $\dot{J}_4$ equal to about 611% of the value given in EIGEN-GRACE02S, that is $\dot{J}_4 = 6.11 \times (-1.41 \cdot 10^{-11}) \simeq -8.61 \cdot 10^{-11}$, by fitting the residuals with a straight line plus a parabola plus the main periodic terms, we again obtained the same measured value of frame-dragging, i.e., $\mu \cong 0.98$, see [13]. Therefore, also in this case the measured value of frame-dragging differs by about 1% *only* with respect to the case of using $\dot{J}_4 = -1.41 \cdot 10^{-11}$ without fitting a quadratic curve, i.e., the result reported in [13]. These analyses clearly show that even in the cases of -100% and $+511\%$ variations between the value of $\dot{J}_4$ of Mother Nature and the value used in the orbital estimation with EIGEN-GRACE02S, it is possible to fit for the $\dot{J}_4^{Effective}$ effect using a quadratic and recover from the fit most of the simulated variation in $\dot{J}_4$ and demonstrate that, in the case of the analysis with EIGEN-GRACE02S, the measured value of the Lense–Thirring effect can only be affected at a level of the order of 1% by the $\dot{J}_{2n}$ uncertainties. Indeed, in these three cases corresponding to different values of $\dot{J}_4$ (zero, the EIGEN-GRACE02S value, and 611% of this EIGEN-GRACE02S value), we always obtained $\mu \cong 0.98$ (see [110]), whereas without fitting any quadratic we obtained $\mu \cong 0.99$.

In conclusion, in the case of the analysis with EIGEN-GRACE02S, this 1% variation, gives the order of magnitude of the estimated error due to the $\dot{J}_{2n}$ uncertainties in our measurement of μ, as reported in [13].

Another way to look at the interdependency between the Lense–Thirring effect and the value of $\dot{J}_4$ is through a χ^2 plot. Indeed in [110] we discuss and give the ellipse representing a confidence level for a 99% probability, see Fig. 15. This ellipse, in the case of EIGEN-GRACE02S, shows that the uncertainty in the Lense–Thirring parameter arising from the uncertainty of $\dot{J}_4$ is just of the order of 0.75% of its value (Fig. 12).

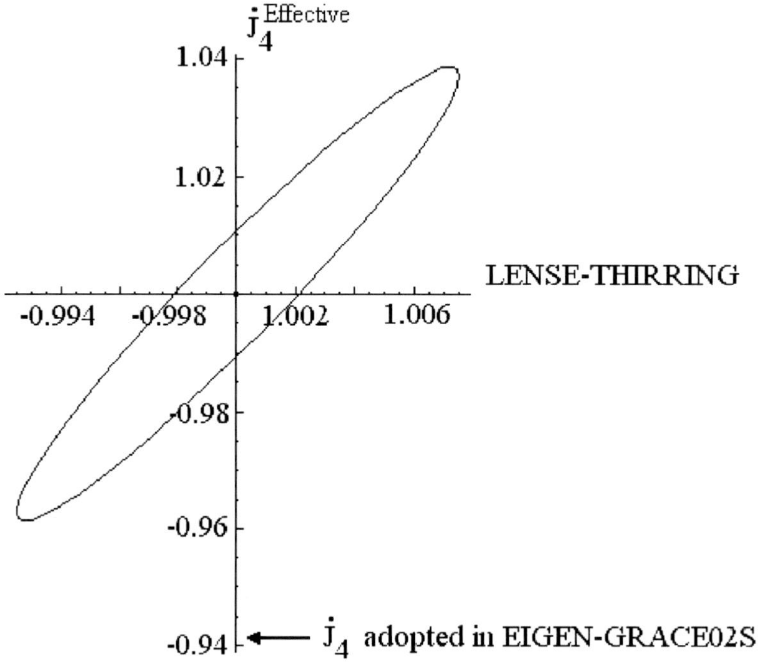

Fig. 12 Confidence Ellipse. On the x axis is the Lense–Thirring effect; the origin, at $x = 1$, corresponds to the fitted value of frame-dragging. On the y axis is $\dot{J}_4^{Effective}$; the origin, at $y = 1$, corresponds to the fitted value $\dot{J}_4^{Effective} \cong -1.51 \cdot 10^{-11}$. This confidence ellipse is tentative since we have assumed an average tentative standard deviation for each combined residual. This figure shows that, in the case of the EIGEN-GRACE02S fits, the uncertainty in the measured Lense–Thirring parameter arising from the uncertainty of $\dot{J}_4$ is of the order of 0.75%

We also stress that in Section 6.3 and in [110] we have presented a worst case error analysis based on the various $\dot{J}_4$ estimates published between 1997 and 2004, and in [15] and [110] the various published values of $\dot{J}_4$ are reported and compared with their mean.

Finally, since the measurements of $\dot{J}_{2n}$ and J_{2n} are independent (as previously explained), the corresponding errors are independent and this allows us to take the RSS of these errors. Nevertheless, even by just adding up these errors we would still get a total error budget of about 10% [129].

In summary, by assuming a total error of about 2% due to the secular variations of the J_{2n} with $2n \geq 4$, obtained by multiplying by a factor 2 the estimated error due to $\dot{J}_4$ in the case of EIGEN-GRACE02S, we have a total effect from tides and other temporal variations in Earth's gravity field:

$$\delta\mu^{\textit{Time variations of even zonals}}_{+ \textit{Tides}} \lesssim 3\%\mu \tag{8}$$

Thus, in conclusion, as far as errors caused from uncertainties in the gravitational perturbations, for the models EIGEN-GRACE02S and EIGEN-GRACE03S, by multiplying by a factor 2 (see Section 6) the error obtained by propagating the GRACE models uncertainties in the nodes combination (5), we have:

$$\delta\mu^{EIGEN-GRACE02S^{Static\ even\ zonals}} \sim 8\%\mu \tag{9}$$

$$\delta\mu^{EIGEN-GRACE03S^{Static\ even\ zonals}} \sim 12\%\mu \tag{10}$$

$$\delta\mu^{Time\ variations\ of\ even\ zonals}_{+\ Tides} \lesssim 3\%\mu \tag{11}$$

Whereas, in the case of JEM03G, the calibrated, i.e., including systematic errors, uncertainties were *not* available to us and we have estimated a grand total error for JEM03G between 10% and 18%.

8.2 Non-Gravitational Forces

Let us now study the main non-gravitational perturbations acting on the LAGEOS satellites: atmospheric drag, direct solar radiation pressure, Earth's albedo radiation pressure, anisotropic thermal radiation (or Yarkovsky–Schach effect) and anisotropic thermal radiation from Earth's infrared radiation (Earth–Yarkovsky or Yarkovsky–Rubincam effect). The effect of these and other perturbations on the node of LAGEOS type satellites has already been extensively treated in numerous papers and NASA studies, see, e.g. [54, 55, 58, 60, 75–79, 92, 93, 104].

8.2.1 Drag-Like Forces

The neutral particle drag acting on a satellite can be written: $\vec{F} = -\frac{C_D}{2} A\rho v\vec{v}$, where C_D is the drag coefficient which depends on the shape of the satellite and on the manner of reflection of the molecules ($C_D \cong 4$ for LAGEOS), A is the cross sectional area, ρ is the density of the atmosphere and $\vec{v}$ is the velocity of the satellite relative to the atmosphere. At the LAGEOS altitude, the neutral plus charged particle drag force acting on a satellite can approximately be found by multiplying the neutral drag force by a well determined factor. In [55] it was shown that the particle drag does not significantly contribute to the total nodal precession, the corresponding error is at most at the level of 10^{-3} of the Lense–Thirring effect.

8.2.2 Radiation Pressure

The un-modeled effects of the direct radiation pressure on the LAGEOS node were calculated in [55, 92]. The result is that the average error on the node of LAGEOS due to un-modeled direct solar radiation pressure is less than 10^{-3} of the Lense–Thirring effect.

This result was derived by assuming an uncertainty of the order of 1% in the value of the instantaneous force on LAGEOS due to direct solar radiation pressure. The error in the value of the reflectivity of LAGEOS has been estimated (see [55, 56]) to be about 5%, or less, of its modeled value, however, its effect is only about 13% of the total (absorbed plus reflected) direct solar radiation pressure effect, and the error in the value of the solar constant has been evaluated to be of the order of 0.3%. Indeed, the solar constant is estimated to vary by about 0.1%, but the measurement uncertainties of the solar irradiance might reach 0.3% (see [55, 56]). In addition, we observe that, in some of our orbital analyses, for each 15-day arc, we estimated and then adjusted the values of the total (absorbed plus reflected) direct solar radiation pressure coefficients of LAGEOS and LAGEOS 2. Therefore, since the solar radiation pressure has a very distinct signature which allows us to distinguish it well from other forces, the estimation and adjustment of the solar radiation pressure coefficient, for each 15-day arc, allowed a further reduction of the total mean uncertainty in the value of the effects due to direct solar radiation pressure. We therefore assume here a mean uncertainty of about 0.5% in the value of the *total*, absorbed plus reflected, direct solar radiation pressure acceleration, that is equivalent to an uncertainty of about 5% in the satellites' reflectivity coefficients. We note that the main direct solar radiation perturbations on the nodes of the LAGEOS satellites have a period much shorter than our periods of observation of 11 and 13.5 years. Therefore, the direct solar radiation perturbations on the nodes are in part averaged out to zero over our much longer period of observation. Furthermore, it is important to stress that, together with a secular trend and with the largest tidal effects, we have also fitted our residuals with the, a priori, largest radiation pressure signals, corresponding to $\Omega - \lambda_\odot$, $\Omega + \lambda_\odot$ and $\lambda_\odot$ (where Ω is the longitude of the ascending node and $\lambda_\odot$ the ecliptic longitude of the Sun), i.e. corresponding to 560, 271 and 365.25 days for LAGEOS, and 226, 953 and 365.25 days for LAGEOS 2 [55, 92, 93, 104]; in this case there was a very small change of only a few parts in 10^{-3} of the value of the measured Lense–Thirring effect, mainly due to the errors in the orbital effects due to albedo with the same periodicities [55, 92, 93, 104] and thus already included (below) in the uncertainties due to albedo.

Thus, the maximum error in our measurement of μ due to direct solar radiation pressure is:

$$\delta\mu^{Direct\,solar}_{radiation} \lesssim 10^{-3}\mu. \qquad (12)$$

8.2.3 Albedo

In a way similar to the analysis of direct solar radiation pressure, we have studied the effects of Earth albedo radiation pressure using three methods: the first is based on analytical calculations, using the orbital perturbation equations, and on extensive works by Martin and Rubincam [78]. The second method is based on two simulations using GEODYN, with and without modeling Earth albedo pressure in the orbital force model, and subsequently analyzing (with Fourier transform) the orbital residuals obtained by taking the difference between the corresponding two sets of simulated orbital elements. The third method is the determination of the maximum deviation of the secular trend of our combination of observables (5) when together with a trend we fitted the residuals with the a priori largest albedo signals, corresponding to $\Omega - \lambda_{\odot}$, $\Omega + \lambda_{\odot}$ and $\lambda_{\odot}$, i.e. corresponding to 560, 271 and 365.25 days for LAGEOS, and 226, 953 and 365.25 days for LAGEOS 2 [92, 93, 104]; using this third method, whether including or not these frequencies in the fit, there was a very small change of only a few parts in 10^{-3} of the value of the measured Lense–Thirring effect, furthermore it is important to note that this small deviation is also due to other gravitational and non-gravitational perturbations with the same periods.

In summary, for the effects of Earth albedo on the nodes of LAGEOS and LAGEOS 2, our analyses have confirmed the numerous and extensive studies of the LAGEOS III experiment and many related papers, see, e.g. [55, 92, 93, 104] and have shown that the errors in the adopted Earth albedo model affect the node of LAGEOS-type satellites with an uncertainty of the order of:

$$\delta\dot{\Omega}^{Albedo} \lesssim 1\%\dot{\Omega} \tag{13}$$

and, for our combination (5) for μ:

$$\delta\mu^{Albedo\,on\,\dot{\Omega}} \lesssim 1\%\mu. \tag{14}$$

This result has also been confirmed by our orbital simulations, with and without the Earth albedo force in GEODYN, and it is based on an estimated error in the adopted Earth albedo model of the order of 10% [92, 93, 104].

8.2.4 Satellite Eclipses

In regard to what concerns the LAGEOS eclipses by Earth, by assuming an error of about 10^{-2} rad in the determination of the boundary of the shadow region generated by Earth, the maximum error in the LAGEOS nodal precession due to mismodeling of the satellite eclipses by Earth is [55, 92, 93, 104]:

$$\delta\dot{\Omega}^{Eclipses} \lesssim 10^{-3}\dot{\Omega}^{Lense-Thirring}. \tag{15}$$

We then have:

$$\delta\mu^{Eclipses} \lesssim 10^{-3}\mu. \tag{16}$$

This estimation of the error due to the mismodeling of the boundary of Earth's shadow was performed by simulating the time evolution of the orbital elements of LAGEOS 2 in two cases having a difference of about 10^{-2} rad in the boundary of the shadow region at the LAGEOS 2 altitude and by then analyzing the effects of this difference on the nodal rate of LAGEOS 2.

8.2.5 Yarkovsky–Schach and Yarkovsky–Rubincam Effects

Finally, let us consider the uncertainty in our measurement of μ due to anisotropic thermal radiation from LAGEOS, a phenomenon known as Yarkovsky or Yarkovsky–Schach effect, and the uncertainty due to the thermal relaxation time of the LAGEOS retro-reflectors re-emitting anisotropically thermal radiation after having absorbed Earth's infrared radiation, a phenomenon known as Yarkovsky–Rubincam or Earth–Yarkovsky effect. This effect [75–77] (see below) shows the extremely high level of accuracy reached in observing, studying and modeling the tiniest perturbations acting on LAGEOS-type satellites.

The electromagnetic radiation from the Sun and the radiation from Earth each instantaneously heat one hemisphere of LAGEOS. Because of the finite heat conductivity of the body, there is an anisotropic distribution of temperature on the satellite. Thus, according to the Stefan–Boltzmann law, there is an anisotropic flux of energy ($\sim T^4$) and momentum from the surface of the satellite giving rise to its acceleration. However, if the satellite is spinning fast enough, the anisotropy in the satellite temperature distribution is mainly latitudinal.

The acceleration, $a_{\Delta T}$, due to this "thermal thrust", acting along the satellite spin axis, is thus: $a_{\Delta T} \sim \frac{4\varepsilon\pi r_L^2 \, \sigma \, T^3 \, \Delta T}{c m_L}$ where $\varepsilon \cong 0.4$ is the emissivity coefficient of LAGEOS, $r_L = 30$ cm its radius, $m_L = 4.1 \times 10^5$ g its mass, $\sigma = 5.67 \times 10^{-8}$ W m^{-2} K^{-4} Stefan's constant, T the average equilibrium temperature of LAGEOS and ΔT the temperature gradient between the two LAGEOS hemispheres. For example, for $T \cong 280$ K and $\Delta T = 5$ K, we get an acceleration $a_{\Delta T}$ of the order of: $a_{\Delta T} \sim 2 \cdot 10^{-9}$ cm/sec^2, constant in direction along one LAGEOS orbit (assuming that the satellite spin axis is substantially constant along one orbit). We insert this value in the equation for the rate of change of the nodal longitude:

$$\frac{d\Omega}{dt} = \frac{1}{nasin\, I} \left(1 - e^2\right)^{-\frac{1}{2}} f \, W \, \frac{r}{a} \, sin\,(\nu + \omega),$$

where $n = 2\pi/P$ is the orbital mean motion, $P = 3.758$ h the LAGEOS orbital period, $a = 12{,}270$ km its orbital semimajor axis, $I \cong 109°.94$ its orbital inclination, $e = 4 \times 10^{-3}$ its orbital eccentricity, f the magnitude of the external force per unit satellite mass, W the direction cosine of the force f along the normal to the orbital plane, ν the true anomaly and ω the argument of the pericenter. We then integrate this expression over one orbital period, $P = 3.758$ h, and neglecting for the moment the satellite eclipses by Earth and Earth radiation, we have (in this calculation W is assumed to be constant along one orbit):

$$\dot{\Omega}^{\Delta T_{\odot}} \lesssim 2 \cdot 10^{-3} \, \dot{\Omega}^{Lense-Thirring}. \tag{17}$$

Therefore, over one orbit, the secular nodal precession due to the LAGEOS thermal anisotropy from the Sun radiation is negligible when the satellite eclipses and Earth radiation are not considered. In regard to the node of LAGEOS 2, considering that its eccentricity is larger than that of LAGEOS ($e_I = 4 \times 10^{-3}$ and $e_{II} = 0.014$) and that the nodal rate of LAGEOS 2 enters with the factor 0.545 in our combination (5) for μ, we get, in our measurement of μ, an error from the node of LAGEOS 2 of the same order of magnitude of the error from the node of LAGEOS. When the satellite orbit is partially in the shadow of Earth, the nodal precession due to the satellite thermal anisotropy is no longer reduced, over one orbit, by a factor of the order of the orbital eccentricity. However, even by considering the satellite eclipses, over our period of observation of about 11 years, we have [55, 92, 93, 104]:

$$\delta\mu^{\Delta T_{\odot}} \lesssim 1\%\mu. \tag{18}$$

Let us finally discuss the Earth–Yarkovsky or Yarkovsky–Rubincam effect. Rubincam discovered that the infrared radiation from Earth plus the thermal inertia of the LAGEOS retro-reflectors can cause a force on the satellite. Infrared radiation from Earth is absorbed by the LAGEOS retro-reflectors, therefore, due to their thermal inertia and to the rotation of the satellite, there is a LAGEOS latitudinal temperature gradient. The corresponding thermal radiation causes an acceleration with an along-track component opposite to the satellite motion. This anisotropic thermal radiation may cause an acceleration of LAGEOS of the order of 10^{-10} cm/s^2, variable in direction, and this thermal, drag-like, force may explain about 70% of the observed, anomalous, LAGEOS semimajor axis decrease of 1.2 mm/day, due to a peculiar along-track acceleration of about -3×10^{-10} cm/s^2 [75–77]. The remainder is probably due to neutral and charged particle drag.

Let us then analyze the Yarkovsky–Rubincam effect on the node. We have: $\vec{a}_{Rubincam} = (k'' \cos\psi)\,\hat{l}$, where $k'' = 6.8 \times 10^{-10}$ cm/s^2, $\hat{l}$ is the spin unit-vector of LAGEOS 2 that, for the moment, we assume to have constant direction over one orbital period of about 3.8 h and $\cos\psi = \hat{l}\cdot\hat{r}_{\alpha}$, where $\hat{r}_{\alpha}$ is the radial unit-vector including the correction for the thermal lag angle of the satellite due to the thermal inertia of the silica retro-reflectors. By integrating the nodal rate equation over one orbital period, we have: $\dot{\Omega} \cong [W/(na \sin I)] \sin\lambda_{\odot}$. Considering the secular effects on $\dot{\Omega}$ only, we then have: $< \dot{\Omega} >^{secular}_{Rubincam} \cong (k''/4na)\, \cos\theta\, \cos I(3{l_z}^2 - 1)$, where $\frac{k''}{4na} \cong 1.9$ mas/yr and the thermal lag angle, θ, is assumed to be: $\theta = 55°$ [92, 93, 104]. For a worst case analysis we thus set $l_z \equiv 1$ and we get on LAGEOS 2: $< \dot{\Omega} >^{Max\ secular}_{Rubincam} \cong 1.3$ mas/year. The periods of the LAGEOS 2 nodal perturbations due to the Yarkovsky–Rubincam effect are: $P(\Omega) \cong 569$ days, with amplitude 0.25 mas/year, and $P(2\Omega) \cong 285$ days, with amplitude 0.02 mas/year. Therefore, these components of the nodal precession associated with the Yarkovsky–Rubincam effect have the period of the node and half of the nodal period. Thus, they give a negligible uncertainty when averaged over the period of the node. Furthermore, in some of the fits of the combined residuals and in the corresponding determinations of μ, we have also fitted for a signal with a period of 569 days, with the result of no substantial change in the value of the secular trend, i.e. of the Lense–Thirring effect.

We then finally get the maximum error $\delta\mu$, in our measurement of μ, due to the Yarkovsky–Schach and Yarkovsky–Rubincam effects:

$$\delta\mu^{Yarkovsky} \lesssim 1\%\mu. \tag{19}$$

In summary, due to the non-gravitational perturbations, we have:

$$\delta\mu^{\textit{Direct solar radiation}} \lesssim 10^{-3}\mu;\ \delta\mu^{\textit{Earth albedo}} \lesssim 1\%\mu;$$
$$\delta\mu^{\textit{Satellite eclipses}} \lesssim 10^{-3}\mu;\ \delta\mu^{\textit{Yarkovsky}} \lesssim 1\%\mu;$$

thus, by just summing these errors, the maximum total error due to the non-gravitational perturbations is:

$$\delta\mu^{\textit{Non-gravitational perturbations}} \lesssim 2.2\%\mu. \tag{20}$$

8.3 Other Error Sources

As far as other conceivable error sources, a possible bias in our measurement of μ might be due to some "imprint" or "memory" of the a priori value of the Lense–Thirring effect used in the determination of the gravity model (see [60]).

Most of the GRACE gravity field models that we used in our analysis have been obtained using the observations of the GRACE spacecraft *only* and by using a set of models and parameters to describe the orbital perturbations. For example, the GFZ models EIGEN-GRACE02S and EIGEN-GRACE03S have been obtained without the Lense–Thirring effect in the data analysis. Thus, the GRACE gravitational field models that we used in our experiment to measure μ, may contain some kind of "imprint" of the a priori, theoretical, value of the Lense–Thirring effect, i.e., zero in the case of the GRACE models that we used in our measurement of μ.

In particular, in the present analysis we are only concerned about a conceivable imprint on the Earth's even zonal harmonics J_{2n}. In the GRACE models the gravitational potential of Earth is measured by observation of the rate of change of the GRACE inter-satellite distance [65], i.e., the gravitational field is measured by observation of the relative acceleration of two test particles (the equation linking relative acceleration and gravitational field is, in general relativity, the geodesic deviation equation [3]). However, the effect of the gravitomagnetic field on the acceleration vector of a satellite [3], decomposed in out-of-plane direction, $\perp$ (orthogonal to the satellite orbital plane); along track, $\hat{\mathbf{v}}$ (along the satellite velocity vector); and radial direction, $\hat{\mathbf{r}}$ (along $\mathbf{r}$), is given by $\mathbf{a} = \mathbf{v} \times \mathbf{H}$, where $\mathbf{H} = \frac{2\,[\mathbf{J} - 3\,(\mathbf{J}\cdot\hat{\mathbf{x}})\,\hat{\mathbf{x}}]}{|\mathbf{x}|^3}$ is the gravitomagnetic field and $\mathbf{v} = v\hat{\mathbf{v}}$. For a polar satellite with a circular orbit we then have $a_{\perp}^{L-T} \sim \sin\phi$, where ϕ is the latitude, $a_{\hat{\mathbf{r}}}^{L-T} = 0$ and $a_{\hat{\mathbf{v}}}^{L-T} = 0$. However, the effect of the even zonal harmonics on the acceleration vector of a satellite is given by $\mathbf{a} = \mathbf{\nabla U}$ and the terms of the potential U generated by the mass, M, of the central body and by the even zonal harmonics, J_{2n}, are [103]: $U = \frac{M}{r}\,[1 - J_2\,(\frac{R_\oplus}{r})^2 P_{20}(\sin\phi) - J_4(\frac{R_\oplus}{r})^4 P_{40}(\sin\phi) - ...]$, where J_2 and J_4 are the un-normalized even zonal harmonic coefficients of degree 2 (quadrupole) and 4, $R_\oplus$ is Earth's equatorial radius and $P_{20}(\sin\phi)$ and $P_{40}(\sin\phi)$ are the Legendre associated functions. For a polar satellite with a circular orbit we then have $a_{\perp}^{J_{2n}} = 0$,

$a_{\hat{\mathbf{r}}}^{J_{2n}} \neq 0$ and $a_{\hat{\mathbf{v}}}^{J_{2n}} \neq 0$. Then, for a polar satellite with a circular orbit the satellite acceleration generated by the even zonal harmonics is just orthogonal to satellite acceleration generated by the gravitomagnetic field (i.e. the Lense–Thirring effect). In conclusion, the static even zonal harmonics determined by a polar satellite with a circular orbit are independent of the a priori value used for the Lense–Thirring effect.

Indeed, the orbit of the GRACE satellites is nearly circular with eccentricity <0.005 and basically polar with inclination $= 89.5°$, thus the values of the even zonal harmonics determined from the GRACE orbital perturbations are substantially independent of the a priori value of the Lense–Thirring effect. The small deviation from a polar orbit of the GRACE satellite, that is 1.7×10^{-2} rad, gives only rise, *at most*, to a very small correlation with a factor 1.7×10^{-2}. For the quasi-polar satellite CHAMP this correlation is, *at most*, $\sim 4.7 \times 10^{-2}$. In addition, since our results are independent of the error in the quadrupole coefficient J_2, any bias in J_2 due to a Lense–Thirring "imprint" is removed in our combination (5). Furthermore, the Lense–Thirring effect depends on the third power of the inverse of the distance from the central body, i.e., $(1/r)^3$, and the J_2, J_4, J_6 ... effects depend on the powers $(1/r)^{3.5}$, $(1/r)^{5.5}$, $(1/r)^{7.5}$ ... of the distance; then, since the ratio of the semimajor axes of the GRACE satellites to the LAGEOS' satellites is $\sim \frac{6780}{12270} \cong \frac{1}{1.8}$, any conceivable "Lense–Thirring imprint" on the spherical harmonics at the GRACE altitude becomes quickly, with increasing distance, a negligible effect, especially for higher harmonics of degree $l \geq 4$. Therefore, any conceivable "Lense–Thirring imprint" is negligible at the LAGEOS' satellites altitude. Furthermore in [60], it is proved with several simulations that by far the largest part of this "imprint" effect is absorbed in the by far largest coefficient J_2.

In conclusion, any error due to a conceivable "imprint" of the Lense–Thirring effect on Earth's even zonal harmonics of the GRACE models (and of the other CHAMP- and GRACE-only models) is negligible on the combination (5) of the nodes of LAGEOS and LAGEOS 2 and therefore in our measurement of the Lense–Thirring effect (see also [129]).

The secular nodal rate of a satellite is produced by the Earth's even zonal harmonics (and by the Lense–Thirring effect) and is described by a nodal rate equation that is a function of the Earth's mass, radius and even zonal harmonics, and of the satellite orbital parameters a, semimajor axis, I, inclination and e, eccentricity (see Eq. (4)), therefore any error in the determination of the satellite orbital parameters will produce an error in the calculated nodal longitude of the satellite. The nodal rate is especially sensitive to errors in the satellite inclination. The error in the nodal rate due to the bias in the variations of the inclinations of LAGEOS and LAGEOS 2, used in our data reduction will be treated in the next section.

We finally observe that the small changes of the semimajor axes of LAGEOS and LAGEOS 2 are absorbed in the empirical, 15-day, along-track acceleration, evaluated and adjusted in our analysis for each 15-day arc; however, this constant acceleration has no effect on the node of a satellite that is not affected by along-track accelerations. Similarly, the uncertainty in the eccentricity has a negligible effect on the nodal rate. We finally note that any uncertainty in the GM value of Earth is removed, using our combination (5), in the main J_2 terms of the equations for the nodal rate (4) and is very small, in comparison with the Lense–Thirring effect, in the other J_{2n} terms with $2n \geq 4$.

8.4 Summary of Error Budget

In summary, for EIGEN-GRACE02S, EIGEN-GRACE03S and JEM03G, we have:

$\delta\mu^{Even\ zonal\ harmonics-EIGEN-GRACE02S} \sim 8\%\,\mu;$
$\delta\mu^{Time\ variations}_{in\ gravitational\ field\ and\ Tides} \sim 3\%\,\mu$
$\delta\mu^{Non\ gravitational}_{perturbations} \lesssim 3\%\,\mu;$
$\delta\mu^{Other\ Error\ Sources} \lesssim 2\ \%\mu.$

Therefore, by taking into account all the error sources, for EIGEN-GRACE02S we finally find a grand total RSS error:

$$(\delta\mu/\mu)^{EIGEN-GRACE02S} \sim 9\%\,\mu\ . \tag{21}$$

Whereas for EIGEN-GRACE03S we have:

$\delta\mu^{Even\ zonal\ harmonics-EIGEN-GRACE03S} \sim 12\%\,\mu;$ and we finally find a grand total RSS error for EIGEN-GRACE03S:

$$(\delta\mu/\mu)^{EIGEN-GRACE03S} \sim 13\%\,\mu\ . \tag{22}$$

Whereas, in the case of JEM03G, the calibrated, i.e., including systematic errors, uncertainties were *not* available to us and we have estimated a grand total error for JEM03G between 10% and 18%.

In conclusion, using the latest GRACE gravity models our measurement of the Lense–Thirring effect has a total error, including systematic errors, of the order of 10%. This error analysis is in very good agreement with that of [15] for the latest GRACE models that is summarized in Table 9.

9 Appendix 2: Some Comments on the Error Analysis and Error Budget of the Gravitomagnetism Measurements with LAGEOS and LAGEOS 2

A single author has claimed in a number of papers (see, e.g., [131–134]) that the error analysis and error budget of the measurement of gravitomagnetism using LAGEOS and LAGEOS 2, and using the forthcoming LARES satellite, has been underestimated. We demonstrate here that these various claims of are ill-founded. These claims are mainly based on the following arguments: (a) the effect of the

Table 9 CSR-UT Estimated Error Budget for Lense–Thirring Test [15]

Error Source	% of Lense–Thirring
Scatter due to method (linear fit w/wo tidal lines)	1
Solar radiation pressure, Earth albedo, thermal reradiation effects	3
Zonal rates (quadratic effect; after mapping to mean epoch)*	1
C_{40} (estimated from scatter of GRACE gravity models)**	10
C_{60} (estimated from scatter of GRACE gravity models)**	5
$\dot{C}_{40}$ (20% uncertainty in mapping to mean epoch)***	3
$\dot{C}_{60}$ (50% uncertainty in mapping to mean epoch)***	2
RSS % of Lense–Thirring	12

* Epoch of GRACE gravity models typically ~2004.0–2005.0; mean epoch of SLR data ~2000.0.
** Assigned sigmas typically too small; used C_{40} scatter $1.3 \cdot 10^{-11}$, C_{60} scatter $1.0 \cdot 10^{-11}$.
*** $\dot{C}_{40}$ uncertainty is estimated to be 20% of $4.7 \cdot 10^{-12}$/year and for $\dot{C}_{60}$ 50% of $1.7 \cdot 10^{-12}$/year.

uncertainties in the measured rate of change of J_4 and of higher even zonal harmonics, (b) the unmodeled changes in the inclination due to atmospheric drag, (c) the reliability of the published uncertainties of the GRACE gravity field models and (d) in the case of the LARES experiment, the inclusion of the even zonal harmonics of degree higher than 20 in the calculation of the total uncertainty.

These claims are based on, at least, the following misunderstandings or miscalculations (as explained in details in [15, 110, 129, 135, 136]). In regard to the claims (a), this author is claiming (see the rebuttal in [110]) that the uncertainties in the measured rate of change of J_4 and of higher even zonal harmonics would introduce a large bias in the measurement of the Lense–Thirring effect, however several independent authors [15, 110, 129, 136] have proved that this type of error, in the LAGEOS and LAGEOS 2 measurements, is in fact at the level of a few percent only of the Lense–Thirring effect; we have discussed this in Sections 6.3 and 8.1.3. In regard to the claim (b), in order to measure gravitomagnetism using the nodal rate of an Earth satellite, the orbital inclination of a satellite needs only to be accurately *measured* but *not* accurately *predicted* in its temporal behavior; this is explained here below and in [135]. Then, in Section 10, we show that the orbital inclination of LAGEOS-type satellites is very accurately *measured* by the technique of laser ranging and thus, the corresponding error in the determination of the Lense–Thirring effect due to the *measurement* errors in the inclination of LAGEOS, LAGEOS 2 and LARES is at most at a level of a fraction of 1% of the Lense–Thirring effect. In regard to the claims (c), here below and in [135], we stress and show that different Earth gravity field models with intrinsically different accuracies cannot be compared or they *must* be compared in the appropriate way, i.e., the accuracy of models derived with a larger GRACE data set or with the standard, accurate, GRACE techniques to derive Earth gravity models *cannot* be estimated by taking their difference with less accurate models, derived with a smaller GRACE data set or based on less accurate techniques or on other less accurate satellite observations. Finally, in regard to the claims (d), in the paper on the LARES experiment [16], elsewhere in this book, we have proven that the error due to the even zonal harmonics of degree higher than 26 is negligible in the LARES experiment, contrary to a number of miscalculations by the same author [132, 133].

- *Orbital inclination determination*

As already stressed, what is critical for the measurement of the Lense–Thirring effect is that the *modeling* of the classical *node* precession (4) (i.e., the prediction of its behavior on the basis of the available physical models) must be accurate, at the level of a few milliarcsec or less [54]). What is critical is *not* that all the quantities entering this equation, i.e., the Earth parameters and the orbital parameters, and in particular the Earth spherical harmonic coefficients and the semimajor axis and the inclination, must be *predicted* in their variations, but instead what is critical is that they must be *determined* with sufficient accuracy via satellite laser ranging and other techniques (such as the GRACE spacecraft to determine the Earth gravity field). For example if the variations of the *inclination* and of the *semimajor axis* of LAGEOS are not well modeled because the effect of particle drag (i.e., "atmospheric drag") is not known with sufficient accuracy, this is *not* critical for our measurement of the Lense–Thirring effect because we are able to *measure* the variations of inclination and semimajor axis accurately enough with satellite laser ranging (see next Section 10 on the uncertainty in the determination of the orbital inclination due to the atmospheric refraction modeling errors) and thus we are able to precisely quantify the effect of these variations on the nodal rate, Eq. (4) (with the orbital estimators GEODYN, EPOS-OC and UTOPIA). Indeed, in the next Section 10 we estimate that the average measurement error in the inclinations of LAGEOS and LAGEOS 2 (that are roughly constant) is respectively about 30 μarcsec (i.e., 0.03 milliarcsec) for LAGEOS and about 10 μarcsec for LAGEOS 2, that, when propagated in the nodal rate, Eq. (4), corresponds to respectively about 0.6% and 0.36% of the Lense–Thirring effect.

The difference between *modeling* and *determining* an orbital element may seem trivial but is critical to understand one of the misunderstandings published in [131] about the error induced in our measurement of the Lense–Thirring effect by the unmodeled inclination variations; indeed this paper [131] concludes: "The atmospheric drag, both in its neutral and charged components, will induce a non-negligible secular decrease of the inclination of the new spacecraft yielding a correction to the node precession of degree $l = 2$ which amounts to 3–9% year^{-1} of the total gravitomagnetic signal. Such a corrupting bias would be very difficult to be modeled". However, as stressed above, these variations in the inclination are very accurately determined with satellite laser ranging (see Section 10) even though they are not modeled (i.e., predictable) at a comparable level of accuracy; nevertheless, the measurement of the inclination variations is what is only needed to accurately model the nodal rate, Eq. (10).

- *How to compare different Earth gravity field models*

The GRACE mission provided a leap in the accurate determination of models for the gravitational field and its temporal variations. Nonetheless, not all models derived from GRACE data are of equal or even nearly equal quality, simply because they are based on the highly accurate GRACE observations.

There are several reasons behind this fact which is easily asserted from an examination of the error spectra for nearly every such model available, at the site of the Int.

Center for Global Earth Models (ICGEM) at: http://icgem.gfz-potsdam.de/ICGEM. The first and most obvious reason is that most of these models are based on an entirely different set of data, from as little as one month to as much as several years. Some of them are derived rigorously from standard reduction techniques and others use innovative approximation techniques that rely on the accurate position and velocity determination of the orbits from the GPS tracking data. When one additionally considers the fact that GRACE's primary mission is to observe the temporal change of the field, one can realize that all of these models are de facto associated with a mean epoch as far as the representation of the static part of the field and since they span quite diverse time periods, an additional source of diversity is the way that temporal change was handled during their creation, i.e. whether it was accounted for and if so, what rates and for which part of the model they were applied. These models are not referenced to a fixed epoch, and this implies that especially for what concerns the very low degrees, users must apply very carefully the re-referencing (mapping) of these coefficients to a fixed epoch for all of them, prior to attempting any comparisons. This task is difficult to accomplish rigorously for all fields, since there are several other background models used during the data reduction process that affect these rates, each of which could be handled in various ways in the different software and by the different groups that generated these fields. Even for the tightly controlled models produced by the three groups that comprise the GRACE project office, UT/CSR, GFZ/Potsdam and NASA/JPL, the direct comparison at the single coefficient level is problematic, although comparison of the RMS coefficient differences by degree or by order, do seem to provide much more reliable comparison statistics and give a more robust way of assessing the relative accuracy of these models (see Fig. 9).

A recommended approach to compare these diverse models is to examine their "by degree" and "per degree" spectral differences with respect to a single benchmark model and after a careful mapping of the coefficients to the same epoch. Such a comparison is available already online at the ICGEM site. What we can surmise from an inspection of those spectra is that the models are very diverse, even when they come from the same group (e.g. GGM02 and GGM03). In conclusion, we must stress that one should not use each coefficient differences between non-equivalent models as a substitute for the accuracy estimates of any particular GRACE gravity model. The proper calibration of their errors is accomplished with the use of ground truth data sets as for example a global set of GPS-derived geoid undulations over the solid Earth surface and oceanographic observations over the oceans.

Let us further discuss how to correctly compare different Earth gravity field models and some previous misleading comparisons published in [134], in this paper the author is comparing different models, some of which already obsolete in 2007 and obtained with the use of less accurate space missions before GRACE, such as CHAMP. For example the author is comparing two state of the art models in 2004, EIGEN-GRACE02S and GGM02S, with the preliminary, older and obsolete models EIGEN-GRACE01S and GGM01S. Of course, the author is getting a large uncertainty as a result, indeed in [134] he concludes "it turns out that the systematic error $\delta\mu$ in the Lense–Thirring measurement is quite larger than in the evaluations so far published based on the use of the sigmas of one model at a time separately, amounting up to 37–43% for the pairs GGM01S/GGM02S and EIGEN-GRACE01S/EIGEN-GRACE02S". However, in order to assess the uncertainty of the newer, much more accurate models, this is simply wrong; for example whereas the uncertainty associated with the measurement using EIGENGRACE02S is about 10% of the Lense–Thirring effect, the uncertainty using the older models is much larger, e.g., for GGM01S was already published [129] to be as large as 24%. This explains why the author gets a large percent "error" in comparing EIGEN-GRACE02S and GGM02S with EIGEN-GRACE01S and GGM01S.

One cannot evaluate the accuracy of the latest GRACE gravity models by comparing them with previous obsolete models obtained using shorter periods of GRACE observations or obtained with different and less accurate techniques. One can see the error spectra for the various models following the links under each model in the tables available at: http://www.gfz-potsdam.de/.

Furthermore, different models cover different periods of time (thus they refer to a different mean epoch). For example, when any of these models state that they have a "zero tide" J_2, this means that their quadrupole coefficient J_2 is different from the J_2 of most of the other models by 4.173×10^{-9} in fully normalized coefficient space. If one does not remove that "kind of reference frame bias", taking the difference of J_2 for two such models, this would make them look as extremely different and it will generate a bias in the error calibration of the accuracy estimates of the coefficients. Similarly, in regard to other even zonal harmonics such as J_4, J_6, etc.

In a more recent paper [133] by the same author, the claims [132] of an error of 1,000% in the measurement the Lense–Thirring effect was resized down by almost two orders of magnitude, with a new claim of an error that, in the worst possible case, can be as large as 33% (by comparing the models AIUB-GRACE01S and ITG-Grace03s with EIGEN-GRACE02S). However, even this more recent comparison is affected by at least four misunderstandings and miscalculations: (a) as just explained, one cannot simply take the difference between the corresponding even zonal harmonics coefficient of two different models, because, *among other things*, they refer to different mean epochs and they have been obtained with different $\dot{C}_{lm}$ corrections; (b) in [133] the author is comparing more recent models, however, the author is still taking the difference between models with different intrinsic accuracies, for example in Table 2 he is comparing the model GGM02S with ITG-Grace03s, however, GGM02S has a lower intrinsic accuracy than ITG-GRACE03S; (c) the author is often comparing the difference of each even zonal coefficient of two different models with their *formal* error to conclude that the differences "ΔC_{l0} are always larger than the linearly added sigmas, apart from l = 12 and l = 18", however, it is well known and rather trivial that the formal error, i.e., the error that does *not* include the *systematic* errors, is usually much smaller than the true error (which includes systematic errors); (d) one cannot take the difference between the even zonal coefficients of different models and then sum the *absolute* values of each difference, called "SAV" in [133], (or similarly take the root sum square of these differences) and then compare this "SAV" sum with the Lense–Thirring effect

(48.2 milliarcsec/yr) in order to more or less implicitly imply that the results of the measurement of the Lense–Thirring effect with these two models would differ by this "SAV" sum. In fact, one *must* perform a *real* data reduction to compare the results with two different models and in the real data reduction with a model, some even zonal coefficients will be larger and some smaller with respect to the other model, in other words in order to compare the results with two different models, the differences with their plus or minus signs have to be considered. For example, in the more recent paper [133], in Tables 3 and 10, the author calculates a difference between GGM02S and GGM03S of 24%, however by doing the real data reduction, Ries et al. [15] obtained a difference between the two models of 13% only.

The appropriate way of comparing different Earth gravity models is manifest from Fig. 9. In this figure different gravity field models are compared by comparing the difference of the corresponding coefficients for each degree (by including the sum over every corresponding *order*, see above) with their published uncertainties. For example the gravity field model GGM01S, a preliminary GRACE model, less accurate and based on a much shorter period of observation of GRACE, is compared with GGM03S, one of the best models today available. Clearly the difference between GMM01S and GGM03S, the solid line in light blue, is of comparable size with the accuracy of the less accurate model, i.e., GGM01S, the dashed line in light blue. This difference cannot be interpreted as the uncertainty in the much more accurate GGM03S model.

We further elucidate this point with a very simple example. Suppose we wish to discuss the accuracy in the measurement of the gravitational constant G, we then take one of its latest measured values, let us say, 6.674215×10^{-11} with a relative uncertainty of 0.0015%, whereas the G value measured in the eighteen century was, let us say, about 6.74×10^{-11} with a relative uncertainty of a few percent. Now, in order to evaluate the accuracy of the latest value of G (and this is exactly what the author of [132, 133] is doing by comparing in [132, 133] even zonal coefficients of different gravity field models with different intrinsic accuracies), we take the difference between these two values. In this way we obtain a relative difference of about 1%. What can one reasonably conclude from this? One may conclude that the older estimate of the uncertainty in G was of the correct order of magnitude but certainly one *cannot* conclude that the 0.0015% uncertainty of the modern measurement of G is wrong by a factor 1000. One cannot seriously difference the older value of the gravitational constant with its newer value to assess the accuracy of its recent measurement. Similarly one cannot seriously compare some obsolete and preliminary Earth gravity models, such as GGM01 and EIGEN-GRACE01S (as this author is doing in [132]) with some of the most accurate models obtained with GRACE (see: http://icgem.gfz-potsdam.de), such as the latest EIGEN-GRACE and GGM models, used for the Lense–Thirring measurement. Furthermore one cannot seriously compare models obtained with GRACE only with models that use GRACE together with other less accurate observations; nevertheless, this is exactly what this author is doing in [132, 133].

10 Appendix 3. Inclination Errors and Atmospheric Delay Modeling Uncertainties in SLR

Here we analyze the uncertainty in the determination of the Lense–Thirring effect due to the measurement error in the inclination of the LAGEOS satellites. This measurement error is mainly due to the atmospheric refraction. Of course, the corresponding uncertainty in the determination of the Lense–Thirring effect using the LAGEOS satellites is induced by the error in the *measurement* of the inclination and *not* by the uncertainty in the *modeling* of the inclination (i.e., the uncertainty in the prediction of its behavior).

Atmospheric refraction is an important accuracy-limiting factor in the use of Satellite Laser Ranging (SLR) for high-accuracy science applications. In most of these applications, and particularly for the establishment and monitoring of the Terrestrial Reference Frame, of great interest is the stability of its scale and its implied height system. The modeling of atmospheric refraction in the analysis of SLR data is based on the determination of the delay in the zenith direction and on the subsequent projection to a given elevation angle using a mapping function. Mendes et al. [139] pointed out some limitations in the Marini–Murray model used in SLR since its introduction in 1973, namely, the modeling of the elevation dependence (the mapping function component of the model), and a <1 mm bias in the computation of the zenith delay. The mapping functions developed by Mendes et al. [139] represent a significant improvement over the built-in mapping function of the Marini–Murray model and other known mapping functions. The new mapping functions can be used in combination with any zenith delay model. Mendes et al. [139] concluded that current zenith delay models have errors at the millimeter level, which increase significantly at $0.355\,\mu$m, reflecting inadequacy in the dispersion formulae incorporated in these models. A more accurate zenith delay model was developed, applicable to the range of wavelengths used in modern SLR instrumentation (0.355–1.064 μm) [140]. Using a three-dimensional ray-tracing procedure based on globally distributed satellite data from the Atmospheric Infrared Sounder (AIRS) instrument on NASA's AQUA platform, as well as three-dimensional analysis fields from the European Center for Medium Weather Forecasting (ECMWF), Mendes et al. assessed the new zenith delay models and mapping functions both spatially and temporally. They also looked at the magnitude of the horizontal gradient contribution to the total delay by ray-tracing and using a parametric model. Ray-tracing does not depend on any models or mapping functions, it uses a three-dimensional spherical grid that covers Earth from its surface to the top of the atmosphere and generates the atmospheric delay value by following the (non-planar) path of a light ray from the tracking station to the satellite and back. The path of the ray is governed by the local refractive index computed on the basis of the conditions within each three dimensional cell of the grid (with horizontal and vertical resolution $\sim$50 km).

Mendes et al. used meteorological data sets from NASA's AIRS (see Fig. 13) in order to validate the new zenith delay and mapping function models, and to develop new models that include variations in horizontal refractive indices. The

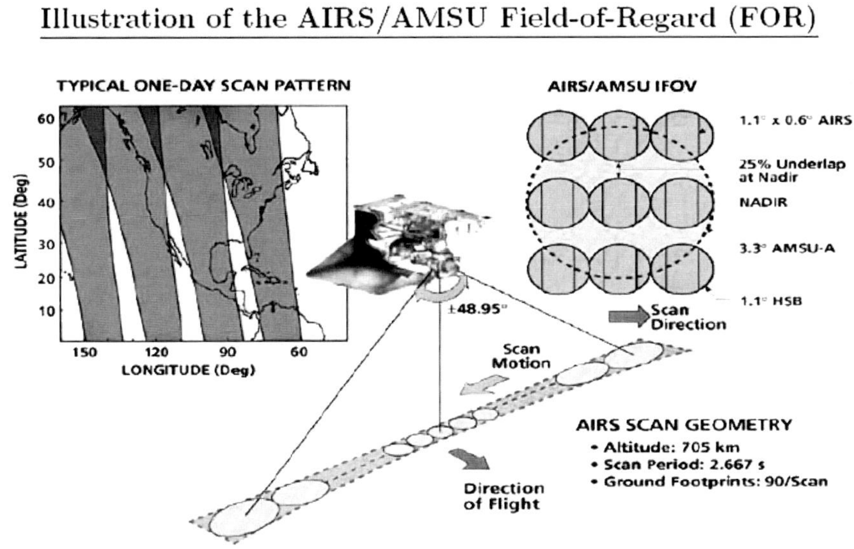

Fig. 13 AIRS concept of operation

AIRS Level-2 support product gives profiles of temperature, pressure and water vapor from the surface to the top-of-the-atmosphere in 100 standard pressure levels. The pressure levels extend from 1.100 up to 0.1 mb.

AIRS obtains temperature profiles to an accuracy of 1° K for every 1 km in the troposphere and 1° K for every 4 km in the stratosphere up till 40 km. The temperature profile accuracy in the troposphere matches that achieved by radiosonde launches from ground stations. The advantage of AIRS is that it provides rapid and global coverage of the earth. AIRS also obtains humidity profiles to an accuracy of 10% in 2 km layers in the lower troposphere and ~50% in the upper stratosphere. The AIRS data is retrieved in the form of "granules". One granule contains 6 min of data and is approximately 1,600 (E–W) × 2,300 (N–S) km in spatial extent with a 50 km resolution within the granule. One day of data yields 240 granules.

Mendes et al. developed a ray-tracing algorithm specifically tailored for AIRS and ECMWF data in order to calculate the atmospheric delay by directly integrating all the values through which the ray traverses, independent of any mapping function. They used a new formulation for the group refractivity based on formulas by [137] that include both hydrostatic and non-hydrostatic components of the group refractivity. In order to perform the ray tracing, the data is first processed and grouped into 10 × 10 degree latitude/longitude grids up to 0.1 mb in order to build three dimensional atmospheric profiles around each operational ILRS-SLR tracking station. AIRS-based Ray-Tracing (ART) and ECMWF-based Ray-Tracing (ERT) were used to calculate both the total atmospheric delay as well as the delay due to horizontal gradients in refractivity. Given the independent accuracy assessment of the simple, closed-formula models in use today, (e.g. Marini–Murray), the effect of the horizontal gradients (which is not accounted for in these models), is the largest remaining error in refraction delay modeling today. If we can quantify the level of that unaccounted error induced on the orbit of LAGEOS due to neglecting the horizontal gradients, we can then place an upper bound on the maximum error in our Lense–Thirring estimate.

Using the three dimensional ray-tracing approach which fully accounts for the *total* atmospheric delay including the gradients, we analyzed a few years of LAGEOS and LAGEOS 2 data as a test case. Comparing the results of these reductions to those obtained using the standard model (Mendes–Pavlis without gradients), we showed that the most significant result was a reduction in the variance of the observation residuals up to 25%, with only random and insignificant differences in the orbits. This was the most significant outcome of implementing the new approach in calculating the atmospheric delay using the three dimensional ray-tracing methodology. With insignificant, random and no secular differences in the orbits, the effect on the estimated Lense–Thirring parameter is also equally insignificant and of no further concern for the Lense–Thirring experiment.

Since refraction does not enter the dynamical model being a media propagation effect on the tracking data, one would expect a priori that the effects on the orbit would be very small, much smaller than the actual variations seen in the tracking data residuals that absorb the majority of refraction errors.

From Fig. 14 we can estimate that on LAGEOS, on the average, the effect on inclination is at the 30 μarcsec (0.03 milliarcseconds) level with a comparable

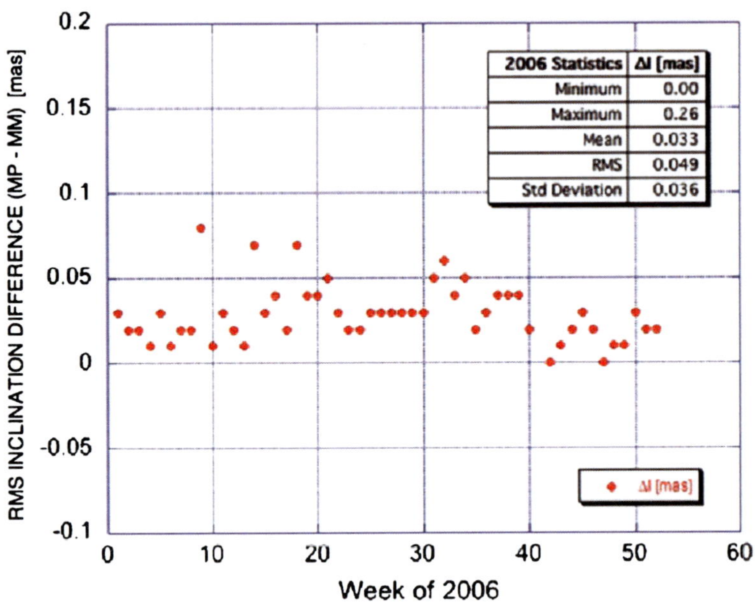

Fig. 14 Weekly arc RMS variation (RMS Difference, Marini–Murray minus Mendes–Pavlis, in mas (milliarcsec)) in the inclination fits of LAGEOS data reduced with two different atmospheric delay models, the Marini–Murray and the Mendes–Pavlis

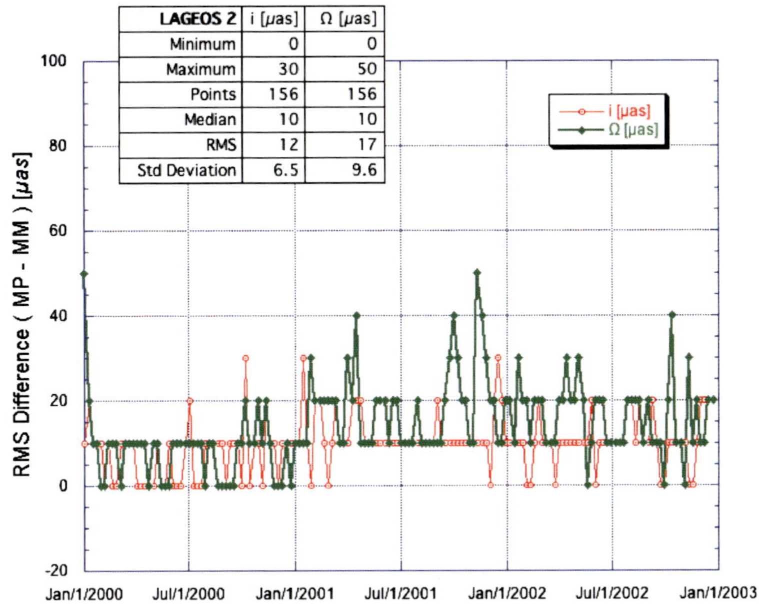

Fig. 15 Weekly arc RMS variation (RMS Difference, Marini–Murray minus Mendes–Pavlis, in μas (microarcsec)) in the inclination and node fits of LAGEOS 2 data reduced with two different atmospheric delay models, the Marini–Murray and the Mendes–Pavlis

standard deviation. For comparison purposes notice that 30 μarcsec at LAGEOS altitude are only <1.5 mm. On the LAGEOS orbit, a 30 μarcsec measurement uncertainty in the inclination corresponds (from the nodal rate (2) of LAGEOS due to the even zonal harmonics) to a nodal rate of 0.6% of the Lense–Thirring effect. Similar conclusions are reached from inspection of Fig. 15, displaying the results of the analysis of several years of older data taken on LAGEOS 2. We notice here that from the analysis of three years of data we again see a very minor effect at the level of about 10 μarcsec on the average in either the inclination or the node of LAGEOS 2, with an even smaller standard deviation than the one for LAGEOS. This may be explained by the fact that in recent years (after 2004) the network has allowed tracking at lower elevations below the original 20° minimum and all the way down to 10°. Refraction errors drop significantly at higher elevations and since the preponderance of the LAGEOS 2 data in these early years is at elevations above 20°, the discrepancy between the two orbits is less affected by the much smaller refraction errors. On the LAGEOS 2 orbit, a 10 μarcsec measurement uncertainty in the inclination corresponds (from the nodal rate eq. (4) of LAGEOS 2 due to the even zonal harmonics) to a nodal rate of 0.36% of the Lense–Thirring effect.

Final tests were performed with the direct application of three dimensional atmospheric ray-tracing as detailed in [138], where two years of data were reduced with atmospheric delay corrections that are obtained with this approach and are thus free of any error in the zenith delay or the mapping function used in the model. Furthermore, three dimensional AIRS-based Ray-Tracing includes automatically the effect of the horizontal gradients as explained ibidem, and what these tests demonstrated is that indeed, as expected, the errors in the atmospheric delay modeling are absorbed by the residuals of the individual stations. The use of the superior 3D ART approach results in significant variance reduction in the station residuals, whether the meteorological information comes from the AIRS global fields or the ECMWF assimilation fields, which in recent years begun using the AIRS data as part of their standard input for their assimilation scheme.

In conclusion, on the combination of the nodes of LAGEOS and LAGEOS 2 (to measure the Lense–Thirring effect) the error due to inclination measurement uncertainties is of the order of 0.5% and similarly for the LARES+LAGEOS+LAGEOS 2 experiment.

Acknowledgement We gratefully acknowledge the support of ASI, the Italian Space Agency, grants I/043/08/0 and I/016/07/0.

References

1. Misner, C.W., Thorne, K.S., and Wheeler, J.A., *Gravitation* (Freeman, San Francisco, 1973).
2. Weinberg, S., *Gravitation and Cosmology: Principles and Applications of the General Theory of Relativity* (Wiley, New York, 1972).
3. Ciufolini, I., and Wheeler, J.A., *Gravitation and Inertia* (Princeton University Press, Princeton, New Jersey, 1995).
4. Will, C.M., *Theory and Experiment in Gravitational Physics*, 2nd edn (Cambridge Univ. Press, Cambridge, UK, 1993).
5. Will, C.M., The confrontation between general relativity and experiment. *Living Rev. Rel.* **9**, 3 (2006); http://www.livingreviews.org/lrr-20063.
6. Riess, A. et al., Observational evidence from supernovae for an accelerating universe and a cosmological constant. *Astron. J.* **116**, 1009–1038 (1998).
7. Perlmutter, S. et al., Measurements of Ω and Λ from 42 high-redshift supernovae. *Astrophys. J.* **517**, 565–586 (1999).
8. Perlmutter, S., Supernovae, dark energy, and the accelerating universe. *Phys. Today*, **56**, 53–59 (2003).
9. Caldwell, R.R., Dark energy. *Phys. World*, **17**, 37–42 (2004).
10. Adelberger, E., Heckel, B. and Hoyle, C.D., Testing the gravitational inverse-square law. *Phys. World*, **18**, 41–45 (2005).
11. Amelino-Camelia, G., Ellis, J., Mavromatos, N.E., Nanopoulos, D.V. and Sarkar, S. Potential sensitivity of gamma-ray burster observations to wave dispersion in vacuo. *Nature*, **393**, 763–765 (1998).
12. Dvali, G., Filtering gravity: modification at large distances? Infrared Modification of Gravity. In Nobel Symp. on Cosmology and String Theory and Cosmology, *Proc. of Nobel Symposium 127*, Sigtuna, Sweden, 2003 (eds Danielsson, U., Goobar, A. and Nilsson, B.) (World Scientific, Singapore, 2005). (Sigtuna, Sweden, August 2003); preprint at http://arXiv.org/hep-th/0402130 (2004).
13. Ciufolini, I. and Pavlis, E.C., A confirmation of the general relativistic prediction of the Lense-Thirring effect. *Nature*, **431**, 958–960 (2004).
14. Ciufolini, I., Dragging of inertial frames. *Nature*, **449**, 41–47 (2007).
15. Ries, J.C., Eanes, R.J. and Watkins. M.M., Confirming the frame-dragging effect with satellite laser ranging, *16th International Workshop on Laser Ranging*, 13–17 October 2008, Poznan, Poland (2008).

16. Ciufolini, I. et al., The LARES space experiment: LARES orbit, error analysis and satellite structure. In this book: *General Relativity and John Archibald Wheeler*, eds. Ciufolini, I. and Matzner, R. (Springer Verlag, 2010).
17. Einstein, A., Letter to Ernst Mach. Zurich, 25 June 1913, in ref. [1] p. 544.
18. Mach, E., *Die Mechanik in Ihrer Entwicklung Historisch Kritisch-Dargestellt* (Brockhaus, Leipzig, 1912); transl. *The Science of Mechanics* (Open Court, La Salle, Illinois, 1960).
19. Barbour, J. and Pfister, H., eds., *Mach's Principle. From Newton's Bucket to Quantum Gravity* (Birkhauser, Boston, 1995).
20. For implications of Mach principle and frame-dragging in cosmology see, e.g., chapter 4 of ref. [3] and Schmid, C., Cosmological gravitomagnetism and Machs principle. *Phys. Rev. D*, **74**, 044031–1–18 (2006).
21. Friedländer, B. and Friedländer, I., *Absolute und Relative Bewegung?* (Simion-Verlag, Berlin, 1896).
22. Föppl, A., Überreinen Kreiselversuch zur messung der Umdrehungsgeschwindigkeit der Erde. *Sitzb. Bayer. Akad. Wiss.* **34**, 5–28 (1904) *Phys. Z.* **5**, 416; see also Föppl, A. Über Absolute und Relative Bewegung. *Sitzb. Bayer. Akad. Wiss.* **34**, 383–95 (1904).
23. de Sitter, W., On Einstein's theory of gravitation and its astronomical consequences. *Mon. Not. Roy. Astron. Soc.* **76**, 699–728 (1916)
24. Lense, J. and Thirring, H., Über den Einfluss der Eigenrotation der Zentralkorper auf die Bewegung der Planeten und Monde nach der Einsteinschen *Phys. Z.*, **19**, 156–163 (1918). See also English translation by Mashhoon, B., Hehl, F.W., Theiss, D.S. *Gen. Relativ. Gravit.*, **16**, 711–750 (1984).
25. Zeldovich, Ya.B. and Novikov, I.D., *Relativistic Astrophysics, Vol. I, Stars and Relativity* (Univ. Chicago Press, Chicago, 1971).
26. Landau, L.D. and Lifshitz, E.M., *The Classical Theory of Fields*, 3rd rev. English edn. (Pergamon, London, 1971).
27. Ciufolini, I. and Ricci, F., Time delay due to spin and gravitational lensing. *Classical and Quantum Gravity*, **19**, 3863–3874 (2002).
28. Ciufolini, I. and Ricci, F., Time delay due to spin inside a rotating shell. *Classical and Quantum Gravity*, **19**, 3875–3881 (2002).
29. Ciufolini, I., Ricci, F., Kopekin, S. and Mashhoon, B. On the gravitomagnetic time delay. *Phys. Lett. A*, **308**, 101–109 (2003).
30. Kerr, R.P., Gravitational field of a spinning mass as an example of algebraically special metrics. *Phys. Rev. Lett.*, **11**, 237–238 (1963).
31. Pugh, G.E., Proposal for a satellite test of the coriolis prediction of general relativity. *Weapons Systems Evaluation Group Research Memorandum* N. 11 (The Pentagon, Washington, 1959).
32. Schiff, L.I., Motion of a gyroscope according to Einstein's theory of gravitation. *Proc. Nat. Acad. Sci.*, **46**, 871–82 (1960) and Possible new test of general relativity theory. *Phys. Rev. Lett.*, **4**, 215–7 (1960).
33. Bardeen, J.M. and Petterson, J.A., The Lense-Thirring effect and accretion disks around Kerr Black Holes. *Astrophysical J.*, **195**, L65–7 (1975).
34. Thorne, K.S., Price, R.H. and Macdonald, D.A., *The Membrane Paradigm* (Yale Univ. Press, NewHaven, 1986).
35. Schäfer, G., Gravitomagnetic effects. *J. Gen. Rel. Grav.*, **36**, 2223–2235 (2004).
36. de Sitter, W., On Einstein's theory of gravitation and its astronomical consequences. *Mon. Not. R. Astron. Soc.*, **76**, 699728 (1916).
37. Ashby, N. and Shahid-Saless, B., Geodetic precession or dragging of inertial frames? *Phys. Rev. D*, **42**, 1118–22 (1990).
38. O'Connel, R.F., A Note on frame dragging. *Class. Quant. Grav.*, **22**, 3815–16 (2005).
39. Ciufolini, I., Frame-dragging, gravitomagnetism and lunar laser ranging, *New Astronomy*, **15**, 332–337 (2010); see also Pavlis, E. and Ciufolini, I., *Proc. of 15th International Laser Ranging Workshop*, Camberra, Australia, October 16–20 (2006).
40. Bertotti, B., Ciufolini, I. and Bender, P.L., New test of general relativity: measurement of de Sitter geodetic precession rate for lunar perigee. *Phys. Rev. Lett.*, **58**, 1062–1065 (1987).

41. Williams, J.G., Turyshev, S.G. and Boggs, D.H., Progress in lunar laser ranging tests of relativistic gravity. *Phys. Rev. Lett.*, **93**, 261101–1–4 (2004).
42. Williams, J.G., Newhall, X.X. and Dickey, J.O., Relativity parameters determined from lunar laser ranging. *Phys. Rev. D*, **53**, 6730–6739 (1996).
43. GRAVITY PROBE-B update at: http://einstein.stanford.edu/
44. Weisberg, J.M. and Taylor, J.H., General relativistic geodetic spin precession in binary pulsar B1913+16: mapping the emission beam in two dimensions. *Astrophys. J.*, **576**, 942–949 (2002).
45. Stairs, I.H., Thorsett, S.E. and Arzoumanian, Z., Measurement of gravitational spin-orbit coupling in a binary-pulsar system. *Phys. Rev. Lett.*, **93**, 141101–1–4 (2004).
46. Murphy, T.W. Jr., Nordtvedt, K. and Turyshev, S.G., Gravitomagnetic influence on gyroscopes and on the lunar orbit. *Phys. Rev. Lett.*, **98**, 071102–1–4 (2007).
47. Kopeikin, S.M., Comment on "Gravitomagnetic Influence on Gyroscopes and on the Lunar Orbit". *Phys. Rev. Lett.*, **98**, 229001 (2007).
48. Murphy, T.W. Jr., Nordtvedt, K. and Turyshev, S.G., Murphy, Nordtvedt, and Turyshev Reply. *Phys. Rev. Lett.*, **98**, 229002 (2007).
49. Barker, B.M. and O'Connel, R.F., The gravitational interaction: Spin, rotation, and quantum effects. A review. *Gen. Rel. Grav.*, **11**, 149–175 (1979).
50. Khan, A.R. and O'Connell, R.F., Gravitational analogue of magnetic force. *Nature* **261**, 480–481 (1976).
51. Ciufolini, I., Gravitomagnetism and status of the LAGEOS III experiment. *Class. Quantum Grav.*, **11**, A73–A81 (1994).
52. The curvature invariants have been calculated using MathTensor, a system for doing tensor analysys by computer, by Parker, L. and Christensen, S.M. (Addison-Wesley, Boston, 1994).
53. Nordtvedt, K., Lunar laser ranging: a comprehensive probe of post-Newtonian gravity. In: *Gravitation: from the Hubble Length to the Planck Length, Proc. I SIGRAV School on General Relativity and Gravitation*, Frascati, Rome, September 2002 (IOP, 2005) p 97–113.
54. Ciufolini, I., Measurement of the Lense-Thirring drag on high-altitude laser-ranged artificial satellites. *Phys. Rev. Lett.* **56**, 278–281 (1986).
55. Ciufolini, I., A comprehensive introduction to the Lageos gravitomagnetic experiment: from the importance of the gravitomagnetic field in physics to preliminary error analysis and error budget. *Int. J. Mod. Phys. A*, **4**, 3083–3145 (1989); see also: [56].
56. Tapley, B., Ries, J.C., Eanes, R.J., and Watkins, M.M., *NASA-ASI Study on LAGEOS III*, CSR-UT publication n. CSR-89-3, Austin, Texas (1989), and Ciufolini, I., et al., ASI-NASA Study on LAGEOS III, CNR, Rome, Italy (1989). See also: I. Ciufolini et al., *INFN study on LARES/WEBER-SAT* (2004).
57. Ries, J.C., Simulation of an experiment to measure the Lense-Thirring precession using a second LAGEOS satellite, Ph.D. dissertation. The University of Texas, Austin (1989).
58. Ciufolini, I., On a new method to measure the gravitomagnetic field using two orbiting satellites. *Nuovo Cimento A*, **109**, 1709–1720 (1996).
59. Ciufolini, I. et al., Measurement of dragging of inertial frames and gravitomagnetic field using laser-ranged satellites. *Nuovo Cimento A*, **109**, 575–590 (1996).
60. Ciufolini, I., Chieppa, F., Lucchesi, D. and Vespe, F., Test of Lense-Thirring orbital shift due to spin. *Class. and Quantum Grav.*, **14**, 2701–2726 (1997). See also: Ciufolini, I., Lucchesi, D., Vespe, F., and Chieppa, F., Measurement of gravitomagnetism. *Europhys. Lett.*, **39**, 359–364 (1997).
61. Ciufolini, I., Pavlis, E.C., Chieppa, F., Fernandes-Vieira, E. and Perez-Mercader, J., Test of general relativity and measurement of the Lense-Thirring effect with two Earth satellites. *Science*, **279**, 2100–2103 (1998).
62. Cohen, S.C. and Dunn, P.J., eds., LAGEOS Scientific Results. *J. Geophys. Res.*, **90** (B11), 9215 (1985).
63. Bender P. and Goad, C.C., The use of satellites for geodesy and geodynamics, in: Veis, G., Livieratos, E. (Eds.), *Proceedings of the Second International Symposium on the Use of Artificial Satellites for Geodesy and Geodynamics*, Vol. II. National Technical University of Athens, p. 145 (1979).

64. Reigber, Ch., Schwintzer, P., Neumayer, K.-H., Barthelmes, F., König, R., Förste, Ch., Balmino, G., Biancale, R., Lemoine, J.-M., Loyer, S., Bruinsma, S., Perosanz, F. and Fayard, T., The CHAMP-only Earth Gravity Field Model EIGEN-2. *Advan. Space Res.*, **31**(8), 1883–1888 (2003), doi: 10.1016/S0273–1177(03)00162–5.
65. Reigber, C., Schmidt, R., Flechtner, F., Koenig, R., Meyer, U., Neumayer, K.H., Schwintzer, P. and Zhu, S.Y., An Earth gravity field model complete to degree and order 150 from GRACE: EIGEN-GRACE02S. *Geodynamics*, **39**, 1–10 (2005). The EIGEN-GRACE02S gravity field coefficients and their calibrated errors are available at: http://op.gfz-potsdam.de/grace/index_GRACE.html
66. Förste, C., Flechtner, F., Schmidt, R., Stubenvoll, R., Rothacher, M., Kusche, J., Neumayer, K.-H., Biancale, R., Lemoine, J.-M., Barthelmes, F., Bruinsma, J., Koenig, R., Meyer, U., EIGEN-GL05C – A new global combined high-resolution GRACE-based gravity field model of the GFZ-GRGS cooperation. General Assembly European Geosciences Union, Vienna, Austria, 2008, *J. Geophys. Res. Abstracts*, **10**, No. EGU2008-A-06944 (2008).
67. Förste, C., Schmidt, R., Stubenvoll, R., Flechtner, F., Meyer, U., Konig, R., Neumayer, H., Biancale, R., Lemoine, J.-M., Bruinsma, S., Loyer, S., Barthelmes, F. and Esselborn, S., The GeoForschungsZentrum Potsdam/Groupe de Recherche de Gèodésie Spatiale satellite-only and combined gravity field models: EIGEN-GL04S1 and EIGEN-GL04C. *J. Geodesy*, **82**, 6, 331–346 (2008).
68. Förste, C., Flechtner, F., Schmidt, R., Meyer, U., Stubenvoll, R., Barthelmes, F., Köenig, R., Neumayer, H., Rothacher, M., Reigber, Ch. Biancale, R., Bruinsma, S., Lemoine, J.M., Raimondo, J.C., A New High Resolution Global Gravity Field Model Derived From Combination of GRACE and CHAMP Mission and Altimetry/Gravimetry Surface Gravity Data. *EGU General Assembly 2005*, Vienna, Austria, 24–29, April 2005.
69. Tapley, B., Ries, J., Bettadpur, S., Chambers, D., Cheng, M., Condi, F., Gunter, B., Kang, Z., Nagel, P., Pastor, R., Pekker, T., Poole, S., and Wang, F., GGM02 An improved Earth gravity field model from GRACE. *J. Geod.*, **79**, 467–478 (2005). The GGM02 gravity model is available at: http://www.csr.utexas.edu/grace/gravity/
70. Tapley, B., Ries, J., Bettadpur, S., Chambers D., Cheng, M., Condi, F., Poole, S., The GGM03 Mean Earth Gravity Model from GRACE. *Eos Trans. AGU* 88(52), Fall Meet.Suppl., Abstract G42A-03 (2007).
71. Mayer-Guerr, T., Eicker, A., Ilk, K.H., ITG-Grace02s: A GRACE Gravity Field Derived from Short Arcs of the Satellites Orbit. *Proc. of the 1st International Symposium of the International Gravity Field Service "Gravity Field of the Earth"*, Istanbul (2007).
72. Mayer-Guerr, T., ITG-Grace03s: The latest GRACE gravity field solution computed in Bonn. presentation at GSTM+SPP, 15–17 Oct 2007, Potsdam.
73. The JEM models were provided by JPL-Caltech.
74. International Earth Rotation Service (IERS) Annual Report, 1996. Observatoire de Paris, Paris (July 1997).
75. Rubincam, D.P., Yarkovsky thermal drag on LAGEOS. *J. Geophys. Res.*, **93** (B11), 13803–13810 (1988).
76. Rubincam, D.P., Drag on the LAGEOS satellite. *J. Geophys. Res.*, **95** (B11), 4881–4886 (1990).
77. Rubincam, D.P., and Mallama, A. Terrestrial atmospheric effects on satellite eclipses with application to the acceleration of LAGEOS. *J. Geophys. Res.*, **100** (B10), 20285–20990 (1995).
78. Martin, C.F., and Rubincam, D.P., Effects of Earth albedo on the LAGEOS I satellite. *J. Geophys. Res.*, **101** (B2), 3215–3226 (1996).
79. Andrès, J.I. et al., Spin axis behavior of the LAGEOS satellites. *J. Geophys. Res.*, **109**, B06403–1–12 (2004).
80. Gross, R.S., Combinations of Earth orientation measurements: SPACE94, COMB94, and POLE94. *J. Geophys. Res.*, **101** (B4), 8729–8740 (1996).
81. Pavlis, D.E. et al., *GEODYN operations manuals*, contractor report, Raytheon, ITSS, Landover MD (1998).

82. Ries, J.C., Eanes R.J. and Tapley, B.D., Lense-Thirring precession determination from laser ranging to artificial satellites. In: *Nonlinear Gravitodynamics, the Lense-Thirring Effect*, Proc. *III William Fairbank Meeting* (World Scientific, Singapore, 2003) pp. 201–211.
83. Ciufolini, I., Frame-dragging and its measurement. In: *Gravitation: from the Hubble Length to the Planck Length, Proc. I SIGRAV School on General Relativity and Gravitation*, Frascati, Rome, September 2002 (IOP, 2005) pp. 27–69.
84. Ciufolini, I., Theory and experiments in general relativity and other metric theories, *Ph.D. Dissertation*, Univ. of Texas, Austin (Pub. Ann Arbor, Michigan, 1984).
85. Peterson, G.E., Estimation of the Lense-Thirring precession using laser-ranged satellites. *Ph. Dissertation*, Univ. of Texas, Austin, (1997).
86. Ries, J.C., Eanes, R.J., Tapley, B.D. and Peterson, G.E., Prospects for an improved Lense-Thirring test with SLR and the GRACE gravity mission. In: *Toward Millimeter Accuracy* Proc. *13th Int. Laser Ranging Workshop*, Noomen, R., Klosko, S., Noll, C. and Pearlman, M. eds., (NASA CP 2003212248, NASA Goddard, Greenbelt, MD, 2003).
87. Pavlis, E.C., Geodetic contributions to gravitational experiments in space. In: *Recent Developments in General Relativity, Genoa 2000*, R. Cianci, et al., eds. (Springer-Verlag, Milan, 2002) pp. 217–233.
88. Rubincam, D.P., General relativity and satellite orbits: the motion of a test particle in the Schwarzschild metric. *Celest. Mech.*, **15**, 21–33 (1977).
89. Cugusi, L. and Proverbio, E., Relativistic effects on the motion of Earth's artificial satellites. *Astron. Astroph.*, **69**, 321–325 (1978).
90. Yilmaz, H., Proposed test of the nature of gravitational interaction. *Bull. Am. Phys. Soc.*, **4**, 65 (1959).
91. Van Patten, R.A., Everitt, C.W.F., Possible Experiment with two counter–orbiting drag–free satellites to obtain a new test of Einstein's general theory of relativity and improved measurements in geodesy. *Phys. Rev. Lett.*, **36**, 629–32 (1976).
92. Ciufolini, I., Paolozzi, A., et al., *LARES phase A study for ASI* (1998).
93. Lucchesi, D.M., Reassessment of the error modelling of non–gravitational perturbations on LAGEOS 2 and their impact in the Lense–Thirring determination. Part I. *Planet. Space Sci.*, **49**, 447–463 (2001).
94. Pavlis, E.C. and Iorio, L., The impact of tidal errors on the determination of the Lense-Thirring effect from satellite laser ranging. *Int. J. Mod. Phys. D*, **11**, 599–618 (2002).
95. Lemoine, F.G., Kenyon, S.C., Factor, J.K., Trimmer, R.G., Pavlis, N.K., Chinn, D.S., Cox, C.M., Klosko, S.M., Luthcke, S.B., Torrence, M.H., Wang, Y.M., Williamson, R.G., Pavlis, E.C., Rapp, R.H. and Olson, T.R., The Development of the Joint NASA GSFC and the National Imagery and Mapping Agency (NIMA) Geopotential Model EGM96, NASA/TP-1998-206861, Goddard Space Flight Center, Greenbelt, Maryland (July 1998).
96. NRC, *Satellite Gravity and the Geosphere: Contributions to the Study of the Solid Earth and Its Fluid Envelope* (National Academy Press, Washington D.C., 1997).
97. Pavlis, E.C., Improvements to geodesy from gradiometers and drag-free satellites. In: Proc. *1st W. Fairbank Memorial Conference on Gravitational Relativistic Experiments in Space*, M. Demianski and C.W.F. Everitt, eds., (World Scientific, 1993).
98. Reigber, Ch., Flechtner, F., Koenig, R., Meyer, U., Neumayer, K., Schmidt, R., Schwintzer, P., and Zhu, S., GRACE Orbit and Gravity Field Recovery at GFZ Potsdam – First Experiences and Perspectives. *Eos. Trans. AGU*, **83**(47), Fall Meet. Suppl., Abstract G12B-03 (2002).
99. Perosanz, F., Loyer, S., Lemoine, J.M.L., Biancale, R., Bruinsma, S. and Vales, N., CHAMP accelerometer evaluation on two years mission. *Geophys. Res. Abstracts*, **5** (CD), Abstract EAE03-A-06989 (2003).
100. Tapley, B.D., The GRACE mission: status and performance assessment. *Eos. Trans. AGU*, **83**(47), Fall Meet. Suppl., Abstract G12B-01 (2002).
101. Watkins, M.M., Yuan, D., Bertiger, W., Kruizinga, G., Romans, L. and Wu, S., GRACE gravity field results from JPL. *Eos. Trans. AGU*, **83**(47), Fall Meet. Suppl., Abstract G12B-02 (2002).
102. Rummel, R., GOCE – its status and promise. *Geophys. Res. Abstracts*, Vol. **5** (CD), Abstract EAE03-A-09628 (2003).

103. Kaula, W.M., *Theory of Satellite Geodesy*, (Blaisdell, Waltham, 1966).
104. Lucchesi, D.M., Reassessment of the error modelling of non-gravitational perturbations on LAGEOS 2 and their impact in the Lense–Thirring determination. Part II. *Planet. Space Sci.*, **50**, 1067–1100 (2002).
105. Yoder, C.F., Williams, J.G., Dickey, J.O., Schutz, B.E., Eanes, R.J. and Tapley, B.D., Secular variations of Earth's gravitational harmonic J2 coefficient from Lageos and nontidal acceleration of Earth rotation. *Nature*, **303**, 757–62 (1983).
106. Cheng, M.K., Shum, C.K. and Tapley, B., Determination of long-term changes in the Earth's gravity field from satellite laser ranging observations. *J. Geophys Res.*, **102** (B10), 22377–22390 (1997).
107. Cheng, M.K. and Tapley, B.D., Temporal variations in J2 from analysis of SLR data. In: *Proc. 12th International Workshop on Laser Ranging* (2000).
108. Cheng M.K. and Tapley B.D., Variations in the Earth's oblateness During the Past 28 years. *J. Geophys. Res.*, **109**, B09402 (2004) doi: 10,1029/2004JB003028.
109. Cheng, M.K., Tapley, B.D., Secular variations in the low degree zonal harmonics from 28 years of SLR data. *Eos Trans. AGU*, **85**(47), Fall Meet. Suppl., Abstract G31C-0801.
110. Ciufolini, I. and Pavlis, E.C., On the Measurement of the Lense-Thirring effect using the nodes of the LAGEOS satellites, in reply to "On the reliability of the so-far performed tests for measuring the Lense-Thirring effect with the LAGEOS satellites" by L. Iorio. *New Astr.*, **10**, (8), 636–651 (2005).
111. Cox, C.M., Klosko, S.M. and Chao, B.F., Changes in Ice-Mass Balance inferred from time variation of the geopotential observed through SLR and Doris tracking. In: *Gravity, Geoid and Geodynamics 2000, International Association of Geodesy (IAG) Symposia* Vol. 123, M.G. Sideris, ed. (Springer, 2000).
112. Devoti, R., Luceri, V., Rutigliano, P., Sciappreta, C. and Bianco, G., Time series of low degree zonals obtained analyzing different geodetic satellites. *Bollettino di Geofisca Teorica ed Applicata*, **40**, 353–358, (1999).
113. Pavlis, E.C., Dynamical Determination of Origin and Scale in the Earth System from Satellite Laser Ranging. In: *Vistas for Geodesy in the New Millennium*, Proc. *2001 International Association of Geodesy Scientific Assembly*, Budapest, Hungary, September 2–7, 2001, J. Adam and K.P. Schwarz, eds. (Springer-Verlag, New York, 2002) pp. 36–41.
114. Schmidt, R., Flechtner, F., Koenig, R., Meyer, U., Neumayer, K.H., Schwintzer, P., Zhu, S.Y., An Earth gravity field model complete to degree and order 150 from GRACE: EIGEN-GRACE02S. *J. Geodyn.*, **39**, 1–10 (2005).
115. Tapley, B.D. et al., Lageos Laser Ranging Contributions to Geodynamics, Geodesy and Orbital Dynamics. In: *Contributions of Space Geodesy to Geodynamics: Earth Dynamics. Geodyn. Ser.*, **24**, 147–174, D.E. Smith and D.L. Turcotte, eds., AGU Washington, D.C. (1993).
116. Tapley, B.D., Chambers, D.P., Bettadpur, S., and Ries, J.C., Large scale ocean circulation from the GRACE GGM01 geoid. *Geophys. Res. Lett.*, **30** (22), 2163 (2003)–1–4 doi: 10.1029/2003GL018622.
117. Cazenave, A., Geogut, P. and Ferhat, G., Secular variations of the gravity field from Lageos 1, Lageos2 and Ajisai. In: *Global Gravity Fields and its temporal variations, Int. Assoc. of. Geod. Symp.*, Vol. 116 (Springer-Verlag, New York, 1996) p 141–151.
118. Cheng, M.K., Eanes, R.J., Shum, C.K., Schutz, B.E. and Tapley, B.D., Temporal variation in low degree zonal harmonics from starlette orbit analysis. *Geophys. Res. Lett.*, **16**, 393–396, (1989)
119. Cheng, M.K., Eanes, R.J., Shum, C.K., Time-varying gravitational effects from analysis off measurements from geodetic satellite. *EoS Trans. AGU*, **74**(43), Fall Meet., Suppl., 196 (1993)
120. Eanes, R.J. and Battadpur, S., Temporal variability of Earth's gravitational field from satellite laser ranging observations. In: *Global Gravity Field and its Temporal Variations, Int. Assoc. of Geod. Symp.*, Vol. 116, 30–41 (Springer-Verlag, New York, 1996).
121. Gegout, P., and Cazenave, A., Geodynamic parameters derived from 7 years of laser data on Lageos. *Geophys. Res. Lett.*, **18**, 1729–1742 (1991).

122. Ivins, E.R., Sammis, C.G. and Yoder, C.F., Deep mantle viscous structure with prior estimate and satellite constraint. *J. Geophys. Res.*, **98**, 4579–4609 (1993).
123. Mitrovica, J.X. and Peltier, W.R., Present-day secular variations in the zonal harmonics of Earth's geopotential. *J. Geophys. Res.*, **98**, 4509–4526 (1993).
124. Nerem, R.S., and Klosko, S.M., Secular variations of the zonal harmonics and polar motion as geophysical constraints. In: *Global Gravity Field and its Temporal Variations, Int. Assoc. of Geod. Symp.* Vol. *116*, 152–163 (Springer-Verlag, New York, 1996).
125. Nerem, R.S., Chao, A.Y., Chan, J.C., Klosko, S.M., Pavlis, N.K. and Williamson, R.G., Temporal variations of the Earth's gravitational field from satellite laser ranging to Lageos. *J. Geophys. Res. Lett.*, **20**, 595–598 (1993).
126. Rubincam, D.P., Postglacial rebound observed by Lageos and the effective viscosity of the Lower mantle. *J. Geophys. Res.*, **89**, 1077–1087 (1984).
127. Trupin, A., Meier, M.F. and Wahr, J., Effects of the melting glaciers on the Earth's rotation and gravity field: 1965–1984. *Geophys. J. Int.*, **108**, 1–15 (1992).
128. Cox, C.M. and Chao, B., Detection of a large-scale mass redistribution in the terrestrial system since 1998. *Science*, **297**, 831 (2002).
129. Ciufolini, I., Pavlis, E.C. and Peron, R., Determination of frame-dragging using Earth gravity models from CHAMP and GRACE. *New Astron.*, **11**, 527–550 (2006).
130. Eanes, R.J., A study of temporal variations in Earth's gravitational field using Lageos-1 laser ranging observations. Ph. D. dissertation, Univ. of Texas at Austin (1995)
131. Iorio, L., On the impact of the atmospheric drag on the LARES mission, arXiv:0809.3564v2, see also: arXiv:0809.3564v1 (2008).
132. Iorio, L., Will the recently approved LARES mission be able to measure the Lense-Thirring effect at 1 %? *Gen. Relativ. Gravit.* **41**, 1717–1724 (2009) doi: 10.1007/s10714-008-0742-1; see also: arXiv:0803.3278v5 [gr-qc] (2008).
133. Iorio, L., An assessment of the systematic uncertainty in present and future tests of the Lense-Thirring effect with satellite laser ranging. *Space Sci. Rev.* **148**, 363–381 (2009); see also: arXiv:0809.1373v2 [gr-qc] (2008).
134. Iorio, L., On some critical issues of the LAGEOS/LAGEOS II Lense-Thirring experiment. arXiv:0710.1022v1 [gr-qc].
135. Ciufolini, I., Paolozzi, A., Pavlis, E.C., Ries, J.C., Koenig, R., Matzner, R.A., Sindoni G., and Neumayer, H. Towards a one percent measurement of frame-dragging by spin with satellite laser ranging to LAGEOS, LAGEOS 2 and LARES and GRACE gravity models. *Space Sci. Rev.*, **148**, 71–104 (2009).
136. Lucchesi, D.M., The impact of the even zonal harmonics secular variations on the Lense-Thirring effect measurement with the two Lageos satellites. *Int. J. of Mod. Phys. D*, **14**, 1989–2023 (2005).
137. Ciddor, P.E., Refractive index of air: New equations for the visible and near infrared. *Applied Optics*, **35**, 1566–1573 (1996).
138. Hulley, G.C. and Pavlis, E.C., A ray-tracing technique for improving Satellite Laser Ranging atmospheric delay corrections, including the effects of horizontal refractivity gradients. *J. Geophys. Res.*, **112**, B06417–1-19 (2007), doi:10.1029/2006JB004834.
139. Mendes, V.B., Prates, G., Pavlis, E.C., Pavlis, D.E. and Langley, R.B., Improved mapping functions for atmospheric refraction correction in SLR. *Geophysical Res. Lett.*, **29**, 1414–1-4 (2002), doi:10.1029/2001GL014394.
140. Mendes, V.B. and Pavlis, E.C., High-accuracy zenith delay prediction at optical wavelengths. *Geophys. Res. Lett.*, **31**, L14602–1-5 (2004), doi:10.1029/2004GL020308.
141. Tapley, B.D., Bettadpur, S., Watkins, M., and Reigber, C., The gravity recovery and climate experiment: Mission overview and early results. *Geophysical Res. Lett.*, **31**, L09607 (2004) doi: 10.1029/2004GL019920.

The Relativity Mission Gravity Probe B, Testing Einstein's Universe

Saps Buchman for the GP-B Collaboration

Abstract The Relativity Mission, Gravity Probe B (GP-B) is the most complex gravitational space experiment to date and the first controlled experiment in which a General Relativistic effect is the main feature of the data. Launched on April 20, 2004 it has successfully taken science data from August 29, 2004 to August 15, 2005. Using four high precision gyroscopes GP-B measures the relativistic precessions of the frame of reference in a 642 km polar orbit. The two precessions are predicted in General Relativity to be the geodetic effect, 6.6 arcsec/year, and the frame dragging effect, 0.042 arcsec/year. Comprehensive pre and post mission calibrations have been performed between April 20 and August 27, 2004 and between August 15 and September 29, 2005. GP-B is the first experiment to see General Relativity directly as the main effect of the measurement and has demonstrated that very complex physics experiments can perform flawlessly in space. We present engineering performance data and general operational information for the major systems of the mission. Lessons learned from the GP-B experiment and their impact on the development of technology for the Laser Interferometer Space Antenna (LISA) and the Space Test of the Equivalence Principle (STEP) are discussed. Technology progress for the design of the next generation of drag-free sensors is presented. Preliminary results set the measurement accuracy at ±97 marcs/year or about ±1.5% of the geodetic effect. The GP-B science team is continuing to refine the analysis of the data and further improvements in accuracy are expected. Final science results will be released in December 2007.

1 Introduction

At a banquet honoring Albert Einstein in 1930 George Bernard Shaw aptly summarized the quest for understanding the Universe in the last 2,000 years:

S. Buchman (✉)
Hanson Experimental Physics Laboratory, Stanford University, Stanford, CA 94305
e-mail: sasha@relgyro.stanford.edu

I. Ciufolini and R.A. Matzner (eds.), *General Relativity and John Archibald Wheeler*, Astrophysics and Space Science Library 367, DOI 10.1007/978-90-481-3735-0_18,

> ...there is an order of men who are makers of Universes. Ptolemy made a Universe which lasted 1,400 years. Newton also made a Universe which has lasted 300 years. Einstein has made a Universe and I can't tell you how long that would last.

This progression exemplifies beautifully the approach to the truth through consecutive approximations. Ptolemy's empirical model used eighty epicycles to explain the motions of the Sun, the Moon, and the five known planets. It matched visual observation to their level of accuracy and was "the Universe" for 13 centuries. Copernicus's heliocentric system (1543) and Keppler's three laws (1618) were the start of the modern description of planetary motion. Newton's Universe, based on the equivalence of the inertial and the gravitational mass is in many ways the beginning of modern science. The limit on the equivalence principle is presently about 10^{-13}.

In 1915 Einstein formulated General Relativity based on the equivalence of gravitational and acceleration fields, describing gravity as the curvature of space, and establishing the equality between the space–time Einstein tensor $G_{\mu\nu}$ and stress-energy tensor $T_{\mu\nu}$ (Eq. 1)

$$G_{\mu\nu} = \frac{8\pi G}{c^4} T_{\mu\nu} \quad G_{\mu\nu} \equiv R_{\mu\nu} - \frac{1}{2} R g_{\mu\nu} + \Lambda g_{\mu\nu} \tag{1}$$

where $R_{\mu\nu}$ is the Ricci tensor, R the Ricci scalar, and $g_{\mu\nu}$ the metric tensor, and Λ Einstein's cosmological constant.

General Relativity remains the longest lived modern physical theory. Early on it has resolved the 43 arcs/century discrepancy in the perihelion shift of Mercury, it has correctly predicted the deflection of light by the Sun (1.75 arcs during the 1919 eclipse) and the redshift in a gravitational field. Its predictions, including the once believed implausible black holes, have all been confirmed by measurements and observations. General Relativity remains an elegant and the most widely accepted theory of gravitation, that agrees with experimental results to better than 10^{-4}. In the fast moving modern science of the last century General Relativity has lasted amazingly long and has been spectacularly successful.

Why then test Einstein? The main question remains the problem of unified physics. General Relativity is not quantizable and thus incompatible with the Grand Unification theories that unite the electromagnetic, strong nuclear, and weak nuclear forces. Grand Unification theories predict violations of General Relativity and the promising areas to discover discrepancies are in rotational effects, below the 10^{-5} level, and in the equivalence principle, below the 10^{-14} level.

Section 2 discusses the rational for performing gravitational experiments in space, and describes the GP-B mission. Section 3 summarizes the performance of the main GP-B systems and the results to date. The LISA and STEP mission concepts and their common technologies with GP-B are summarized in Section 4. Section 5 discusses the lessons learned from GP-B and their applications to LISA and STEP. Section 6 describes advances in the area of drag-free technology with applications to future gravitational space missions.

2 The GP-B Experiment

"No mission could be simpler than Gravity Probe B. It's just a star, a telescope, and a spinning sphere." – William Fairbank.

GP-B is measuring two rotational effects predicted by gravitational theories; the geodetic precession due to the distortion of space by a massive body, and the frame dragging precession, a twisting of space caused by the rotation of the body [2, 10]. Earth is the massive body, with the experiment in a 642 km polar orbit. Figure 1 shows this very simple concept of GP-B. The frame of reference in the field of the Earth, measured by gyroscopes, is compared to the reference frame of distant stars determined by a telescope locked on a guide star. The guide star is HR8703 (IM Pegasi) a binary with optical and radio frequency emissions. Its proper motion with respect to Quasars is determined by VLBI measurements. Equation 2 gives the two precessions as computed in General Relativity and expanded in the Parameterized Post Newtonian (PPN) formalism used for metric theories in weak gravitational fields. The geodetic precession in the plane of the polar orbit, first term, and the frame dragging orthogonal to the orbit plane, second term, are proportional respectively to the mass and the moment of inertia of the Earth.

$$\overline{\Omega} = \left(\gamma + \frac{1}{2}\right) \frac{GM}{c^2 R^3} \left(\overline{R} \times \overline{v}\right) + \left(\gamma + 1 + \frac{\alpha_1}{4}\right) \frac{GI}{2c^2 R^3} \left[\frac{3\overline{R}}{R^2} \cdot \left(\overline{\omega}_e \cdot \overline{R}\right) - \overline{\omega}_e \right]. \tag{2}$$

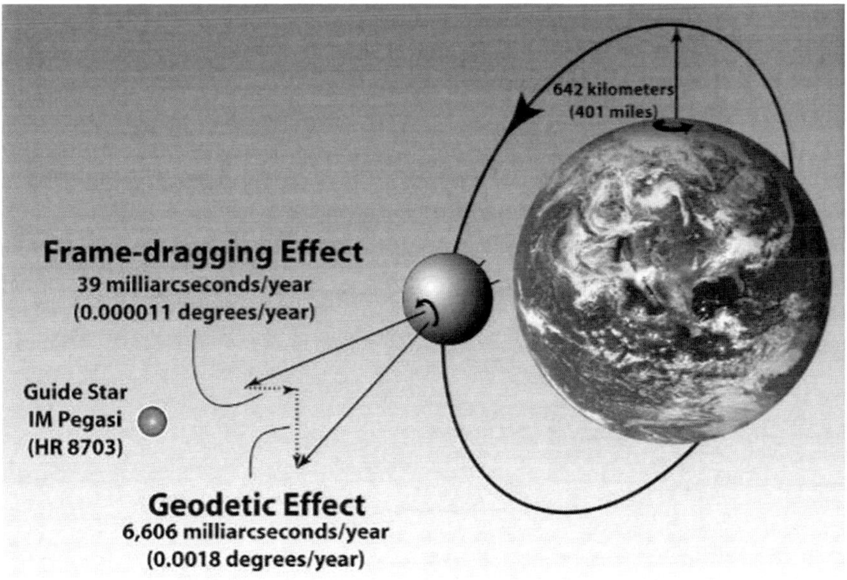

Fig. 1 The relativity mission, gravity probe B concept

GP-B is a unique gravitational measurement in space, as a controlled physics experiment with an apparatus specifically designed for the purpose with fully determined parameters. While based on a simple concept, the precision requirements make GP-B a very complex experiment based on cutting edge technology, never before used in space. Its superb performance demonstrates that such complex experiments can and will work and greatly increases the confidence in the success of future experiments in space.

Figure 2 shows the preliminary analysis results of the direct measurement of the geodetic effect, the north-south inertial orientation of the four gyroscopes to $\pm 1.5\%$ ($\pm$97 marcsec/year).

Present measurement limitations (April 2007) are mainly due to incomplete torque modeling, interruptions in the data stream, roll phase uncertainties, and systematic read-out errors. The SQUID readout noise is contributing less than 1marcs/year to the error. Note: deg/h are the units habitually used for gyroscope drift: $1\,\mathrm{marcs/year} = 3.2\times 10^{-11}$ deg/h. Fragmentation of the year-long data stream, caused by anomalies, torques, and data analysis approaches, will reduce the nominal measurement accuracy $\delta\Omega$ to $\delta\Omega_n$ proportionally to the number n of segments (considered equal for this example); Eq. 3.

$$\delta\Omega = L \cdot T^{-3/2} \qquad \delta\Omega_n = L \cdot T^{-1/2} \cdot t^{-1} = \delta\Omega \cdot n \tag{3}$$

where L is the proportionality constant, T is the nominal science duration, t the segment length and $n \equiv T/t$.

The spacecraft rolls phase locked around the line of sight to the guide star (telescope axis) with a period of 77.5 s. This results in the spectral shifting of the science signal from zero frequency to 12.9 mHz (the gyroscope 1/f read-out noise is reduced) as well as in the averaging of the body fixed disturbance torques acting on the gyroscopes. The averaging is proportional to the angle of about 20 arcs between the telescope axis and the spin axes of the gyroscopes, equivalent to a reduction in disturbance by 10^4. The roll phase is determined by star trackers and used to separate Einstein's predicted gyroscope spin-axis drifts.

Deciding on performing a space experiment requires the careful balancing of the advantages and disadvantages of space. Such experiments are costly and of very long duration, require survivability in the severe launch environment, are a single opportunity endeavor with any major anomaly potentially causing total mission loss, and are handicapped by limitations in communications. However, low seismic noise in the frequency region below 1 Hz, the low gravity environment, long interfer ometry baselines, and long integration times make space the only possible way of achieving the requirements for a number of experiments including GP-B, LISA, and STEP. Table 1 summarizes the main advantages and disadvantages of space experimentation, while Fig. 3 shows the rational behind performing GP-B in orbit. Only the very low gravity of space allows the gyroscope performance required for the measurement of the relativistic precessions. GP-B is the first experiment to directly measure gravitational relativity as the main measured effect.

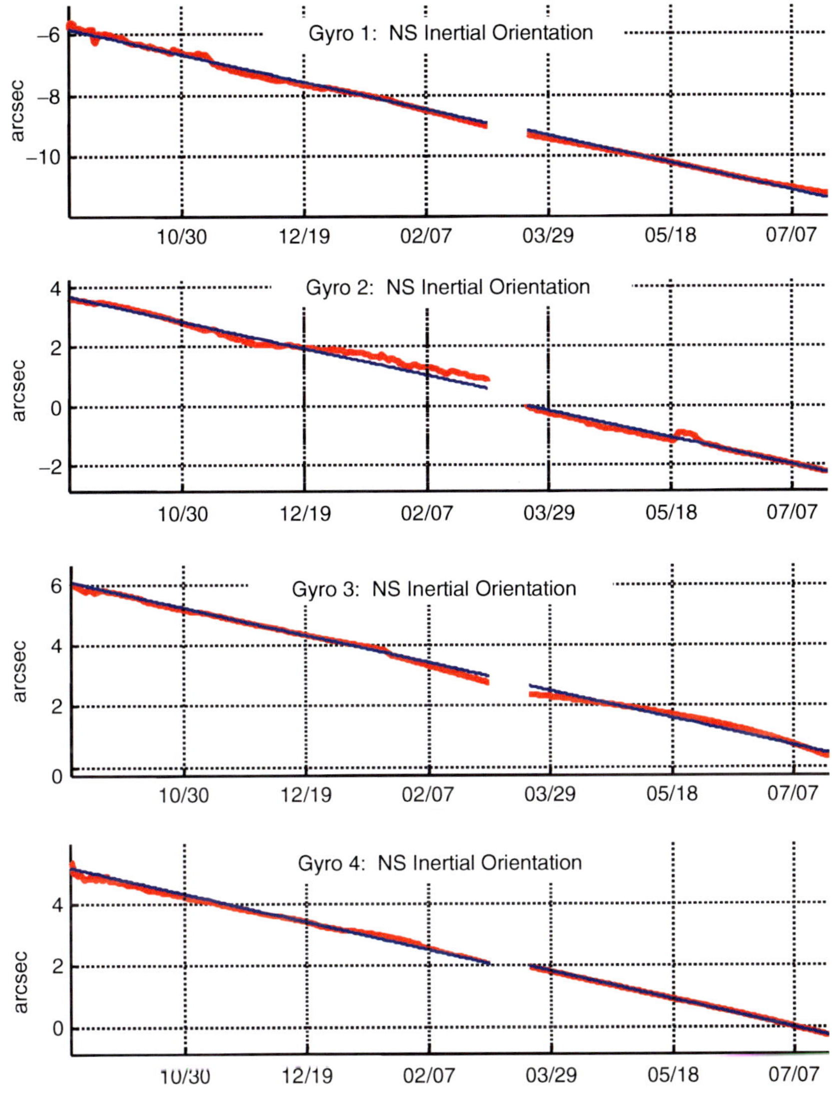

Fig. 2 Geodetic effect seen directly; raw data in *red*, with torque analysis, *blue*

Table 1 Main advantages and disadvantages of space experiments

Experimentation in space	
Advantages	Disadvantages
Low seismic noise below 1 Hz (LISA)	High acceleration launch
Low acceleration; $<10^{-10}$ m/s^2; GP-B, LISA, STEP	High cost and long development
Long baselines; 5×10^6 km for LISA	Single opportunity: high reliability req.
Long measurement times (GP-B, LISA, STEP)	Restricted communications band

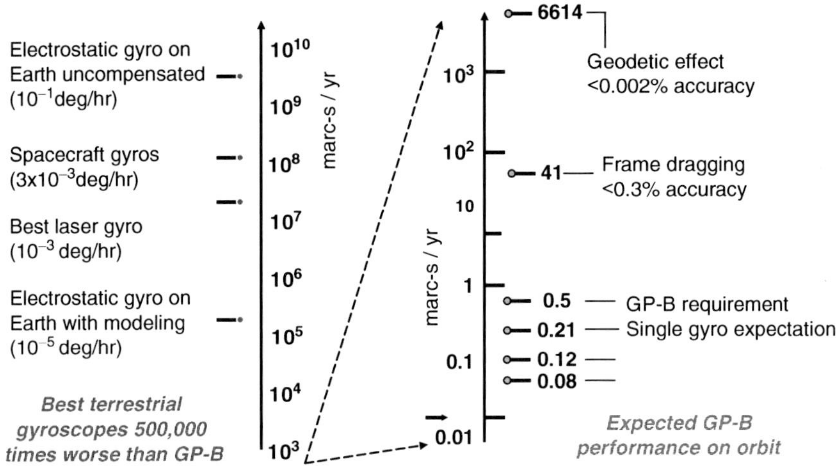

Fig. 3 GP-B gyroscopes in space versus ground based gyroscopes

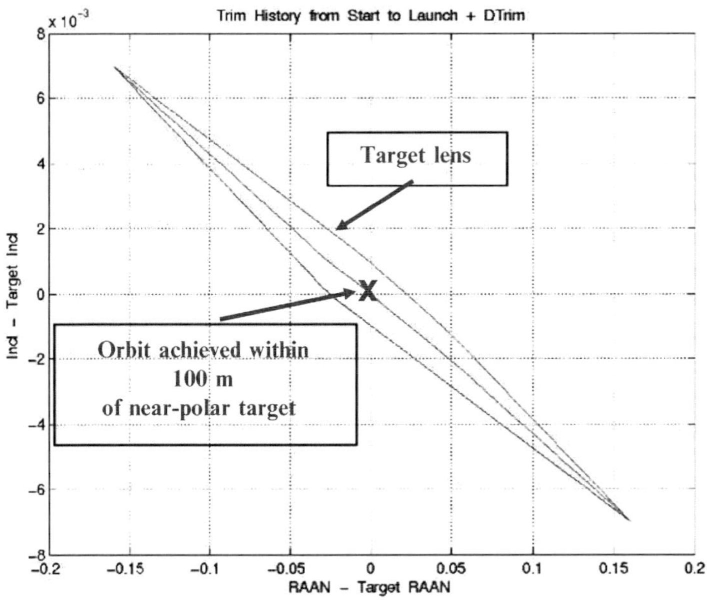

Fig. 4 GP-B orbit insertion

3 GP-B Performance

GP-B was launched on April 20, 2004. The Delta II rocket placed the space vehicle in within 100 m of the required orbit. Figure 4 shows the nearly perfect insertion point in relation of the "target lens" the required final orbit area. The quality of the insertion made orbit adjustment unnecessary and therefore allowed for simplified experiment initialization.

Until Science Mission start, on August 29, 2004, the experiment underwent a period of 128 days of initialization that included instrument calibrations, guide-star acquisition, spacecraft rolling, drag-free set-up, spin-up of gyroscopes and their alignment with the roll axis, and low-temperature bake-out. The 128 days was triple the expected duration of 40 days and more than twice the 60 days initialization allocation with contingency margin, notwithstanding the excellent communication bandwidth provided by the ground and space NASA networks. Thus, the first GP-B lesson learned is to allow for a factor of three contingency in the planned experiment initialization period. The science data collection phase lasted for 353 days until August 15, 2005. In addition to the collection of science data, redundancy and internal cross-checks of these data were performed. The 46 day calibration phase, ending with the depletion of the liquid helium on September 29, 2005, consisted of repeat calibrations while deliberately enhancing the disturbing effects.

The GP-B, Science Instrument Assembly (SIA), contains four electrostatically suspended cryogenic gyroscopes, four dc SQUID read-outs, and a 14 cm aperture 380 cm focal length Cassegranian telescope, all mounted in a quartz block that ensures accurate and stable alignment (Fig. 5). Each gyroscope is sufficient for the GP-B experiment, thus the local frame is measured with quadruple redundancy while the telescope has double redundancy for measuring the far-stars reference frame.

The SIA is mounted in the cryogenic probe, that is in turn inserted in the dewar. The spacecraft systems are mounted around the dewar that supplies the mechanical core of the space vehicle (Fig. 6).

We present the GP-B instrumental performance in terms of the cutting edge technologies for the mission that were never before achieved in space or, in some cases, even on the ground. Table 2 lists the ten "Near-zeroes and Superlatives", their performance requirements and their actual measured performance in orbit. All requirements were achieved or surpassed.

The gyroscopes are at the core of the GP-B mission. The reference to the universal frame, the telescope, is required to perform 10^3 times better than the best prior

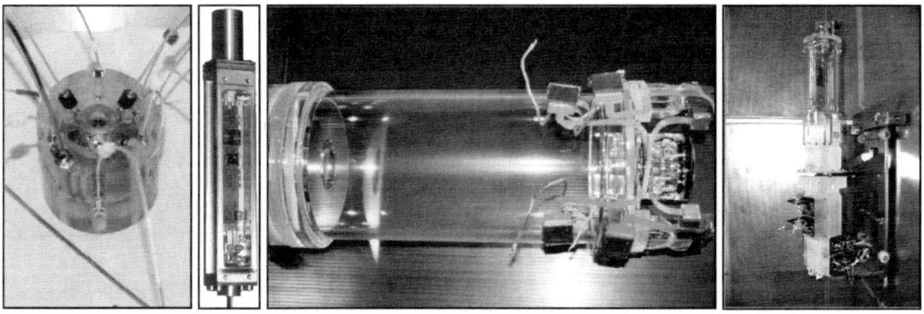

Fig. 5 Assembled gyroscope, dc SQUID, telescope, and SIA

Fig. 6 Cryogenic probe, dewar, space vehicle in faring, GP-B launch April 20, 2004

Table 2 GP-B critical "near-zeros" and superlatives: requirements and performance in orbit

#	Systems	Requirement	In orbit performance
1	Rotor asphericity	<50 nm	<10 nm
2	Rotor inhomogeneity	<25 nm	<7 nm
3	Electric charge	<15 pC	<15 pC
4	Residual acceleration	$<10^{-12}$ m/s^2 at roll	$<10^{-12}$ m/s^2 at roll
5	Gas pressure	$<10^{-11}$ torr	$<10^{-14}$ torr
6	Magnetic field	<9 μG in rotor	<3 μG in rotor
7	Low-noise SQUID	<200 marcs/$\sqrt{\text{Hz}}$at roll	<190 marcs/$\sqrt{\text{Hz}}$atroll
8	Sensitive star-tracker	0.1 marcs	0.1 marcs
9	Largest flight dewar	2,500 l, 16.5 month, 1.8 K	2,500 l, 17.3 month, 1.8 K
10	Star proper motion	<0.3 marcs/year	<0.15 marcs/year

star trackers, while the reference to the local frame, the gyroscopes, are required to have drifts up to 10^7 times smaller than the best modeled previous instruments. Most of the 10 "near-zeroes" and superlatives are therefore technologies required to achieve the gyroscope performance. The main features of the GP-B gyroscopes are: electrical suspension and position measurement, Helium gas spin-up, magnetic read-out, and cryogenic operation. Sphericity uniformity is critical to the performance of the gyroscopes by minimizing the torques caused by the electrostatic fields. The gyroscopes are 3.9 cm diameter fused quartz spheres coated with 1μm thick Niobium film of 2% uniformity. Quartz substrate sphericity is $2 \cdot 10^{-7}$ of diameter or less than 10 nm peak-to-valley; gyroscope housing uniformity is about one magnitude less. Figure 7 shows the color coded surface profile of a flight rotor mapped to 1 nm precision and its equivalent surface roughness if expanded to the size of the Earth.The centered gyroscopes are spaced at 32 μm from the suspension electrodes with a maximum clearance of 20 μm from other housing features. In order to reduce precession the center of mass coincides to the center of geometry (mass unbalance) to $2 \cdot 10^{-7}$ of diameter. Figure 8 shows the measurement method of the mass unbalance using the maximum of the range of positions of the gyroscope as swept by its polhoding motion; on the right is the schematic of the forces moving the mass unbalance δr with respect to the spin axis. Table 3 shows

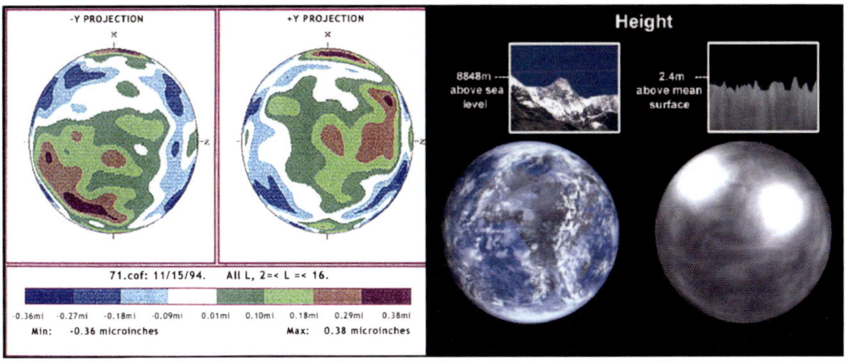

Fig. 7 Surface profile of GP-B flight rotor (spherical to 0.2 ppm) and its equivalent scaling to earth size

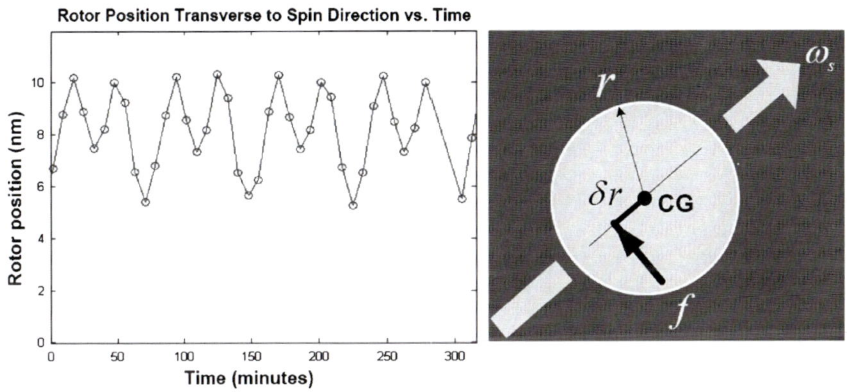

Fig. 8 Gyroscope mass unbalance δr measured by maximum variation of position transverse to spin direction

Table 3 Mass unbalance: pre-launch estimate and on-orbit data

Gyro #	Pre-launch estimate (nm)	On-orbit data (nm)
1	18.8	10.1
2	14.5	4.8
3	16.8	5.4
4	13.5	8.2

the pre-launch estimates of the mass unbalance versus the measurements on orbit. The mass unbalance values are well below the requirement of 50 nm.

The GP-B gyroscopes were spun up in orbit with Helium gas warmed to 6.5 K, below the superconducting transition of their Niobium coating. Spin-up proceeded by bringing the gyroscopes close to the channel through which Helium flows at 750 cm^3/min 95% of the gas is being exhausted directly to space through dedicated pumping lines, while the remaining 5% are pumped through the instrument. The leakage gas limits the flow rate of gas for spin-up and causes some slow-down of

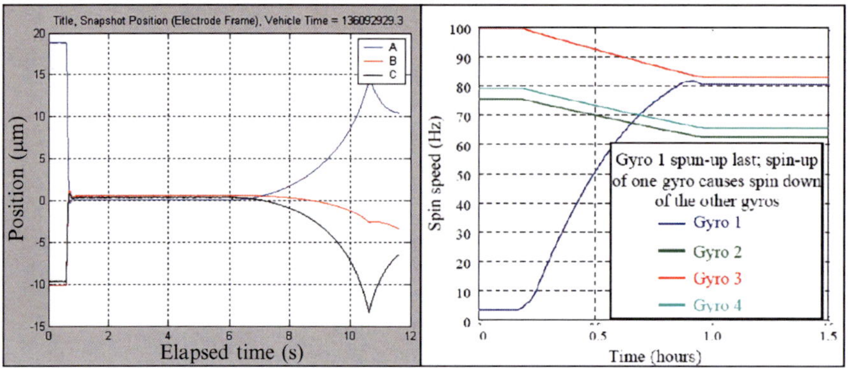

Fig. 9 First gyroscope suspension and GP-B gyroscope spin-up

Table 4 Spin speed and spin-down rates

Gyro #	Spin-spee d (Hz)	Spin-down rate (μHz/h)
1	79.4	0.57
2	61.8	0.52
3	82.1	1.30
4	64.8	0.28

the other gyroscopes. Figure 9 shows the first gyroscope suspension and the spin-up of gyroscope #1. Gyroscope #1 was spun-up last; its spin-up causes some spin down of the other gyroscopes. Table 4 gives the science mission spin-speed and average spin-down rate for each of the four gyroscopes.

The gyroscope charge was controlled to less than 15 pC; equivalent to a potential of 15 mV for the 1 nC total gyroscope capacitance. Charge management was achieved by UV generated photoelectrons from the gyroscope and housing, which were moved onto or away from the gyroscope by the appropriate electrostatic bias [1]. The charge was measured to better than 2 pC by applying out of phase 20 mV potentials to opposite electrodes at about 50 mHz.

As expected, two main charging mechanisms required management. The largest by far, was the one time charge of between 200 and 500 pC generated by the separation of dissimilar metals during initial gyroscope levitation. In order to achieve science electrostatic suspension mode this charge needed to be reduced to less than 15 pC (Fig. 9). During science mission the charging was caused by cosmic radiation at a daily rate of 0.1–0.15 pC. During a major solar flare (sun spot 720) the charging rates increased to between 0.6 and 1.2 pC/day. As calculated prior to launch, the cosmic radiation charging was always positive and caused mainly by protons stopping in thc gyroscope. In order to maintain the charge below 15 pC only two discharge operations were required during the mission, at 3 months and 7 months after science start. Figure 10a shows the discharging of gyroscope #1 after levitation. The initial charge from levitation was 450 pC and the discharge rate was 70 pC/h. Figure 10b shows the charge history of the four gyroscopes, including the two discharge operations; scale is 5 mV/division vertical and 50 days/division horizontal.

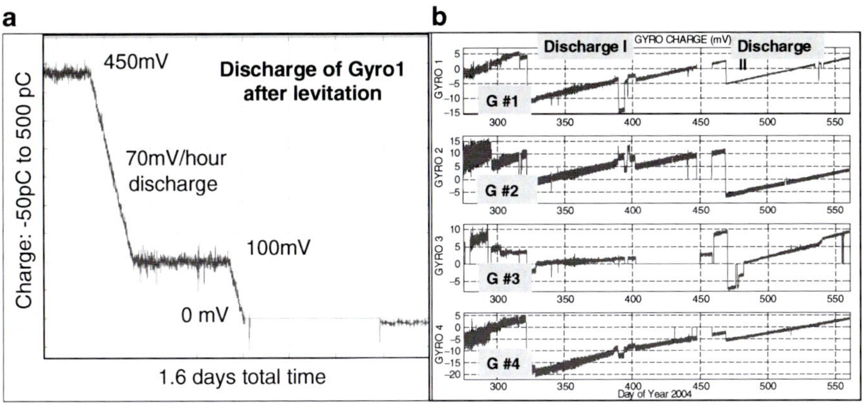

Fig. 10 (**a**) Discharge of gyro 1 after levitation. (**b**) Charging and discharging of the GP-B gyroscopes; 5 mV/div vertical, 50 days/div horizontal

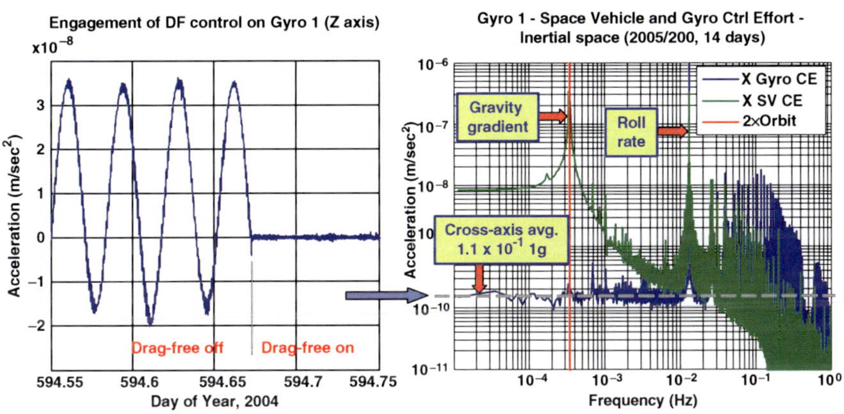

Fig. 11 Acceleration of drag-free gyro: drag free off and on

Gyroscope performance is dependent on the level of drag-free the control systems are reaching [4]. By controlling the position of the spacecraft around a test mass, in the case of GP-B a gyroscope, drag-free technology reduces the disturbing acceleration effects in orbit as caused principally by radiation pressure and residual gas. The average acceleration transverse to roll axis was approximately 10^{-10} m/s^2. Figure 11 shows the acceleration level with drag-free off, left, and drag free on, right.

Cutting edge technology, which can only be tested by analysis and simulation on the ground, requires maximum flexibility in the modes of operations, incorporating as many software 'hooks' as practical. An example is provided by the three modes of operation of the electrostatic suspension of the GP-B gyroscopes. In the accelerometer mode the position of the gyroscope is the input for the control provided by the electrostatic suspension. This mode is used for science in three of the four gyroscopes as well as for levitation, calibrations, spin-up, and spin alignment.

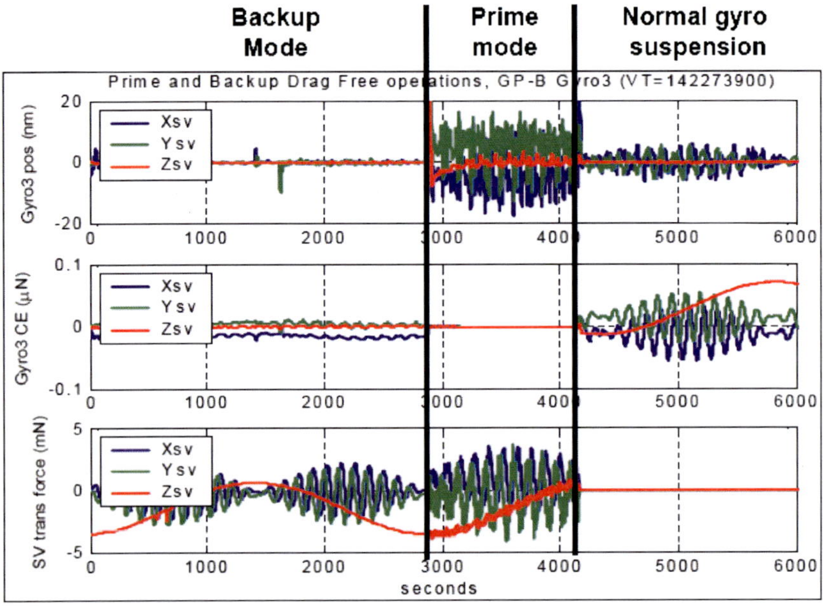

Fig. 12 The three suspension modes of the GP-B gyroscopes

The drag-free mode uses the gyroscope position measurements as input for the control system of the space vehicle thrusters while the gyroscope electrostatic suspension is turned off. This is a possible science mission mode for the drag-free sensor gyroscope. For flexibility, GP-B has also provided a third mode, the back-up drag-free. It uses the space vehicle thrusters as primary actuators to minimize the gyroscope suspension control effort and the electrostatic suspension for fine tuning of the gyroscope position. The performance of the back-up drag free system has been shown in orbit to be better and more reliable then the primary drag-free, and was therefore used during the entire science mission. Figure 12 shows the three suspension modes; the three axes in the space vehicle coordinates for the position and control effort of gyroscope #3 and the space vehicle forcing.

GP-B's attitude and control (ATC) and drag-free systems pose a 9 degrees of freedom (DOF) problem with complex dynamics and coupling. The integrated system actively controls 9 interacting DOF: 3 in attitude of spacecraft to track guide star and maintain roll phase (using thrusters), 3 in translation: drag-free about the geometric center of the gyroscope housing (using thrusters), and 3 in translation of the gyroscope with respect to its housing (using the gyroscope suspension system) (Fig. 13). Figure 14 shows the three main mechanisms used by the space vehicle for ATC and drag-free control; from left to right (not to scale) a proportional Helium thruster cut-out, a star tracker, and a mass trim mechanism used to adjust the center of mass and the axes of inertia of the space vehicle. Including the redundant systems, the GP-B space vehicle has a total of 16 thrusters 2 star trackers, and 7 mass trim mechanisms. Orbital position and velocity were determined by a GPS system,

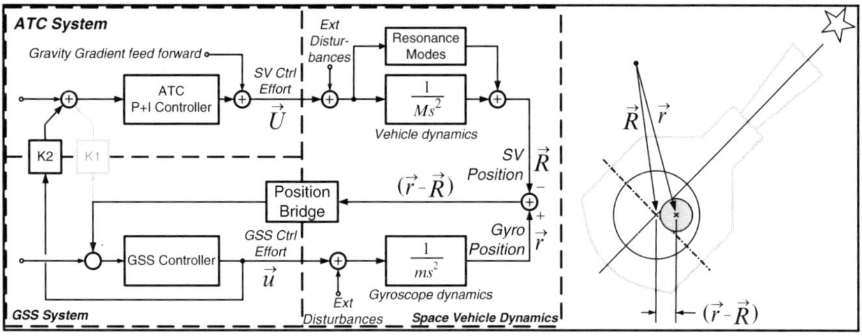

Fig. 13 Schematic of the GP-B 9 DOF drag-free and ATC control systems

Fig. 14 Helium thruster, star tracker, mass trim mechanism (not to scale)

a commercial system modified by Stanford. The modifications included hardware upgrades for space use and sophisticated new software that allowed the system to track the GPS constellation while the space vehicle was both orbiting Earth and rolling with a 77.5 period around its axis. The GPS was the main data source for orbit determination, with a secondary back-up of laser ranging from corner cubes. One of the two fully redundant sets of receivers with four antennas each provided position, velocity, and time data every 10 s; more than 5,000 points/day. The GP-B rms position and velocity accuracy requirements were 25 m and 7.5 cm/s. The actual achieved accuracies were 2.5 m rms and 2.2 mm/s rms; 10 and 35 times better than required.

Table 5 gives the details of the control systems used for GP-B, the first precision controlled six degrees of freedom spacecraft. Abbreviations: DOF = degrees of freedom, GS = guide star, ST = science telescope, STr = star tracker, MT = micro thruster, SC = spacecraft, I = moment of inertia of spacecraft, CM = center of mass of spacecraft, MTM = mass trim mechanism.

Since the gyroscopes are not at the center of mass of the space vehicle, the small differences in gravity due to the differences in the distance to Earth, the gravity gradients, can be seen as forces on the gyroscope of about 10^{-6} m/s^2. Figure 15 shows the increased control effort of the suspension system due to the gravity gradient while the gyroscopes operate as accelerometers. The demonstrated accelerometer performance is better than 10^{-12} m/s^2 from 0 to 1 Hz. The gravity gradient varies at twice orbital frequency.

Table 5 Six degrees of freedom spacecraft control

Controller	DOF	Requirement	Sensor	Actuator	Control computation
GS pointing	2	20 marcs	ST	MT	SC computer
Drag-free	3	10^{-12}m/s^2 rms	GSS	MT	SC computer
Other gyros		0.3 nm at 13 mHz	GSS	GSS	GSS
SC roll	1	20 arcs rms	STr	MT	SC computer
Orbit		<500 m from pole	GPS	MT	SC computer
SC max I		MT capability	MT demand	MTM	On ground
SC CM		0.3 mm	GSS	MTM	On ground

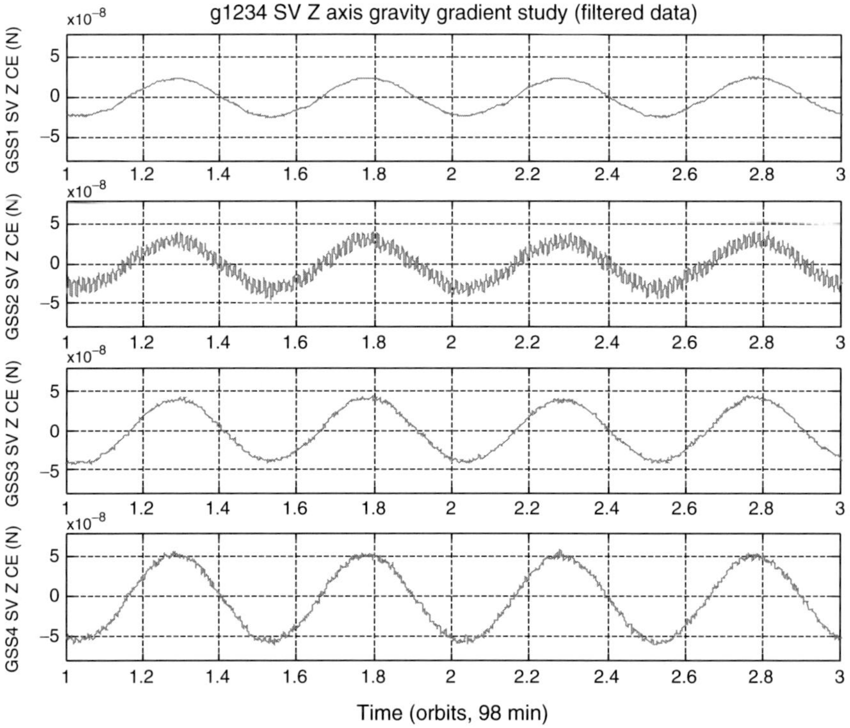

Fig. 15 Gravity gradient for the gyroscopes in accelerometer mode

The gyroscopes are spun with Helium gas in orbit. To remove the remaining Helium, the entire apparatus is heated to 6 K, evacuated to space, ($5{\cdot}10^{-9}$ to 10^{-8} torr), isolated from the environment by closing all exhaust valves, and finally cooled down to the approximately 1.8 K science working temperature. The pressure in the apparatus at 1.8 K, as determined from experiments at higher temperatures is about 10^{-14} torr. Table 6 gives the average spin-down periods before and after bake-out. Figure 16 shows the spin-down rates and implied pressures before, during, and after the bake-out. The spin-down rates after bake-out are not dominated by gas differential damping.

Table 6 Spin-down periods (years) before and after bake-out

Spin-down periods on orbit (years)		
Gyro #	Before bake-out	After bake-out
1	75	15,900
2	62	13,600
3	60	7,200
4	46	26,400

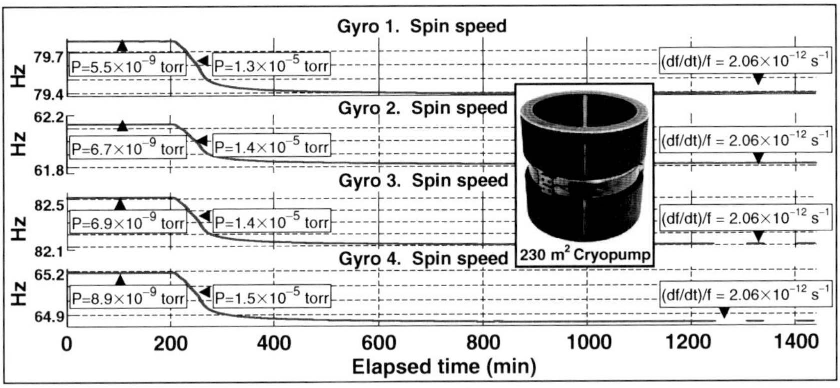

Fig. 16 Spin-down rates and pressures before, during, and after bake-out. Insert cryopump picture

Table 7 Trapped field in gyroscopes

Gyro #	Trapped field (μgauss)
1	3.0
2	1.3
3	0.8
4	0.2

GP-B's magnetic readout requires both ultra-low magnetic field at the instrument location and very high magnetic shielding of outside sources. Taking advantage of the conservation of magnetic flux in a superconducting loop, magnetic fields of less than 10^{-7} gauss are achieved by the expansion of a series of nested lead bags [7]. A system of magnetic shields, including the lead bag, attenuates the external field by 120 db. Table 7 gives the trapped field in the gyroscopes superconducting coating as measured in orbit. Note that the trapped field decreases with the depth of the gyroscope location inside the lead bag.

The spinning superconducting gyroscopes develop magnetic dipoles, also called the London moment, aligned with the spin axis and, for the uniform spherical gyroscopes, with the angular momentum. Figure 17 shows the schematic configuration of the GP-B read-out, based on the coupling of the London moment flux through a superconducting read-out coil to a dc SQUID. Equation 4 gives magnetic field in gauss as a function of the angular spin frequency ω and the electron mass and charge. The readout sensitivity is 190 marcs/$\sqrt{\text{Hz}}$ (5×10^{-11} G/$\sqrt{\text{Hz}}$) at 12.9 mHz roll frequency (Fig. 18). Table 8 gives the SQUID noise in $\mu\Phi_0/\sqrt{\text{Hz}}$ and in

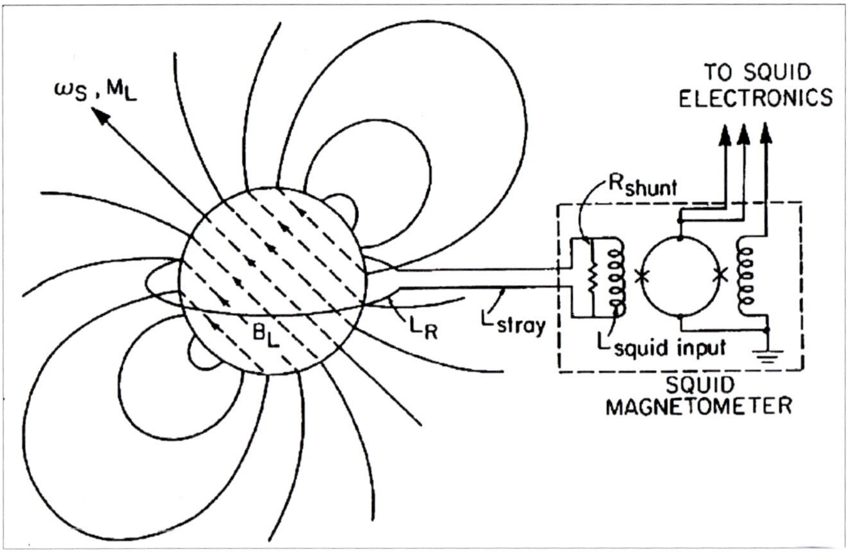

Fig. 17 Schematic of GP-B London moment readout

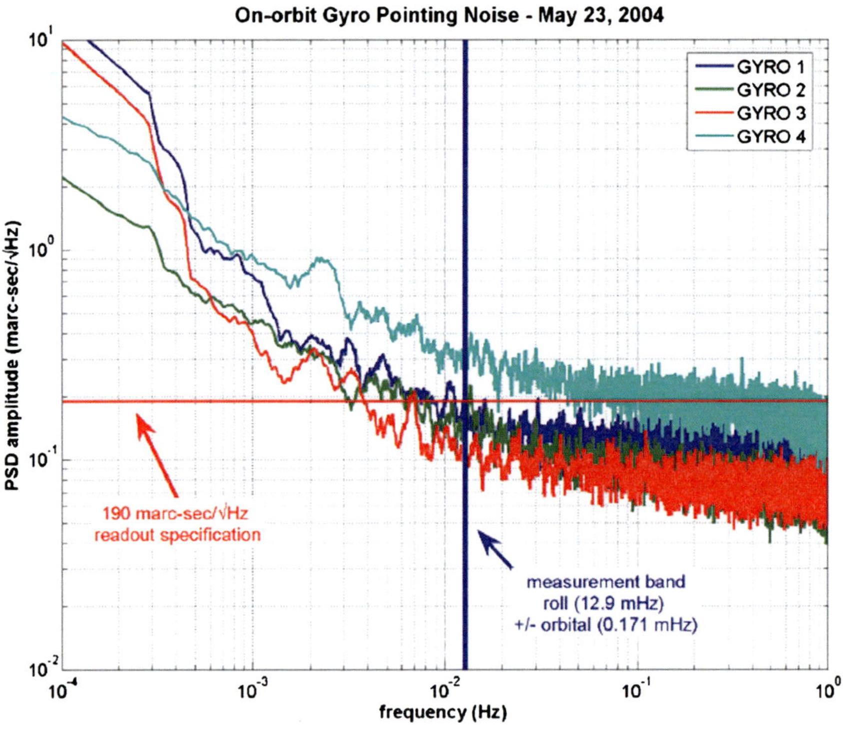

Fig. 18 On orbit gyroscope read-out noise

Table 8 SQUID performance in orbit

SQUID #	Noise at 12.9 mHz ($\mu\Phi_0/\sqrt{Hz}$) SPEC = 47.5 $\mu\Phi_0/\sqrt{Hz}$	Noise at 12.9 mHz (marcs/$\sqrt{Hz}$) SPEC = 190 marcs/$\sqrt{Hz}$	SQUID read-out limit (marcs/year)
1	28	168	0.198
2	20	157	0.176
3	25	140	0.144
4	24	344	0.348

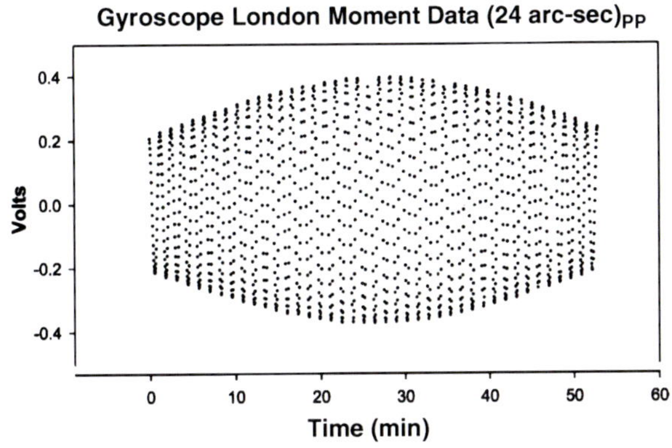

Fig. 19 GP-B half orbit SQUID read-out

marcs/$\sqrt{\text{Hz}}$ as well as the read-out limit for a year of integration. Figure 19 shows the SQUID output data for the guide star valid portion of the orbit; the roll frequency output is modulated by the orbital aberration.

$$\bar{B}_L = -\frac{2m_e c}{e}\overline{\omega}_s = -1.14 \cdot 10^{-7}\overline{\omega}_s \text{ (gauss)} \quad (4)$$

The GP-B telescope, used as a very high quality star tracker, has a folded Cassegranian configuration. For instrument geometry optimization a tertiary mirror returns the image to the front end. Image dividers split the beam in orthogonal components with double redundancy read-out. The sensors are silicon detectors working at 72 K; with two sensors per detector assembly (Fig. 20). The telescope points to the guide star (GS) during about half the orbit, GS valid, and is obscured by the Earth during the other half, GS invalid. Figure 22a shows the telescope pointing data during GS valid. The star is acquired in less than 1 min; September 9, 2004, 54 min total science data. The telescope measurement precision is 0.1 marcs, with its pointing of 34.5 marcs/$\sqrt{\text{Hz}}$ determined by the Helium thruster noise.

Calibration and matching of the scale factors of the telescope and gyroscope is done by two methods. Firstly a slow (29 s and 34 s for the x and y axes) of 60 marcs amplitude was applied with the ATC system allowing scale factor matching for accurate subtraction. Secondly the orbital and annual aberration of the guide star

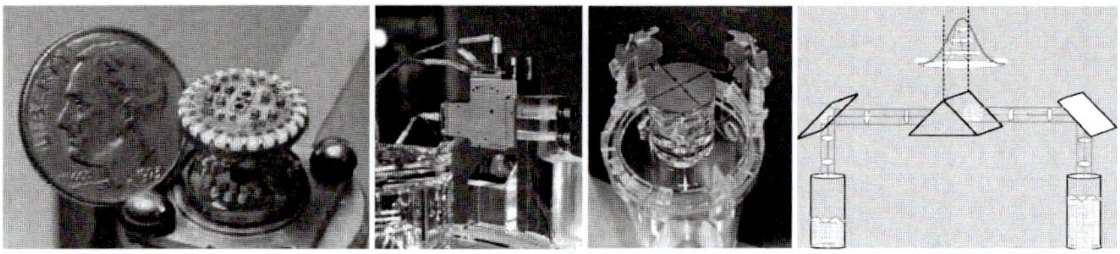

Fig. 20 Telescope detector, detector assembly and image divider

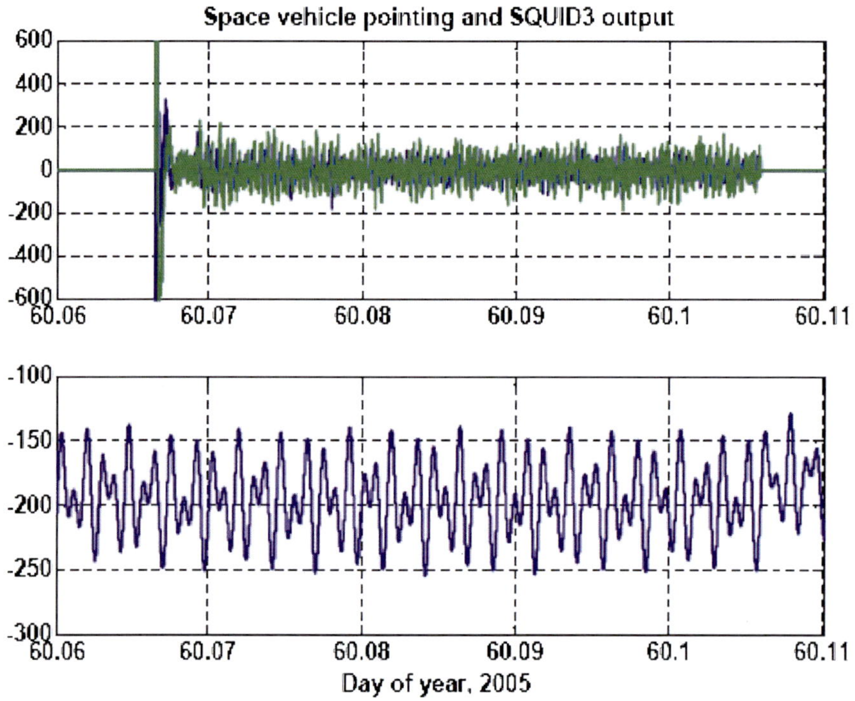

Fig. 21 Telescope pointing (*top*) and SQUID read-out (*bottom*) for scale factor matching

was used to provide the absolute calibration of gyroscope read-out. The motion of the Earth around the Sun causes an annual aberration of 20.4958 arcs, while the motion of the space vehicle around the earth causes an orbital aberration of 5.1856 arcs (at the 97.5 min. orbital period). Figure 21 shows the space vehicle pointing and the SQUID #3 read-out for a guide star valid period on March 1, 2005.

The proper motion of the guide star HR8703 (IM Pegasus) was determined to better than 0.3 marcs/year by Very Large Baseline Interferometry using the radio emission of this binary system. IM Peg's identification was confirmed by acquiring neighboring stars in known configuration with respect to it, as well as measuring its well established binary period. Figure 22b and c show respectively the stars acquired and the light curve , measured from the ground and in orbit.

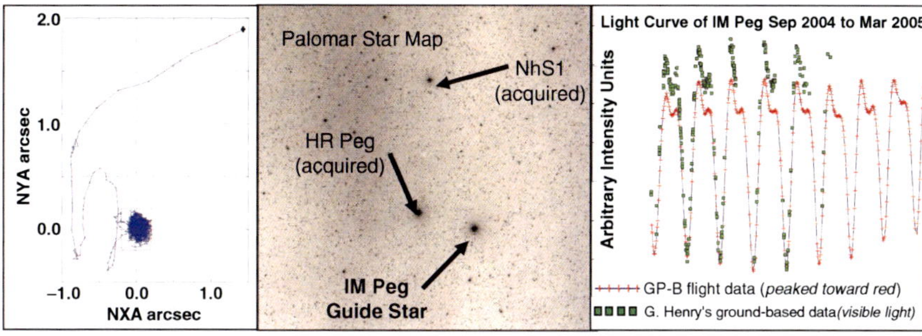

Fig. 22 (**a**) GS valid, 54 min. (**b**) Stars acquired near IM Peg. (**c**) Ground and orbit light curves

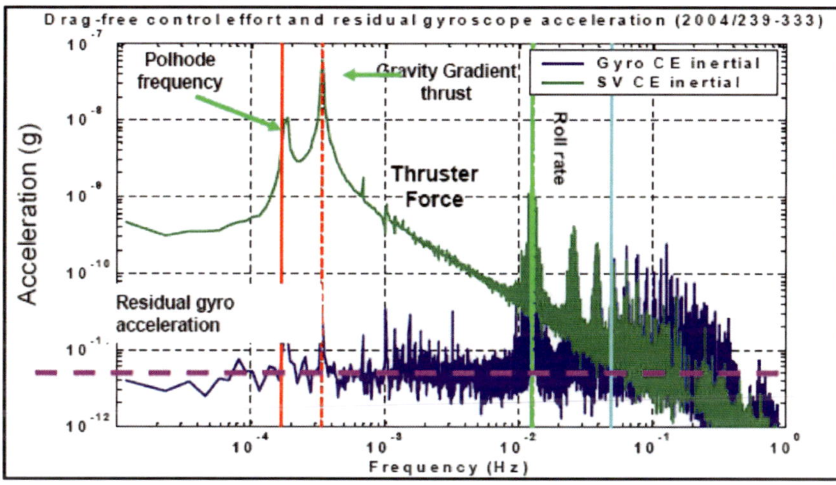

Fig. 23 Space vehicle z axis modulation at polhode frequency of drag free sensor

The main unexpected experimental observation was the spatial extent of the patch effect on both gyroscope and housing. Patch effects are spatial variations in surface potential caused by non-uniform dipole layers. Ground microscopy and data indicated that the electrostatic patches are of micron size and fraction of volt amplitude. In the low gravity in orbit it was determined that size of the patches was up to the order of centimeters. The cause for the patches on the gyroscope was most likely the coherence of the crystalline structure of the thin films over large numbers of single crystals. Misalignment torques, polhode damping and spin-down were the main results of the patch effect.

The experimental observations of excess gyroscope into suspension system coupling were:

(a) The 10^{-9} m/s^{2}space vehicle z (telescope) axis bias modulation at the polhode frequency of the drag free gyroscope (Fig. 23).
(b) Control effort modulation at the 1.3 Hz gyroscope spin speed of 2×10^{-7} N, about 30% of total suspension force (Fig. 24).

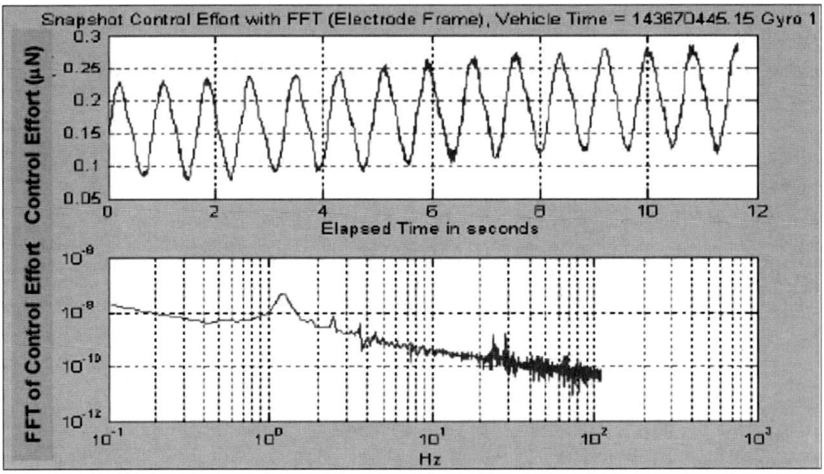

Fig. 24 Control effort modulation with gyroscope spinning at 1.3 Hz

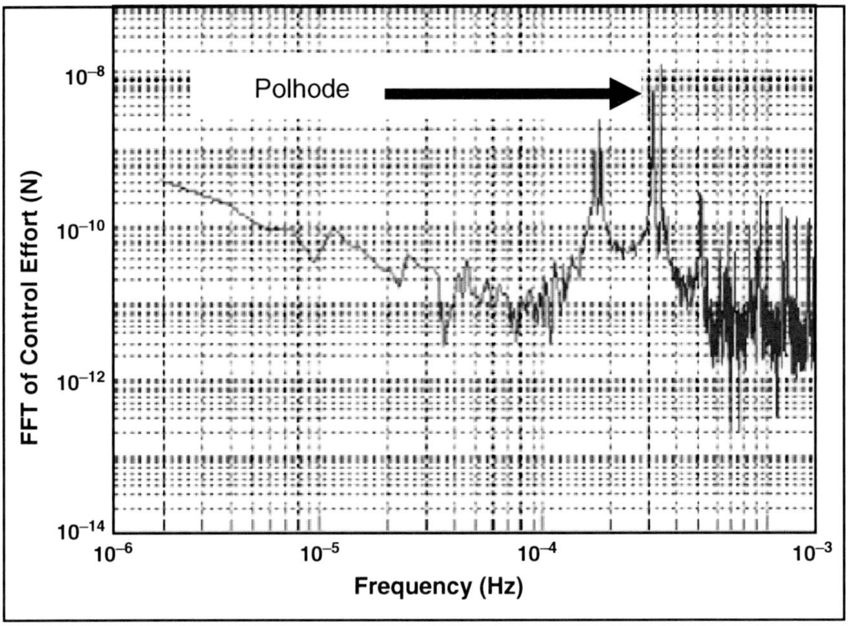

Fig. 25 Control effort modulation at polhode frequency with gyroscope spinning at 80 Hz

(c) Position and suspension voltage modulation at the 1.3 Hz gyroscope spin speed of 60 nm_{pp}.
(d) Control effort modulation of 10^{-8} N, at polhode frequency, for 80 Hz gyroscope spin speed, about 30% of total suspension force (Fig. 25).

Rotor geometry cannot account for the magnitude of observed effects. Mass unbalance and surface waviness are of the order of 3×10^{-3} of the gap, and would produce effects of the same order of magnitude, about 100 times smaller than the 3×10^{-3} effects observed. Multiple calculations of the interaction of the trapped flux in the rotor with the gyroscope housing have shown that these are orders of magnitude smaller than the measured coupling. However electrostatic patches of 50–100 mV on both rotor and housing are consistent with all observations and predictions.

Misalignment torques cause precession orthogonal to the plane of the misalignment; defined by the gyroscope spin axis and the spacecraft roll axis. Only relativity effects will contribute to angular momentum changes in the misalignment plane. Therefore the accurate measurement of the misalignment angle, including guide star occulted periods makes it possible to separate the precessions due to misalignment torques from those caused by relativity (Fig. 26a). The misalignment during occultation is measured using the science gyroscopes themselves.

Using the property that misalignment torques are orthogonal to the misalignment angle, the plot of drift rate versus misalignment phase gives a measure of the present consistency between the four gyroscopes, and therefore a measure of the quality of the torque model (Fig. 27). The schematic in Fig. 27 shows schematically the method. This analysis approach is called the 'geometric method' by the GP-B team.

Damping of the polhode period changes the trapped flux through the read-out loop and therefore causes variation in the scale factor of the gyroscope. In order to account for this effect in the data analysis high precision knowledge of polhode period and phase is required (Fig. 26b). These are used as known parameters in the modeling of the gyroscope motion.

The gyroscope spin-down rates (Table 4) are equivalent to a gas dissipation pressure of about 10^{-11} torr. However, thermal measurements indicate that the Helium

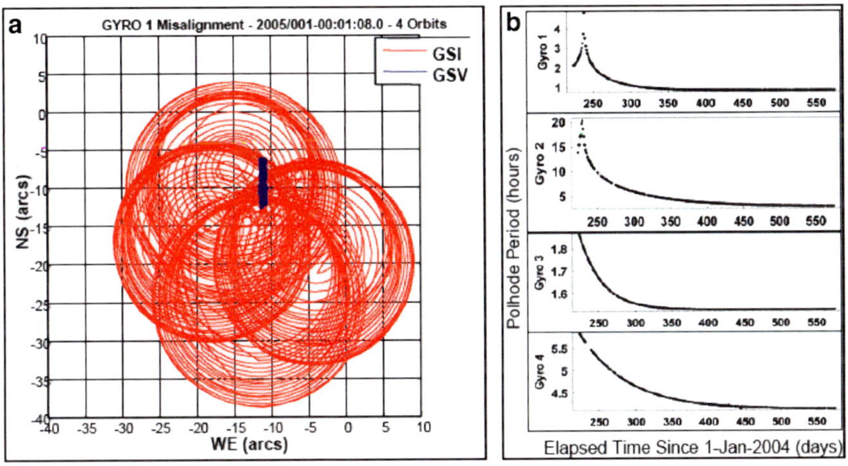

Fig. 26 (**a**) Misalignment angle GS valid and invalid. (**b**) Polhode period variation

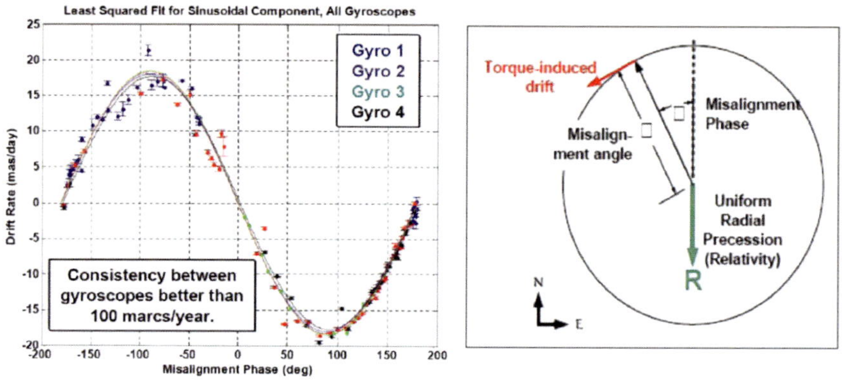

Fig. 27 Drift rate versus misalignment phase; the geometric method

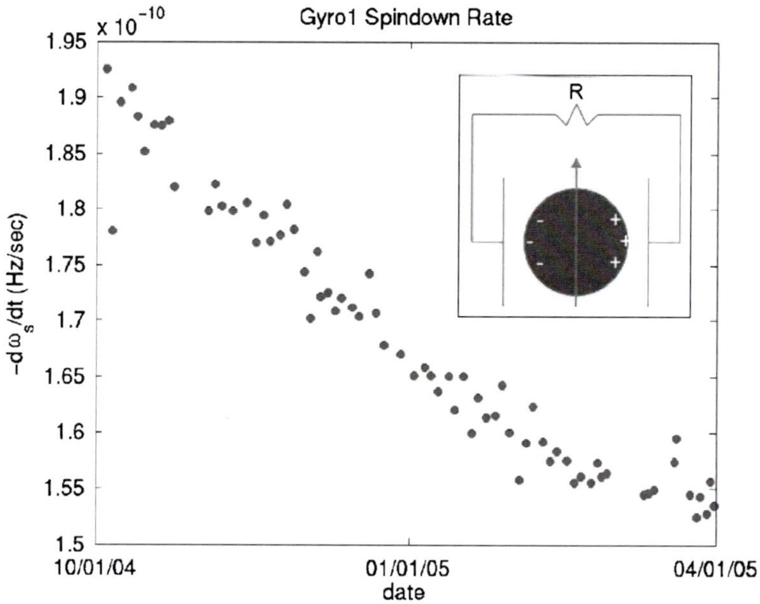

Fig. 28 Spin-down rate of gyroscope #1. Insert is a schematic of the spin-down dissipation mechanism

pressure is less than 10^{-14} torr. The excess spin-down is caused by image charges of the gyroscope flowing through the resistors that ground the housing coatings. Patch effect potentials of 40–80 mV account for the measured spin-down rates. Note that the spin-down rates, independent of their source, meet the GP-B requirement of about 1 μHz/h. Figure 28 gives the variation of the spin down rate of gyroscope #1 for 6 months of the mission, with the inset showing a schematic representation of the spin-down dissipation mechanism. Note that the spin-down rate is determined to about 5×10^{-12} Hz/s.

4 GP-B, LISA, STEP

GP-B, LISA, and STEP are three ultra untypical physics experiments that can meet their precision requirements only by using the quiet environment of space. In particular, the significant technologies overlap of GP-B, LISA [6], and STEP [11] present us with the opportunity to greatly enhance the probability of success for LISA and STEP by judiciously using the lessons learned from GP-B. The top three of these lessons are the implementation of the simplest most robust design, the use of the most advanced technology available, and the employment of innovative operation methods.

LISA is a space gravitational wave observatory based on the interferometric measurement of the distances between three spacecraft placed in an equilateral triangle with five million kilometers side. The formation center is twenty degrees behind the Earth on its solar orbit and inclined sixty degrees to the ecliptic (Fig. 29a). LISA's bandwidth is 3×10^{-5} to 1 Hz, complimentary to the ground gravitational wave detectors (Fig. 29b). LISA takes advantage of the low gravitational environment and the very long baselines available in space. The main GP-B heritage technologies are the gravitational reference sensor (GRS), the high stability optical benches and telescopes, the drag free technology, and the multiple degrees of freedom controls systems. STEP will improve the measurement of the equivalence principle by five orders of magnitude; from 10^{-13} to 10^{-18}. The experiment compares the rate of fall of different materials and relies on the low gravity and the long integration times in a 450 km orbit. The science signal is modulated at orbital frequency (Fig. 30a). GP-B technologies used for STEP are the cryogenic, magnetic-shielding, drag-free, multiple degrees of freedom controls, and cold gas thrusters systems, the dc SQUID read-out, and the high stability optical bench. Figure 30b shows the history of improvements of the equivalence principle measurement.

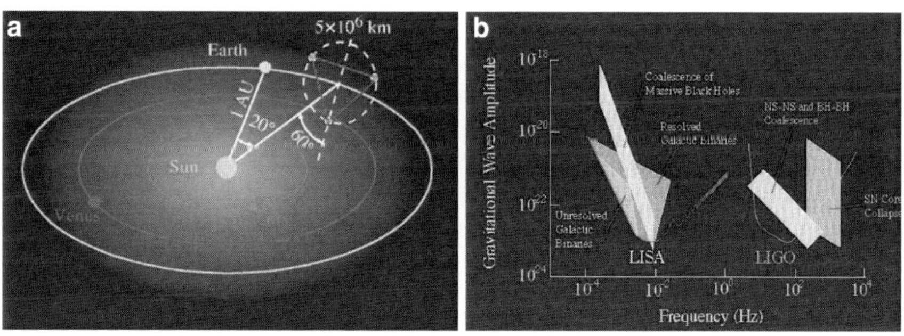

Fig. 29 (**a**) LISA concept and (**b**) LISA performance complementary to the LIGO ground detector

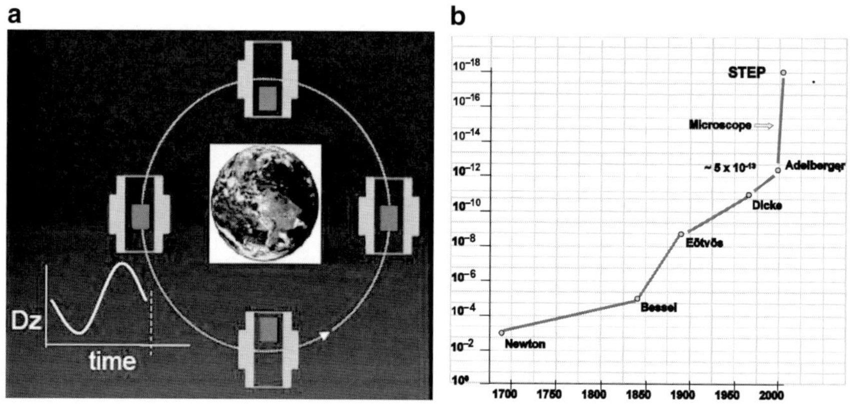

Fig. 30 (**a**) STEP concept. (**b**) History of improvements of the equivalence principle measurements

Table 9 Comparison of significant features of GP-B, STEP and LISA

	GP-B	STEP	LISA
Measurements/controls	6	13	16×3
Communications	Ground + TDRSS	Ground + TDRSS	DSN
Drag-free sensors	1	2	6
Formation flying	No (1 satellite)	No (1 satellite)	Yes (3 satellites)
Frequency band	13 mHz	0.18 mHz	$3{\cdot}10^{-5}$Hz to 1 Hz
Drag-free level in band	10^{-12}m/s^2	10^{-11}m/s^2	10^{-11}m/s$^2/\sqrt{\text{Hz}}$
Cryogenics	Yes	Yes	No

5 Lessons Learned

GP-B, STEP, and LISA have in common sophisticated drag-free attitude and control systems and their payload and vehicles are integrated in a single instrument. Table 9 compares some of their most significant features. Note that while the degrees of control equal the number of independent measurements, each degree of control is activated by more than one actuator, adding further complexity to the experiments.

Lessons learned fall into three main categories; the team, the instrument, and the unexpected. A successful team requires a combination of skills, in this case a close collaboration of physicists and engineers. These instruments are in fact the payload, spacecraft, and control system combined. The approach of designing and building the science payload and the spacecraft separately and then integrating them is unworkable in these experiments were all systems are contributing directly to the experimental error and are expected to perform well beyond the standards of existing space equipment. The lesson is therefore to build a science and engineering collaboration from the start; engineers with space experience must learn the fundamentals of the science and the scientists need to understand the requirements of work in space. Space experienced personnel are indispensable from the earliest stages of the project in order to include the required knowledge in the design and development; experimenting in space is really very different then working in the laboratory. Students, graduate and undergraduate, are essential to the foundation of the team. Their enthusiasm, hard work, and willingness to undertake 'unsolvable' problems are at the core of the team's performance. And of course they are the future generation of engineers and scientists. Lastly it is imperative to keep the core team for the operations and data analysis stages; it is the team that has build and tested the instrument and its members are the only ones who will be able to operate it successfully in orbit and understand the acquired data. The GP-B core team, including the developers of the instrument, cryogenic systems, and attitude and control system were an integral and critical part of the operations team, and remain at the center of the data analysis team.

Table 9 illustrates not only the similarities but also the difference in technical challenge faced by the three experiments. STEP has the advantage that most technologies involved have been demonstrated by GP-B to the required performance level. LISA has the advantage of not requiring cryogenics. However, in its present configuration, it faces the technical challenges of the many degrees of control for the six drag-free sensors, a restricted communications channel, the need for formation flying, wide frequency band measurement, three orders of magnitude improved drag-free level and complex optical metrology.

The experiment must be instrumented as extensively as possible. Space tests are truly a one-shot opportunity and the experimenter must have maximum opportunity to understand data, anomalies, and calibrations. For the high accuracy tests discussed all system interact to some extent and therefore all need monitoring at five to ten times their maximum operating frequency. Restricting the data telemetry to twice or even a few times the frequency (Nyquist) of the high end of the measurement band is insufficient. Periodical and anomaly-triggered snapshots of the parameters of higher frequency sub-systems are a practical way of understanding their impact on the science data. The GP-B science data was at 13 mHz. Snapshots of the critical electronics, gyroscopes, SQUIDs, and telescope, were collected periodically and for anomalies at sampling frequencies ranging from 220 to 2,200 Hz. These snapshots have proven critical to data analysis and anomaly resolution.

Incremental prototyping is the ideal approach in the development these instruments allowing testing and de-bugging of increasingly complex configurations. The integrated GP-B system went though four incrementally complete system tests. Ground testing should follow the principle "Test it like you fly it". This includes, again in incremental prototyping stages; the instrument, the environment as practical, the flight software, and the telemetry and ground commanding. A high fidelity simulator with hardware in the loop is an essential development, testing, flight operations, and data analysis tool. Development of the simulator starts with a software simulation with hardware prototypes included and ends prior to launch with high fidelity simulations of the instrument using flight hardware software. The simulator includes an identical copy of the mission operations system, making it a virtual science instrument. Following are two, of many, examples of the use of the simulator for anomaly resolution during GP-B operations.

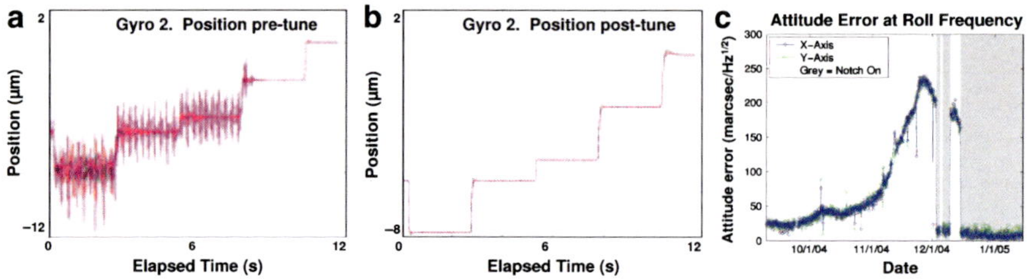

Fig. 31 (**a**) Suspension stability pre and post tune. (**b**) Attitude control error at roll frequency; notch filter on in gray areas

Figure 31a shows the optimization of the in-orbit performance of gyroscope's #2 suspension system. Control loop instability was observed for gyroscope positions more than 2 μm from center; total spacing to suspension electrodes is 32.5 μm. With no risk to the flight instrument, the simulator was used to reproduce the anomaly, find and test a solution, and verify the software and telemetry modifications on the ground before applying them to the experiment in orbit. The simulator accurately reproduced both the anomalous and the optimized performance; the solution was the increase in modulation frequency of the suspension system from 25 to 30 Hz. Note that the anomaly was detected and corrected by the use of gyroscope suspension snapshots, 220 Hz for 12 s. The continuous 0.5 Hz data would have clearly been inadequate to the task.

Figure 31b shows the roll frequency component of the attitude control error. The large errors at roll frequency were induced by an anomalous large sensitivity to thermal variations of the platform supporting the navigation instrumentation; the increase in December 2004 coincides with the GP-B spacecraft moving into full Sun illumination. A notch filter at roll frequency, designed and tested with the simulator, was up-loaded into the flight attitude control system, permanently solving the problem.

The principal technical lessons learned from GP-B including the requirement for a charge control system with wide dynamic range and the importance of the understanding and characterization of surfaces are described in detail in previous sections. Gyroscope charge at levitation was more then ten times larger than accumulated cosmic charging during the entire mission and needed to be removed in hours. The patch effect was the main unexpected disturbance significantly complicating the data analysis.

There are two types of the "unexpected". Real surprises and standard but consistently underestimated difficulties. Rigorous planning can largely mitigate the second category. Required data and commanding rates will be significantly higher than planned for in the early stages of the experiment design; the standard 50% early development contingency should be increase to as much as 100–200%. In particular the experiment set-up stages and anomaly resolution periods will require significantly increased data rates. GP-B data was collected both through the space TDRSS network, ˜12 contacts/day, 20–40 min/contact, 1–2 Kbits/s data rate, as well as the ground based network of NASA ground stations, 4 contacts/day, 10–12 min/contact, 32 Kbits/s data rate. The total stored data, including derived monitors and raw and processed data streams is 1.5 Tbytes.

GP-B initial set-up duration was 128 days, versus the planned 40 days plus 20 days contingency, which was expected to be unnecessary. More than 100 anomalies occurred during the mission, including the failure of two thrusters, multiple bit upset events (MBE) at ten times the expected rate, computer reboots, antennas and star sensors degradation. A large percentage of anomalies occurred during initialization, a period that required more than 10,000 commands to the space vehicle for experimental set-up and anomaly resolution. Regarding LISA two significant points of reference are the GP-B 128 days commissioning duration and the 5 year required by the ground based LIGO to achieve design performance. An aggravating factor is the narrow DSN communication bandwidth as compared to the much faster GP-B telemetry and the non-limiting LIGO instrumental communications capability. LISA requires simulation work at much higher level than any previous mission as well as the development of advanced autonomous software.

Extensive calibrations have proven invaluable for GP-B and have both accurately determined the instrument performance parameters and revealed anomalous behavior. For this class of instruments, designed for minimum sensitivity to disturbance, two methods of calibration are used. Firstly, instrumental performance in planned operating conditions sets an upper limit to their sensitivity. Secondly, operations in a highly degraded controlled environment allow the actual measurement of the effect of the disturbance on the instrument and thus the extrapolation to performance numbers.

Data analysis will always be more complex and time consuming than the most conservative ground estimate. Again contingency scheduling at factors of two to three beyond what is considered "reasonable" on the ground is a very good move. One cannot overemphasize the need for the experience of the core team for data analysis.

A positive aspect of space experimentation is that most systems will work better in this quiet environment. The most complex and apparently risky systems seem to be the best behaved in space; GP-B gyroscope, SQUID's telescope, cryogenics. Experimental plans cannot be based on this fact but in most cases a positive margin of performance is available; and very helpful. The "unknown unknowns", the real surprises, will occur and will by definition be the least expected effects. Preparing for the unexpected is possible through the use of extensive instrumentation and comprehensive data and with the help provided by the experience of the core team.

6 Improving Drag-Free: The Modular Gravitational Reference Sensor

The Gravitational Reference Sensor (GRS), by physically shielding a proof mass from stochastic non-gravitational disturbances, provides the basis for following an inertial trajectory. By itself, or when integrated with a complete drag-free control system, the GRS opens new frontiers in Earth imaging and space science; autonomous maintenance and prediction of spacecraft orbits, satellite constellation maintenance without ground tracking, disturbance free inertial trajectories, new observations of the universe through gravity waves, new understanding of the physics of the Universe, precision formation flying for satellite constellation interferometers and imagers, geodesy and surface imaging, integrated precision inertial sensing of translation and rotation, autonomous spacecraft rendezvous, guidance and space-based target tracking [5].

A Disturbance Reduction System (DRS) is composed of the GRS,the actuating thrusters and the attitude and translation control algorithms. The first drag-free system, called DISCOS, was developed by Stanford University and launched on the U.S. Navy's TRIAD drag-free spacecraft in 1974 [4]. Recently, GRS technology found application in space science, lying at the heart of the recently launched NASA Gravity Probe B mission to test Einstein's Theory of Relativity [10]. The GRS will again be crucial to the Laser Interferometer Space Antenna (LISA) [3], part of the NASA SSE Strategic Plan; the LISA GRS will require disturbance reduction improvements of at least four orders of magnitude over current performance.

TRIAD I was launched September 2, 1972 with its Disturbance Compensation System (DISCOS) built by Stanford. Discos had three axis translation control and demonstrated an acceleration below 5×10^{-11} m/s^2 averaged over 3 days, corresponding to a residual acceleration of 2.7×10^{-8} m/s$^2/\sqrt{\text{Hz}}$.at 5 μHz [4]. GP-B was launched April 20, 2004 with drag-free control of all degrees of freedom, three translations and three rotations, and had a residual acceleration of 10^{-11}m/s$^2\sqrt{\text{Hz}}$ at 13 mHz.

The three main principles we applied in the design of a modern drag-free sensor are: (a) minimize the forces on the test mass (TM), (b) optimize the TM center of mass determination, and (c) use modern technology; including the optical read-out of the TM position, charge management with UV LEDs, diffraction grating technology, optimized algorithms for TM position determination. In order to practically eliminate TM forcing we use a non-supported spinning spherical TM. The spin, at about 10 Hz, spectrally shifts the TM measurement noise, (including spin, precession, and polhode) above the science band; 1 Hz for LISA. A gap equal or larger than the TM diameter significantly reduces surface patch effect disturbances. A double sided grating on the sensor housing isolates the GRS from external measurements. The distance from the grating to the TM center of mass is measured directly, making the GRS modular, MGRS, and thus insulating it from other experimental systems [9]. With the TM center of mass measured optically no active electrostatic systems are disturbing the TM. UV LEDs for charge control, gratings for non-transmissive optics advanced interferometry methods, and laser frequency doubling to green are some of the technologies that contribute to solving the challenges of advanced drag-free systems.

Optical sensing offers a high-resolution method of sensing across a large gap while maintaining low disturbances. The sensing element is a low-finesse Fabry-Perot 1.6 cm cavity formed between a Littrow mounted 900 lines/mm diffraction grating and the surface of the proof-mass. The sensor reached 10 pm/$\sqrt{\text{Hz}}$

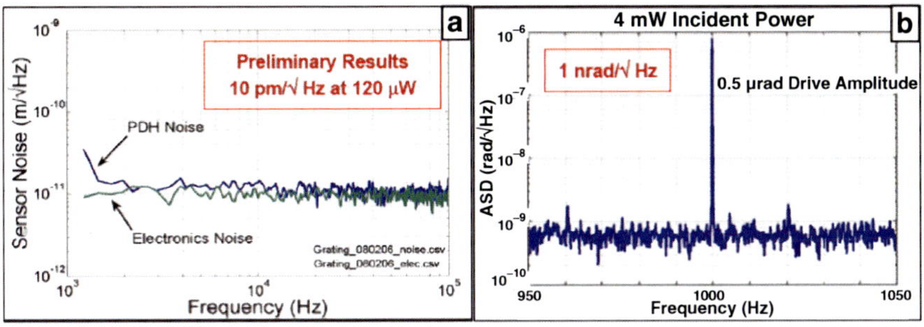

Fig. 32 Performance of optical sensors: (**a**) linear, (**b**) angular

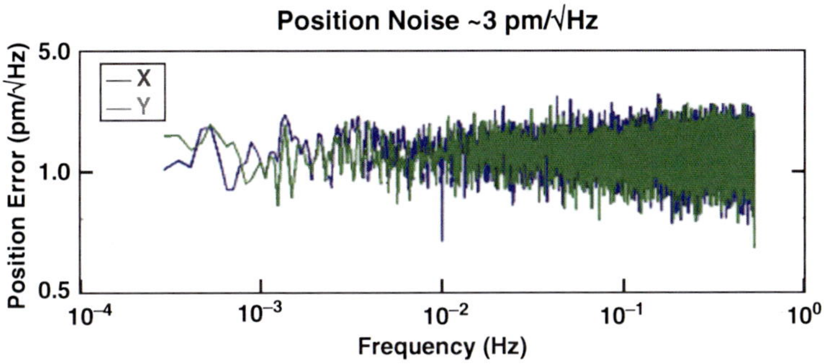

Fig. 33 Numerical simulation of optical read-out of 10 Hz spinning sphere

sensitivity with 120 μW of optical power (Fig. 32a). Further improvements should achieve 1 pm/$\sqrt{\text{Hz}}$ sensitivity for 20 μW optical power in the LISA science band (3×10^{-5} to 1 Hz).

High precision angular sensing is needed in measuring the TM orientation, telescope steering, and point-ahead-angle control. We have demonstrated the use of grating diffraction orders as angular sensing signal beams, taking advantage of the grating angular magnification. The angular sensitivity achieved is better than 1 nrad/$\sqrt{\text{Hz}}$ with a 6 cm working distance (Fig. 32b).

A three dimensional numerical simulation of a sphere spinning at 10 Hz has been performed in order to demonstrate the feasibility of measuring the TM center of mass. Six optical sensors, with two sensors in the 60 degrees configuration of the LISA gratings are being used in the model. The position noise and surface roughness of the sphere are 50 nm. The results indicate that the TM center of mass can be measured to 3 pm/$\sqrt{\text{Hz}}$. Further work incorporating additional disturbances is in progress (Fig. 33).

Since the tip/tilt sensor will require a grating atop the proof mass, we have demonstrated several ways of fabricating gratings on dielectric and gold surfaces: electron-beam lithography, mechanical transfer imprinting, and ion-beam writing. The grating patterns display only small irregularities caused mainly by the step size

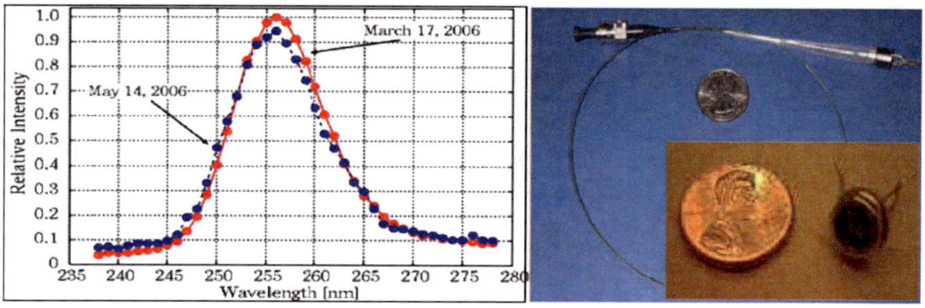

Fig. 34 UV LED spectral stability, in TI05 packaging and coupled to optical fiber

of the ion beam movement. Measured diffraction efficiencies meet requirements for the MGRS tip/tilt and displacement sensing.

Charge management using photoelectrons has been demonstrated in the GP-B experiment using Hg discharge sources for the generation of the UV [1]. Deep UV LED charge management systems have the advantages of high dynamic range, low disturbance, low power consumption, and light weight. We have demonstrated AC charge management using a UV LED, and have started the space qualification of these devices A UV LED has operated for more than 8,000 h under AC charge management working conditions. Additional measurements show power and spectral stability in the UV LED emission over this period of time. Figure 34 shows the UV LED spectral stability over 10 months and photographs of the LED in its TI05 packaging and coupled to an optical fiber.

A major noise source for the MGRS is due to spatial and temporal variations in surface potential (or patch effect) across the surfaces of the test mass and its housing. Such variations will lead to force gradients which will result in a significant acceleration noise term. We have used a Kelvin probe to make spatial and temporal measurements of contact potential differences for a selection of materials (Au/Pt, beryllia, alumina, titanium) and coatings (gold, diamond-like carbon, indium tin oxide, titanium carbide). The data showed evidence of variations in the patch potentials related to pressure and contamination effects. Temporal variation measurements were limited by the current accuracy of the instrument [8]. Figure 35 shows examples of Kelvin probe spacial scans and a photograph of the system.

7 Conclusions

GP-B, the first controlled experiment in which a General Relativity effect is the main feature of the data, has demonstrated that complex physics experiments do work in space. Preliminary results, in April 2007, set the measurement accuracy at ±97 marcs/year or about 1.5% of the geodetic effect. Further improvement in accuracy is expected before the end of 2007. LISA and STEP are space physics experiments significantly similar in many features to GP-B. The main lessons that

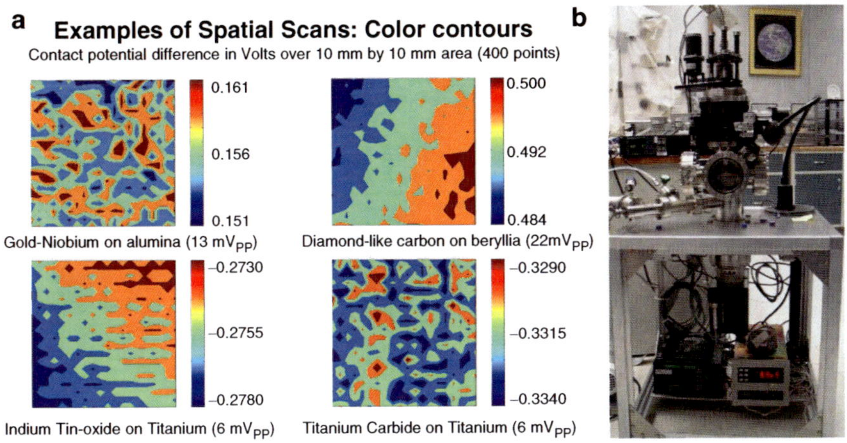

Fig. 35 Kelvin probe: (**a**) spacial scans of coatings and (**b**) apparatus

GP-B can convey are: (a) a strong core team of scientists and engineers with space experience that works from design to data analysis, (b) a flexible, comprehensively monitored and diagnosed instrument, (c) significant longer schedule contingency for operations and data analysis, and (d) extensive calibrations to ensure results credible to the science community and the experimental team.

Based on the experience and lessons from GP-B, advanced drag-free technology is being developed. The MGRS makes possible the achievement of drag-free performances appropriate for the future generation of space experiments. Among its design features are the complete elimination of electrostatic forcing and sensing, its isolation from other experimental systems, a spherical test mass spinning above the science measurement band, optical linear and angular sensing and gaps larger than the size of the sphere. Technologies that will ensure the performance and reliability of space-based gravitational wave detectors include the MGRS, in-field laser beam pointing, the use of gratings as non-transmissive optics, lasers with frequency doubled to green, and UV LED charge management.

Acknowledgments This research supported by NASA on contract NAS8-39225.

References

1. S. Buchman, et al, *Rev.Sci.Instrum.* **66**, 120 (1995).
2. S. Buchman, et al, *Advances in Space Research*; vol. **25**, no. 6, 1177–80 (2000).
3. K. Danzmann and A. Rudiger, *Class. Quantum Grav.* **20** (10), S1 (2003).
4. D.B. DeBra, *APL Technical Digest*; vol. **12**, no. 2, p. 14–26 (April–June 1973).
5. D. B. DeBra, *Class. Quantum Grav.* **14** (6), 1549 (1997).
6. J. E. Faller, et al, *Proceedings of the Colloquium on Kilometric Optical Arrays in Space*, Cargese, Corsica, 23–25 October (1984).

7. J. C. Mester, et al, *Advances in Space Research*; vol. **25**, no. 6, 1185–8 (2000).
8. N. A. Robertson, et al, *Class. Quantum Grav.* **23** (7), 2665 (2006).
9. K. X. Sun, et al, *Class. Quantum Grav.* **22** (10), S287 (2005).
10. J. P. Turneaure, et al, *Advances in Space Research* (UK), vol **9**, 29–38 (1989).
11. P. Worden, A cryogenic test of the equivalence principle *Thesis (Ph.D.)* - Dept. of Physics, Stanford University (1976).

The LARES Space Experiment: LARES Orbit, Error Analysis and Satellite Structure

Ignazio Ciufolini, Antonio Paolozzi, Erricos Pavlis, John Ries, Rolf Koenig, Richard Matzner, and Giampiero Sindoni

Abstract The LARES space experiment, by the Italian Space Agency (ASI), is based on the launch of a new laser ranged satellite, called LARES (LAser RElativity Satellite), using the new launch vehicle VEGA (Veicolo Europeo di Generazione. Avanzata, provided by ESA). LARES will have an altitude of about 1,450 km, orbital inclination of about 71.5° and nearly zero eccentricity. The LARES satellite together with the satellites LAGEOS (LAser GEOdynamics Satellite launched by NASA) and LAGEOS 2 (built by ASI and launched by NASA and ASI) and with improved GRACE (Gravity Recovery and Climate Experiment, a NASA/DLR, German Space Agency, mission) Earth's gravity field models will allow a measurement of the Earth's gravitomagnetic field and of Lense–Thirring effect with an uncertainty of a few percent. After a description of the LARES experiment and

I. Ciufolini (✉)
Dip. Ingegneria dell'Innovazione, Università del Salento and INFN Sezione di Lecce, via Monteroni, 73100 Lecce, Italy
e-mail: ignazio.ciufolini@unisalento.it

A. Paolozzi and G. Sindoni
Scuola di Ingegneria Aerospaziale, Sapienza Università di Roma, via Salaria 851/881, 00138 Roma, Italy
e-mail: giampiero.sindoni@uniroma1.it; antonio.paolozzi@uniroma1.it

E. Pavlis
Joint Center for Earth Systems Technology, University of Maryland, Baltimore County, 1000 Hilltop Circle, Baltimore, Maryland 21250, USA
e-mail: epavlis@umbc.edu

J. Ries
Center for Space Research, University of Texas at Austin, Austin, Texas 78759, USA
e-mail: ries@csr.utexas.edu

R. Koenig
GFZ German Research Centre for Geosciences, Potsdam, Germany
e-mail: rolf.koenig@gfz-potsdam.de

R. Matzner
Center for Relativity, University of Texas at Austin, Austin, Texas 78712, USA
e-mail: matzner2@physics.utexas.edu

I. Ciufolini and R.A. Matzner (eds.), *General Relativity and John Archibald Wheeler*, Astrophysics and Space Science Library 367, DOI 10.1007/978-90-481-3735-0_19,

of the orbit of LARES, we present an analysis of the main error sources affecting the measurement of gravitomagnetism; these are due to the uncertainties in the Earth's gravitational field, and in particular to the Earth's even zonal harmonics, to the time dependent Earth's gravitational field, and in particular to $\dot{J}_6$ and to the K_1 tide. We also discuss the effect of particle drag and the error due to the uncertainties in the measurement of the orbital inclination. We finally describe some technical and engineering aspects of the LARES mission, and in particular: the laser ranging technique, the cube corner reflectors and the satellite body. We conclude with a brief discussion of LARES separation system and the selected launcher.

1 Introduction

In 1984, the use of the nodes of two laser ranged satellites of LAGEOS type to measure the Lense–Thirring effect was proposed [1,2]. The idea was to orbit a laser-ranged satellite of LAGEOS-type (called LAGEOS III and later on LARES), with an inclination supplementary to the one of LAGEOS (launched in 1976 by NASA, National Aeronautics and Space Administration), in order to cancel out all the secular effects on the nodes of the two laser-ranged satellites due to the deviations of the Earth's gravitational field from spherical symmetry and in particular due to the Earth's even zonal harmonics. All the other orbital parameters of LARES/LAGEOS III were proposed to be equal to the ones of LAGEOS, in particular the semimajor axis was proposed to be approximately equal to 12,270 km. The mass of LAGEOS and of the proposed LAGEOS III satellite is about 400 kg. Several papers [3], international studies [4, 5], proposals [6] and Ph.D. dissertations [7, 8] analyzed the LAGEOS III proposal.

Unfortunately, even though such an orbit of LARES would have allowed a complete cancellation of the static Earth's spherical harmonics secular effects, in order to measure the much smaller Lense–Thirring effect, the weight of the proposed LARES satellite of about 400 kg and especially the high altitude of its orbit implied an expensive launch vehicle. For this reason a LARES satellite of only about 100 kg of weight was later designed [6], nevertheless the high altitude of LARES was still somehow expensive to achieve.

Nevertheless, three new factors have changed the need of such a high altitude orbit for LARES: (a) the idea to use the nodes of N laser-ranged satellites to measure the Lense–Thirring effect and to cancel the uncertainty due to the first N-1 even zonal harmonics [3, 9, 10], (b) the launch of the GRACE spacecraft in 2002 and the publication of a new generation of very accurate Earth's gravity field models using the GRACE observations [11–15] and (c) the possibility to launch the LARES satellite using the much less expensive qualifying flight of the new VEGA launcher, built by ESA (European Space Agency) and ELV (ASI, Italian Space Agency, and AVIO), however at a much lower altitude than originally planned.

(a) The idea to use the nodes of N satellites of LAGEOS type to cancel the effect of the first N-1 Earth even zonal harmonics and to measure the Lense–Thirring effect was published in 1989 [3] (see also [9], on page 336) as a possible alternative to the concept of the supplementary inclination satellites. This technique and in particular the idea to use the two nodes of the satellites LAGEOS and LAGEOS 2, together with the perigee of LAGEOS 2, was described in details, together with the corresponding formula, in 1996 [16]. In the 1989 paper (see also [9]), in order to measure the Lense–Thirring effect and to cancel the even zonal harmonics uncertainties, it was proposed: "For J_2, this corresponds from formula (3.2), to an uncertainty in the nodal precession of 450 milliarcsec/year, and similarly for higher J_{2n} coefficients. Therefore the uncertainty in $\dot{\Omega}^{Lageos}$ is more than ten times larger than the Lense–Thirring precession. A solution would be to orbit several high-altitude, laser-ranged satellites, similar to LAGEOS, to measure J_2, J_4, J_6 etc, and one satellite to measure $\dot{\Omega}_{Lense-Thirring}$". At that time the error due to the even zonal harmonics was quite large due to the much less accurate Earth gravity models (available at that time) and the LAGEOS 2 satellite was not yet launched. (It was launched in 1992 by NASA and ASI.)

This technique to use N observables to cancel the effect of the first N-1 even zonal harmonics was explicitly described in [16] (see also the explicit calculations about the use of the nodes of N laser-ranged satellites in [8]) and led to an observation of the Lense–Thirring effect using the LAGEOS and LAGEOS 2 satellites and the gravity field model EGM96 [17, 18] (using the nodes of LAGEOS and LAGEOS 2 and the perigee of LAGEOS 2 in order to cancel the error in the first two even zonal harmonics by using three "observables", including the perigee, which however introduced relatively large errors due to its unmodelled non-gravitational perturbations) and to the 2004–2008 measurements [10, 19–21] (with accuracy of the order of about 10%) of the Lense–Thirring effect using the LAGEOS satellites and the accurate Earth's gravity field models obtained by using the GRACE spacecraft. In these measurements we used the same technique of the 1998 paper but we did not use the perigee of LAGEOS II (that the authors of [10] tried for a long time to avoid since the publishing of their 1997–1998 papers) thanks to the new generation of GRACE Earth's gravity models. The 2004-measurement is just the case of $N = 2$ described in the above mentioned 1989 paper and it uses the nodes of the two laser ranged satellites LAGEOS and LAGEOS 2 in order to cancel the effect of the first even zonal harmonic coefficient J_2 of Earth and to measure the Lense–Thirring effect (the explicit expression of this combination was also given in [22]).

(b) The 2004 accurate measurement of the Lense–Thirring effect was possible thanks to the launch of the GRACE satellite and the publication of its accurate gravity field models by GFZ German Research Centre for Geosciences, Center for Space Research (CSR) of the University of Texas at Austin and by other groups [11, 13–15]. The use of the GRACE-derived gravitational models, when available, to measure the Lense–Thirring effect with accuracy of a few percent was, since almost a decade ago, a well-known possibility to all the researchers in this field and was presented (and published) at several meetings by Ries et al. [23], Pavlis [24], etc.

(c) In 2004, Paolozzi of "Scuola d'Ingegneria Aerospaziale" of the University of Rome "La Sapienza" discovered the possibility to use the qualifying flight of the new launcher VEGA to orbit LARES [25]. However, this launch for LARES will be at a much lower altitude than the originally planned satellite at about 6,000 km. The altitude achievable with this qualifying launch is of about 1,450 km. In 2005, one of us [26] informed the other members of the LARES team that CSR had done some simulations supporting the possibility of using a lower orbit laser-ranged satellite to measure the Lense–Thirring effect, this possibility was later on discussed also in [27, 51].

In the following we describe the orbit of the LARES satellite in relation to the error in the measurement of the Lense–Thirring effect.

2 A New Laser-Ranged Satellite at a Lower Altitude Than LAGEOS and LAGEOS 2

The simplest conceivable orbit in order to cancel the effect of all the even zonal harmonics on the node of a satellite would be a polar orbit, indeed for such an orbit the effect of the even zonal harmonics on the satellite node would be zero and, however, the node of the satellite would still be perturbed by the Earth's gravitomagnetic field, i.e., would be affected by the Lense–Thirring effect.

Unfortunately, as pointed out in the 1989 LAGEOS III NASA/ASI study [4, 5] and explicitly calculated by Peterson (1997) (Chap. 5 of [8]), the uncertainty in the K_1 tide (tesseral, $m = 1$, tide) would make such an orbit unsuitable for the Lense–Thirring measurement. Indeed, a polar satellite would have a secular precession of its node whose uncertainty would introduce a large error in the Lense–Thirring measurement. In addition, it would be quite demanding to launch LARES with the requirement of a small orbital injection deviation from a polar orbit (in order to cancel the error due to the uncertainties in the static Earth gravity field, for a very accurate measurement of the Lense–Thirring effect at an altitude lower than the one planned for LAGEOS III, the deviation from a polar orbit should be less than a tenth of degree).

Furthermore, a quasi-polar orbit would have a nodal precession, due to its departure from 90° of inclination, and thus one could simply fit for the effect of the K_1 tide using a periodical signal exactly at the nodal frequency. Such frequency (with the period of the LAGEOS satellites node) is indeed observed in the LAGEOS and LAGEOS 2 analyses already mentioned [10, 17] and is the largest periodical amplitude observed in the combined residuals.

Furthermore, in regard to the effect of the static even zonal harmonics, by using the technique explained in [3, 16] and by using the nodes of the satellites LARES, LAGEOS and LAGEOS 2, we would be able to cancel the uncertainties due to the first two even zonal harmonics, C_{20} and C_{40}, and our measurement will only be affected by the uncertainties due to the even zonal harmonics with degree strictly higher than 4. In the next section we present the results of the error analysis due to the uncertainties in the Earth even zonal harmonics.

By solving the system of the three equations for the nodal precessions of LAGEOS, LAGEOS 2 and LARES in the three unknowns, J_2, J_4 and Lense–Thirring effect, we have a combination of three observables (the three nodal rates) which determines the Lense–Thirring effect independently of any uncertainty δC_{20} and δC_{40} in the first two even zonal harmonics. This same technique was applied in [17] using the nodes of LAGEOS and LAGEOS 2 and the perigee of LAGEOS 2 and in [10] using the nodes of LAGEOS and LAGEOS 2 only.

It turns out [51] that some values of the inclination of LARES would minimize the error in the measurement of the Lense–Thirring effect since they would minimize the error due to the uncertainty in the largest (not cancelled using the combination of the three observables) even zonal harmonic C_{60}.

In Fig. 1 we have plotted the error in the measurement of the Lense–Thirring effect, using LARES, LAGEOS and LAGEOS 2, as a function of the inclination and of the semimajor axis. The range of the altitude of LARES is between 1,000 and 2,000 km and of the inclination between 0° and 360°; of course for a satellite in a nearly polar orbit the use of the LAGEOS and LAGEOS 2 satellites would not be anymore useful in order to reduce the error budget (and would indeed only introduce an additional error), since the effect of the even zonal harmonics on the node of LARES would be nearly zero. However, as previously remarked, the measurement of the Lense–Thirring effect using a polar orbit would be substantially affected by the uncertainty in the K_1 tide.

In Fig. 2 we have plotted the error in the measurement of the Lense–Thirring effect as a function of the inclination by assuming an altitude of LARES of 1,450 km, i.e., a LARES semimajor axis of about 7,830 km.

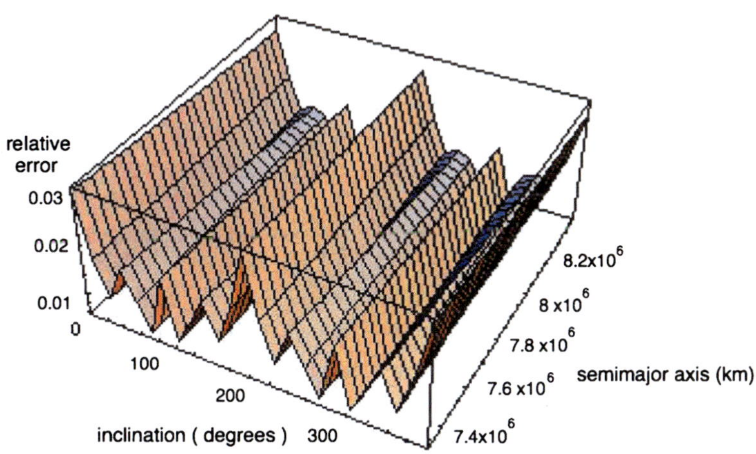

Fig. 1 Uncertainty in the measurement of the Lense–Thirring effect, due to the even zonal harmonics uncertainties, as a function of the inclination and of the semimajor axis of LARES, using LARES, LAGEOS and LAGEOS 2 [51]. The range of the altitude of LARES is between 1,000 and 2,000 km and of the inclination between 0° and 360°

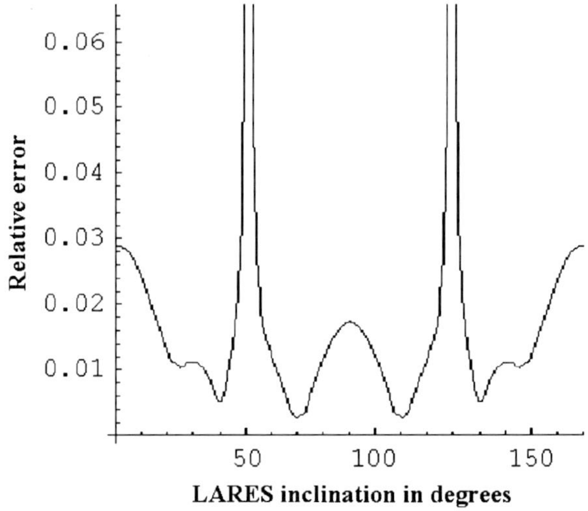

Fig. 2 Uncertainty in the measurement of the Lense–Thirring effect, due to the even zonal harmonic uncertainties, as a function of the inclination of LARES, using LARES, LAGEOS and LAGEOS 2 [51]. The altitude of LARES is here 1,500 km and the range of the inclination between 0° and 180°

From Fig. 2 we can see that any inclination from 60° to 86° and from 94° to 120° would be suitable for a measurement of the Lense–Thirring effect with accuracy of a few percent. An inclination of LARES of about 110° or 70° would minimize the error. In deriving this result, we have assumed: (a) zero eccentricity for the LARES orbit, (b) we have only considered the effect of the first 5 even zonal harmonics: C_{20}, C_{40}, C_{60}, C_{80} and C_{100} and (b) we have considered the uncertainties in the even zonal spherical harmonics C_{60}, C_{80} and C_{100} equal to those of the EIGEN-GRACE02S Earth's gravity model [11], i.e., we have assumed $\delta C_{60} = 0.2049 \cdot 10^{-11}$, $\delta C_{80} = 0.1479 \cdot 10^{-11}$, $\delta C_{100} = 0.2101 \cdot 10^{-11}$. However, by including higher degree even zonal harmonics, the results of Figs. 1 and 2 would only slightly change. Indeed, in the next section and in Figs. 3 to 6, we calculate and show the error in the measurement of the Lense–Thirring effect, with LARES, LAGEOS and LAGEOS 2, due to the even zonal harmonics up to degree 70, corresponding to the LARES orbit of 1,450 km of altitude and 71.5° of inclination. The results described in this paper and those shown in Figs. 1 to 6, are based on the calibrated uncertainties (i.e., including systematic errors) of the EIGEN-GRACE02S model (used in [10]); even though the real error in these EIGEN-GRACE02S coefficients would probably be about two or three times larger than these published uncertainties, EIGEN-GRACE02S was a preliminary 2004 model and by the time of the launch of LARES and of its data analysis (about 2011–2017), much improved Earth's gravity field models based on much longer data set of GRACE observations would be available, with real errors even less than the EIGEN-GRACE02S uncertainties we used in the analysis of this paper.

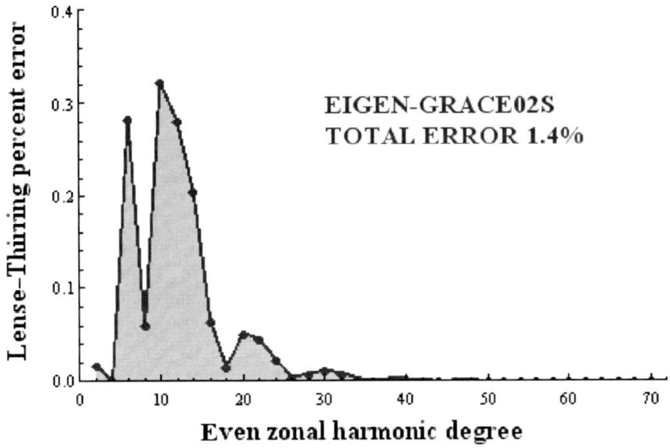

Fig. 3 Percent error in the measurement of the Lense–Thirring effect using LARES, LAGEOS and LAGEOS 2 as a function of the uncertainties of each even zonal harmonic. The model used is EIGEN-GRACE02S (GFZ Potzdam 2004) and the uncertainties in this model include systematic errors. Using EIGEN-GRACE02S, the total error in the measurement of the Lense–Thirring effect due to the even zonal harmonics is 1.4%. An improvement in the Earth gravity field by about an order of magnitude with respect to this 2004 model is expected at the time of the LARES data analysis

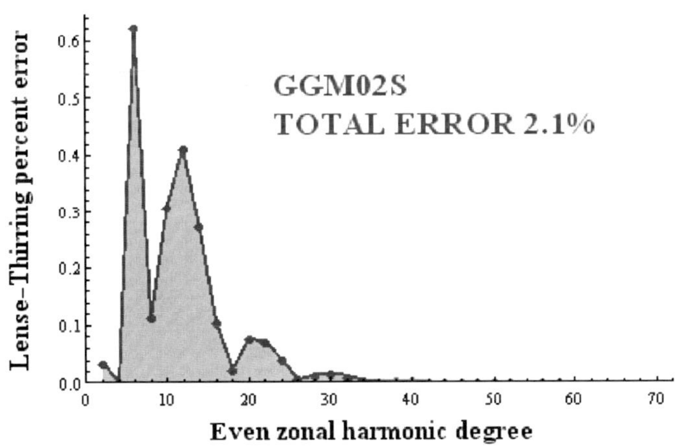

Fig. 4 Percent error in the measurement of the Lense–Thirring effect using LARES, LAGEOS and LAGEOS 2 as a function of the uncertainties of each even zonal harmonic. The model used is GGM02S (CSR 2004) and the uncertainties in this model include systematic errors. Using GGM02S, the total error in the measurement of the Lense–Thirring effect due to the even zonal harmonics is 2.1%. An improvement in the Earth gravity field by about an order of magnitude with respect to this 2004 model is expected at the time of the LARES data analysis

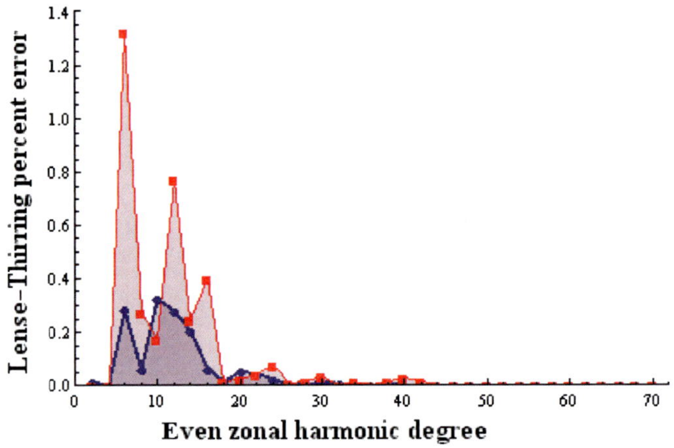

Fig. 5 Percent error in the measurement of the Lense–Thirring effect using LARES, LAGEOS and LAGEOS 2 as a function of the uncertainty due to each even zonal harmonic. The points in blue are the errors obtained using the model EIGEN GRACE02S (i.e., Fig. 1 rescaled) and the points in red are the errors obtained using, as uncertainty of each coefficient, the difference between the value of this coefficient in the two different models EIGEN-GRACE02S and GGM02S. The total error in the measurement of the Lense–Thirring effect using EIGEN-GRACE02S is 1.4% and by using as uncertainties the differences between the coefficients of the two models is 3.4%. However, at the time of the LARES data analysis an improvement in the Earth gravity field of about one order of magnitude has to be taken into account with respect with these 2004 models that were based on less than 365 days of observations of the GRACE spacecraft

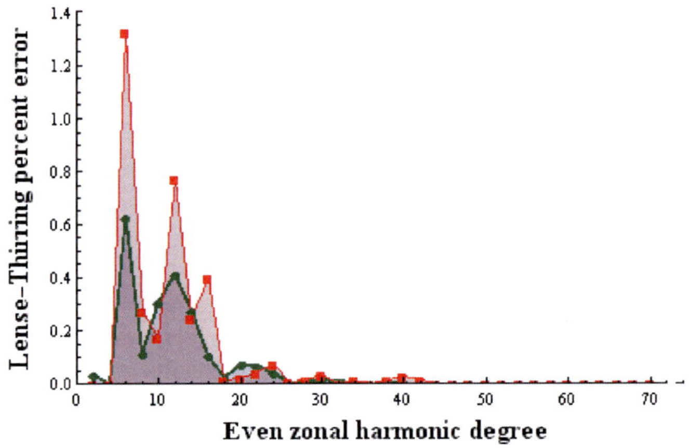

Fig. 6 Percent error in the measurement of the Lense–Thirring effect using LARES, LAGEOS and LAGEOS 2 as a function of the uncertainty due to each even zonal harmonic. The points in green are the errors obtained using the model GGM02S (i.e., Fig. 1 rescaled) and the points in red are the errors obtained using, as uncertainty of each coefficient, the difference between the value of this coefficient in the two different models EIGEN-GRACE02S and GGM02S. The total error in the measurement of the Lense–Thirring effect using GGM02S is 2.1% and by using as uncertainties the differences between the coefficients of the two models is 3.4%. However, at the time of the LARES data analysis an improvement in the Earth gravity field of about one order of magnitude has to be taken into account with respect with these 2004 models that were based on less than 365 days of observations of the GRACE spacecraft

In regard to the other orbital perturbations that affect the LARES experiment we briefly discuss here the tidal effects, particle drag and thermal drag; for a detailed treatment of other perturbations we refer to [3–5, 19, 21]. In regard to the orbital perturbations of the LARES experiment due to the time dependent Earth's gravity field, we observe that the largest tidal signals are due to the zonal tides with $l = 2$ and $m = 0$, due to the Moon node, and to the K_1 tide with $l = 2$ and $m = 1$ (tesseral tide). However, the medium and long period zonal tides ($l = 2$ and $m = 0$) will be cancelled using the combination of the three nodes together with the static C_{20} uncertainty (also the uncertainty in the time-dependent secular variations $\dot{C}_{20}$, $\dot{C}_{40}$ will be cancelled using this combination of three observables). Furthermore, the tesseral tide K_1 will be fitted for over a period equal to the LARES nodal period, as explained above (see [4, 5] and Chap. 5 of [8]) and this tide would then introduce a small uncertainty in our combination. In regard to the non-gravitational orbital perturbations, we observe here that the unmodelled thermal drag perturbations on the LARES orbit would be reduced with respect to the LAGEOS satellites thanks to the much smaller (by a factor of about 0.38) cross-section to mass ratio of LARES (see the Sections 4 and 5 on the LARES structure); furthermore, accurate measurements of the thermal properties of the LARES and of its retro-reflectors should be performed. We point out that the neutral and charged particle drag on the LARES node at an altitude of about 1,450 km would be a negligible effect for an orbit with very small eccentricity, even by assuming that the exosphere would be co-rotating with Earth at 1,450 km of altitude and assuming density variations in the exosphere. Indeed, as calculated in [3] for the LAGEOS III satellite, in the case of zero orbital eccentricity $e = 0$ the total drag effect on the LARES node would be zero; indeed the nodal rate of a satellite due to particle drag is a function of $sin\ \nu \cdot cos\ \nu$ (ν is the true anomaly) and the total nodal shift is then zero over one orbit; in the case of a small orbital eccentricity, the total shift would be proportional to the eccentricity and it would still be a small effect as calculated in [3].

3 Gravitational Uncertainties and Even Zonal Harmonics

According to what is shown in Figs. 1 and 2, and to what can be achieved with the first VEGA flight, the orbital parameters of LARES will be: inclination equal to about 71.5°, semimajor axis equal to about 7,830 km and eccentricity nearly zero.

Using the node of the new laser ranged satellite LARES and the nodes of LAGEOS and LAGEOS 2, thanks to the accurate Earth gravity field models obtained with the GRACE spacecraft, it is then possible to achieve a measurement of the Lense–Thirring effect and of the Earth gravitomagnetic field with an accuracy of the order of 1%. The error sources of gravitational origin, i.e., those due to the uncertainties in the Newtonian gravitational field, are by far larger than the uncertainties of non-gravitational origin, i.e. radiation pressure, both from Sun and Earth, thermal thrust and particle drag; indeed the LAGEOS satellites and especially the LARES satellite are extremely dense spherical satellites with very small cross-sectional-to-mass ratio [3, 50]. In particular LARES is made of Tungsten alloy and has a radius of only 18.2 cm making it the densest known single object in the solar system (see Sections 4 and 5). Among the gravitational perturbations of the orbit of the LARES and LAGEOS satellites by far the largest ones are due to the non-sphericity of the Earth gravity field, both its static and time dependent part, described by the expansion of the Earth potential in spherical harmonics [3, 30]. In particular, the only secular effects affecting the orbit of an Earth satellite are due to the axially symmetric deviations from spherical symmetry of the Earth gravitational potential described by the even zonal harmonics, i.e., the spherical harmonics of the Earth gravitational potential of even degree and zero order, e.g., the C_{20} harmonic describing the well known quadrupole moment of Earth.

In Figs. 3, 4, 5 and 6 we display the error in the LARES experiment due to each even zonal harmonic up to degree 70. Once again we stress that the large errors due to the uncertainties of the first two even zonal harmonics, i.e., of degree 2 and 4, are eliminated using the three observables, i.e. the 3 nodes. Figures 3, 4, 5 and 6 clearly display that the error due to each even zonal harmonic of degree higher than 4 is considerably less than 1% and in particular that the error is substantially negligible for the even zonal harmonics of degree higher than 26. In Fig. 3 we show the percent errors in the measurement of the Lense–Thirring effect in the LARES experiment due to each individual uncertainty of each even zonal harmonic corresponding to the model EIGEN-GRACE02S [11]. In Fig. 4 we show these percent errors for the model GGM02S [12]. In Fig. 5 we display the maximum percent errors, due to each even zonal harmonic, obtained by considering as uncertainty for each harmonic the difference between the value of that harmonic in the EIGEN-GRACE02S model minus its value in the GGM02S model; this is a technique used in space geodesy to estimate the reliability of the published uncertainties of a model. Of course, in order to use this technique, one must difference models of comparable accuracy, i.e., models that are indeed comparable, or use this technique only to assess the errors of the less accurate model.

In conclusion, using EIGEN-GRACE02S and GGM02S, the total error in the measurement of the Lense–Thirring effect due to the even zonal harmonics is respectively 1.4% and 2.1%. However, an improvement in the gravity field models by about an order of magnitude is expected at the time of the LARES data analysis. Indeed, these two models, EIGEN-GRACE02S and GGM02S, have been obtained with a relatively small amount of observations of the GRACE spacecraft (launched in 2002) over less than 365 days and therefore a substantial factor of improvement of these GRACE models, of about one order of magnitude, has to be taken into account at the time of the LARES data analysis (between 2011 and 2017), thanks to longer GRACE observational periods. As GOCE [28] is going to improve mainly the short wavelength part of the gravity field models, i.e., the higher degrees, the GRACE models will still play a basic role for gravitomagnetism measurements.

We finally point out that a single author, see, e.g., [29], has claimed that the inclusion of the even zonal harmonics of degree higher than 20 would largely increase the total uncertainty of the LARES experiment. However, Figs. 3, 4, 5 and 6 demonstrate that these claims are ill-founded and that the error due to the even zonal harmonics of degree higher than 26 is negligible in the LARES experiment, contrary to a number of miscalculations by the same author, e.g., [29]. Indeed, the contributions of higher degree spherical harmonics of the Earth's gravity field to a satellite orbital motion, and in particular their contributions to the satellite nodal precession, decrease as the inverse power of the semimajor axis to the degree of the even zonal harmonic (see the LAGEOS paper [21] elsewhere in this book), and thus quickly *decrease* with the degree and does *not increase* as claimed in various papers by the same author: the expansion of the Earth's gravity field in spherical harmonics is indeed an expansion with terms that decrease as the inverse power of the semimajor axis to the degree of the harmonics [30]. Figures 3, 4, 5 and 6 show that the total error in the LARES experiment, due to the uncertainties in the even zonal harmonics of degree strictly higher than four, is of the order of one percent only. These figures prove indeed that the claims (published, e.g., in [29]) that by considering the uncertainties in the even zonal harmonics higher than degree 20, i.e., from degree 20 to degree 70, the error largely increases, are obviously misleading and wrong by a large factor.

Indeed, the results shown in Figs. 3, 4, 5 and 6 have been obtained both by precise analytical propagation of the error of each coefficient in the nodal rate equation (using Mathematica) and confirmed by orbital propagation (with GEODYN). However, these results can also be easily understood and derived, in order of magnitude, in the following simple way: (a) the uncertainties of each even zonal harmonic published with a GRACE gravity field model are very roughly constant, from degree 12 to degree 60; they may at most change by a factor 5 (see http://icgem.gfz-potsdam.de), so it is the difference between different Earth gravity models of comparable accuracy (see Fig. 9 of the LAGEOS paper elsewhere in this book [21] that displays the difference between a number of gravity field models as a function of the degree of each harmonic and see also there the discussion about the appropriate way of comparing different Earth gravity field model); (b) in the nodal rate equation [21], the size of each even zonal coefficient of degree l roughly decreases as the inverse, $l + 1.5$, power of the satellite semimajor axis, a (or better of the ratio between the Earth radius, $R_\oplus$, and the satellite semimajor axis, a), where the 1.5 power comes from the common coefficient $\frac{n}{a} = \frac{2\pi}{Pa}$ in the nodal rate equation and P is the satellite orbital period. For example, for $l = 2$ the error corresponding to the term $l = 2$ is proportional to $(R_\oplus/a)^{3.5}$, however for $l = 60$ is proportional to $(R_\oplus/a)^{61.5}$. Now, since the error corresponding to each even zonal harmonic uncertainty can be calculated by multiplying each even zonal uncertainty (very roughly constant from $l = 12$ to $l = 60$) for the corresponding even zonal coefficient in the nodal rate equation (that goes as the inverse, $l + 1.5$, power), the total error due to each even zonal harmonic uncertainty is roughly proportional to the inverse, $l + 1.5$, power of the satellite semimajor axis a. This simply explains the results displayed in Figs. 3, 4, 5 and 6.

4 Technical and Engineering Aspects of LARES Mission

Lares satellite was first proposed in the phase A study financed by ASI [31] in 1998, even though a first idea on how to reduce surface perturbations on a laser ranged satellite was given in Ref. [32]. In the following years some other studies have been conducted on the LARES satellite Refs. [33–39, 51]. In February 2008, ASI selected Carlo Gavazzi Space to be the prime contractor of the LARES system. In the previous sections it has been described how to circumvent the problem of classical gravitational perturbations, while in the following sections are reported design solutions to minimize the non gravitational perturbations such as particle drag, interaction with the magnetic field of Earth and thermal thrust (induced by thermal gradients on the satellite surface).

4.1 Laser Ranging

The first aspect to analyze is the technique used to accurately track a passive satellite such as LAGEOS or LARES: the laser ranging technique. Laser ranging can be assimilated to an optical radar. In fact distances are measured by accurately measuring the time of flight of a laser pulse. Laser pulses are very short, typically of the order of nanoseconds. The space resolution corresponding to the pulse length is not sufficient and for this reason the measurement takes into account the shape of the function energy vs time. This allows to reach accuracies of the order of 1 cm. From some stations it is possible to reach accuracies of few millimeters. Another font of uncertainty is due to the presence of the atmosphere, whose composition (humidity for instance) varies with whether with consequent variation of the speed of light. One technique developed to reduce this uncertainty is the two color technique which employs lasers at two different wavelengths. The principle of positioning a satellite in space is very simple and shown in Fig. 7: knowing the distance from at least three different stations it is possible to locate the satellite in space. It is however possible to determine the position of the satellite also from one single station. In fact besides the distance it is possible to obtain from the laser station also the other two coordinates of the pointing device (for example a mirror with a two degrees of freedom mounting positioned in front of the telescope used to direct the laser light).

4.2 Cube Corner Reflectors

The second aspect concerns the Cube Corner Reflectors (CCRs). It is extremely important to have on the surface of the satellite CCRs that are designed and manufactured very accurately to allow to redirect the laser pulse back to the ground station. In Fig. 8 it is reported a scheme showing in two dimensions the principle of a CCR. In three dimensions there will be three reflections on the back faces of the CCR

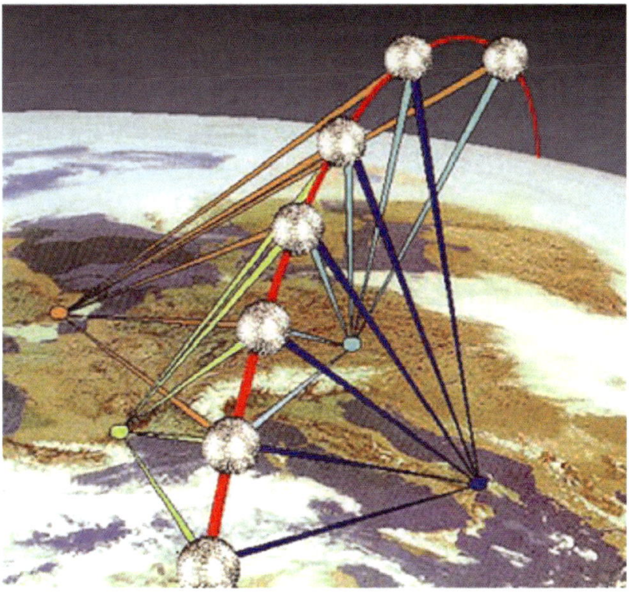

Fig. 7 Satellite positioning with laser ranging stations

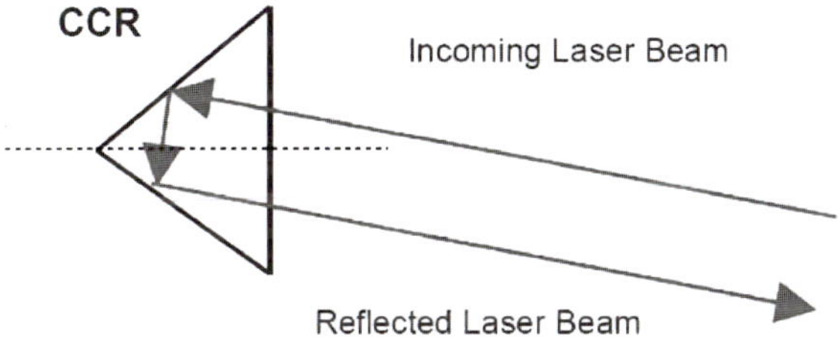

Fig. 8 Principle of a CCR

before the signal comes back on the opposite direction. In Fig. 9 is reported a CCR that will be used for the LARES satellite. It has to be considered that those CCRs can operate at distances from hundreds to several thousands of kilometers and up to the Moon distance even though only few stations are capable of performing today lunar laser ranging. For these reasons the manufacturing tolerances are quite demanding. There are several designs of CCRs that basically differs in the way used to mount them on the satellite surface, but the common characteristics is that they have a front face and three back faces at 90° one from the other. They can in fact, as the name recalls, be obtained by cutting an edge of a cube. The back faces of the CCRs can be either coated or uncoated. In the first case one relies on total internal reflection. Uncoated CCRs were used in the lunar retroreflector experiment carried on the Moon by the crews of Apollo 11, 14 and 15. The reason for this choice is the concern related to the heating of the coating and consequent CCR deformation. In order

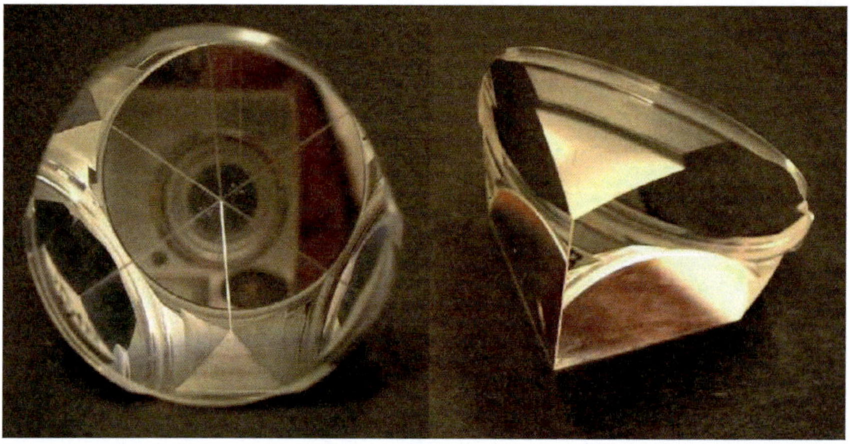

Fig. 9 Photograph of a LARES cube corner reflector from two different points of view

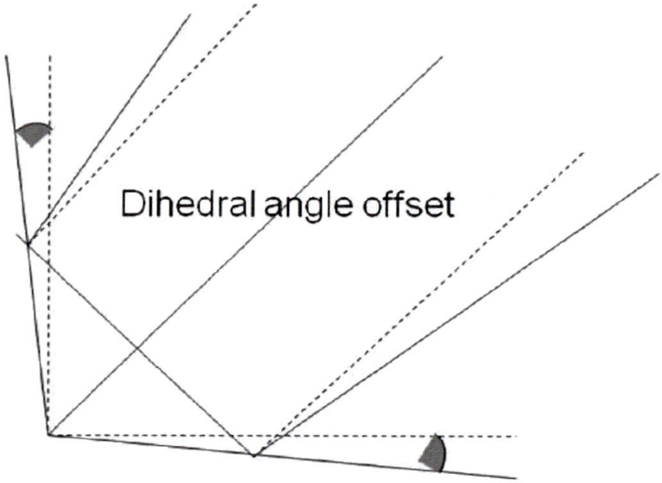

Fig. 10 The increase of the dihedral angle by a small quantity causes the spread of the energy over a bigger area

to receive the signal back at the transmitting station, the dihedral angles of the CCR should be at exactly 90° if the relative motion is negligible. In the case of a satellite CCR, the dihedral angles need to be slightly increased so that, as shown in Fig. 10, the spreading of the returning signal will compensate for the satellite motion. The small angular quantity is called Dihedral Angle Offset (DAO) and for LAGEOS satellite it is equal to 1.25 arcsec and for LARES 1.5 arcsec for each of the three back faces.

The energy distribution at the receiving station is called Far Field Diffraction Pattern (FFDP) and can be calculated theoretically using diffraction theory. The

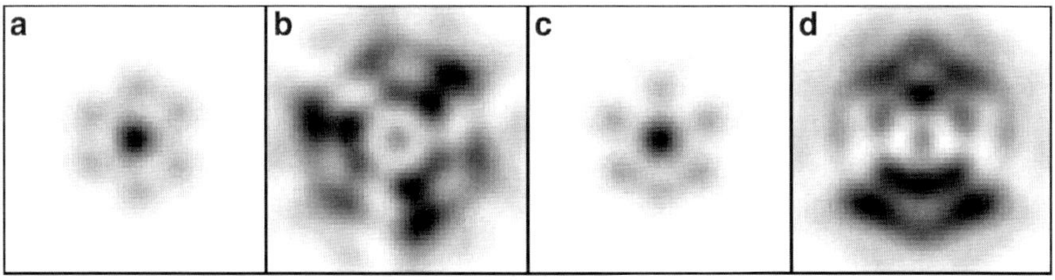

Fig. 11 Calculated FFDP of an uncoated CCR for (**a**) circular polarization, no DAO, (**b**) circular polarization DAO = 1.25 arc seconds, (**c**) linear vertical polarization, no DAO, (**d**) for lineare vertical polarization, DAO = 1.25 arcsec. Darker areas are relevant to higher energy [52]

FFDP of a mirror is an Airy pattern. In Fig. 11 are reported the theoretical FFDPs of uncoated CCRs with and without DAO (see Ref. [40] for more details). In Fig. 11 it is possible to recognize that the energy distribution of the returning laser pulse is not uniform. In the case of no DAO such a distribution resembles the FFDP of a mirror characterized by a central peak followed by an annulus of low intensity and other annuli alternatively of higher and lower intensities rapidly going to zero. In Fig. 11 the mentioned annuli are substituted by lobes that are arranged circumferentially around the central peak. It is important to observe that the application of a DAO allowed to move the energy from the central peak (Figs. 11a, c) to the external lobes (Figs. 11b, d). In fact for a moving CCR the pattern received on the ground will shift in the direction of the satellite velocity and consequently a low energy area will hit the detector on the ground if no DAO is present as in the case reported in Fig. 11a, c. The introduction of a DAO will change the pattern in such a way to reduce the central peak intensity in favor of other lobes that will hit the ground station. For LAGEOS and LARES the manufacturing tolerances of the CCRs, accepted for the DAO is $+/-0.5$ arcsec. Therefore in the case of LARES the DAO of the 92 CCRs will range from 1 arcsec to 2 arcsec. What in practice is measured for qualification and acceptance of the CCRs is the energy in a specified annulus of the FFDP.

Other important characteristics of the CCRs are the surface qualities expressed in fraction of wavelength λ. For the three back faces is $\lambda/10$, for the front face is $\lambda/8$. Since the impinging wavefront will enter the front face, it will be subjected to three back reflections and will pass again through the front face, it is expected that the exiting wavefront will be within $\lambda/4$. All the values are peak-to-valley not RMS, that means that the front face, for instance, should have the surface inside two perfect planes distant $\lambda/8$. The wavelength used for testing the CCRs has been $\lambda = 632$ nm. In practice due to the 0.5 arcsec tolerance on the DAO, the three back faces of each CCR will be offset with a different angle. For example a CCR that has been subjected to a preliminary test had the following DAOs expressed in arcsec: 1.34, 1.19, 1.75. In Fig. 12 (left), darker areas corresponds to higher energy, on the other hand in Fig. 12 (right) brighter areas corresponds to higher energy. It is possible to recognize the absence of a high energy central peak in both theoretical and experimental FFDP. Also the vertical line at the center of the pattern is at low energy in both the experimental and calculated FFDP.

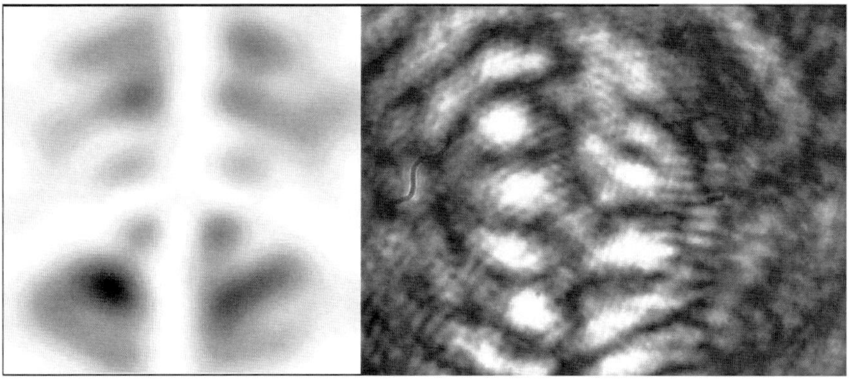

Fig. 12 Calculated (*left*) and experimental (*right*), preliminary, FFDP of an uncoated CCR with measured DAOs of 1.34, 1.19, 1.75 arcsec

To conclude this section we mention a completly different type of retroreflector that exploits the Luneburg (often spelled Luneberg) lens concept (Ref. [41]). This concept relay on a spherical retroreflector that has been manufactured and tested on board the METEOR-3M satellite, launched on December 10, 2001 [42].

4.3 LARES Satellite

The third item is the satellite body for which an original design has been devised. LARES is a Laser Ranged (LR) satellite, and as such is completely passive and consequently very reliable. All laser ranged satellites are similar since they are basically constituted by an inert mass covered on the surface by retroreflectors. What basically makes one LR satellite different from another is the size, weight, number, distribution and type of retroreflectors. Then there are many small constructive issues that make those satellites different such as: mounting systems of retroreflectors, number, material and shape of the pieces of the main body of the satellite.

There are many laser ranged satellites in orbit (Fig. 13) or ready to fly (Fig. 14). In particular the last satellite is based on the Luneburg concept mentioned above. It is constituted by two concentric glass spheres with index of refraction chosen in such a way to focus a plane wavefront on the back face of the sphere which will reflect back the impinging light (Ref. [43]). But LR satellites are not the only ones that carry CCRs: there are satellites such as the GPS satellites and the new constellation of GALILEO that have arrays of CCRs, also some remote sensing satellites such ERS, carries arrays of CCRs.

As described earlier the LAGEOS satellites have a special importance for the measurement of the Lense–Thirring effect, as mentioned those satellites orbit at about 6,000 km altitude. The satellite body is constituted by three main components:

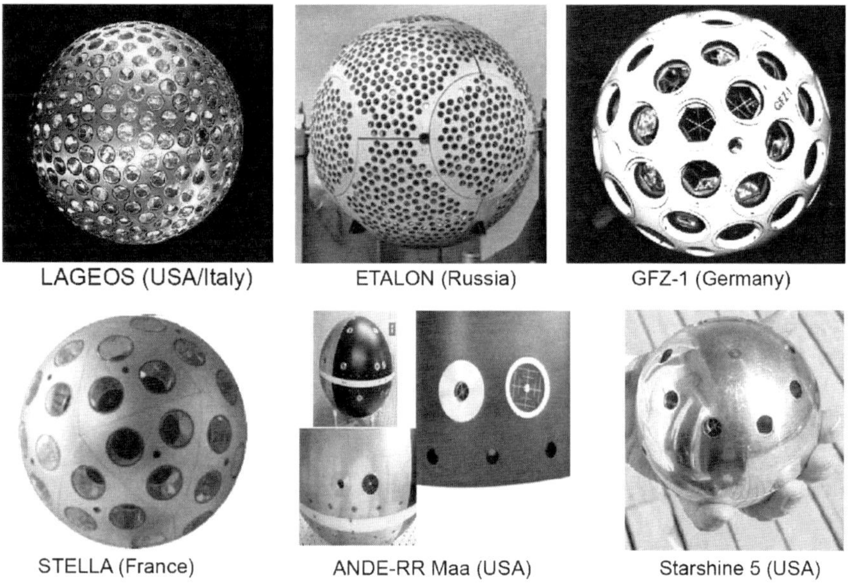

Fig. 13 Some laser ranged satellites already launched

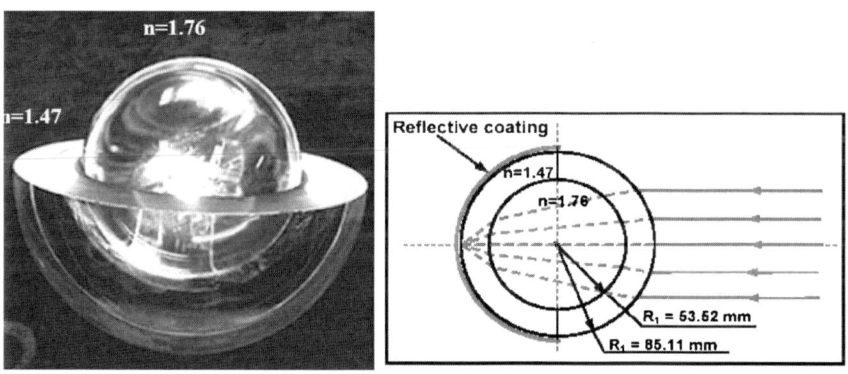

Fig. 14 Spherical retroreflector (diameter 17cm)

two aluminium shells and a cylindrical core of copper alloy, bolted together by a copper-beryllium shaft as shown in Fig. 15.

An important parameter of a LR satellite is the surface-to-mass ratio (A/M). In the hypothesis of a perfectly spherical and homogeneous satellite of radius r and density ρ this parameter is:

$$\frac{A}{M} = \frac{1}{3\rho r} \text{ with the constraint on } M = \rho \frac{4}{3} \pi r^3 \qquad (1)$$

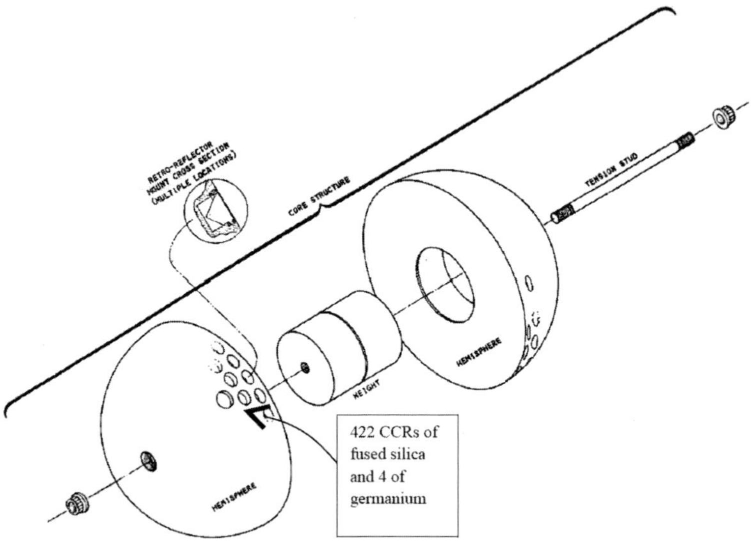

Fig. 15 LAGEOS structural components

Thus, in order to minimize the non-gravitational perturbations, one has to minimize this ratio. It is easy to recognize that this minimization problem is solved maximizing the density. Let us take as a reference the mean density of LAGEOS satellite of about 3,600 kg/m^3, which is a fairly high density for a space structure. Looking at the periodic table of elements one finds out that the highest density is provided by osmium 22,600 kg/m^3, followed by iridium, platinum, rhenium, neptunium, plutonium, tungsten, gold and uranium. Considering the cost, workability and toxicity of the materials just listed, it has been decided to use a tungsten alloy as the best compromise. The nominal density of tungsten is 19,350 kg/m^3, however for best machining performance a tungsten alloy has been adopted for the satellite body, screws and retainer rings. Although slightly higher densities could be obtained with special custom alloys, the small gain in density does not justify the extra cost required. The higher density available by catalogue is 18,500 kg/m^3 with nominal 97% W and 3% Cu, Ni and Fe. We preferred an 18,000 kg/m^3 alloy with nominal 95% W and 5% Cu, Ni. In fact although the former alloy has a higher density, the presence of Fe makes it magnetic and consequently, once the satellite is in orbit it will be subjected to interaction with the magnetic field of Earth. This interaction is very small. However the small gain in density does not justify the bigger complication of dealing with a magnetic material. Other small improvements to decrease the A/M ratio are herein described.

LAGEOS CCR cavities were cylindrical causing a useless satellite mass reduction with consequent increase of A/M. By machining the cavity as a perfect match with the CCR, with say one millimeter gap, would have allowed the maximum mass increase. However the cost and time required for this solution would have been relatively higher. A good compromise has been obtained by manufacturing a conical shape cavity as reported in Fig. 16.

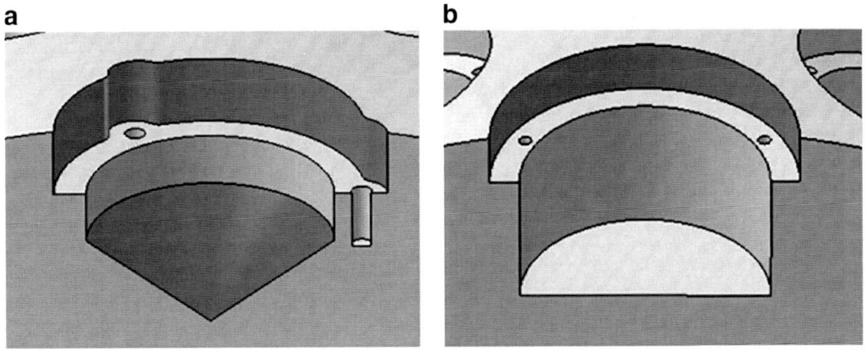

Fig. 16 LARES and LAGEOS cavities

Another reason for not adopting the perfect match CCR cavity, was the tight program schedule for LARES. The CCRs orientation with respect to the axis perpendicular to the CCR front face (CCR azimuth) can be optimized because reflected power is a function of that orientation. The final decision on that optimal orientation was delayed and therefore the final choice was the conical cavity shape which is insensitive to the CCR azimuthal orientation. Particle drag on satellite motion is negligible because of the very small A/M ratio adopted for LARES (about 2.7 times smaller than that of LAGEOS). Thermal contact conductances are the main reason for the onset of thermal gradients on a space structure. On the LAGEOS satellites the three body design introduces not negligible temperature differences on the two hemispheres. Although the main contribution to thermal thrust is given by the CCRs and relevant external part of the mounting system, we adopted a single piece design for the satellite to minimize thermal gradient on the surface of the LARES body. In Fig. 17 is reported a photograph of the LARES satellite during manufacturing. For more details on the mechanical design of LARES please refer to [44].

4.4 Separation System

Another challenging engineering task is represented by the separation system. A first study was conducted in Ref. [45], while some more advanced studies are reported in Refs. [46, 47]. The separation system is required to keep the satellite fixed inside the fairing of the launch vehicle during the launch phases and then release it with the required separation velocity. Due to the similarity with the LAGEOS satellites the first idea was to derive the LARES separation system from that one of LAGEOS (Fig. 18). However the cost and time of qualification required for such a system were not compatible with the program and it was decided to adopt a system described in Ref. [46]. Due to the high forces applied on the satellite by the arms of the separa-

Fig. 17 Demonstration model of the LARES satellite of ASI manufactured with a tungsten alloy at OMPM Angri, Italy, for Carlo Gavazzi Space

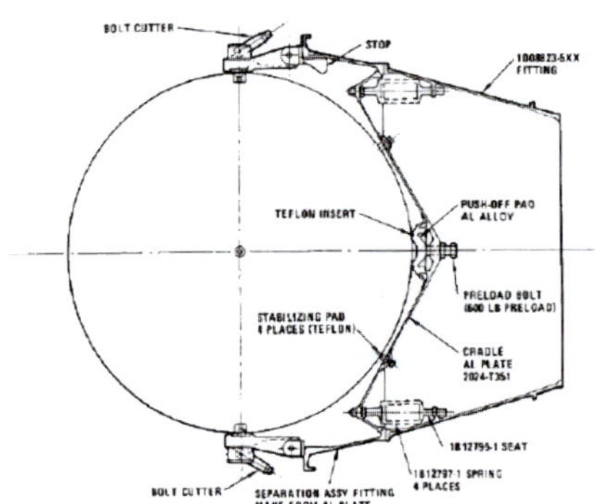

(**a**) LAGEOS I separation system, ref. [48].

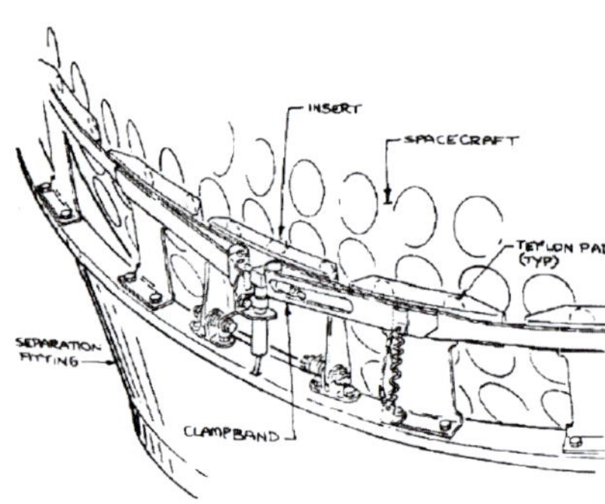

(**b**) Three dimensional sketch of a detail of LAGEOS I separation system, Ref. [48].

Fig. 18 Separation system of LAGEOS I

tion system the current design has slightly changed with respect to what described in the previous references (Fig. 19). This new configuration is under development and offers a higher stiffness with an acceptable mass increase of the arms. Four independent mechanisms are positioned 90° from each other (in the sketch are reported

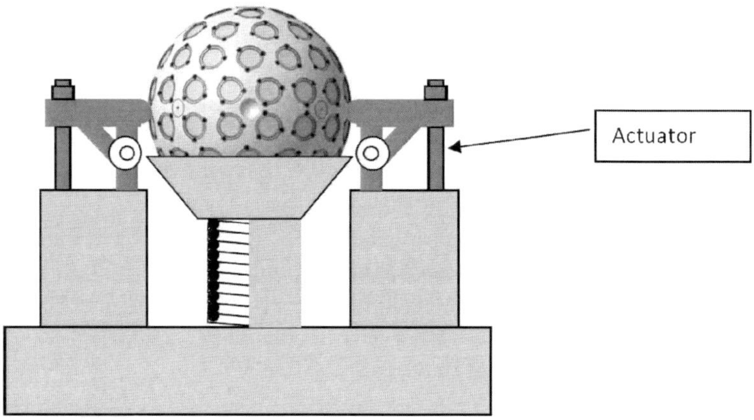

Fig. 19 Sketch of LARES separation system

only two of them). The mechanism is constituted by a hinge with a rotational spring and one actuator. When the four independent actuators will release the brackets, the pre-loaded spring (about 3,500 N) below the satellite will push LARES away from the last stage at the nominal speed of 0.75 m/s. In the LAGEOS separation system the preload of the four springs (only two of those are visible in Fig. 18(a)) was about 500 N each for a total of about 2,000 N. Another difference is in the number of actuators, two redundant for LAGEOS (only one was sufficient to let the clamp open and release the satellite) and four for LARES (at least three need to work for releasing the satellite see Ref. [46]).

4.5 *Launch Vehicle*

LARES is a low cost mission for the measurement of the Lense–Thirring effect. In the year 2008, the qualification launch of VEGA accepted LARES as the main passenger considering that the low cost of the LARES system makes acceptable the higher risk associated with the qualification launch. VEGA is a four stage launch vehicle. The first three stages are solid rocket motors. The fourth stage is an Attitude Vernier Upper Module (AVUM) that uses liquid propellant. This liquid stage can be restarted up to three times in the qualification flight. The launch vehicle will put LARES into an elliptical orbit with the apogee (the point of the orbit more distant from the centre of Earth) at 1,450 km. The second burn of the AVUM will circularise the orbit (1,450 × 1,450 km). The third and last (for VEGA commercial flights the restarts can be up to five) burn will deorbit the AVUM at (350 × 1,450 km) for the release of the secondary payloads and for the fulfilment of the space debris mitigation procedures. LARES will be put at about 71.5 degrees of inclination. VEGA with its equatorial launch base, located in Kourou, French Guiana, can inject satellites

Table 1 Main characteristics of VEGA launch vehicle

Typical length	Typical diameter	Typical liftoff mass	Payload capacity
29.9 m	3.025 m	137 tons	1,500 kg into a 700 km circular polar orbit

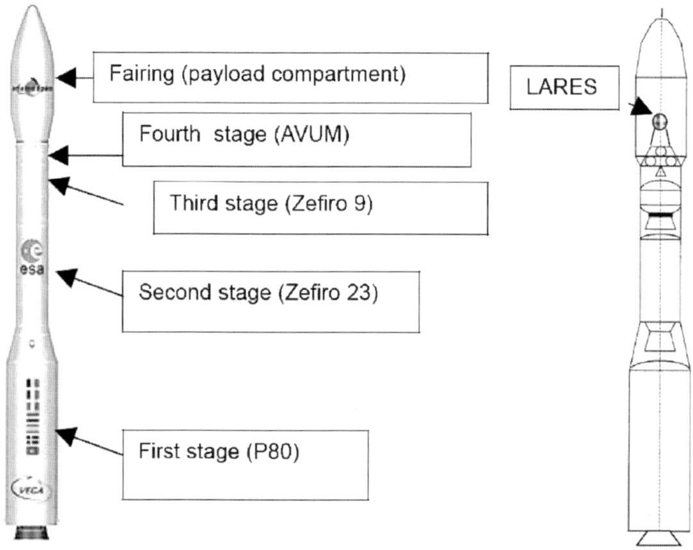

Fig. 20 VEGA Launch vehicle

at orbital inclination from 5.2° to sun synchronous (i.e., 600–800 km in altitude, and inclinations of about 98°), and masses up to 2,500 kg in low Earth orbit. Main characteristics are reported in Table 1 [49].

In Fig. 20 is reported a drawing of the VEGA launch vehicle. The qualification of VEGA main components is in progress. All the solid rocket motors have been tested. In Fig. 21 are reported two pictures showing the third stage (Zefiro 9) being tested.

5 Conclusions

A subpolar orbit for LARES at an altitude of about 1,450 km, together with the LAGEOS satellites and with the GRACE models, will provide a measurement of Earth gravitomagnetism with an accuracy of a few percent. In order to minimize the measurement error induced by the uncertainties in the Earth even zonal harmonics and however by necessarily taking into account the VEGA qualifying flight launch constraints, the orbital parameters finally chosen for LARES are: inclination about 71.5°, semimajor axis about 7,830 km and eccentricity nearly zero; a small orbital

Fig. 21 Third stage (Zefiro 9) mounted at the Inter-force Test Range in Salto di Quirra, Sardinia (Italy) (*left*), Firing test (*right*)

injection error from these values is quite acceptable for the measurement of gravitomagnetism. The weight and radius of the LARES satellite will be about 390 kg and 18 cm respectively. LARES will have 92 CCRs housed in conical cavities. The satellite has been designed to minimize the non-gravitational perturbations and be as nearly as possible a test-particle freely falling in the Earth gravitational field. This will allow the measurement of tiny relativistic perturbations such as the Lense–Thirring effect.

Acknowledgement We gratefully acknowledge the support of ASI, the Italian Space Agency, grant I/043/08/0.

References

1. Ciufolini, I., Theory and Experiments in General Relativity and other Metric Theories. *Ph. Dissertation*, Univ. of Texas, Austin, Ann Arbor, Michigan (1984).
2. Ciufolini, I., Measurement of the Lense-Thirring drag on high-altitude laser-ranged artificial satellites. *Phys. Rev. Lett.*, **56**, 278–281 (1986).
3. Ciufolini, I., A comprehensive introduction to the Lageos gravitomagnetic experiment: from the importance of the gravitomagnetic field in physics to preliminary error analysis and error budget. *Int. J. Mod. Phys. A* **4**, 3083–3145 (1989); see also: [4, 5].
4. Tapley, B., Ries, J.C., Eanes, R.J., and Watkins, M.M., *NASA-ASI Study on LAGEOS III*, CSR-UT publication n. CSR-89-3, Austin, Texas (1989).
5. Ciufolini, I. et al., ASI-NASA Study on LAGEOS III, CNR, Rome, Italy (1989). See also: I. Ciufolini et al., INFN study on LARES/WEBER-SAT (2004).
6. Ciufolini, I., Paolozzi, A., Pavlis, E.C. et al., LARES phase A study for ASI (1998).

7. Ries, J.C., Simulation of an experiment to measure the Lense-Thirring precession using a second LAGEOS satellite. *Ph. Dissertation*, Univ. of Texas, Austin (1989).
8. Peterson, G.E., Estimation of the Lense-Thirring precession using laser-ranged satellites. *Ph. Dissertation*, Univ. of Texas, Austin (1997).
9. Ciufolini, I. and Wheeler, J.A., *Gravitation and Inertia* (Princeton University Press, Princeton, New Jersey, 1995).
10. Ciufolini, I. and Pavlis, E.C., A confirmation of the general relativistic prediction of the Lense-Thirring effect, *Nature*, **431**, 958–960 (2004).
11. Reigber, C., Schmidt, R., Flechtner, F., Koenig, R., Meyer, U., Neumayer, K.H., Schwintzer, P. and Zhu, S.Y., An Earth gravity field model complete to degree and order 150 from GRACE: EIGEN-GRACE02S. *J. Geodyn.*, **39**, 1–10 (2005). The EIGEN-GRACE02S gravity field coefficients and their calibrated errors are available at: http://op.gfz-potsdam.de/grace/index_GRACE.html
12. Tapley, B., Ries, J., Bettadpur, S., Chambers, D., Cheng, M., Condi, F., Gunter, B., Kang, Z., Nagel, P., Pastor, R., Pekker, T., Poole, S., and Wang, F., GGM02 An improved Earth gravity field model from GRACE. *J. Geod.*, **79**, 467–478 (2005). The GGM02 gravity model is available at: http://www.csr.utexas.edu/grace/gravity/
13. Reigber, Ch., Flechtner, F., Koenig, R., Meyer, U., Neumayer, K., Schmidt, R., Schwintzer, P. and Zhu, S., GRACE Orbit and Gravity Field Recovery at GFZ Potsdam – First Experiences and Perspectives. *Eos. Trans. AGU*, **83**(47), Fall Meet. Suppl., Abstract G12B-03 (2002).
14. Tapley, B.D., The GRACE mission: status and performance assessment. *Eos. Trans. AGU*, **83**(47), Fall Meet. Suppl., Abstract G12B-01 (2002).
15. Watkins, M.M., Yuan, D., Bertiger, W., Kruizinga, G., Romans, L., and Wu, S., 2002. GRACE gravity field results from JPL. *Eos. Trans. AGU*, **83**(47), Fall Meet. Suppl., Abstract G12B-02.
16. Ciufolini, I., On a new method to measure the gravitomagnetic field using two orbiting satellites. *Nuovo Cimento A*, **109**, 1709–1720 (1996).
17. Ciufolini, I., Pavlis, E.C., Chieppa, F., Fernandes-Vieira, E. and Perez-Mercader, J., Test of general relativity and measurement of the Lense-Thirring effect with two Earth satellites. *Science*, **279**, 2100–2103 (1998).
18. Ciufolini, I., Chieppa, F., Lucchesi, D. and Vespe, F., Test of Lense-Thirring orbital shift due to spin. *Class. Quantum Grav.*, **14**, 2701–2726 (1997).
19. Ciufolini, I., Pavlis, E.C. and Peron, R., Determination of frame-dragging using Earth gravity models from CHAMP and GRACE. *New Astron.*, **11**, 527–550 (2006).
20. Ries, J.C., Eanes, R.J. and Watkins. M.M., Confirming the frame-dragging effect with satellite laser ranging. *16th International Workshop on Laser Ranging*, 13–17 October 2008, Poznan, Poland (2008).
21. Ciufolini, I, Pavlis, E.C., Ries, J., Koenig, R., Sindoni, G., Paolozzi, A. and Newmayer, H.K., Gravitomagnetism and its measurement with laser ranging to the LAGEOS satellites and GRACE Earth gravity models. in this book: *General Relativity and John Archibald Wheeler*, eds. Ciufolini, I., and Matzner, R. (Springer Verlag, 2010).
22. Ciufolini, I., Frame-dragging and its measurement. In: *Gravitation: from Hubble Length to the Planck Length, Proc. I SIGRAV School on General Relativity and Gravitation*, Frascati, Rome, September 2002 (IOP, 2005) pp. 27–69.
23. Ries, J.C., Eanes, R.J., Tapley, B.D. and Peterson, G.E., Prospects for an improved Lense-Thirring test with SLR and the GRACE gravity mission, in: *Toward Millimeter Accuracy Proc. 13th Int. Laser Ranging Workshop*, Noomen, R., Klosko, S., Noll, C. and Pearlman, M. eds., (NASA CP 2003212248, NASA Goddard, Greenbelt, MD, (2003).
24. Pavlis, E.C., Geodetic contributions to gravitational experiments in space. In: *Recent Developments in General Relativity, Genoa 2000*, R. Cianci, et al., eds. (Springer-Verlag, Milan, 2002) pp. 217–233.
25. Paolozzi, A., Private communication (2004).
26. Ries, J., Private communication (2005).
27. Iorio, L., The impact of the new Earth gravity models on the measurement of the Lense-Thirring effect with a new satellite. *New Astron.*, **10**, 616–635 (2005).

28. ESA (2006): ESA's Gravity Mission – GOCE Brochure. Available at http://esamultimedia.esa.int/docs/BR209web.pdf
29. Iorio, L., Will the recently approved LARES mission be able to measure the Lense-Thirring effect at 1 %? *Gen. Relativ. Gravit.*, **41**, 1717–1724 (2009) doi: 10.1007/s10714-008-0742-1; see also: arXiv:0803.3278v5 [gr-qc] (2008).
30. Kaula, W.M., *Theory of Satellite Geodesy* (Blaisdell, Waltham, 1966).
31. Ciufolini, I., Paolozzi, A. et al. LARES, Laser Relativity Satellite, for the study of the Earth gravitational field and general relativity measurement. *Phase A report, studio piccole missioni ASI* (1998).
32. Ciufolini, I., Paolozzi, A., and Peroni I. A Preliminary Study for a Satellite Minimizing Solar Radiation Perturbation. In: Atti *XIV Congresso Nazionale AIDAA*, 521–530 (1997).
33. Ciufolini, I., and Paolozzi, A. LARES: A New Laser-ranged satellite for fundamental physics and general relativity. *Actual Problems of Aviation and Aerospace Systems*, **1**, 61–73 (1999).
34. Ciufolini, I., Paolozzi, A., Peroni, I., and Gabrielli, A. A study for a laser ranged relativity satellite. In: *Applied Mechanics in the Americas, Proc. of the PACAM VI*, vol. **6**, 467–470 (1999).
35. Cimino, C., Paolozzi, A., and Peroni, I. Thermal Analysis of a Satellite Retroreflector Using MSC/NASTRAN. In: *Conferenza Utenti MSC*, Roma, Ottobre 1999, pp. 9.1–12 (1999).
36. Paolozzi, A., A new laser ranged satellite for testing general relativity. In: *Ninth Marcel Grossmann Meeting on General Relativity*, Roma, Italy, 2–8 July 2000 (World Scientific, 2002) pp. 1766–1770.
37. Lucchesi, D.M. and Paolozzi, A. A Cost Effective Approach for LARES Satellite. *XVI AIDAA*, Palermo, Sept. 2001, paper 111 (2001) pp. 1–14.
38. Ciufolini, I., Currie, D.G., and Paolozzi, A. The LARES Mission for Testing the Dynamics of General Relativity. *IEEE Aerospace Conference*, Montana, USA, March 2003.
39. Ciufolini, I., Currie, D.G., Paolozzi, A., and Pavlis, E.C. LARES/WEBERSAT, frame-dragging and fundamental physics. In: *Proc. Frontier Science 2004, Physics and Astrophysics in Space*, Frascati, 14–19 June, 2004 (INFN, Frascati, 2004).
40. Paolozzi, A., Ciufolini, I. , Lucantoni, A., and Arnold, D., Optical Design of LARES Satellitte. *XX AIDAA Congress*, Milano, Italy, June 29–July 3, 2009.
41. Luneburg, R.K., *Mathematical Theory of Optics* (Providence, Rhode Island: Brown University, 1944) pp. 189–213.
42. Burmistrov, V.B., Parkhomenko, N.N., Roy, Y.A., Shargorodsky, V.D., Vasiliev, V.P., Degnan, J.J., Habib, S., Glotov, V.D., and Sokolov, N.L., Spherical Retroreflector with an Extremely Small Target Error: International Experiment in Space, available at: http://cddis.nasa.gov/lw13/docs/papers/target_vasiliev_1m.pdf
43. Shargorodsky, V.D., Vasiliev, V.P., Belov, M.S., Gashkin, I.S., and Parkhomenko, N.N., Spherical Glass Target Microsatellite, Institute for Precision Instrument Engineering, Moscow, Russia. Available at: http://cddis.gsfc.nasa.gov/lw15/docs/document/15_TargetSignalsSession.pdf, pp. 21–25.
44. Paolozzi, A., Ciufolini, I., Vendittozzi, C., and Peroni, I., Mechanical Design of LARES Satellite. *XX AIDAA Congress*, Milano, Italy, June 29–July 3 2009.
45. Paolozzi, A., Ciufolini, I., Onorati, F.M., Acquaroli, L., Scolamiero, L., Sindoni, G., Paris, C., Vendittozzi, C., Ramiconi, M., Preli, N., Lucantoni, A., Passeggio, F., and Berardis, S., Fibre optic sensors for the validation of the numerical simulation on the breadboard of the LARES separation system. In: *IAC-08*, Glasgow, Scotland, October 2008.
46. Paolozzi, A., Ciufolini, I., Paris, C., Sindoni, G., Ramiconi, M., Onorati, F.M., and Scolamiero, L., Design of LARES Separation System. *XX AIDAA Congress*, Milano, Italy, June 29– July 3, 2009.
47. Paolozzi, A., Ciufolini, I., Paris, C., Acquaroli, L., Piersigilli, P., and Gabrielli, A., Tests on LARES separation system components using fiber optic sensors. *XX AIDAA Congress* Milano, Italy, June 29–July 3, 2009.
48. NASA technical memorandum NASA TM X-64915 LAGEOS phase B technical report (NASA), Prepared by the Science and Engineering Directorate of Marshall Space Flight Center, Alabama, February 1975. Available at: http://ntrs.nasa.gov/archive/nasa/casi.ntrs.nasa.gov/19750011251_1975011251.pdf

49. VEGA User's Manual, Issue 3/ Revision 0 March 2006, Arianespace, at http://www.arianespace.com/launch-services-vega/VEGAUsersManual.pdf
50. Ciufolini, I., Paolozzi, A., Pavlis, E.C., Ries, J.C., Koenig, R., Matzner, R.A., Sindoni, G., and Neumayer, H., Towards a one percent measurement of frame-dragging by spin with satellite laser ranging to LAGEOS, LAGEOS 2 and LARES and GRACE gravity models. *Space Sci. Rev.*, **148**, 71–104 (2009).
51. Ciufolini I., On the orbit of the LARES satellite (2006), arXiv:gr-qc/0609081v1
52. Paolozzi, A., Ciufolini, I., Lucantoni, A., Arnold, D., Optical design of LARES satellite. *XX AIDAA Congress*, Milano, Italy, June 29–July 3, 2009.

The History of the So-Called Lense–Thirring Effect, and of Related Effects

Herbert Pfister

Abstract Some historical documents, especially the Einstein–Besso manuscript from 1913, an extensive notebook by H. Thirring from 1917, and the correspondence between Thirring and Einstein in the year 1917 reveal that most of the merit for the so-called Lense–Thirring effect of general relativity belongs to Einstein. Besides telling this "central story" of the effect, I give a short "prehistory", with contributions by E. Mach, B. and I. Friedlaender, and A. Föppl, followed by the later history of the problem of a correct centrifugal force inside a rotating mass shell, which was resolved only relatively recently.

1 The History of the So-Called Lense–Thirring Effect

The idea that rotating bodies may exert not only the static (Newtonian) gravitational force but also an additional "dragging force" on test particles deflecting them in the direction of the rotation, was first formulated in 1872 by E. Mach in a small booklet [1], and then in more detail in his famous book on mechanics from 1883 [2]: "The principles of mechanics can, presumably, be so conceived that even for relative rotations centrifugal forces arise. Newton's experiment with the rotating vessel of water simply informs us that the relative rotation of the water with respect to the sides of the vessel produces no noticeable centrifugal forces, but that such forces are produced by its relative rotation with respect to the mass of the earth and other celestial bodies. No one is competent to say how the experiment would turn out if the sides of the vessel increased in thickness and mass until they were ultimately several leagues thick." And although Mach did not provide a concrete extension of Newton's laws of inertia and gravitation, e.g., by adding velocity-dependent forces, and although he did not perform any "dragging experiments", Mach's mechanics was a decisive stimulus for other physicists to do such things.

H. Pfister (✉)
Institute for Theoretical Physics, University of Tübingen, Auf der Morgenstelle 14, D-72076 Tübingen, Germany
e-mail: herbert.pfister@uni-tuebingen.de

I. Ciufolini and R.A. Matzner (eds.), *General Relativity and John Archibald Wheeler*, Astrophysics and Space Science Library 367, DOI 10.1007/978-90-481-3735-0_20,

In 1896 the Friedlaender brothers published a very interesting but not well known booklet [3]. They formulated, in extension of Mach's work, interesting, in some cases even prophetic theoretical ideas: "It seems to me that the correct form of the law of inertia will only then have been found when *relative inertia* as an effect of masses on each other and *gravitation*, which is also an effect of masses on each other, have been derived on the basis of a *unified law*" (already a hint to an equivalence principle!), with the footnote "In this connection it is greatly to be desired that the question of whether Weber's law is to be applied to gravitation and also the question of the propagation velocity of gravitation should be resolved." (Weber's law of 1846 modified Coulomb's law by a factor $1 - (1/2c^2)(dr/dt)^2 + (r/c^2)d^2r/dt^2$, hereby combining Coulomb's law with Ampere's law, and with Faraday's law of induction.) Friedlaender's concrete – but of course unsuccessful – experiment searched for a possible influence of a rapidly rotating, heavy fly-wheel on a torsion balance mounted above the fly-wheel.

A quite different, and, in principle, more promising experiment was performed in 1904 by A. Föppl [4]. Here the rotating source was the whole earth, and the test system was a gyroscope, consisting of two heavy fly-wheels, rotating with angular velocity up to 2,300 rpm. Föppl tested whether the rotating earth induced a Coriolis-type "dragging force" on the gyroscope axis, and he found that such an effect was less than 2% of the angular velocity of the earth. Today we know that a positive measurement of dragging effects due to the rotating earth requires an experimental accuracy better than 10^{-9} (ratio between Schwarzschild and geometric radius of the earth) what seems to be completely out of reach for a Föppl-type gyroscope, even with present technology.

As is well known, Einstein started his search for a relativistic gravitation theory in 1907 [5], where he introduced the equivalence principle, and derived therefrom a gravitational redshift and a light deflection. In 1912, Einstein formulated a scalar, relativistic gravitation theory [6]. Within this theory, Einstein performed the first concrete calculation of a Machian dragging effect [7]. He introduced the model of an infinitely thin, spherical mass shell (mass M, radius R), a model which is very useful still today in general relativity, because (a) it represents the optimal substitute for the Newtonian mass point which is forbidden in general realtivity due to the collapse phenomenon, (b) it allows the study of mass effects by solving only the vacuum Einstein field equations (in the interior and exterior of the mass shell). Einstein then considered a test mass m at the center of this shell and derived that the presence of the mass shell M induces an increase of m by a factor $(1+M/R)$. On the basis of this result he calculated that if an external force exerts a linear acceleration Γ on the mass shell, the test mass m is dragged along with the acceleration $\gamma = (3M/2R)\Gamma$. (Here, and in the following, units are chosen such that the gravitational constant and the light velocity have value 1.)

Very soon after the papers [6, 7], still in 1912, Einstein reached decisive new conclusions regarding a relativistic theory of gravitation: it should be based on a non-Euclidean (pseudo-Riemannian) geometry with metric tensor $g_{\mu\nu}$, and it should be a tensorial theory with the whole energy–momentum tensor $T_{\mu\nu}$ as source of the gravitational field. The first such theory was the so-called Entwurf theory of

Einstein and Grossmann from May 1913 [8] which, however, was not yet generally covariant. Whereas the Entwurf paper contains no direct application of the new gravitational theory, such applications were performed soon afterwards, by Einstein with his friend M. Besso in June 1913, in the so-called Einstein–Besso manuscript [9]. The main objective of this manuscript was the perihelion advance of Mercury which at that time was the only observation in conflict with Newton's theory of gravitation. However, besides the (incorrect) result for the perihelion shift, the Einstein–Besso manuscript contains some other interesting and future-oriented results: they derive a Coriolis force inside a spherical, rotating mass shell, and calculate the resulting "dragging" of test particles: for the ratio d between the induced angular velocity of the test particles and the angular velocity of the mass shell they get $d = 4M/3R$, half the value which Thirring derived in 1918 in general relativity [10]. This is the only part of the manuscript entering Einstein's great talk in September 1913 at the Naturforscher–Versammlung in Vienna [11], where he also remarks that "unfortunately the expected effect is so small that we cannot hope to verify it in terrestrial experiments or in astronomy". Einstein and Besso also calculate the motion of the nodes of planets in the field of the rotating sun. If one compares their result with the later calculation of Lense and Thirring [12] in general relativity, it can be seen that the effect in the Entwurf theory is only $1/4$ of the effect in general relativity.

Hans Thirring started his work on dragging effects by rotating bodies in 1917. Originally he tried to set up an experiment for measuring centrifugal-type forces inside a heavy, rotating hollow cylinder. This is somewhat strange: whereas Friedlaender and Föppl can be pardoned for not having estimated the order of magnitude of the effect searched for, in the year 1917, 12 years after the publication of Einstein's special relativity, it should have been clear that such a relativistic correction to Newton's theory of gravitation can be at most of the order M/R, i.e., below the ridiculously small value 10^{-24} for all conceivable laboratory systems ($M < 100$ kg, $R > 10$ cm). Furthermore, this value was explicitly contained in Einstein's paper [11], and e.g. in the Schwarzschild solution from 1916.

The history of the origin and development of the two central papers of Thirring [10] and Lense and Thirring [12] is well documented because the estate of Hans Thirring, stored at the central physics library in Vienna contains a 156-page notebook "Wirkung rotierender Massen" (the effect of rotating masses) [13]. This notebook, written mainly in the time April–December 1917, is a remarkable document, and it allows a detailed view into the workshop of a theoretical physicist of that time, but it is somewhat deficient in physical insight and a sense for reality and measurability. The first third of the notebook contains, besides obsolete and partly crossed out pages, calculations (within the weak field limit of general relativity) for a rotating sphere and for a rotating mass shell (with angular velocity ω), but Thirring confines himself to the diagonal metric components whose deviations from the Minkowski metric are of order ω^2. With date July 17, the notebook contains the draft of a letter to Einstein (also published in [14]) in which Thirring tells his results so far, the problems he has with them (e.g., the wrong axial component of the so-called centrifugal force), and he asks Einstein for his advice, and whether he could think of an experimental confirmation of such a centrifugal effect on the innermost moon of Jupiter. Einstein's answer of August 2, 1917 [15] is quite short, but it exposes the weak points in Thirring's work in an admirably clear and concise way: "To your example of the hollow sphere it is only to be added that, besides the centrifugal field ... also a Coriolis field results which corresponds to the components g_{41}, g_{42}, g_{43} of the potential, and which is proportional to the first power of ω. This field acts orthogonally deflecting on moving masses, and produces, e.g., a rotation of the pendulum plane in the Foucault experiment. I have calculated this dragging for the earth; it stays far below any measurable amount. Such a Coriolis field is produced also by the rotation of the sun and of Jupiter, and it causes secular changes of the orbital elements of the planets (respectively the moons) which, however, stay far below the measurement error. ... Nevertheless, the Coriolis field seems to be accessible to measurement more easily than your correction terms to g_{44} because the latter have the same symmetry properties as the field distortion due to oblateness."

The first entries in Thirring's notebook after the receipt of Einstein's letter deal with topics he has never considered before: "Calculation of g_{14}, g_{24}, and g_{34} for the rotating spherical shell", and "Determination of the Coriolis force from g_{24}". Later pages from December 1917 contain a first draft of the article [10]. On December 3, 1917 Thirring writes a second letter to Einstein. He says that he is preparing two articles [10, 12] for publication. He then addresses a problem, also appearing at the end of [10]: by transforming the interior of the rotating mass shell to a coordinate system rotating with an appropriate angular velocity $\tilde{\omega}$, one can eliminate the Coriolis force. But, according to Thirring's results, the centrifugal force does not vanish in the same rotating system. In this connection it is interesting that very recent historical research by M. Janssen [16] has revealed that already Besso (and Einstein) realized this wrong centrifugal force (in their work on the rotating spherical mass shell in the Entwurf theory) in August 1913, and also Ehrenfest mentions this problem in a letter of August 1913 to Lorentz. Einstein's immediate answer (of December 7, 1917) to Thirring is, however, not illuminating in this respect. The pages of the notebook, dated January 25–28, 1918, contain the draft of §§1 – 2 of the paper [12]. Here Thirring omits the – partly very involved – expressions of order ω^2, worked out in the notebook, and confines himself to the terms of first order in ω. Thirring's notebook contains no details of §§3 – 4 (transformation of the equations of motion from Cartesian coordinates to the orbital elements used in astronomy, and application to the planets and moons of the solar system) of the paper [12]. It is therefore plausible that these (and only these) parts were calculated and formulated by J. Lense. This supposition is supported by the sentence "My colleague, Prof. Lense, ... has then taken over the task to compare the results with observations on stars" in Thirring's later article [17].

From this analysis of Thirring's notebook (for a more detailed analysis of this notebook see [18]), and of other historical documents, I come to the following conclusion concerning the respective merits of Einstein, Thirring, and Lense for the so-called Lense–Thirring effect: Einstein surely was the first to have the idea that such a dragging effect should result from a relativistic, tensorial theory of gravity. He introduced the extremely useful model of a rotating mass shell into the game, and he calculated (in the Entwurf theory) the Coriolis force inside a rotating shell [11], and the motion of the nodes of planets due to the sun's rotation [9]. Herein he achieves with a minimum of calculational expense (of only first order in ω) a maximum of physical insight and practical results. Since the calculated effects were, however, below measurability at that time, and because Einstein presumably realized that the effects in general relativity would not differ qualitatively from his results in the Entwurf theory, he did not repeat the calculations after November 1915. But it was Einstein who, through his letter of August 2, 1917, put Thirring on the track of the dragging effects (of first order in ω) in general relativity, known today as "Lense–Thirring effects". Possibly due to the fact that Mach's discussion of Newton's bucket is centered on centrifugal effects, and that also in daily life centrifugal effects usually dominate over Coriolis effects, Thirring is so fixed on these centrifugal effects that, until Einstein's intervention, he never considers the Coriolis terms of order ω. Obviously, Thirring never estimated the order of magnitude of the calculated centrifugal effects, which are far below measurability under realistic circumstances, even with today's technology. Furthermore, Thirring overlooked many centrifugal effects active in a rotating mass shell, and therefore never got a correct centrifugal force field. (I will come back to this topic in Section 2.) Nevertheless, even Thirring's published article [10] puts more weight on the centrifugal effects than on the Coriolis effects, notwithstanding Einstein's suggestion to the contrary. And even if one bears in mind that at the time Einstein's and Thirring's papers were published, the standards of quoting references and acknowledging contributions by colleagues, differed from today's practice, it appears very strange that in papers [10] and [12] neither is Einstein's paper [11] quoted, nor are the decisive stimuli provided by his letter [15] acknowledged. As already mentioned, the contributions made by Lense obviously consist (only) of the astronomical evaluations of Thirring's formulas. Therefore I argue that the so-called Lense–Thirring effect should, in order to be historically correct and fair, be called the Einstein–Thirring–Lense effect.

In the oral presentation at Erice I gave a detailed mathematical derivation of the so-called Lense–Thirring effect. However, since this derivation can be found in the textbook literature (see, e.g., Chap. 6 of [19]), I will here confine myself to the final formula for this effect: in Cartesian coordinates, the acceleration (relative to the asymptotic "fixed stars") of exterior, free test bodies, caused by the angular momentum $\vec{J}$ of a (spherical) body, is given by

$$\frac{d^2\vec{x}}{dt^2} = \frac{d\vec{x}}{dt} \times \vec{H}, \qquad \text{with} \qquad \vec{H} = \frac{2}{r^3}\left[\vec{J} - 3\frac{(\vec{J}\vec{r})\vec{r}}{r^2}\right].$$

Whereas this formula is valid for all $r > R$, Thirring's derivation was limited to the far field $r \gg R$, and therefore would be applicable neither to the Gravity Probe B experiment ($r/R \approx 1.10$) nor to the LAGEOS satellites ($r/R \approx 1.92$).

2 Thirring's Work on the Rotating Mass Shell, and the Problem of a Correct Centrifugal Acceleration

I now like to come to a problem of a more theoretical and fundamental character, which was initiated by Mach, Einstein and Thirring: the question of relativity of rotation, and in particular the dragging of inertial frames inside a rotating mass shell, representing a very idealized model for the fixed stars, galaxies, and quasars in our real universe. Thirring calculated this dragging effect again only in the weak field limit, but now up to the second order in ω. Thirring presumed a mass shell (mass M, radius R) of spherical shape and mass distribution, consisting of dust particles. After correction of an integration error, observed by M. Laue and W. Pauli, Thirring [20] got for the acceleration of test particles inside the mass shell

$$\begin{aligned} \ddot{x} &= -\frac{8M}{3R}\omega\dot{y} + \frac{4M}{15R}\omega^2 x \\ \ddot{y} &= \frac{8M}{3R}\omega\dot{x} + \frac{4M}{15R}\omega^2 y \\ \ddot{z} &= -\frac{8M}{15R}\omega^2 z, \end{aligned}$$

where his result is only derived for $r \ll R$, and would receive corrections of order $M\omega^2 r^2/R^2$ for $r \not\ll R$. The terms of order ω obviously represent a Coriolis-type acceleration $2\vec{\omega} \times \dot{\vec{r}}$, with "dragging factor" $d = 4M/3R$, which is very small for all laboratory – or earth-size mass shells. The terms of order ω^2 were discussed by Thirring and others as centrifugal accelerations but obviously these terms also contain a (wrong) axial z-component, which Thirring tried to explain as a result of the (special relativistic) mass increase of the rotating equatorial parts of the mass shell. Whereas this interpretation contains a grain of truth, Thirring's analysis of the rotating mass shell in order ω^2 suffers from many deficiencies and inconsistencies which were realized and corrected only very gradually.

The first essential deficiency of Thirring's model was found and corrected by K. Lanczos in 1923 [21]: it is clear already from Newtonian physics that dust particles can move on circular orbits of radius R around the mass center M only for the Keplerian frequency $\omega_K = \sqrt{MG/R^3}$, what is never realizable for all particles of the spherical shell. Indeed, Thirring's model violates the "conservation law" $T^{\mu\nu}{}_{;\nu} = 0$ so that the calculated gravity field does (in order ω^2) not really solve Einstein's field equations (in the shell). To make the model consistent, one has, according to Lanczos, to introduce appropriate stresses in the shell material, compensating for the centrifugal forces. This has (mathematically) the consequence that the spatial components T_{ik} have to vanish, whereby all factors in front of the ω^2-terms in the accelerations of test particles (again only calculated for $r \ll R$) are reduced to half their values. But the problem with the axial component of the "centrifugal acceleration" persists. More than 30 years later, Bass and Pirani [22] partly repeat Lanczos' arguments but present them in more detail, and generalize Thirring's model to a latitude-dependent mass–density

$\rho(\vartheta) = \rho_0(1 + N\omega^2 R^2 \sin^2\vartheta)$ of the shell, with constants ρ_0 and N. Bass and Pirani call the choice $N = -1$ "the most interesting case": "This corresponds to a mass distribution which just compensates for the special-relativistic increase in density, and represents a uniform mass distribution in the reference frame in which the shell is rotating. This choice leads to the annihilation of both the radial and the axial 'centrifugal' forces, leaving the 'Coriolis' force intact." In the introduction of [22] it is mentioned that in Thirring's model the self-interaction of the shell (being proportional to M^2) is neglected, but no attempt is made to extend the model beyond the weak field approximation of first order in M. Another argument, calling for a treatment of the rotating mass shell at least up to order M^2, was presented by C. Soergel-Fabricius [23]: As already discussed by Thirring [10], it is possible to eliminate the Coriolis acceleration inside the rotating mass shell by a transformation to a reference frame back-rotating with angular velocity $\tilde{\omega} = -\frac{4M}{3R}\omega$. However, the centrifugal acceleration can vanish in the same reference system, as it should according to Mach's demand for relativity of rotation, at best if it is of order $\tilde{\omega}^2 r \sim M^2\omega^2 r/R^2$, instead of the order $M\omega^2 r/R$ in [10].

A treatment of the rotating mass shell even exactly in M was then started in 1966 by Brill and Cohen [24], by considering a rotational perturbation not of Minkowski spacetime but of the Schwarzschild solution. However, they confined themselves to the first order in the angular velocity ω, and derived (now for the whole, flat interior of the shell) a Coriolis-type acceleration, with dragging factor

$$d = \frac{4\alpha(2-\alpha)}{(1+\alpha)(3-\alpha)},$$

with $\alpha = M/2R$, and where R denotes the shell radius in isotropic coordinates. In the weak field limit $M \ll R$, this dragging factor coincides of course with Thirring's result $d = 4M/3R$. But the central new result of Brill and Cohen is that in the collapse limit $R \to M/2$ the dragging factor attains the value $d = 1$. This signifies – within the model class of rotating mass shells, and up to first order in ω – a complete realization of the Machian postulate of relativity of rotation: in the collapse limit the interior of the shell ties off (as a type of separate universe) from the exterior space–time, and interior test bodies and inertial frames are dragged along with the full angular velocity ω of the shell.

An extension of the Brill and Cohen results to higher orders in ω, and in particular the long-standing problem of the gravitational induction of a correct centrifugal force by rotating masses, had to wait for another 19 years to be solved [25]. The solution is based on two "new" observations which could and should have been made already in Thirring's time, but which, for inexplicable reasons, were overlooked by all authors before 1985:

(a) Any physically realistic, rotating body will suffer a centrifugal deformation in orders ω^2 and higher, and cannot be expected to keep its spherical shape.
(b) If we aim and expect to realize in the interior of the rotating mass shell quasi-Newtonian conditions with "correct" Coriolis and centrifugal forces – and no other forces! – the interior of the mass shell obviously has to be a flat piece of

space–time. In first order of ω, this flatness is more or less trivial because the only non-Minkowskian metric component $g_{t\varphi}$ is constant there, i.e., we have a constantly rotating Minkowski metric, and therefore a structurally correct Coriolis force. In contrast, in order ω^2 this flatness is by no means trivial, and it is indeed violated for Thirring's solution, due to the axial component of his "centrifugal force".

With these observations, the problem of a correct centrifugal force inside a rotating mass shell boils down to the question whether it is possible to connect a flat "rotating" interior metric through a mass shell (with, to begin with, unknown geometrical and material properties) to the non-flat but asymptotically flat exterior metric of a rotating body. In full generality, this would represent a mathematically quite intricate free-boundary-value problem for the stationary and axisymmetric Einstein equations. However, if we confine ourselves to a perturbation expansion in the angular velocity ω, all metric functions can be expanded in spherical harmonics, i.e., due to the axial symmetry, just in Legendre polynomials $P_l(\cos\vartheta)$, where in order ω^n the index l is limited by $l \leq n$. In this way, the Einstein equations reduce to a system of ordinary differential equations for the functions $f_l^{(i)}(r)$ multiplying $P_l(\cos\vartheta)(i = 1, \ldots, 4$, for the four different metric coefficients describing the stationary, axisymmetric space–time in the exterior of the mass shell).

In order ω^2 the shell geometry is then given by $r_S = R(1 + \omega^2 c_2 P_2(\cos\vartheta))$, with a constant c_2, and with corresponding corrections in higher (even) orders ω^{2n}. Furthermore, it turns out [26] that in order ω^3 the flatness of the interior space–time can only be maintained if the shell material rotates differentially: $\omega_S = \omega(1 + \omega^2 e_2 P_2(\cos\vartheta))$, with a constant e_2, and with corresponding corrections in higher (odd) orders ω^{2n+1}. Surprisingly, the flatness condition enforces a prolate form of the shell: invariant equatorial circumference smaller than the invariant polar circumference. The conditions that the exterior metric (written, e.g., in the isotropic coordinate r) is asymptotically flat, and joins, at $r = r_S$, continuously to the interior flat metric, lead (for given M and R) to a unique determination of the constants c_{2n} and e_{2n}, and of the functions $f_l^{(i)}(r)$. The energy–momentum tensor $T_{\mu\nu}$ of the shell material results then uniquely from the discontinuities of the radial derivatives of $f_l^{(i)}(r)$ at $r = r_S$, and $T_{\mu\nu}$ is indeed ϑ-dependent in orders ω^2 and higher (compare [22]). Let me close this section with a quotation from Brian Greene's recent popular book "The Fabric of the Cosmos" [27], which, surprisingly, devotes many pages to Machian questions. On p. 417 one can read: "Calculations initiated in 1912 by Einstein (even before he completed general relativity), which were significantly extended in 1965 by Dieter Brill and Jeffrey Cohen, and finally completed in 1985 by the German physicists Herbert Pfister and K. Braun, showed that space inside the hollow sphere would be dragged by the rotational motion and set into a whirlpool-like spin." And I may add: these calculations correctly produce (in the whole interior $r < R$ of the shell) all known inertial forces (Coriolis, centrifugal, and no others), and herewith completely realize Mach's idea of relativity of rotation.

3 Generalizations to Cosmological Models, and to Linearly Accelerated Mass Shells

Here I like to present two comments concerning generalizations of these "Machian" results for rotating mass shells. The first concerns their cosmological relevance: it is clear that Mach envisaged a realization of his ideas about relativity of rotation, if any, only in a cosmological context. Therefore, the work of Einstein, Thirring, and others, which confirmed some aspects of this 'relativity of rotation' in the model class of isolated rotating mass shells, was often criticized for the asymptotic flatness of the exterior solution, instead of using cosmological boundary conditions. It turns out, however, that the essential dragging results for rotating bodies and mass shells in an asymptotically flat background carry over with only minor changes to cosmological boundary conditions. As shown by my former PhD-student C. Klein [28], it is possible to embed a slowly rotating mass shell with flat interior also in a (rotationally perturbed) Friedmann universe, and the resulting dragging factor compares reasonably well with the results of Thirring [10] and Brill and Cohen [24], but depends of course also on the type of the Friedmann cosmos ($k = 0, \pm 1$), and on its mass density. More recently, analyses of rotational perturbations of pure FRW-cosmologies (without the somewhat unrealistic mass shell), and of their Machian impact, have appeared, by Bičák, Lynden–Bell, and Katz [29], and by C. Schmid [30].

The second and final comment concerns dragging phenomena due to linearly accelerated masses or mass shells, a question initiated already in 1912 by Einstein [7]. Only very few papers (in the time 1980–2000) have taken up this problem, and, in my judgement, they compare only quite poorly to the rotating systems considered by Thirring, Brill–Cohen, and others. A quite general and severe problem with linearly accelerated bodies is that they need – in contrast to rotating bodies – a perpetual supply of energy, in order to maintain the acceleration. (The constructors of linear accelerators for elementary particles painfully know this!) And since in general relativity the equations of motion of bodies are already contained in the field equations, the energy source (or the motor of the accelerated system) has to be included in the considered system in order to deal with a self-consistent problem. Very recently we have (for the first time) constructed such a self-consistent model [31]: a charged, spherical mass shell (mass M, charge q, radius R) is linearly accelerated by a weak, axisymmetric, dipolar, and asymptotically vanishing charge distribution $\lambda\sigma(r)\cos\vartheta$ ($\lambda \ll 1$) in the exterior of the shell. We analyzed this system only up to order t^2 around the moment of time symmetry $t = 0$. It can be proven that the interior of the (Reissner–Nordström-like) shell stays flat. In analogy to the rotating mass shell, we choose (up to first order of λ) the shell as spherical and rigidly accelerating, we set the correction terms to energy–density and "pressure" to zero, and we choose the shell material as electrically isolating, and the system free of magnetic fields (up to order t^2). Hereby the (asymptotically flat) solutions of the Einstein–Maxwell equations are uniquely determined, and therefore also the linear acceleration g of neutral test particles inside the shell, the linear acceleration b of the shell as a whole, and the dragging factor $d = g/b$. For the simplest power law charge distribution $\sigma(r) \sim r^{-5}$, having a finite dipole moment, we get in the weak field limit $M/R \ll 1$, and $q/R \ll 1$ (by chance) exactly the Thirring result $d = 4M/3R$, and for arbitrary M/R a result very similar to the Brill–Cohen result. In particular, in the collapse limit we get $d = 1$, i.e., total dragging, for arbitrary charge distributions $\sigma(r)$.

Altogether, it is now clear that for (first-order) linear and rotational accelerations of a mass shell, the interior of this shell can be kept flat, and the dragging effects in this shell exactly mimic the corresponding well-known 'inertial forces' in accelerated reference systems in Newtonian physics. Since general accelerations can (in principle) be constructed from appropriate linear and circular accelerations, we have now in hand very good arguments for the validity of a 'quasiglobal equivalence principle' in general relativity [25]: "If some finite laboratory (a flat region of space–time) is in arbitrary (weak) accelerated motion relative to the fixed stars, then all motions of free particles and all physical laws, measured from laboratory axes, are modified by inertial forces. It is argued that exactly the same modified motions and laws can be induced (at least for some time) at all places of a laboratory at rest relative to the fixed stars, by suitable and suitably accelerated masses outside the laboratory (e.g., in a mass shell)." In this connection it may be remarked that already in the dawn of general relativity, in the years 1912–1913, similar ideas arose in discussions of Einstein with Ehrenfest [32] and Mie [11]. But at that time the participants were quite sceptical about such a "macro-equivalence".

References

1. Mach, E. (1872). *Die Geschichte und die Wurzel des Satzes von der Erhaltung der Arbeit.* (Calve, Prag).
2. Mach, E. (1883). *Die Mechanik in ihrer Entwicklung. Historisch-kritisch dargestellt.* (Brockhaus, Leipzig).
3. Friedlaender, B., and I. (1896). *Absolute oder relative Bewegung?* (Simion, Berlin).
4. Föppl, A. (1904). *Sitzb. Bayer. Akad. Wiss., 34,* 5.
5. Einstein, A. (1907). *Jahrb. Radioakt. Elektronik, 4,* 411 (Chap. V, pp. 454–462).
6. Einstein, A. (1912). *Ann. Phys., 38,* 355, and 443.
7. Einstein, A. (1912). *Vierteljahrschr. gerichtl. Med. öffentl. Sanitätswesen, 44,* 37.
8. Einstein, A., and Grossmann, M. (1913). *Entwurf einer verallgemeinerten Relativitätstheorie und einer Theorie der Gravitation.* (Teubner, Leipzig).
9. Klein, M.J., et al. (eds.) (1995). *The Collected Papers of Albert Einstein, Vol. 4, pp. 344–473.* (Princeton University Press).
10. Thirring, H. (1918). *Phys. Zs., 19,* 33.
11. Einstein, A. (1913). *Phys. Zs., 14,* 1249.
12. Lense, J., and Thirring, H. (1918). *Phys. Zs., 19,* 156.
13. Thirring, H. (1917). *Wirkung rotierender Massen.* (Österr. Zentralbibl. Phys., Wien).
14. Schulmann, R., et al. (eds.) (1998). *The Collected Papers of Albert Einstein, Vol. 8, Doc. 361.* (Princeton University Press).
15. Schulmann, R., et al. (eds.) (1998). *The Collected Papers of Albert Einstein, Vol. 8, Doc. 369.* (Princeton University Press).
16. Janssen, M. (2007). *The Genesis of General Relativity (ed. by J.Renn), Vol. 2, pp. 785–837.* (Springer, Berlin).

17. Thirring, H. (1966). *Almanach Österr. Akad. Wiss., 166,* 361.
18. Pfister, H. (2007). *Gen. Rel. Grav., 39,* 1735.
19. Ciufolini, I., and Wheeler, J.A. (1995). *Gravitation and Inertia.* (Princeton University Press).
20. Thirring, H. (1921). *Phys. Zs., 22,* 29.
21. Lanczos, K. (1923). *Zs. Phys., 14,* 204.
22. Bass, L., and Pirani, F.A.E. (1955). *Phil. Mag., 46,* 850.
23. Soergel-Fabricius, C. (1960). *Zs. Phys., 159,* 541, and *161,* 392.
24. Brill, D., and Cohen, J.M. (1966). *Phys. Rev., 143,* 1011.
25. Pfister, H., and Braun, K.H. (1985). *Class. Quant. Grav., 2,* 909.
26. Pfister, H., and Braun, K.H. (1986). *Class. Quant. Grav., 3,* 335.
27. Greene, B. (2004). *The Fabric of the Cosmos.* (Knopf, New York).
28. Klein, C. (1993). *Class. Quant. Grav., 10,* 1619.
29. Bičák, J., Lynden-Bell, D., and Katz, J. (2004). *Phys. Rev. D 69,* 064011.
30. Schmid, C. (2006). *Phys. Rev. D 74,* 044031.
31. Pfister, H., Frauendiener, J., and Hengge, S. (2005). *Class. Quant. Grav., 22,* 4743.
32. Klein, M.J. et al. (eds.) (1993). *The Collected Papers of Albert Einstein, Vol. 5, Docs. 409 and 411.* (Princeton University Press).

Part V
Miscellaneous

Atom Interferometers and Optical Clocks: New Quantum Sensors Based on Ultracold Atoms for Gravitational Tests in Earth Laboratories and in Space

Guglielmo M. Tino

Abstract Cooling and trapping of neutral atoms has been one of the most active fields of research in physics in recent years. Several methods were demonstrated to reach temperatures as low as a few nanokelvin allowing, for example, the investigation of quantum degenerate gases. The ability to control the quantum degrees of freedom of atoms opens the way to applications for precision measurement of fundamental physical quantities. Here, experiments we are performing using atom interferometry to determine the gravitational constant G and test the Newtonian gravitational law at micrometric distances will be presented. Other experiments in progress, planned or being considered using atom interferometers and new optical atomic clocks in ground laboratories and in space are also discussed.

1 Introduction

The field of laser cooling and manipulation of atoms has been one of the most active in physics in recent years. Atoms from a room-temperature vapour can be cooled to temperatures as low as a few nanokelvin. At such low temperatures, the wave properties of the atoms become relevant and this gives rise to completely new phenomena such as Bose–Einstein condensation and allows to perform experiment where the matter waves interfere just as usual waves do. In 1997, the Nobel Prize in physics was awarded to C. Cohen-Tannoudji, S. Chu and W.D. Phillips for their contribution in this field [1]. In 2001, the Nobel Prize was given to E.A. Cornell, W. Ketterle and C.E. Wieman for the observation of Bose–Einstein condensation in ultracold dilute atomic gases [2]. In 2005, J.L. Hall and T.W. Hänsch received the Nobel Prize for the invention of a method that will lead to new clocks based on optical transitions of ultracold atoms [3].

G.M. Tino (✉)
Dipartimento di Fisica and LENS – Università di Firenze, Istituto Nazionale di Fisica Nucleare, Sezione di Firenze, via Sansone 1, I-50019 Sesto Fiorentino, Florence, Italy
e-mail: guglielmo.tino@fi.infn.it

I. Ciufolini and R.A. Matzner (eds.), *General Relativity and John Archibald Wheeler*, Astrophysics and Space Science Library 367, DOI 10.1007/978-90-481-3735-0_21,

Atomic quantum sensors based on cold atom physics have already demonstrated their potential for precision measurements. Today, atomic clocks approach a precision of few parts in 10^{17} in the measurement of time and frequency; on ground, atom interferometers promise sensitivities of $10^{-10}\mathrm{g}/\sqrt{\mathrm{Hz}}$ for acceleration measurements and of $10^{-9}\mathrm{rad}/s\sqrt{\mathrm{Hz}}$ for the detection of tiny rotations. The field is now mature both from the point of view of the understanding of the basic physics underlying laser cooling and laser manipulation of atoms and for the development of a solid technology for the experimental implementation of new quantum devices.

Here I describe some gravitational physics experiments in progress or under development based on atom interferometers and new atomic clocks.

2 Precision Gravity Measurements by Atom Interferometry: Measurement of G and Test of Newtonian Law at Micrometric Distances

Recent advances in atom interferometry led to the demonstration of different schemes for fundamental physics experiments and for applications: atom interferometry was used for precision measurements of gravity acceleration [4], Earth's gravity gradient [5, 6], rotations [7, 8] and $\hbar$/m [9, 10]. An overview of basic principles and seminal theoretical and experimental work can be found in Ref. [11]. Atom interferometers are promising sensors for the investigation of the gravitational interaction such as equivalence principle tests [12, 13], $1/r^2$ law test [14, 15], gravitational waves detection [16–19] and for possible applications in geophysics [4, 6, 20]. Quantum devices based on ultracold atoms show extraordinary features in terms of sensitivity and spatial resolution, which are important for studies of surfaces, Casimir effects [21], and searches for deviations from Newtonian gravity predicted by theories beyond the standard model [22, 23].

In Section 2.1, I describe the operation of an atom interferometer conceived for measuring the gravitational constant G. In Section 2.2, I discuss how using laser-cooled strontium atoms in optical lattices, persistent Bloch oscillations are observed for about 10 s, and gravity is determined with ppm sensitivity on micrometer scale. This method can improve the sensitivity in the search of deviations from Newtonian gravity in the micrometer distance range.

2.1 Measurement of G

After Cavendish first measurement, more than 300 experiments have been performed to measure G, but the results are not in agreement. In 2002 the recommended CODATA [24] value ($G = 6.6742(10) \times 10^{-11}$ $\mathrm{m^3kg^{-1}s^{-2}}$) uncertainty was reduced by one order of magnitude down to 150 ppm compared to the previous one (CODATA 1998) but this is still much higher than the uncertainty of any other physical constant. Problems in measuring G with high accuracy arise from the weakness of the gravitational force, from the impossibility of shielding it and from the difficulty of realizing well-defined masses and positioning them at well-known distances.

We have applied Raman interferometry techniques with Rb atoms to determine the Newtonian gravitational constant G [25, 26]. We implemented a new measurement scheme aiming to get rid of, or at least to better identify, such systematic effects. In our experiment freely falling microscopic bodies (atoms) are used as probes of the gravitational field induced by heavy and well-characterized source masses. The vertical acceleration is simultaneously measured in two vertically separated positions with two atomic samples that are launched in rapid sequence with a juggling method. From the differential acceleration measurements, and from the knowledge of the added mass distribution, we determine the value of G. The result of another conceptually similar experiment was recently reported in Ref. [27].

The experimental apparatus, described in detail in Ref. [26], consists of a Raman interferometer used as a gravity-gradiometer and two sets of heavy source masses (SM). Rb atoms are laser-cooled and trapped and launched upwards into a 1 m long, magnetically shielded tube where the interferometer sequence takes place. While falling down, they are detected at their passage through the central vacuum chamber. The two sets of SM are symmetrically arranged around the tube and can be vertically moved with high precision.

The gradiometer requires two clouds of cold atoms moving with the same velocity at the same time, but vertically displaced. A vertical separation of 35 cm for atoms launched 60 and 95 cm above the MOT results in a launch delay between the two clouds of about 100 ms. The two atomic clouds are prepared using the juggling technique [28]. During the ballistic flight of the first cloud of atoms, a second cloud is loaded. Just before the first cloud falls down in the MOT region, the second one is launched. Then the first cloud, used as a cold and intense source of atoms, is recaptured, cooled and launched upwards within less than 50 ms. In our experimental sequence, the first cloud is launched 60 cm upwards, which leads to a loading time of 650 ms for the second cloud. In this way, the number of atoms launched in each of the two clouds used in the gradiometer is 5×10^8. After the launch, the atoms are selected both in velocity and by their m_F state. The selection procedure uses vertical beams so that the state preparation can take place simultaneously on both clouds. After the selection sequence, the atoms end up in the $F = 1$, $m_F = 0$ state with a horizontal temperature of 4 μK and a vertical temperature of 40 nK, corresponding to velocity distribution widths (HWHM) respectively of 3.3 v_{rec} an 0.3 v_{rec}.

A sequence of three vertical velocity-selective Raman pulses is used to realize the interferometer. The first ($\pi/2$ pulse) splits the atomic wave packet, the second (π pulse) induces the internal and external state inversion and the third ($\pi/2$ pulse) recombines the matter waves after their different space–time evolution. Stimulated Raman transitions are driven by two extended cavity phase-locked diode lasers, with a relative frequency difference equal to the ^{87}Rb ground state hyperfine splitting frequency ($\nu_{hf\,^{87}Rb} = 6.835$ GHz) and amplified by a single tapered amplifier. A detailed description of the laser locking system can be found in Ref. [29].

To compensate for the Doppler shift of the atomic resonance during the atomic free fall trajectory, the Raman beams frequency difference is linearly swept. The interferometric sequence is defined in such a way that the π pulse is sent 5 ms before the atoms reach the top of their trajectory, when their velocity is still high enough to discriminate between upwards and downwards propagating Raman beams. For a Raman beam intensity of 30 mW/cm^2, the π pulse lasts 100 μs.

The interferometric phase shifts are detected using the relative phase of the Raman beams as a reference. To scan the interferometric fringes, a controlled phase jump ϕ_L is applied after the π pulse to the rf signal generated by the low phase noise reference oscillator. The population of the two hyperfine sublevels of the ground state after the interferometric sequence is measured using normalized fluorescence detection. With a typical number of 5×10^4 detected atoms per cloud per state, the SNR is 60/1. The main interferometer phase term is the one induced by Earth's gravity

$$\phi(g) = k_{eff}\, g\, T^2, \tag{1}$$

with $\hbar k_{eff}$ being the momentum transferred to the atoms during each Raman pulse. A gravity gradient determination consists of two vertically separated acceleration measurements within the interferometer region. If g_{DW} and g_{UP} are the gravity acceleration values at the height of the lower and upper interferometers the following relative phase shift is observed

$$\phi(\Delta g) = k_{eff}\, (g_{DW} - g_{UP})\, T^2. \tag{2}$$

A simultaneous realization of these measurements overcomes the stringent limit set by the phase noise through common mode rejection. The Raman sequence interval T, as well as the gradiometer sensitivity, can then be increased up to the limit set by the size of the apparatus.

For the determination of G, in the double differential scheme, the measurements are repeated twice in the same point. Two sets of masses are used to generate a well-known gravitational field. Each set is made of 12 identical cylinders, symmetrically arranged in a hexagonal configuration around the vertical axis of the atomic fountain. The cylinders have a diameter of 100 mm and a height of 150 mm. The two sets of masses are placed on two large titanium rings, which in turn are held by a mount specifically designed for the experiment. A vertical translation mechanism allows to independently move the two sets of SM with a fine control of the position on the order of 5 μm. The SM can be placed at a relative distance ranging between 4 and 50 cm. SM have been positioned close to the atomic trajectories in the gradiometer configuration. The change of the local acceleration due to the added gravitational potential can be measured, thus allowing to determine the gravitational constant G, once the SM density distribution and their positions are well-known.

In a first step the turning point of the upper cloud is located above the two sets of SM, and the acceleration induced on the atoms is in the $-z$ direction. The opposite happens for the lower cloud. The differential phase term is then determined for a different position of the SM; moving them to the external positions with respect to the atoms clouds, the sign of the induced acceleration is inverted. By evaluating the difference of such consecutive measurements a reduction of systematic effects [30] is achieved, due for instance to spatially inhomogeneous spurious accelerations, which are constant on the time scale of SM repositioning. Among these effects, the Earth's gravity gradient g' is the most important. Other minor contributions are due to inhomogeneous electric and magnetic fields as well as to inertial forces.

Considering the differences between two consecutive measurements, the resulting phase shift from the whole data set is 144(5) mrad, which corresponds to a sensitivity of $3 \times 10^{-9} g$ and a relative uncertainty of 4×10^{-2} in the measurement of G.

The cylinders for the final G measurement are made of a non-magnetic tungsten alloy. The proposed accuracy of $\Delta G/G = 10^{-4}$ for the final measurement of G can be reached only optimizing all the parameters so far considered [26, 31]. Atomic motion's initial parameters also will be critical for the final accuracy. The sensitivity to initial atomic position and velocity can be dramatically reduced by choosing the optimum combination of the two SM configuration and atoms' trajectories. The interferometers will be realized in regions where the acceleration is stationary in both SM configurations. By optimizing the atoms–masses relative position in this way $\Delta G/G = 10^{-4}$ can be reached with an initial position uncertainty of 1 mm and an initial velocity uncertainty of 5 mm/s.

2.2 *Testing the Newtonian Gravity Law*

The confinement of ultracold atoms in optical lattices, regular structures created by interfering laser beams where the atoms are trapped by the dipole force, provides clean model systems to study quantum physics problems [32]. For example, Bloch oscillations, predicted for electrons in a periodic crystal potential in presence of a static electric field [33] but not observed in natural crystals, were directly observed using atoms in an optical lattice [34].

In our experiment, laser-cooled ^{88}Sr atoms are trapped in a 1-dimensional vertical optical lattice. The insensitivity to stray fields and collisions makes Sr in optical lattices, a candidate also for future clocks [35], a unique sensor for small-scale forces. The combination of the periodic optical potential and the linear gravitational potential gives rise to Bloch oscillations at frequency ν_B given by

$$\nu_B = \frac{mg\lambda_L}{2h} \tag{3}$$

where m is the atomic mass, g is the acceleration of gravity, λ_L is the wavelength of the light producing the lattice, and h is Plancks constant. Since both λ_L and m are well known, the overall force along the lattice axis can be determined by measuring the Bloch frequency ν_B. In order to do a force measurement with given interrogation time, the atomic wavefunction has to undergo a coherent evolution on the same time timescale. The most common effects limiting the coherence time for ultracold atoms are perturbations due to electromagnetic fields and atom–atom interactions. ^{88}Sr is in this respect a good choice because in the ground state it has zero orbital, spin and nuclear angular momentum that makes it insensitive to stray electric and magnetic fields that otherwise need to be shielded. In addition, ^{88}Sr has remarkably small atom–atom interactions [36]; this prevented so far the achievement of Bose–Einstein condensation for this atom [36, 37] but becomes an important feature in experiments where collisions lead to a loss of coherence limiting the measurement time and the potential sensitivity.

Given the small extension of ultracold Sr atoms confined in optical lattice potential, and its insensitivity to stray fields and elastic collisions, Sr in optical lattices results to be a unique sensor for small-scale forces with better performances and reduced complexity compared to proposed schemes using degenerate Bose [38] or Fermi [39] gases. This improves the feasibility of new experiments on gravity in unexplored regions. The method used to produce ultracold Sr atoms was already described in Ref. [40]. The experiment starts with trapping and cooling $\sim 5 \times 10^7$ ^{88}Sr atoms at 3 mK in a magneto-optical trap (MOT) operating on the $^1S_0$-$^1P_1$ blue resonance line at 461 nm. The temperature is then further reduced by a second cooling stage in a red MOT operating on the $^1S_0$-$^3P_1$ narrow transition at 689 nm and finally we obtain $\sim 5 \times 10^5$ atoms at 400 nK. After this preparation phase, that takes about 500 ms, the red MOT is switched off and a one-dimensional optical lattice is switched on adiabatically in 50 μs. The lattice potential is originated by a single-mode frequency-doubled Nd:YVO4 laser ($\lambda_L = 532$ nm) delivering up to 350 mW on the atoms with a beam waist of 200 μm. The beam is vertically aligned and retro-reflected by a mirror producing a standing wave with a period $\lambda_L/2 = 266$ nm. The corresponding photon recoil energy is $E_R = h^2/2m\lambda_L^2 = k_B \times 381$ nK. As expected from band theory [41], the amplitude of the oscillation in momentum space decreases as the lattice depth is increased. This suggests that in order to measure the Bloch frequency with maximum contrast the intensity of the lattice laser should be reduced. On the other hand, reducing the intensity results in a loss in the number of trapped atoms because of the smaller radial confinement. For this reason, we used a lattice depth of 10 E_R. For a lattice potential depth corresponding to 10 E_R, the trap frequencies are 50 kHz and 30 Hz in the longitudinal and radial direction, respectively. Before being transferred in the optical lattice, the atom cloud in the red MOT has a disk shape with a vertical size of 12 μm rms. In the transfer, the vertical extension is preserved and we populate about 100 lattice sites with 2×10^5 atoms with an average spatial density of $\sim 10^{11}$ cm^{-3}. After letting the atoms evolve in the optical lattice, the lattice is switched off adiabatically and we measure the momentum distribution of the sample by a time-of-flight measurement, after a free fall of 12 ms, using a resonant probe laser beam and absorption imaging on a CCD camera. The atoms confined in the optical lattice can be seen performing Bloch oscillations due to the combined effect of the periodic and gravitational potential. The average force arising from the photon recoils transferred to the atoms compensates gravity. The images obtained by absorption imaging are integrated along the horizontal direction and fitted with the sum of two Gaussian functions. From each image, two quantities are extracted : the first is the vertical momentum distribution of the lower peak. The second is the width of the atomic momentum distribution (i.e., the second momentum of the distribution). We find that the latter is less sensitive against noise-induced perturbations to the vertical momentum. We observed ~4,000 Bloch oscillations in a time $t = 7$s. During this time, about 8,000 photon momenta are coherently transferred to the atoms. Oscillations continue for several seconds and the measured damping time of the amplitude is $\tau \sim 12$ s. To our knowledge, the present results for number of Bloch oscillations, duration, and the corresponding number of coherently transferred photon momenta, are by far the highest ever achieved experimentally in any physical system.

From the measured Bloch frequency $\nu_B = 574.568(3)$ Hz we determine the gravity acceleration along the optical lattice $g = 9.80012(5)$ ms^{-2}. The overall estimated sensitivity is 5×10^{-6} g and, neglecting the 500 ms preparation of the atomic sample, we have a sensitivity of 4×10^{-5} g at 1 s. We expect that a sensitivity of 10^{-7} g can be achieved using a larger number of atoms, and reducing the initial temperature of the sample. Apart from collisional relaxation, which should contribute to decoherence on a minute timescale, the main perturbation to quantum evolution is represented by vibrations of the retro-reflecting mirror [42]. Minor contributions to decoherence may come from the axial momentum dispersion of the lattice at 10^{-6} due to its radial extension.

The micrometric spatial extension of the atomic cloud in the vertical direction, and the possibility to load it into the optical potential at micrometric distance from a surface, makes the scheme we demonstrated particularly suitable for the investigation of forces at small spatial scales. The possibility of investigating the gravitational force at small distances by atomic sensors was proposed in Ref. [14], discussed in detail in Ref. [43], and preliminarily demonstrated in Ref. [44]. Deviations from the Newtonian law are usually described assuming a Yukawa-type potential

$$V(r) = -G\frac{m_1 m_2}{r}(1 + \alpha e^{-r/\lambda}) \tag{4}$$

where G is Newton gravitational constant, m_1 and m_2 are the masses, r is the distance between them. The parameter α gives the relative strength of departures from Newtonian gravity and λ is its spatial range. Experiments searching for possible deviations have set bounds for the parameters α and λ. Recent results using microcantilever detectors lead to extrapolated limits $\alpha \sim 10^4$ for $\lambda \sim 10$ μm and for distances ~1 μm it was not possible to perform direct experiments so far [22, 23].

The small size and high sensitivity of the atomic probe allows a direct, model-independent measurement at distances of a few μm from the source mass with no need for modeling and extrapolation as in the case of macroscopic probes. This allows to directly access unexplored regions in the $\alpha - \lambda$ plane. Also, in this case quantum objects are used to investigate gravitational interaction.

Our results indicate that our Sr atoms when brought close to a thin layer can be used as probe for the gravitational field generated by the massive layer [45]. If we consider, in fact, a material of density ρ and thickness d, the added acceleration of gravity in proximity of the source mass is $a = 2\pi G\rho d$ so that when $d \sim 10$ μm and

$\rho \simeq 10$ g/cm^3 as for tungsten crystals the resulting acceleration is $a \sim 4 \times 10^{-11}$ ms^{-2}. Measuring ν_B at a distance $\sim$4 μm away from the surface would allow to improve the constraint on α by two orders of magnitude at the corresponding range $\lambda \sim 4$ μm. Spurious non-gravitational effects (Van der Waals, Casimir forces), also present in other experiments, can be reduced by using an electrically conductive screen and performing differential measurements with different source masses placed behind it. Moreover, by repeating the same experiment with the four stable isotopes (three bosons, one fermion, with atomic mass ranging from 84 to 88), we can further discern among gravitational and other forces.

3 Optical Atomic Clocks

Although present microwave atomic clocks reached remarkable performances, a new type of clocks based on optical atomic transitions promises dramatic improvements. In an optical atomic clock, a laser in the visible region of the electromagnetic spectrum is used to induce a forbidden atomic transition. The resonance signal is detected by measuring the population in the two clock levels and used to keep the laser frequency tuned on the atomic transition By using optical frequencies ($\nu_0 \sim 10^{15}$ Hz) rather than microwave frequencies ($\nu_0 \sim 10^{10}$ Hz), an optical clock operates with a much smaller unit of time. This leads to an enormous improvement in stability and also to a higher accuracy since several key frequency shifts are fractionally much smaller in the optical domain [3].

The measurement of optical frequencies has recently been made practical by the development of the femtosecond-laser frequency-combs [46, 47]. Combined with narrow linewidth lasers, this has made possible the first generation of optical atomic frequency standards and clocks, based on cold trapped neutral atoms, such as Sr or Yb, and ions, such as Sr^+, Yb^+, Al^+, Hg^+. Optical clocks have already demonstrated stabilities below 1 part in 10^{14} at 1 s integration time and 5 parts in 10^{17} at 20,000 s and have the potential to reach a systematic fractional frequency uncertainty approaching 10^{-18} level.

In space, where the weightlessness and the extremely quiet environment ensure the ideal conditions for detecting very narrow signals, these performances can be improved even further. Clocks in space represent unique tools to test fundamental laws of physics at an unprecedented level of accuracy and to develop applications in time and frequency metrology, universal time scales, global positioning and navigation, geodesy, and gravimetry. Measurements of the constancy and isotropy of the speed of light can be performed by continuously comparing a space clock to a ground clock. Tests based on this technique have already been performed in 1997 by comparing clocks on-board GPS satellites to a hydrogen maser [48]. Optical clocks orbiting around the Earth, combined to a time and frequency transfer link not degrading the clock performances can improve present results by about four orders of magnitude. Optical clocks could measure the Einstein's gravitational red-shift with a relative uncertainty of few parts in 10^8, demonstrating a new efficient way of mapping the Earth gravity field and define the shape of the geoid at the centimeter level. At the same time, the comparison of two optical clocks based on appropriately chosen transitions during free flight in a varying gravitational potential can test the universality of the gravitational red-shift at the same accuracy level. As direct consequence of Einstein's equivalence principle, general relativity and other metric theories of gravitation forbid any time variation of non-gravitational constants. Today, optical clocks offer the possibility to test time variations of fundamental constants at a high accuracy level. Such measurements complement the tests of the local Lorentz invariance and of the universality of the free fall to experimentally establish the validity of Einstein's equivalence principle.

Third generation navigation systems will benefit from the technology development related to optical clocks. New concepts for global positioning systems based on a reduced set of ultra-stable space clocks in orbit associated with simple transponding satellites could be studied and, as already discussed, a new kind of geodesy based on the precise measurement of the gravitational red-shift can be envisaged. At present, the most advanced project on microwave atomic clocks based on cold atoms for space is ACES (*Atomic Clock Ensemble in Space*). New ESA studies will demonstrate the high potential of the emerging technology of optical clocks both for fundamental physics studies and applications. Optical clock laboratory demonstrators will be realized, characterized, and compared preparing the background for the development of optical clocks for space applications.

References

1. S. Chu, C. Cohen-Tannoudji, and W.D. Phillips, *Nobel Lectures in Physics 1997*, Rev. Mod. Phys. **70**, 685 (1998).
2. E.A. Cornell and C.E. Wieman, *Nobel Lectures in Physics 2001*, Rev. Mod. Phys. **74**, 875, 1131 (2002); W. Ketterle, *ibid.*, 1131 (2002).
3. J. Hall and T. Haensch, *Nobel Lectures in Physics 2005*, Rev. Mod. Phys. **78**, 1279 (2007).
4. A. Peters, K.Y. Chung, and S. Chu, Metrologia **38**, 25 (2001).
5. M.J. Snadden, J.M. McGuirk, P. Bouyer, K.G. Haritos, and M.A. Kasevich, Phys. Rev. Lett. **81**, 971 (1998).
6. J.M. McGuirk, G.T. Foster, J.B. Fixler, M.J. Snadden, and M.A. Kasevich, Phys. Rev. A **65**, 033608 (2002).
7. T.L. Gustavson, P. Bouyer, and M. Kasevich, Phys. Rev. Lett. **78**, 2046 (1997).
8. T.L. Gustavson, A. Landragin, and M.A. Kasevich, Class. Quantum Grav. **17**, 2385 (2000).
9. D.S. Weiss, B.C. Young, and S. Chu, Phys. Rev. Lett. **70**, 2706 (1993).
10. R. Battesti, P. Cladé, S. Guellati-Khélifa, C. Schwob, B. Grémaud, F. Nez, L. Julien, and F. Biraben, Phys. Rev. Lett. **92**, 253001 (2004).
11. P.R. Berman, ed., *Atom interferometry* (Academic press, Chestnut Hill, 1997).
12. S. Fray, C.A. Diez, T.W. Haensch, and M. Weitz, Phys. Rev. Lett. **93**, 240404 (2004).
13. S. Dimopoulos, P. Graham, J. Hogan, and M. Kasevich, PRL **98**, 111102 (2007).
14. G.M. Tino, in *2001: A Relativistic Spacetime Odyssey - Proceedings of JH Workshop, Firenze, 2001* (I. Ciufolini, D. Dominici, L. Lusanna eds., World Scientific, 2003). Also, Tino G. M., Nucl. Phys. B **113**, 289 (2003).
15. G. Ferrari, N. Poli, F. Sorrentino, and G.M. Tino, Phys. Rev. Lett. **97**, 060402 (2006).
16. Chiao, Y. Raymond, Speliotopoulos, and D. Achilles, Journal of Modern Optics **51(6-7)**, 861 (2004).

17. A. Roura, D. Brill, B. Hu, C. Misner, and W. Phillips, Phys. Rev. D **73**, 084018 (2006).
18. P. Delva, M.-C. Angonin, and P. Tourrenc, Phys. Lett. A **357**, 249 (2006).
19. G. Tino and F. Vetrano, Class. Quantum Grav. **24**, 2167 (2007).
20. N. Sneeuw, R. Rummel, and J. Müller, Class. Quantum Grav. **13**, A113 (1996).
21. M. Antezza, L.P. Pitaevskii, and S. Stringari, Phys. Rev. Lett. **95**, 113202 (2005).
22. J.C. Long, H.W. Chan, A.B. Churnside, E.A. Gulbis, M.C.M. Varney, and J.C. Price, Nature **421**, 922 (2005).
23. S.J. Smullin, A.A. Geraci, D.M. Weld, J. Chiaverini, S. Holmes, and A. Kapitulnik, Phys. Rev D **72**, 122001 (2005).
24. P.J. Mohr and B.N. Taylor, Rev. Mod. Phys. **77-1**, 42 (2005).
25. J. Stuhler, M. Fattori, T. Petelski, and G.M. Tino, J. Opt. B: Quantum Semiclass. Opt. **5**, S75 (2003).
26. A. Bertoldi, G. Lamporesi, L. Cacciapuoti, M.D. Angelis, M. Fattori, T. Petelski, A. Peters, M. Prevedelli, J. Stuhler, and G.M. Tino, Eur. Phys. J. D **40**, 271 (2006).
27. J.B. Fixler, G.T. Foster, J.M. McGuirk, and M. Kasevich, Science **315**, 74 (2007).
28. R. Legere and K. Gibble, Phys. Rev. Lett. **81**, 5780 (1998).
29. L. Cacciapuoti, M. de Angelis, M. Fattori, G. Lamporesi, T. Petelski, M. Prevedelli, J. Stuhler, and G.M. Tino, Rev. Sci. Instrum. **76**, 053111 (2005).
30. J. Schurr, F. Nolting, and W. Kündig, Phys. Rev. Lett. **80**, 1142 (1998).
31. M. Fattori, G. Lamporesi, T. Petelski, J. Stuhler, and G. Tino, Phys. Lett. A **318**, 184 (2003).
32. I. Bloch, Nat. Phys. **1**, 253001 (2005, and references therein).
33. F. Bloch, Z. Phys. **52**, 555 (1929).
34. M. Raizen, C. Salomon, and Q. Niu, Physics Today **50** (1997, and references therein).
35. M. Takamoto, F.-L. Hong, R. Higashi, and H. Katori, Nature **435**, 321 (2005).
36. G. Ferrari, R.E. Drullinger, N. Poli, F. Sorrentino, and G. Tino, Phys. Rev. A **73**, 23408 (2006).
37. T. Ido, Y. Isoya, and H. Katori, Phys. Rev. A **61**, 061403(R) (2000).
38. B.P. Anderson and M.A. Kasevich, Science **282**, 1686 (1998).
39. G. Roati, E. de Mirandes, F. Ferlaino, H. Ott, G. Modugno, and M. Inguscio, Phys. Rev. Lett. **92**, 230402 (2004).
40. N. Poli, R.E. Drullinger, G. Ferrari, J. Léonard, F. Sorrentino, and G.M. Tino, Phys. Rev. A **71**, 061403(R) (2005).
41. N. Ashcroft and N. Mermin, *Solid State Physics* (Saunders, 1976).
42. Independent measurements with an accelerometer at the level of the retro-reflecting mirror indicate a seismic noise consistent with the observed damping time.
43. S. Dimopoulos and A.A. Geraci, Phys. Rev. D **68**, 124021 (2003).
44. D.M. Harber, J.M. Obrecht, J.M. McGuirk, and E.A. Cornell, Phys. Rev. A **72**, 033610 (2005).
45. S. Kuhr, W. Alt, D. Schrader, M. Mueller, V. Gomer, and D. Meschede, Science **293**, 278 (2001).
46. T. Udem, J. Reichert, R. Holzwarth, and T.W. Haensch, Phys. Rev. Lett. **82**, 3568 (1999).
47. S.A. Diddams, D.J. Jones, J. Ye, S.T. Cundiff, J.L. Hall, J.K. Ranka, R.S. Windeler, R. Holzwarth, T. Udem, and T. Haensch, Phys. Rev. Lett. **84**, 5102 (2000).
48. P. Wolf and G. Petit, Phys. Rev. A **56**, 4405 (1997).

The York Map and the Role of Non-inertial Frames in the Geometrical View of the Gravitational Field*

Luca Lusanna

Abstract The role of non-inertial frames in a class of models of general relativity is clarified by means of Dirac's theory of constraints. The identification of a York canonical basis allows to give the interpretation of the gauge variables as generalized inertial effects and to identify the Dirac observables of the gravitational field with generalized tidal effects. York time is the gauge variable controlling the clock synchronization convention. Differently from special relativity, the instantaneous 3-spaces are dynamically determined.

Dirac's constraint theory [1] is the natural language to describe both gauge theories and gravitational physics. Even if the Hamiltonian approach lies at the heart of the Faddeev–Popov measure of the path integral and of the BRST method, both particle physicists and general relativists tend to prefer the configurational one due to its manifest covariance and to the possibility of avoiding to face the problem of what is *time* in a relativistic theory. However, this is an illusion, because the problem cannot be eluded when we try to establish a well-posed Cauchy problem for the field equations, in absence of which we cannot use the existence and uniqueness theorem for the solutions of partial differential equations and speak of predictability. As a consequence, the basic problem is how to separate the gauge degrees of freedom from the gauge-invariant genuine dynamical variables (the Dirac observables, DO). While the former are (modulo some restriction) completely arbitrary, the latter have to satisfy deterministic hyperbolic partial differential equations in every completely fixed gauge. Now, only the Hamiltonian formalism has a natural (even if still heuristic from the point of view of mathematical rigor) tool to implement this separation: the Shanmugadhasan canonical transformations [2] adapted to the first- and second-class constraints of the model and to the second Noether theorem underlying their existence due to the local invariances of the action functional. This type of transformations were introduced by Dirac for the electromagnetic field in a seminal

*Talk at the International School on *Astrophysical Relativity* John Archibald Wheeler, Erice June 1–7, 2006.

L. Lusanna (✉)
Sezione INFN di Firenze, Polo Scientifico, Via Sansone 1, 50019 Sesto Fiorentino (FI), Italy
e-mail: lusanna@fi.infn.it

I. Ciufolini and R.A. Matzner (eds.), *General Relativity and John Archibald Wheeler*, Astrophysics and Space Science Library 367, DOI 10.1007/978-90-481-3735-0_22,

paper [3], in which its DO were identified with the transverse vector potential and electric field of the radiation gauge. This work has been extended to Yang–Mills theory in Ref. [4] and applied in Ref. [5] to the classical version of the standard model of elementary particles. See Ref. [6] for a review.

However, these results *hold only in inertial frames in Minkowski space–time* and are possible because gauge transformations act on an *inner space*: the redundant gauge variables are present only to implement the manifest covariance of the model under the action of the kinematical Poincare' transformations connecting inertial frames and under local *inner* Lie groups. As a consequence, in every formulation both at the classical and quantum level the standard model of elementary particles and its extensions are a chapter of the theory of representations of the Poincare' group in inertial frames in Minkowski space–time (the non-dynamical container of the fields).

In special relativity the structure of the light-cones is an absolute non-dynamical object [7]: they are the only information (the conformal structure) given by the theory to an (either inertial or accelerated) observer in each point of her/his world-line. There is no notion of instantaneous 3-space, of spatial distance and of one-way velocity of light between two observers.[1] The light postulates say that the two-way (or round trip) velocity of light c (only one clock is needed in its definition) is constant and isotropic. For an ideal inertial observer Einstein's convention for the synchronization of distant clocks[2] selects the constant time hyper-planes of the inertial frame having the observer as time axis as the instantaneous Euclidean 3-spaces, with their Euclidean 3-geodesic spatial distance and with the one-way and two-way velocities of light equal.

But this convention does not work for realistic accelerated observers, because coordinate singularities are produced in the attempt (the $1+3$ point of view) to build (either Fermi or rotating) 4-coordinates around the observer world-line. They must use the more complex conventions arising from the introduction of an extra structure: a global $3+1$ splitting of Minkowski space–time (a choice of *time*, the starting point of the Hamiltonian formalism). Each space-like leaf of the associated foliation is *both a Cauchy surface for the field equations and a convention* (different from Einstein's one) *for clock synchronization*. If we introduce *Lorentz–scalar observer-dependent radar 4-coordinates* $x^\mu \mapsto \sigma^A = (\tau; \sigma^r)$, where x^μ are Cartesian

---

[1] By contrast in Newton physics there are distinct absolute notions of *time* and *space*, so that we can speak of absolute simultaneity and of instantaneous Euclidean 3-spaces with the associated Euclidean spatial distance notion. This non-dynamical chrono-geometrical structure is formalized in the so called Galilei space–time. The Galilei relativity principle assumes the existence of preferred inertial frames with inertial Cartesian coordinates centered on inertial observers, connected by the kinematical group of Galilei transformations. In Newton gravity the equivalence principle states the equality of inertial and gravitational mass. In non-inertial frames inertial (or fictitious) forces proportional to the mass of the body appear in Newton's equations.

[2] The inertial observer A sends a ray of light at x_i^o to a second accelerated observer B, who reflects it towards A. The reflected ray is reabsorbed by the inertial observer at x_f^o. The convention states that the clock of B at the reflection point must be synchronized with the clock of A when it signs $\frac{1}{2}(x_i^o + x_f^o)$.

coordinates, τ is an arbitrary monotonically increasing function of the proper time of the observer and σ^r curvilinear 3-coordinates having the observer world-line as origin, this leads to the definition of a *non-inertial frame* centered on the accelerated observer [8]. Every $3+1$ splitting, satisfying certain Møller restrictions (to avoid coordinate singularities) and with the leaves tending to hyper-planes at spatial infinity (so that there are asymptotic inertial observers to be identified with the fixed stars), gives a conventional definition of instantaneous 3-space (in general a Riemannian 3-manifold), of 3-geodesic spatial distance and of one-way velocity of light (in general both point-dependent and anisotropic). The inverse coordinate transformation $\sigma^A \mapsto x^\mu = z^\mu(\tau, \sigma^r)$ defines the *embedding* of the simultaneity surfaces Σ_τ into Minkowski space–time. The $3+1$ splitting leads to the following induced 4-metric (a functional of the embedding): ${}^4g_{AB}(\tau, \sigma^r) = \frac{\partial z^\mu(\sigma)}{\partial \sigma^A}\, {}^4\eta_{\mu\nu}\, \frac{\partial z^\nu(\sigma)}{\partial \sigma^B} = {}^4g_{AB}[z(\sigma)]$.

Parametrized Minkowski theories [7, 9] allow to give a description of every isolated system (particles, strings, fields, fluids), in which the transition from a $3+1$ splitting to another one (i.e., a change of clock synchronization convention) is a *gauge transformation*. Given any isolated system admitting a Lagrangian description, one makes the coupling of the system to an external gravitational field and then replaces the 4-metric ${}^4g_{\mu\nu}(x)$ with the induced metric ${}^4g_{AB}[z(\tau, \sigma^r)]$ associated to an arbitrary admissible $3+1$ splitting. The Lagrangian now depends not only on the matter configurational variables but also on the embedding variables $z^\mu(\tau, \sigma^r)$ (whose conjugate canonical momenta are denoted $\rho_\mu(\tau, \sigma^r)$). Since the action principle turns out to be invariant under *frame-preserving diffeomorphisms*, at the Hamiltonian level there are four first-class constraints $\mathcal{H}_\mu(\tau, \sigma^r) = \rho_\mu(\tau, \sigma^r) - l_\mu(\tau, \sigma^r)\, T^{\tau\tau}(\tau, \sigma^r) - z^\mu_s(\tau, \sigma^r)\, T^{\tau s}(\tau, \sigma^r) \approx 0$ in strong involution with respect to Poisson brackets, $\{\mathcal{H}_\mu(\tau, \sigma^r), \mathcal{H}_\nu(\tau, \sigma^r_1)\} = 0$. Here $l_\mu(\tau, \sigma^r)$ are the covariant components of the unit normal to Σ_τ, while $z^\mu_s(\tau, \sigma^r) = \frac{\partial z^\mu(\tau, \sigma^r)}{\partial \sigma^s}$ are the components of three independent vectors tangent to Σ_τ. The quantities $T^{\tau\tau}$ and $T^{\tau s}$ are the components of the energy–momentum tensor of the matter inside Σ_τ describing its energy- and momentum-densities. As a consequence, Dirac's theory of constraints implies that the configuration variables $z^\mu(\tau, \sigma^r)$ are arbitrary *gauge variables*. Therefore, all the admissible $3+1$ splittings, namely all the admissible conventions for clock synchronization, and all the admissible non-inertial frames centered on time-like observers are *gauge equivalent*.

By adding four gauge-fixing constraints $\chi^\mu(\tau, \sigma^r) = z^\mu(\tau, \sigma^r) - z^\mu_M(\tau, \sigma^r) \approx 0$ ($z^\mu_M(\tau, \sigma^r)$ being an admissible embedding), satisfying the orbit condition $det\, |\{\chi^\mu(\tau, \sigma^r), \mathcal{H}_\nu(\tau, \sigma^r_1)\}| \neq 0$, we identify the description of the system in the associated non-inertial frame centered on a given time-like observer. The resulting effective Hamiltonian for the τ-evolution turns out to contain the potentials of the *relativistic inertial forces* present in the given non-inertial frame. As a consequence, the gauge variables $z^\mu(\tau, \sigma^r)$ describe the *spatio-temporal appearances* of the phenomena in non-inertial frames, which, in turn, are associated to *extended* physical laboratories using a metrology for their measurements compatible with the notion of simultaneity (distant clock synchronization convention) of the non-inertial frame (think to the description of the Earth given by GPS). Therefore, notwithstanding mathematics tends to use only coordinate-independent notions, physical metrology forces us to consider intrinsically coordinate-dependent quantities like the non-inertial Hamiltonians. For instance, the motion of satellites around the Earth is governed by a set of empirical coordinates contained in the software of NASA computers: this is a *metrological standard of space–time around the Earth*.

Inertial frames centered on inertial observers are a special case of gauge fixing in parametrized Minkowski theories. For each configuration of an isolated system there is a special $3+1$ splitting associated to it: the foliation with space-like hyper-planes orthogonal to the conserved time-like 4-momentum of the isolated system. This identifies an intrinsic inertial frame, the *rest-frame*, centered on a suitable inertial observer (the covariant non-canonical Fokker–Pryce center of inertia of the isolated system) and allows to define the *Wigner-covariant inertial rest-frame instant form of dynamics* for every isolated system, which allows to give a new formulation of the relativistic kinematics [10] of N-body systems and continuous media (relativistic centers of mass and canonical relative variables, rotational kinematics and dynamical body frames, multipolar expansions, Møller radius) and to find the theory of relativistic orbits. Instead *non-inertial rest frames* are $3+1$ splittings of Minkowski space–time having the associated simultaneity 3-surfaces tending to Wigner hyper-planes at spatial infinity.

Moreover it is now possible to define relativistic and non-relativistic quantum mechanics of particles in non-inertial frames [11], with a multi-temporal quantization scheme in which the gauge variables $z^{\mu}(\tau,\sigma^{r})$ (the appearances) are c-numbers (generalized times) and only the particle degrees of freedom are quantized. What is still lacking is the quantization of a scalar field in non-inertial frames. Torre and Varadarajan [12] have shown that the traditional Tomonaga–Schwinger approach does not lead in general to a unitary evolution. From the $3+1$ point of view we have to restrict the $3+1$ splittings to those whose simultaneity leaves admit an instantaneous Fock space, in which the Bogoliubov transformation between two such leaves of the foliation is Hilbert–Schmidt (unitary evolution). Moreover all such admissible $3+1$ splittings must be unitarily gauge equivalent.

Things change dramatically when gravity is taken into account [7]. In general relativity there is no absolute notion: the full chrono-geometrical structure of space–time is dynamical. The relativistic description of gravity abandons the relativity principle and replaces it with the equivalence principle. Special relativity can be recovered only locally by a freely falling observer in a neighborhood where tidal effects are negligible. As a consequence, *global inertial frames do not exist*.

The general covariance of Einstein's formulation of general relativity leads to a type of gauge symmetry acting also on space–time. The Hilbert action is invariant under coordinate transformations (*passive* off-shell diffeomorphisms as local Noether transformations), whereas the abstract differential geometric formulation is invariant under *active* diffeomorphisms of the space–time 4-manifold extended to tensors (on-shell dynamical symmetries of Einstein's equations; it is assumed that the space of solutions exists according to some notion of integrability). Both in the off-shell and on-shell viewpoints the gauge group is a group of diffeomorphisms acting also on space–time. (The same happens in every model with some type of reparametrization invariance.)

Even if from a mathematical point of view the gauge variables are still arbitrary degrees of freedom not determined by the field equations, they are no more redundant variables of an inner space, but are connected with the *appearances* of phenomena in the various coordinate systems of Einstein's space–times.

In Einstein's geometrical view of the gravitational field the basic configuration variable is the metric tensor over space–time (10 fields), which, differently from every other field, has a double role:

(i) It is the mediator of the gravitational interaction, like every other gauge field.
(ii) It describes the dynamical chrono-geometrical structure of space–time by means of the line element $ds^2 = {}^4g_{\mu\nu}(x)\, dx^\mu\, dx^\nu$. As a consequence, it *teaches relativistic causality* to the other fields: now the conformal structure (the allowed paths of light rays) is point-dependent.

In canonical ADM metric gravity [13] (and in its tetrad gravity extension [14, 15] needed for fermions[3]) we have again to start with the same pattern of $3+1$ splittings, to be able to define the Cauchy and simultaneity surfaces for Einstein's equations. As a consequence, and having in mind the inclusion of particle physics, we must select a family of *non-compact* space–times M^4 with the following properties:

(i) *Globally hyperbolic* and *topologically trivial*, so that they can be foliated with space-like hyper-surfaces Σ_τ diffeomorphic to R^3 ($3+1$ splitting of space–time with τ, the scalar parameter labeling the leaves, as a *mathematical time*).
(ii) *Asymptotically flat at spatial infinity* and with boundary conditions at spatial infinity independent from the direction, so that the *spi group* of asymptotic symmetries is reduced to the Poincare' group with the ADM Poincare' charges as generators. In this way we can eliminate the *super-translations*, namely the obstruction to define angular momentum in general relativity, and we have the same type of boundary conditions which are needed to get well defined non-Abelian charges in Yang–Mills theory, opening the possibility of a unified description of the four interactions with all the fields belonging to same function space [6, 7, 13]. All these requirements imply that the *admissible foliations* of space–time must have the space-like hyper-surfaces tending in a direction-independent way to Minkowski space-like hyper-planes at spatial infinity, which moreover must be orthogonal there to the ADM 4-momentum. Therefore, M^4 is *asymptotically Minkowskian* [16] with the asymptotic Minkowski metric playing the role of an *asymptotic background*. Moreover the simultaneity 3-surfaces (the Riemannian instantaneous 3-spaces) must admit an involution (Lichnerowicz 3-manifolds [17]) allowing the definition of a generalized Fourier transform with its associated concepts of positive and negative energy, so to avoid the claimed impossibility to define particles in curved space–times.

---

[3] This leads to an interpretation of gravity based on a congruence of time-like observers endowed with orthonormal tetrads: in each point of space–time the time-like axis is the unit 4-velocity of the observer, while the spatial axes are a (gauge) convention for observer's gyroscopes.

(iii) All the fields have to belong to suitable *weighted Sobolev spaces* so that; (i) the admissible space-like hyper-surfaces are Riemannian 3-manifolds without asymptotically vanishing Killing vectors [16, 18] (we furthermore assume the absence of any Killing vector); (ii) the inclusion of particle physics leads to a formulation without Gribov ambiguity [4, 19].

In absence of matter the class of Christodoulou–Klainermann space–times [20], admitting asymptotic ADM Poincare' charges and an asymptotic flat metric meets these requirements.

This formulation, the *rest-frame instant form of metric and tetrad gravity*, emphasizes the role of *non-inertial frames* (the only ones existing in general relativity): each admissible $3+1$ splitting identifies a global non-inertial frame centered on a time-like observer. In these space–times each simultaneity surface is the rest frame of the 3-universe, there are asymptotic inertial observers (the fixed stars) and the switching off of the Newton constant in presence of matter leads to a deparametrization of these models of general relativity to the non-inertial rest-frame instant form of the same matter with the ADM Poincare' charges collapsing into the usual kinematical Poincare' generators. This class of space–times is suitable to describe the solar system (or the galaxy), is compatible with particle physics and allows to avoid the splitting of the metric into a background one plus a perturbation.[4] With the addition of suitable asymptotic terms it can probably be adapted to cosmology [23].

The first-class constraints of canonical gravity (8 in metric gravity, 14 in tetrad gravity[5]) imply the existence of an equal number of arbitrary gauge variables and of only $2+2$ genuine physical degrees of freedom of the gravitational field: $r_{\bar{a}}(\tau,\sigma^r)$, $\pi_{\bar{a}}(\tau,\sigma^r)$. It can be shown [13, 14, 24] that the super-Hamiltonian constraint generates Hamiltonian gauge transformations implying the *gauge equivalence* of clock synchronization conventions like it happens in special relativity[6]

[4] This splitting is the basic tool for the linearization of Einstein's equations (see the theory of gravitational waves) and for their replacement with a non-geometric spin-two theory over Minkowski space–time, in which diffeomorphisms acting on space–time are discontinuously replaced with gauge transformations acting on an inner space. However, as shown by Deser [21], the non-geometrical spin-two theory becomes inconsistent if we add the energy–momentum tensor $T_{\mu\nu}$ of dynamical matter as a source: the only way to recover consistency (at the price of loosing an energy conservation law for the gravitational field) is to recover Einstein's theory. Notwithstanding Deser's result, particle physicists prefer to rely on Feynman's statement [22] that *the geometrical interpretation is not really necessary or essential to physics*. The basic reason seems to be the absence of an energy conservation law for the gravitational field, replaced by a coordinate-dependent notion of energy density.

[5] Tetrad gravity has 10 primary first class constraints and four secondary first class ones. Six of the primary constraints describe the extra freedom in the choice of the tetrads. The other four primary (the vanishing of the momenta of the lapse and shift functions) and four secondary (the super-Hamiltonian and super-momentum constraints) constraints are the same as in metric gravity. In Ref. [14] 13 of the 14 constraints were solved: the super-Hamiltonian one can be solved only after linearization [15].

[6] The special relativistic constraints $\mathcal{H}_\mu(\tau,\sigma^r) \approx 0$ are replaced by the super-Hamiltonian and super-momentum ones.

(no Wheeler–DeWitt interpretation of it as a Hamiltonian). As shown in Refs. [24],[7] the gauge variables describe *generalized inertial effects* (the appearances), while the 2 + 2 gauge invariant DO describe *generalized tidal effects*. In Refs. [14, 15] a Shanmugadhasan canonical basis, adapted to 13 of the 14 tetrad gravity first class constraints (not to the super-Hamiltonian one) was found. With its help it can be shown [24] that a completely fixed Hamiltonian gauge is equivalent to the choice of a *non-inertial frame* with its adapted radar coordinates centered on an accelerated observer and its instantaneous 3-spaces (simultaneity surfaces): again this corresponds to an extended physical laboratory.[8]

In the rest-frame instant form of gravity [13, 14], due to the DeWitt surface term the effective Hamiltonian is not weakly zero (*no frozen picture* of dynamics), but is given by the weak ADM energy $E_{ADM} = \int d^3\sigma\, \mathcal{E}_{ADM}(\tau, \sigma^r)$ (it is the analogous of the definition of the electric charge as the volume integral of the charge density in electromagnetism). The ADM energy density depends on the gauge variables, namely it is a coordinate-dependent quantity (the *problem of energy* in general relativity). In a completely fixed gauge, in which the inertial effects are given functions of the DO, $\mathcal{E}_{ADM}(\tau, \sigma^r)$ becomes a well defined function only of the DO's and there is a deterministic evolution of the DO's (the tidal effects) given by the Hamilton equations. A universe M^4 (a 4-geometry) is the equivalence class of all the completely fixed gauges with gauge equivalent Cauchy data for the DO on the associated Cauchy and simultaneity surfaces Σ_τ. In each completely fixed gauge (an off-shell non-inertial frame determined by some set of gauge-fixing constraints determining the gauge variables in terms of the tidal ones) we find the solution for the DO in that gauge (the tidal effects) and then the explicit form of the gauge variables (the inertial effects). As a consequence, the final admissible (on-shell gauge equivalent) non-inertial frames associated to a 4-geometry (and their instantaneous 3-spaces, i.e., their clock synchronization conventions) are *dynamically determined* [24].

A first application of this formalism has been the determination [15] of *post-Minkowskian background-independent gravitational waves* in a completely fixed non-harmonic 3-orthogonal gauge with diagonal 3-metric. It can be shown that the requirements $r_{\bar{a}}(\tau, \sigma^r) \ll 1$, $\pi_{\bar{a}}(\tau, \sigma^r) \ll 1$ lead to a weak field approximation based on a Hamiltonian linearization scheme: (i) linearize the Lichnerowicz equation (i.e., the super-Hamiltonian constraint), determine the conformal factor of the

---

[7] In these papers there is also a solution of Einstein's Hole Argument: in this class of space–times it is possible to identify the point-events of space–time by means of the four tidal degrees of freedom of the gravitational field. In other words, space–time and gravitational field are two faces of the same entity. The previous identification is not valid in spatially compact space–times without boundary, where the Dirac Hamiltonian weakly vanishes and there is a frozen picture of dynamics.

[8] Let us remark that, if we look at Minkowski space–time as a special solution of Einstein's equations with $r_{\bar{a}}(\tau, \sigma^r) = \pi_{\bar{a}}(\tau, \sigma^r) = 0$ (zero Riemann tensor, no tidal effects, only inertial effects), we find [13] that the dynamically admissible 3 + 1 splittings (non-inertial frames) must have the simultaneity surfaces Σ_τ *3-conformally flat*, because the conditions $r_{\bar{a}}(\tau, \sigma^r) = \pi_{\bar{a}}(\tau, \sigma^r) = 0$ imply the vanishing of the Cotton–York tensor of Σ_τ. Instead, in special relativity, considered as an autonomous theory, all the non-inertial frames compatible with the Møller conditions are admissible [8], namely there is much more freedom in the conventions for clock synchronization.

3-metric and then the lapse and shift functions; (ii) find E_{ADM} in this gauge and disregard all the terms more than quadratic in the DO; (iii) solve the Hamilton equations for the DO. In this way we get a solution of linearized Einstein's equations, in which the configurational DO $r_{\bar{a}}(\tau, \sigma^r)$ play the role of the *two polarizations* of the gravitational wave.

In Refs. [25, 26] there is the description of relativistic fluids and of the Klein–Gordon field in the framework of parametrized Minkowski theories. This formalism allows to get the Lagrangian of these matter systems in the formulation of tetrad gravity of Refs. [14, 15]. The resulting first-class constraints depend only on the mass density $\mathcal{M}(\tau, \sigma^r)$ (which is metric-dependent) and the mass-current density $\mathcal{M}_r(\tau, \sigma^r)$ (which is metric-independent) of the matter. For Dirac fields the situation is more complicated due to the presence of second class constraints (see Ref. [27] for the case of parametrized Minkowski theories with fermions). It turns out that the point Shanmugadhasan canonical transformation of Ref. [15], adapted to 13 of the 14 first class constraints is not suited for the inclusion of matter due to its *non-locality*. Therefore the search started for a local point Shanmugadhasan transformation adapted only to 10 of the 14 constraints, i.e., not adapted to the super-Hamiltonian and super-momentum constraints.

The new insight came from the so-called York–Lichnerowicz conformal approach [28–31] to metric gravity in globally hyperbolic (*but spatially compact*) space–times. The starting point is the decomposition ${}^3g_{ij} = \phi^4\, {}^3\hat{g}_{ij}$ of the 3-metric on an instantaneous 3-space Σ_o of a $3+1$ splitting of space–time in the product of a *conformal factor* $\phi = (det\, {}^3g)^{1/12}$ and a *conformal 3-metric* ${}^3\hat{g}_{ij}$ with $det\, {}^3\hat{g}_{ij} = 1$ (${}^3\hat{g}_{ij}$ contains 5 of the 6 degrees of freedom of ${}^3g_{ij}$). The extrinsic curvature 3-tensor ${}^3K_{ij}$ of Σ_o is decomposed in its trace ${}^3K$ (the *York time*) plus the *distortion tensor*, which is the sum of a TT[9] symmetric 2-tensor ${}^3A_{ij}$ (2 degrees of freedom) plus the 3-tensor ${}^3W_{i;j} + {}^3W_{j;i} - \frac{2}{3}\, {}^3g_{ij}\, {}^3W^k{}_{;k}$ depending on a covariant 3-vector ${}^3W_i$ (*York gravitomagnetic vector potential*; 3 degrees of freedom). Having fixed the lapse and shift functions of the $3+1$ splitting and having put ${}^3K = const.$, one assigns ${}^3\hat{g}_{ij}$ and ${}^3A_{ij}$ on the Cauchy surface Σ_o. Then, ${}^3W_i$ is determined by the super-momentum constraints on Σ_o and ϕ is determined by the super-Hamiltonian constraint on Σ_o. Then, the remaining Einstein's equations (see Refs. [18, 28] for the existence and unicity of solutions) determine the time derivatives of ${}^3g_{ij}$ and of ${}^3K_{ij}$, allowing to find the time development from the initial data on Σ_o. However, a canonical basis adapted the previous splittings was never found. The only result is contained in Ref. [32], where it was shown that, having fixed ${}^3K$, the transition from the non-canonical variables ${}^3\hat{g}_{ij}$, ${}^3A_{ij}$, ${}^3W_i$ to the space of the gravitational initial data satisfying the constraints is a canonical transformation, named *York map*.

In Ref. [33] a new parametrization of the original 3-metric ${}^3g_{ij}$ was proposed, which allows to find local point Shanmugadhasan canonical transformation, adapted to 10 of the 14 constraints of tetrad gravity, implementing a York map. The 3-metric ${}^3g_{rs}$ may be diagonalized with an *orthogonal* matrix $V(\theta^r)$, $V^{-1} = V^T$, $det\, V = 1$,

[9] Traceless and transverse with respect to the conformal 3-metric.

depending on 3 Euler angles θ^r. The gauge Euler angles θ^r give a description of the 3-coordinate systems on Σ_τ from a local point of view, because they give the orientation of the tangents to the 3 coordinate lines through each point (their conjugate momenta $\pi_i^{(\theta)}$ are determined by the super-momentum constraints and replace the York gravitomagnetic potential ${}^3W_i$), ϕ is the conformal factor of the 3-metric, i.e. the unknown in the super-Hamiltonian constraint[10] (its conjugate momentum is the gauge variable describing the form of the simultaneity surfaces Σ_τ), while the two independent eigenvalues of the conformal 3-metric ${}^3\hat{g}_{rs}$ (with determinant equal to 1) describe the genuine *tidal* effects $R_{\bar{a}}$, $\bar{a} = 1, 2$, of general relativity (the non-linear "graviton", with conjugate momenta $\Pi_{\bar{a}}$). In the York canonical basis [33] the gauge variable, which describes the freedom in the choice of the clock synchronization convention, i.e., in the definition of the instantaneous 3-spaces Σ_τ, is the trace ${}^3K(\tau, \sigma^r)$ of the extrinsic curvature of Σ_τ. It is both the York time and the momentum conjugate to the conformal factor.

The tidal effects $R_{\bar{a}}(\tau, \sigma^r)$, $\Pi_{\bar{a}}(\tau, \sigma^r)$, are DO *only* with respect to the gauge transformations generated by 10 of the 14 first class constraints. Let us remark that, if we fix completely the gauge and we go to Dirac brackets, then the only surviving dynamical variables $R_{\bar{a}}$ and $\Pi_{\bar{a}}$ become two pairs of *non canonical* DO for that gauge: the two pairs of canonical DO have to be found as a Darboux basis of the copy of the reduced phase space identified by the gauge and they will be (in general non-local) functionals of the $R_{\bar{a}}$, $\Pi_{\bar{a}}$ variables. This shows the importance of canonical bases like the York one: the tidal effects are described by *local* functions of the 3-metric and its conjugate momenta.

Since the conformal factor ϕ and the momenta $\pi_i^{(\theta)}$ conjugate to the Euler angles θ^r are determined by the super-Hamiltonian and super-momentum constraints, the *arbitrary gauge variables* of the York canonical basis are $\alpha_{(a)}$, $\varphi_{(a)}$, θ^i, $\pi_{\tilde{\phi}}$, n and $\bar{n}_{(a)}$. As shown in Refs. [24, 33], they describe the following generalized *inertial effects*:

(a) The angles $\alpha_{(a)}(\tau, \sigma^r)$ and the boost parameters $\varphi_{(a)}(\tau, \sigma^r)$ describe the arbitrariness in the choice of a tetrad to be associated to a time-like observer, whose world-line goes through the point (τ, σ^r). They fix *the unit 4-velocity of the observer and the conventions for the gyroscopes and their transport along the world-line of the observer.*

(b) The angles $\theta^i(\tau, \sigma^r)$ (depending only on the 3-metric) describe the arbitrariness in the choice of the 3-coordinates on the simultaneity surfaces Σ_τ of the chosen non-inertial frame centered on an arbitrary time-like observer. Their choice induces a pattern of *relativistic standard inertial forces* (centrifugal, Coriolis,...), whose potentials are contained in the weak ADM energy E_{ADM}.

[10] The only role of the conformal decomposition ${}^3g_{ij} = \phi^4\, {}^3\hat{g}_{ij}$ is to identify the conformal factor ϕ as the natural unknown in the super-Hamiltonian constraint, which becomes the *Lichnerowicz equation*. See Ref. [13] for a different justification of this result based on constraint theory and the two notions of strong and weak ADM energy.

These inertial effects are the relativistic counterpart of the non-relativistic ones (they are present also in the non-inertial frames of Minkowski space–time).

(c) The *shift* functions $\bar{n}_{(a)}(\tau, \sigma^r)$, appearing in the Dirac Hamiltonian, describe which points on different simultaneity surfaces have the same numerical value of the 3-coordinates. They are the inertial potentials describing the effects of the non-vanishing off-diagonal components ${}^4g_{\tau r}(\tau, \sigma^r)$ of the 4-metric, namely they are the *gravito-magnetic potentials*[11] responsible of effects like the dragging of inertial frames (Lense–Thirring effect) [31] in the post-Newtonian approximation.

(d) $\pi_\phi(\tau, \sigma^r)$, i.e., the York time ${}^3K(\tau, \sigma^r)$, describes the arbitrariness in the shape of the simultaneity surfaces Σ_τ of the non-inertial frame, namely the arbitrariness in the choice of the convention for the synchronization of distant clocks. Since this variable is present in the Dirac Hamiltonian, it is a *new inertial potential* connected to the problem of the relativistic freedom in the choice of the *instantaneous 3-space*, which has no non-relativistic analogue (in Galilei space–time is absolute and there is an absolute notion of Euclidean 3-space). Its effects are completely unexplored. For instance, since the sign of the trace of the extrinsic curvature may change from a region to another one on the simultaneity surface Σ_τ, *the associated inertial force in the Hamilton equations may change from attractive to repulsive in different regions.*

(e) The *lapse* function $N(\tau, \sigma^r) = 1 + n(\tau, \sigma^r)$ describes the arbitrariness in the choice of the unit of proper time in each point of the simultaneity surfaces Σ_τ, namely how these surfaces are packed in the $3 + 1$ splitting.

As a consequence, differently from special relativity, the conventions for clock synchronization and the whole chrono-geometrical structure of M^4 (gravito-magnetism, 3-geodesic spatial distance on Σ_τ, trajectories of light rays in each point of M^4, one-way velocity of light) are *dynamically determined* [24].

The use of Dirac theory of constraints introduces a different point of view on the gauge-fixing and the Cauchy problem. While the gauge fixing to the extra six primary constraints fixes the tetrads (i.e., the spatial gyroscopes and their transport law), the gauge fixing to the four primary plus four secondary constraints follows a different scheme from the one used in the York–Lichnerowicz approach, which influenced contemporary numerical gravity. Firstly one adds the four gauge fixings to the secondary constraints (the super-Hamiltonian and super-momentum ones), i.e., one fixes ${}^3K$, i.e., the simultaneity 3-surface, and the 3-coordinates on it (namely 3 of the 5 degrees of freedom of the conformal 3-metric ${}^3\hat{g}_{ij}$). The preservation in time of these four gauge fixings generates other four gauge fixing constraints determining

[11] In the Post-Newtonian approximation in *harmonic gauges* they are the counterpart of the electro-magnetic vector potentials describing magnetic fields [15, 31]: (A) $N = 1 + n$, $n \stackrel{def}{=} -\frac{4\epsilon}{c^2}\Phi_G$ with Φ_G the *gravito-electric potential*; (B) $n_r \stackrel{def}{=} \frac{2\epsilon}{c^2} A_{G\,r}$ with $A_{G\,r}$ the *gravito-magnetic* potential; (C) $E_{G\,r} = \partial_r\,\Phi_G - \partial_\tau\,(\frac{1}{2}\,A_{G\,r})$ (the *gravito-electric field*) and $B_{G\,r} = \epsilon_{ruv}\,\partial_u\,A_{G\,v} = c\,\Omega_{G\,r}$ (the *gravito-magnetic field*). Let us remark that in arbitrary gauges the analogy with electro-magnetism breaks down.

the lapse and shift functions consistently with the shape of the simultaneity 3-surface and with the choice of 3-coordinates on it (here is the main difference with the conformal approach and most of the approaches to numerical gravity).

The clarification of the interpretational issues allowed by the York canonical basis will allow to face many problems.

1. To understand better the Hamiltonian distinction between inertial and tidal effects, a detailed study of the Post-Newtonian solutions of Einstein's equations adopted by the IAU conventions [34] for the barycentric and geocentric celestial reference frames has begun. In particular it will clarify the mixing of the general relativistic effects like the gravitational redshift with the special relativistic ones like the Doppler effect and the Coriolis forces near the rotating Earth. This is no more an academic research, because in a few years the European Space Agency (ESA) will start the mission ACES [35] about the synchronization of a high-precision laser-cooled atomic clock on the space station with similar clocks on the Earth surface by means of microwave signals. If the accuracy of 5 picosec. will be achieved, it will be possible to make a coordinate-dependent test of effects at the order $1/c^3$, like the second order Sagnac effect (sensible to Earth rotational acceleration) and the general relativistic Shapiro time-delay created by the geoid [36]. It will be important to find the Post-Newtonian deviation from Einstein's convention to be able to synchronize two such clocks and to understand which metrological protocols have to be used for time dissemination at this level of accuracy. Therefore, the problem of clock synchronization is becoming every day more important due to GPS, to the ACES mission of ESA, to the Bepi–Colombo mission to Mercury and to the future space navigation inside the solar system.
2. The geometric vision of space–time will soon be enriched with the Hamiltonian reformulation of the Newman–Penrose formalism, in particular of the 10 Weyl scalars. This will allow (a) to search the Bergmann observables [24] (special DO describing *scalar tidal effects*) and try to understand which inertial effects may have a coordinate-independent form (like gravito-magnetism) and which are intrinsically coordinate-dependent like the ADM energy density; (b) to look for the existence of a closed Poisson algebra of scalars and for Shanmugadhasan canonical bases incorporating the Bergmann observables, to be used to find new expressions for the super-Hamiltonian and super-momentum constraints, hopefully easier to be solved. If the Torre–Varadarajan no-go theorem [12] can be avoided and the scalar field can be quantized in an admissible set of non-inertial frames in Minkowski space–time, it will be possible to arrive at a multi-temporal background- and coordinate-independent multi-temporal quantization (see Ref. [11]) of the gravitational field, in which only the Bergmann observables (the scalar tidal effects) are quantized.
3. The study the 2-body problem in general relativity in various coordinate systems at least in the weak field approximation, with a Grassmann regularization of the self-energies, following the track of Refs. [37] is now possible. In these papers the use of Grassmann-valued electric charges to regularize the Coulomb self-energies allowed to arrive to the Darwin and Salpeter potentials starting

from classical electrodynamics of scalar and spinning particles, instead of deriving them from quantum field theory. The solution of the Lichnerowicz equation would allow to find the expression of the relativistic Newton and gravito-magnetic action-at-a-distance potentials between the two bodies (sources, among other effects, of the Newtonian tidal effects) and the coupling of the particles to the DO of the gravitational field (the genuine tidal effects) in various radar coordinate systems: it would amount to a re-summation of the $1/c$ expansions of the Post-Newtonian approximation. Also the relativistic version of the quadrupole formula for the emission of gravitational waves from the binary system could be obtained and some understanding of how is distributed the gravitational energy in different coordinate systems could be obtained.[12] It would also be possible to study the deviations induced by Einstein's theory from the Keplerian standards for problems like the rotation curves of galaxies, whose Keplerian interpretation implies the existence of dark matter.[13]

4. With more general types of matter (fluids, electro-magnetic field) we could define Hamiltonian numerical gravity (for instance with a post-Minkowskian development in powers of the Newton constant) and try to find strong-field approximations to be used in the gravitational collapse and to find the strong-field deviations from the Newton potential. This last problem is completely open in every approach.

A review of the further developments in this line of research can be found in L. Lusanna, "Post-Minkowskian Gravity: Dark matter as a relativistic inertial effect?", Arxiv 0912.2935.

References

1. P.A.M. Dirac, *Lectures in Quantum Mechanics*, Belfer Graduate School of Science (Yeshiva University, NewYork, 1964).
2. L. Lusanna, *The Shanmugadhasan Canonical Transformation, Function Groups and the Second Noether Theorem*, Inter. J. Mod. Phys. **A8**, 4193–4233 (1993).
3. P. A. M. Dirac, *Gauge Invariant Formulation of Quantum Electrodynamics*, Can. J. Phys. **33**, 650–659 (1955).
4. L. Lusanna, *Classical Yang-Mills Theory with Fermions. I. General Properties of a System with Constraints*, Int. J. Mod. Phys. **A10**, 3531-3579 (1995); *II. Dirac's Observables*, Int. J. Mod. Phys. **A10**, 3675–3757 (1995).

[12] If the ADM energy-momentum pseudo-tensor will be identified, its reformulation as the energy–momentum tensor of a viscous pseudo-fluid will allow to check whether the pressure field is positive-definite or not, namely whether the gravitational energy contributes to dark energy.

[13] Since the relativistic inertial forces are present in the Hamilton equations for the gravitational DO and for the matter in the gauge-dependent instantaneous 3-space, they may be a relativistic alternative to the MOND model [38] (modification of the non-relativistic Newton equations on the acceleration side for slow accelerations). At least part of the dark mattr could be explained by relativistic inertial effects. See Ref. [39] for a possible gravito-magnetic origin of dark matter.

5. L. Lusanna and P.Valtancoli, *Dirac's Observables for the* $SU(3) \times SU(2) \times U(1)$ *Standard Model*, Int. J. Mod. Phys. **A13**, 4605–4690 (1998).
6. L. Lusanna, *Towards a Unified Description of the Four Interactions in Terms of Dirac-Bergmann Observables*, invited contribution to the book *Quantum Field Theory: a 20th Century Profile* of the Indian National Science Academy, ed. A.N.Mitra, foreward F.J. Dyson (Hindustan Book Agency, New Delhi, 2000) (hep-th/9907081).
7. L. Lusanna, *The Chrono-Geometrical Structure of Special and General Relativity: A Re-Visitation of Canonical Geometrodynamics*, lectures at 42nd Karpacz Winter School of Theoretical Physics: Current Mathematical Topics in Gravitation and Cosmology, Ladek, Poland, 6-11 Feb 2006, Int. J. Geom. Methods in Mod. Phys. **4**, 79 (2007). (gr-qc/0604120).
8. D. Alba and L. Lusanna, *Simultaneity, Radar 4-Coordinates and the 3C1 Point of View about Accelerated Observers in Special Relativity* (2003) (gr-qc/0311058); *Generalized Radar 4-Coordinates and Equal-Time Cauchy Surfaces for Arbitrary Accelerated Observers* (2005), Int. J. Mod. Phys. **D16**, 1149 (2007) (gr-qc/0501090).
9. L. Lusanna, *The N- and 1-Time Classical Description of N-Body Relativistic Kinematics and the Electromagnetic Interaction*, Int. J. Mod. Phys. **A12**, 645–722 (1997); *The Chrono-geometrical Structure of Special and General Relativity: towards a Background-Independent Description of the Gravitational Field and Elementary Particles* (2004), in *General Relativity Research Trends*, ed. A.Reiner, Horizon in World Physics vol. 249 (Nova Science, New York, 2005) (gr-qc/0404122).
10. D. Alba, L. Lusanna and M. Pauri, *New Directions in Non-Relativistic and Relativistic Rotational and Multipole Kinematics for N-Body and Continuous Systems* (2005), invited contribution for the book *Atomic and Molecular Clusters: New Research* (Nova Science) (hep-th/0505005).
    D. Alba, H.W. Crater and L. Lusanna, *Hamiltonian Relativistic Two-Body Problem: Center of Mass and Orbit Reconstruction*, to appear in J. Phys. A (gr-qc/0610200).
11. D. Alba and L. Lusanna, *Quantum Mechanics in Non-Inertial Frames with a Multi-Temporal Quantization Scheme: I) Relativistic Particles*, Int. J. Mod. Phys. **A21**, 2781 (2006) (hep-th/0502060); II) *Non-Relativistic Particles*, Int. J. Mod. Phys. **A21**, 3917 (2006) (hep-th/0504060), Int. J. Mod. Phys.
12. C.G. Torre and M. Varadarajan, *Functional Evolution of Free Quantum Fields*, Clas. Quantum Grav. **16**, 2651–2668 (1999).
13. L. Lusanna, *The Rest-Frame Instant Form of Metric Gravity*, Gen. Rel. Grav. **33**, 1579–1696 (2001) (gr-qc/0101048).
14. R. De Pietri, L. Lusanna, L. Martucci and S. Russo, *Dirac's Observables for the Rest-Frame Instant Form of Tetrad Gravity in a Completely Fixed 3-Orthogonal Gauge*, Gen. Rel. Grav. **34**, 877–1033 (2002) (gr-qc/0105084).
    L. Lusanna and S. Russo, *A New Parametrization for Tetrad Gravity*, Gen. Rel. Grav. **34**, 189–242 (2002) (gr-qc/0102074).
15. J. Agresti, R. De Pietri, L. Lusanna and L. Martucci, *Hamiltonian Linearization of the Rest Frame Instant Form of Tetrad Gravity in a Completely Fixed 3-Orthogonal Gauge: a Radiation Gauge for Background-Independent Gravitational Waves in a Post-Minkowskian Einstein Space-Time*, Gen. Rel. Grav. **36**, 1055–1134 (2004) (gr-qc/0302084).
16. Y. Choquet-Bruhat, *Positive-Energy Theorems*, in *Relativity, Groups and Topology II*, eds. B.S. De Witt and R. Stora (North-Holland, Amsterdam, 1984).
17. A. Lichnerowicz, *Propagateurs, Commutateurs et Anticommutateurs en Relativite Generale*, in Les Houches 1963, *Relativity, Groups and Topology*, eds. C. DeWitt and B. DeWitt (Gordon and Breach, New York, 1964).
    C. Moreno, *On the Spaces of Positive and Negative Frequency Solutions of the Klein-Gordon Equation in Curved Space-Times*, Rep. Math. Phys. **17**, 333 (1980).
18. Y. Choquet-Bruhat and J.W. York jr, *The Cauchy Problem*, in *General Relativity and Gravitation*, Vol. 1, pp. 99–172, ed. A. Held (Plenum Press, New York, 1980).
    G. Choquet and Y. Choquet-Bruhat, *Sur un probleme lie' a' la stabilite' des donne'es initiales en relativite' gene'rale*, C.R. Acad. Sc. Paris **A287**, 1047 (1978).

19. V. Moncrief, *Gribov Degeneracies: Coulomb Gauge Conditions and Initial Value Constraints*, J. Math. Phys. **20**, 579 (1979).
20. D. Christodoulou and S. Klainerman, *The Global Nonlinear Stability of the Minkowski Space*. (Princeton University Press, Princeton, 1993).
21. S. Deser, *Self-Interaction and Gauge Invariance*, Gen. Rel. Grav. **1**, 9 (1970) (gr-qc/0411023).
22. R.P. Feynman, *Lectures on Gravitation*, 1962–63, p. 110, Lecture 8 (California Institute of Technology, 1971).
23. S.A. Klioner and M.H. Soffel, *Refining the Relativistic Model for GAIA: Cosmological Effects in the BCRS*, Proc. of the Symposium *The Three-Dimensional Universe with GAIA*, Paris 2004, pp. 305–309 (ESA SP-576, January 2005) (astro-ph/0411363).
24. L. Lusanna and M. Pauri, *The Physical Role of Gravitational and Gauge Degrees of Freedom in General Relativity - I: Dynamical Synchronization and Generalized Inertial Effects; II : Dirac versus Bergmann Observables and the Objectivity of Space-Time*, Gen. Rel. Grav. **38**, 187 and 229 (2006) (gr-qc/0403081 and 0407007). *Explaining Leibniz Equivalence as Difference of Non-Inertial Appearances: Dis-solution of the Hole Argument and Physical Individuation of Point-Events*, talk at the Oxford Conference on Spacetime Theory (2004), Studies in History and Philosophy of Modern Physics **37**, 692 (2006) (gr-qc/0604087). *Dynamical Emergence of Instantaneous 3-Spaces in a Class of Models of General Relativity*, to appear in the book *Relativity and the Dimensionality of the World*, ed. A. van der Merwe, Springer Series *Fundamental Theories of Physics* (gr-qc/0611045).
25. L. Lusanna and D. Nowak-Szczepaniak, *The Rest-Frame Instant Form of Relativistic Perfect Fluids with Equation of State* $\rho = \rho(\eta, s)$ *and of Nondissipative Elastic Materials.*, Int. J. Mod. Phys. **A15**, 4943 (2000).
D. Alba and L. Lusanna, *Generalized Eulerian Coordinates for Relativistic Fluids: Hamiltonian Rest-Frame Instant Form, Relative Variables, Rotational Kinematics* Int. J. Mod. Phys. **A19**, 3025 (2004) (hep-th/020903).
26. L. Lusanna and M. Materassi, *A Canonical Decomposition in Collective and Relative Variables of a Klein-Gordon Field in the Rest-Frame Wigner-Covariant Instant Form*, Int. J. Mod. Phys. A15, 2821 (2000).
27. F. Bigazzi and L. Lusanna, *Dirac Fields on Spacelike Hypersurfaces, their Rest-Frame Description and Dirac Observables*, Int. J. Mod. Phys. **A14**, 1877 (1999).
28. J.W. York jr., *Role of Conformal Three-Geometry in the Dynamics of Gravitation*, Phys. Rev. Lett. **28**, 1082 (1972); *Kinematics and Dynamics of General Relativity*, in *Sources of Gravitational Radiation*, Battelle-Seattle Workshop 1978, ed. L.L. Smarr (Cambridge Univ.Press, Cambridge, 1979).
29. A. Lichnerowicz, *L'inte'gration des e'quations de la gravitation relativiste' et le proble'me des n corps*, J. Math. Pures Appl. **23**, 37 (1944).
30. Y. Choquet-Bruhat, *Cauchy Problem*, in *Gravitation: an Introduction to Current Research*, ed. L. Witten, pp. 138–168 (Wiley, New York, 1962).
31. I. Ciufolini and J.A. Wheeler, *Gravitation and Inertia* (Princeton Univ.Press, Princeton, 1995).
32. J. Isemberg and J.E. Marsden, *The York map is a Canonical Transformation* , J. Geom. Phys. **1**, 85 (1984).
33. D. Alba and L. Lusanna, *The York Map as a Shanmugadhasan Canonical Transformation in Tetrad Gravity and the Role of Non-Inertial Frames in the Geometrical View of the Gravitational Field* (gr-qc/0604086), Gen. Rel. Grav. **39**, 2149 (2007).
34. M. Soffel, S.A. Klioner, G. Petit, P. Wolf, S.M. Kopeikin, P. Bretagnon, V.A. Brumberg, N. Capitaine, T. Damour, T. Fukushima, B. Guinot, T. Huang, L. Lindegren, C. Ma, K. Nordtvedt, J. Ries, P.K. Seidelmann, D. Vokroulicky', C. Will and Ch. Xu, *The IAU 2000 Resolutions for Astrometry, Celestial Mechanics and Metrology in the Relativistic Framework: Explanatory Supplement* Astron. J., **126**, pp. 2687–2706, (2003) (astro-ph/0303376).
35. L. Cacciapuoti, N. Dimarcq and C. Salomon, *Atomic Clock Ensemble in Space: Scientific Objectives and Mission Status*. See the talks at the *SIGRAV Graduate School on Experimental Gravitation in Space*(Firenze, September 25–27, 2006) (http://www.fi.infn.it/GGI-grav-space/egs_s.html) and at the Workshop *Advances in Precision Tests and Experimental*

Gravitation in Space (Firenze, September 28/30, 2006) (http://www.fi.infn.it/GGI-grav-space/egs_w.html).

36. L. Blanchet, C. Salomon, P. Teyssandier and P. Wolf, *Relativistic Theory for Time and Frequency Transfer to Order* $1/c^3$, Astron. Astrophys. **370**, 320 (2000).
37. Crater, H. and Lusanna, L. *The Rest-Frame Darwin Potential from the Lienard-Wiechert Solution in the Radiation Gauge*, Ann. Phys. (N.Y.) **289**, 87–177 (2001) (hep-th/0001046). Alba, D., Crater, H. and Lusanna, L. *The Semiclassical Relativistic Darwin Potential for Spinning Particles in the Rest-Frame Instant Form: Two-Body Bound States with Spin 1/2 Constituents*, Int. J. Mod. Phys. **A16**, 3365–3477 (2001) (hep-th/0103109).
38. M. Milgrom, *A Modification of the Newtonian Dynamics as a Possible Alternative to the Hidden Mass Hypothesis*, Astrophys.J. **270**, 365 (1983); *MOND as Modified Inertia* (astro-ph/0510117).
    J.D. Bekenstein, *The Modified Newtonian Dynamics -MOND- and its Implications for New Physics*, (astro-ph/0701848).
39. F.I. Cooperstock and S. Tieu, *Galactic Dynamics via General Relativity and the Exotic Dark Matter Enigma*, Mod. Phys. Lett. **A21**, 2133 (2006); *Galactic Dynamics via General Relativity: A Compilation and New Developments*, Int. J. Mod. Phys. **A13**, 2293 (2007) (astro-ph/0610370).

Index

F

Q